P9-DUY-975

Understanding Ultra Wide Band Radio Fundamentals

Prentice Hall Communications Engineering and Emerging Technologies Series

Theodore S. Rappaport, *Series Editor*

DI BENEDETTO & GIANCOLA *Understanding Ultra Wide Band Radio Fundamentals*

DOSTERT *Powerline Communications*

DURGIN *Space–Time Wireless Channels*

GANZ, GANZ, & WONGTHAVARAWAT *Multimedia Wireless Networks: Technologies, Standards, and QoS*

GARG *Wireless Network Evolution: 2G to 3G*

GARG *IS-95 CDMA and cdma2000: Cellular/PCS Systems Implementation*

GARG & WILKES *Principles and Applications of GSM*

KIM *Handbook of CDMA System Design, Engineering, and Optimization*

LIBERTI & RAPPAPORT *Smart Antennas for Wireless Communications: IS-95 and Third Generation CDMA Applications*

MURTHY & MANOJ *Ad Hoc Wireless Networks: Architectures and Protocols*

PAHLAVAN & KRISHNAMURTHY *Principles of Wireless Networks: A Unified Approach*

RAPPAPORT *Wireless Communications: Principles and Practice, Second Edition*

RAZAVI *RF Microelectronics*

REED *Software Radio: A Modern Approach to Radio Engineering*

STARR, CIOFFI, & SILVERMAN *Understanding Digital Subscriber Line Technology*

STARR, SORBARA, CIOFFI, & SILVERMAN *DSL Advances*

TRANTER, SHANMUGAN, RAPPAPORT, & KOSBAR *Principles of Communication Systems Simulation with Wireless Applications*

VANGHI, DAMNJANOVIC, & VOJCIC *The cdma2000 System for Mobile Communications*

WANG & POOR *Wireless Communication Systems: Advanced Techniques for Signal Reception*

Understanding Ultra Wide Band Radio Fundamentals

Maria-Gabriella Di Benedetto
University of Rome La Sapienza

Guerino Giancola
University of Rome La Sapienza

PRENTICE HALL
PTR

PRENTICE HALL PTR
UPPER SADDLE RIVER, NJ 07458
WWW.PHPTR.COM

Library of Congress Cataloging-in-Publication Data

A catalog record for this book can be obtained from the Library of Congress

Editorial/production supervision: *Nicholas Radhuber*
Publisher: *Bernard Goodwin*
Cover design director: *Jerry Votta*
Cover design: *Talar Boorujy*
Manufacturing manager: *Maura Zaldivar*
Editorial assistant: *Michelle Vincenti*
Marketing manager: *Dan DePasquale*

Published by Prentice Hall Professional Technical Reference
Pearson Education, Inc.
Upper Saddle River, New Jersey 07458

Prentice Hall books are widely used by corporations and government agencies for training, marketing, and resale.

Prentice Hall offers excellent discounts on this book when ordered in quantity for bulk purchases or special sales. For more information, please contact:
U.S. Corporate and Government Sales
1-800-382-3419
corpsales@pearsontechgroup.com

For sales outside of the U.S., please contact:
International Sales
1-317-581-3793
international@pearsontechgroup.com

Printed in the United States of America

1st Printing

ISBN 0-13-148003-0

Pearson Education LTD.
Pearson Education Australia PTY, Limited
Pearson Education Singapore, Pte. Ltd.
Pearson Education North Asia Ltd.
Pearson Education Canada, Ltd.
Pearson Educación de Mexico, S.A. de C.V.
Pearson Education — Japan
Pearson Education Malaysia, Pte. Ltd.

To Matti,
for his first repunit

To Erika,
my most beautiful

Contents

Preface

The last two years have witnessed an increased interest in both chip manufacturing companies and standardization bodies in Ultra Wide Band (UWB). Appealing features such as flexibility and robustness, as well as high-precision ranging capability, have polarized attention and made UWB an excellent candidate for a variety of applications. Given the strong power emission constraints imposed by the regulatory bodies in the United States, but likely to be adopted by other countries as well, UWB is emerging as a particularly appealing transmission technique for applications requiring either high bit rates over short ranges or low bit rates over medium-to-long ranges. The high bit rate/short range case includes Wireless Personal Area Networks (WPANs) for multimedia traffic, cable replacement such as wireless USB and DVI, and wearable devices, e.g., wireless Hi-Fi headphones. The low bit rate/medium-to-long range case applies to long-range sensor networks such as indoor/outdoor distributed surveillance systems, non-real-time data applications, e.g., e-mail and instant messaging, and in general all data transfers compatible with a transmission rate in the order of 1 Mb/s over several tens of meters. A recent release of the IEEE 802.15.4 standard for low rate WPANs (IEEE 802.15.4-2003, 2003) has increased attention towards the low bit rate case.

The scenarios of applications mentioned above refer to networks that commonly adopt the self-organizing principle, that is, distributed networks. Examples of these networks are ad hoc and sensor networks, i.e. groups of wireless terminals located in a limited-size geographical area, communicating in an infrastructure-free fashion, and without any central coordinating unit or base-station. Communication routes may be formed by multiple hops to extend coverage. This paradigm can be viewed as different in nature from the cellular networking model where typically nodes communicate by establishing single-hop connections with a central coordinating unit serving as the interface between wireless nodes and the fixed wired infrastructure.

The goal of this book is to help understanding UWB. But what is UWB?

The general consensus establishes that a signal is UWB if its bandwidth is large with respect to the carrier or center frequency of the spectrum, that is, if its fractional bandwidth is high. The common adoption of the term UWB, which comes to us from the radar community, is compliant with this definition, and refers to electromagnetic waveforms with an instantaneous fractional bandwidth greater than about 0.20–0.25. These waveforms, because of their large bandwidth, must, at least in principle, friendly coexist with other Hertzian waveforms, which are present in the air interface. The coexistence principle introduces strong limitations over Power Spectral Densities (PSDs), and raises the issue of designing power efficient networks.

Traditionally, UWB signals have been obtained by transmitting very short pulses, rather than continuous waveforms, with typically no Radio Frequencies modulation. This technique has been extensively used in radar applications and goes under the name of Impulse Radio (IR).

Regarding wireless communications, the primal technique for transferring information over the radio medium was based, in fact, on the emission of pulsed signals. As described by (Sobol, 1984) in a milestone review paper, Marconi's first experiments, back in 1894–1896, used spark gap transmitters to transmit Morse Code messages over two miles, and Fessenden transmitted speech as early as in 1900 over one mile using a spark gap transmitter. Technological limitations and commercial pressure for reliable communications strongly favored, however, a shift of research and development towards continuous-wave transmissions, and IR remained relatively confined to the radar field up to recent years. The legacy of Marconi manifested from time to time: In 1946 a remarkable microwave radio relay system was developed by (Black, Beyer, Grieser, and Polkinghorn, 1946). This system was based on the transmission of pulses that were position-modulated and ensured two-way voice transmission over radio links totalling 1600 miles, and one-way over 3200 miles. A complete survey of IR-UWB research in both radar and communication fields is included in the historical perspective presented by (Barrett, 2000). As indicated by Barrett the term UWB was coined by the U.S. Department of Defense in 1989. During the 1990s, a few small and medium-sized enterprises reintroduced the idea of wireless communications based on the UWB concept and developed UWB technology following the IR paradigm, promoting the transmission of virtually carrierless and extremely short pulses.

The most influential milestone in the history of UWB wireless communications was set in April 2002, when the Federal Communications Commission (FCC) approved the first guidelines allowing — at least in the United States — the intentional emission of UWB signals contained within specified emission masks (FCC, 2002). According to the FCC rules, however, the UWB concept is not limited to pulsed transmission, but can be extended to continuous-like transmission techniques, provided that the occupied bandwidth of the transmitted signal is greater than 500 MHz. The effect of the FCC release was twofold. On one hand, the FCC regulation of UWB emissions raised the interest of major chip manufacturers, such as Texas Instruments, Motorola, IBM, and Intel. On the other hand, discussions were triggered around the advantages of the original IR scheme vs. the traditional carrier-based continuous transmission alternative. The above lack of agreement is reflected in the current diatribe on UWB standardization, in particular in the United States in the framework of the IEEE 802.15.3a Task Group. This group was formed in late 2001 with the task of investigating innovative solutions for the development of high-speed and low-power WPANs. Currently (May 2004), two different proposals for a physical layer based on UWB are under consideration: a Multi-Band (MB) approach combining frequency hopping with Orthogonal Frequency Division Multiplexing, or OFDM (Batra et al., 2003), and a second approach using Direct-Sequence UWB, or DS-UWB, which preserves the original UWB pulsed nature (Roberts, 2003).

To evaluate and compare the different physical layer proposals that were submitted to the IEEE, the 802.15.3a Study Group formed a subcommittee devoted to the definition of a standard UWB channel model. In February 2003, a Final Report summarizing the work of the channel modeling subcommittee was released (IEEE 802.15.SG3a, 2003). In this report, a channel model for indoor UWB propagation and related recommendations on how to use the model for evaluating physical layer performance were proposed.

Medium Access Control (MAC) is another flourishing area in the definition of protocols for wireless local area networks (WLANs) and WPANs. Among the several are the IEEE 802.11 and HIPERLAN/2 standards for WLANs up to 54 Mbit/s, the Bluetooth standard for short-range and low bit rate wireless communications, and the most recent IEEE 802.15.3 (IEEE 802.15.3-2003, 2003) for short-range and high bit rate WPANs. The latter defines a MAC protocol for the high bit rate case (11–55 Mb/s) and distances up to 50 m. The protocol, which is TDMA-based, was originally developed based on a traditional, narrowband (15 MHz on-air bandwidth) physical layer in the 2.4 GHz unlicensed band. The sudden and strong interest for UWB caused a rushed adoption of the IEEE 802.15.3 MAC standard also for the UWB physical layer, although this MAC is neither tailored nor optimized to UWB peculiarities.

Regarding the introduction of UWB in low-rate, location-enabled applications, standardization is taking its first steps within the IEEE 802.15.4a Task Group with a first meeting scheduled in May 2004. The main interest is in providing communications with high-precision ranging and localization, low-power emission and consumption, and a low cost.

Outside the United States, and in particular in Europe, a standardization activity for short-range UWB devices is carried out by the TG31A group of the European Telecommunications Standards Institute (ETSI). Currently, the task group is about to deliver a first UWB standard draft. As regards research and promoting activities, the 6th IST European Union Framework Integrated Project PULSERS (*www.pulsers.net*), which started on January 1, 2004, is taking the lead. Project PULSERS gathers over 30 European and international partners. A roadmap for locating information related to currently released standards is included in the appendix of the book.

APPROACH

This book covers the theoretical basis of UWB radio communications and gives practical examples of UWB communication systems and concepts. Both theoretical and practical aspects are treated in each chapter of the book, in correspondence to each of the analyzed topics. Practical aspects are illustrated within the text in specifically highlighted sections which we have called "checkpoints". These checkpoints include MATLAB codes aimed at deepening the understanding of the theory, and also complementing it by introducing the simulation of case examples. The checkpoints should help the reader to fully understand the theoretical material, and also to integrate the theory with practical applications, such as the simulation of specific algorithms, for example, the IEEE 802.15.3a channel model. It is also hoped that training on practical examples will provoke thought and stimulate creative understanding of UWB radio communications. At the end of each chapter, a section titled "Further Reading" has been included, to give the reader suggestions about related literature.

Organization of the Book

This book can be schematically structured into three parts. The first part (Chapters 1 to 6) covers the UWB radio fundamental principles, modulation, and spectral characteristics. The second part (Chapters 7 and 8) is dedicated to channel modeling and reception. The third part (Chapters 9 to 11) moves to the networking aspects. In particular, the book is organized as follows.

Chapter 1 introduces the core concepts of UWB radio communications and sets a definition for the UWB radio signal. The UWB principle is also analyzed in light of recent regulation actions issued by U.S. authorities.

Chapter 2 surveys the different approaches that can be adopted to generate an UWB signal. Impulse radio methods based on the generation of pulses that are very short in time and that are either position-modulated (Pulse Position Modulation, or PPM) or amplitude-modulated (Pulse Amplitude Modulation, or PAM) are described. Data symbol encoding methods, such as time hopping and DS, are addressed. The chapter also discusses the generation of UWB signals using nonimpulsive schemes, such as OFDM, in which the ultra wide bandwidth is produced by a very high data rate.

Chapter 3 derives the PSD for time-hopping UWB signals using PPM. The adopted approach follows the analog PPM theory of the old days and reconciles this well-known modulation method with its digital variant currently in vogue in wireless communications.

Chapter 4 derives the PSD of DS-UWB signals. It also includes the derivation of the PSD for Time-Hopping UWB (TH-UWB) signals using PAM, since this spectrum can be derived in a straightforward manner from the DS case.

Chapter 5 deals with the analysis of spectral properties of a nonimpulsive modulation scheme and in particular of OFDM as proposed in the multi-band proposal to IEEE 802.15.TG3a.

Chapter 6 analyzes the problem of complying with emission masks as set by regulatory bodies. It first analyzes how to read and apply an emission mask. Second, it introduces the methodology for performing a link budget for a point-to-point UWB link.

Chapter 7 discusses the choice of the impulse response of the pulse shaper in impulse radio systems as a function of the PSD of the transmitted signal. It investigates the effect of pulse width variation and differentiation as well as a combination of different waveforms to generate a signal that complies with the power limitations set by the emission masks defined in Chapter 6.

Chapter 8 analyzes the signal at the receiver, after propagation over the radio channel. It first analyzes receiver structures for different modulation formats. It then proceeds in analyzing channel modeling and multi-path fading, presents a survey of traditional approaches, and includes a description and simulation of the UWB channel model proposed by the IEEE 802.15.TG3a. It includes an analysis of the RAKE receiver for multi-path environments, and ends with addressing the problem of synchronization between transmitter and receiver.

Chapter 9 moves up to the design of a multi-user UWB system. It analyzes multi-user interference and extends the performance analysis of Chapter 6 to a multi-user environment.

This chapter also forms the basis for understanding the algorithms that rule access to the medium, which are the object of Chapter 11.

Chapter 10 is devoted to the analysis of ranging and positioning algorithms and protocols, and to understanding how positioning and ranging information can be exploited to design power-aware and location-based routing strategies. Basic principles are reviewed, and a few examples of positioning systems, such as the Global Positioning System (GPS), are presented.

Chapter 11 deals with the MAC module. It first reviews examples of MAC implementations for a few popular wireless networks such as IEEE 802.11b, Bluetooth, and IEEE 802.15.3. It then introduces a proposal for an UWB-tailored MAC that attempts to take into account UWB-specific multi-user interference and synchronization issues, and incorporates the capability of providing a network of nodes with ranging information.

The appendix to the book briefly reviews the current trends in UWB standardization and provides the reader with a roadmap for locating information in the book related to standards.

AUDIENCE AND COURSE USE

This book is targeted to engineering graduate students, postdoctoral scholars, researchers, faculty members, scientists, and engineers in academia, as well as in the public and private sectors in the broad area of wireless communications. The book can serve both as a rapid introduction as well as a reference book to be used by the designer or in classrooms. The book is applicable to different course structures. Table P–1 indicates a few options on which chapters to study in a class, depending on course length and topic.

Table P–1 Book use for courses on UWB of different length and topic

Course description	Material
Short courses	
UWB principles (1 day)	Selected material from Chapters 1, 2, and 6
Ranging and positioning with UWB (1 day)	Chapters 1 and 10
MAC issues in UWB networks (1 day)	Selected material from Chapters 1, 2, 9, and 11
Standards activities of WPANs: IEEE 802.15.3a (3 days)	*Physical layer proposal*: Chapters 1 and 2 *Physical layer: spectral properties*: Chapters 4 and 5 *Channel model*: Section 8.2 and Checkpoint 8–2 *MAC layer*: Chapter 11

UWB principles for the less mathematical-oriented (2 days)	*Introduction to UWB signals and simulation tool MATLAB* *UWB definition and signal*: Chapters 1 and 2 *Performance of UWB link*: Chapter 6 *UWB localization*: Chapter 10
Impulse Radio (4 days)	*IR-UWB definition and IR-UWB signal*: Chapters 1 and 2, except Section 2.3 *IR-UWB spectral properties*: Chapters 3 and 4 *Emission masks, performance of UWB radio link, and Pulse shaper*: Chapters 6 and 7 *Muti-user communications*: Chapter 9
UWB system design (3 days)	*IR-UWB definition and IR-UWB signal*: Chapters 1 and 2, except Section 2.3 *Link budget*: Chapter 6 *Receiver and synchronization*: Chapter 8
UWB channel modeling (2 days)	*UWB definition*: Chapters 1 and 2 *UWB channel model*: Chapter 8
Graduate courses	
UWB communication systems: theory and simulation 1 term= 50 hours	*Introduction to UWB signals and simulation tool MATLAB* *Spectral properties of UWB signals*: Chapters 1 to 5 *Performance analysis for AWGN channel*: Chapters 6 and 7 *UWB channel and receiver*: Chapter 8 *Multi-user communications*: Chapter 9 *Location-aware MAC design* : Chapters 10 and 11

WEB SITE AND USE OF MATLAB

The Web site for this book is: *http://authors.phptr.com/dibenedetto.* It includes all the MATLAB functions introduced within the checkpoints. These functions are provided in the form of MATLAB m-files. The m-files are organized in separate directories corresponding to the checkpoints of the book. To use the functions with MATLAB, the reader must copy the m-files to a directory of the hard disk of a local computer. This directory with all its possible sub-folders must be then added to the "Search Path" of MATLAB (see MATLAB Help for information about this procedure).

All m-files have been tested using version 6 of MATLAB.

Acknowledgments

The writing of this book has been made possible thanks to the help of many people. In particular, we wish to thank the unwavering support and important contribution of Luca De Nardis throughout the book writing process. We are most grateful to Mauro Montanari of Thales Italia S.p.A. — Communications Division for support and encouragement especially at the onset of this book project, and to our colleagues and friends of the INFOCOM Department. We would also like to thank our many partners in European research projects for their collaboration, and in particular our friends of the UCAN and PULSERS projects as well as our project coordinators Hrjehor Mark and Gert Kreiselmeier. The support of our research over the past several years by the European Union is gratefully acknowledged. We also acknowledge our most valuable association with the Center of Excellence DEWS of the University of L'Aquila, in particular with its director, Prof. Maria Domenica Di Benedetto.

Over the years, our work has benefited from the interaction with many former students. They are too numerous to be mentioned, but we wish to thank in particular Maria Stella Iacobucci and Matthias Junk.

Regarding the preparation of the manuscript, we are indebted to Salvatore Falco for his valuable assistance in handling the manuscript and shaping it into a final formatted file. Thanks are also due to Luigi Taglione for his effort in the organization of the reference sections. We would also like to particularly thank our editor, Bernard Goodwin, our series editor, Theodore Rappaport, our production editor, Nicholas Radhuber, as well as the entire staff at Prentice-Hall for providing an outstanding editorial environment.

Finally, we wish to thank our families and friends. In particular, MGDB wishes to acknowledge the support of members of her truly Italian-style family: Pupa Falcone, Marika Di Benedetto, Alberto Sangiovanni-Vincentelli, Matteo Gobbi, Maria Ricci, Marco Sangiovanni-Vincentelli, and especially Felice Di Benedetto for his careful reading of the entire manuscript and for his encouragement. Without them and the friendship of Alessandra Fiumara and Abeer Alwan, this work would have not been completed. GG would like to express his profound gratitude to the two persons who inspired his passion for research: Maria-Gabriella Di Benedetto, his Ph.D. supervisor, and Gennaro Fedele, who, throughout his life, has had an ever lasting impact on his future directions.

MARIA-GABRIELLA DI BENEDETTO

GUERINO GIANCOLA

Rome, Italy
May 2004

Fluctuat Nec Mergitur

REFERENCES

Batra, A., et al., *Multi-band OFDM Physical Layer Proposal for IEEE 802.15 Task Group 3a*, Available at *www.multibandofdm.org/papers/15-03-0268-01-003a-Multi-band-CFP-Document.pdf* (September 2003).

Barrett, T.W., "History of UltraWideBand (UWB) Radar & Communications: Pioneers and Innovators," Proceedings of Progress In Electromagnetics Symposim 2000 (PIERS2000), Cambridge, MA (July 2000).

Black, H.S., J.W. Beyer, T.J. Grieser, and F.A. Polkinghorn, "A multichannel microwave radio relay system," *AIEE Transaction Electrical Engineering*, Volume: 65 (1946), 798–805.

Federal Communications Commission, "Revision of Part 15 of the Commission's rules Regarding Ultra-Wideband Transmission Systems: First report and order," *Technical Report FCC 02-48* (adopted February, 14 2002; released April 22, 2002).

IEEE 802.15.SG3a, "Channel modeling Sub-committee Report Final," IEEE P802.15-02/490r1-SG3a, (February 2003).

IEEE 802.15.3-2003, "IEEE standard for information technology — telecommunications and information exchange between systems — local and metropolitan area networks — specific requirements part 15.3: wireless medium access control (MAC) and physical layer (PHY) specifications for high rate wireless personal area networks (WPANs)," (September 2003).

IEEE 802.15.4-2003, "IEEE standard for information technology — telecommunications and information exchange between systems — local and metropolitan area networks specific requirements part 15.4: wireless medium access control (MAC) and physical layer (PHY) specifications for low-rate wireless personal area networks (LR-WPANs)," (October 2003).

Roberts, R., *XtremeSpectrum CFP Document*, Available at *grouper.ieee.org/groups/802/15/pub/2003/Jul03/03154r3P802-15_TG3a-XtremeSpectrum-CFP-Documentation.pdf* (July 2003).

Sobol, H., "Microwave Communications — An Historical Perspective," IEEE Transactions on Microwave Theory and Techniques, Volume: MTT-32, Issue: 9 (September 1984).

CHAPTER 1

Ultra Wide Band Radio Definition

Ultra Wide Band (UWB) radio is based on the radiation of waveforms, which are characterized by an instantaneous fractional energy bandwidth greater than about 0.20–0.25. In this chapter, we will analyze the fractional energy bandwidth principle and set the definition of UWB radio.

1.1 Fractional Bandwidth

The most common adoption of the term "Ultra Wide Band (UWB)" comes from the UWB radar world and refers to electromagnetic waveforms that are characterized by an instantaneous fractional energy bandwidth greater than about 0.20–0.25. To better understand this definition, we first need to define the energy bandwidth of the waveform. Let E be the instantaneous energy of the waveform; the energy bandwidth is then identified by the frequencies f_L and f_H, which delimit the interval where most of E (say over 90%) falls. We call the width of the interval $[f_L, f_H]$, i.e. $f_H - f_L$ the *energy bandwidth*.

In the radar field, UWB radio is based on the radiation of waveforms formed by a sequence of very short pulses; "very short" refers to a duration of the pulse that is typically about a few hundred picoseconds in communication systems. In these systems, the information to be transferred is represented in digital form by a binary sequence, and the information of each bit (0 or 1) is transferred using one or more pulses in a code-repetition fashion. The repetition of a pulse for representing a single bit increases robustness in the transmission of each bit. The pulsed transmission principle can be applied in a straightforward manner to communications, and forms the basis for the most common UWB transmitter scheme in communication systems. Recent regulations issued by U.S. authorities have extended the UWB concept to continuous transmission techniques, as will be further illustrated in Chapters 2, "The UWB Radio Signal," and 5, "The PSD of MB-UWB

Signals." For now, let us consider a pulsed transmission scheme (Di Benedetto and Vojcic, 2003).

It is important to note that E is indicated as an instantaneous energy in that it must be computed over an interval that corresponds to the duration of a pulse. If the decision concerning a single bit involves the processing of several pulses, as is usually the case, E refers to the overall energy of all the processed pulses engaged in the decision over one single bit. This is an important concept: If several pulses are used for the transmission of one bit, we are interested in specifying the energy of the group of pulses involved in the decision of a single bit since the effect noise at the receiver must be evaluated with respect to the energy of a well-based definition of useful signal energy.

Figure 1–1 illustrates the energy bandwidth concept. Note that if f_L is the lower limit and f_H is the higher limit of the Energy Spectral Density (ESD), then the center frequency of the spectrum is located at $(f_H + f_L)/2$. The fractional bandwidth is defined as the ratio of the energy bandwidth and the center frequency and is expressed as:

$$\text{fractional bandwidth} = \frac{(f_H - f_L)}{\left(\dfrac{f_H + f_L}{2}\right)} \tag{1–1}$$

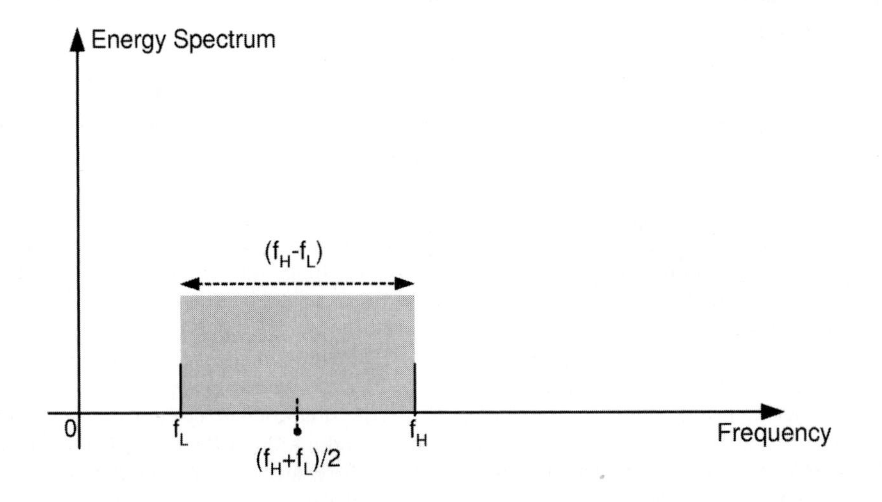

Figure 1–1 Energy bandwidth.

If the fractional bandwidth is greater than 0.20–0.25, we say that the signal is UWB. A signal with an energy bandwidth of 2 MHz, for example, is UWB if the center frequency of its spectrum is lower than 10 MHz. Note that the definition of an UWB signal can be given only relatively to the center frequency.

Often, the term "percent bandwidth" is used. *Percent bandwidth* is simply the fractional bandwidth in percent units. A signal with an energy bandwidth of 1 MHz and a center frequency of 2 MHz, for example, has a percent bandwidth of 50%. It is an UWB signal since its fractional bandwidth is 0.5, which is higher than the 0.20–0.25 lower limit.

Also in use is the relative bandwidth, which is equal to half the value of the fractional bandwidth. The relative bandwidth represents the ratio between half the energy bandwidth and the center frequency.

The lower and higher limits f_L and f_H need to be defined in less general terms. There may be different ways of selecting these frequencies, depending on how stringent the requirements on the bandwidth are set. In a recent release of UWB emission masks in the United States (Federal Communications Commission [FCC], 2002) f_L and f_H are set to the lower and upper frequencies of the -10 dB emission points. The selection of -10 dB over the -20 dB bandwidth established by the Defense Advanced Research Projects Agency (DARPA, 1990) is motivated by the fact that UWB emission is permitted at low power levels, which are close to the noise floor. Under these conditions, the -20 dB emission points cannot be measured reliably. Always in (FCC, 2002), a signal is assumed to be UWB if its bandwidth at -10 dB emission points exceeds 500 MHz, regardless of the fractional bandwidth value. The 500 MHz bandwidth value is lower than the 1.5 GHz minimum bandwidth limit established by DARPA (1990). The reduction is motivated by the use of the -10 dB bandwidth rather than the -20 dB bandwidth adopted in (DARPA, 1990).

The 500 MHz minimum bandwidth limit sets a threshold at 2.5 GHz. Below the threshold signals are UWB if their fractional bandwidth exceeds 0.20, while above the threshold signals are UWB if their bandwidth exceeds 500 MHz.

CHECKPOINT 1–1

In this first checkpoint, we introduce computer simulation for evaluating the bandwidth of a simple reference signal. Two MATLAB functions are considered: The first, Function 1.1, generates a rectangular waveform of unitary amplitude; the second, Function 1.2, evaluates and represents the bandwidth of any input signal.

Function 1.1 (see Appendix 1.A) generates a rectangular waveform with fixed time duration and modulates it at radio frequencies. The user must set the following parameters within the function: the time duration of the rectangular waveform `width`, the number of samples representing the rectangular waveform `points`, and the frequency `f0` of the carrier, which can be eventually introduced to transmit the waveform at radio frequencies. Function 1.1 returns two outputs: the waveform `signal` and the value of the sampling period `dt`.

We will use Function 1.1 for generating two signals. The first, vector `rect_A`, is a base-band rectangular waveform with length t = 100 ms, while the second, vector `rect_B`, is a modulated signal at frequency f_0 = 1 kHz having a rectangular envelope of the same length as `rect_A`. Signal `rect_A` is generated by setting the following parameters in Function 1.1: `width=1e-1; points=1000; f0=0`. The command line for generating the waveform is:

```
[rect_A,dt_A] = cp0101_genrect;
```

This command stores both the signal and the value of the sampling period in memory and produces the plot in Figure 1–2.

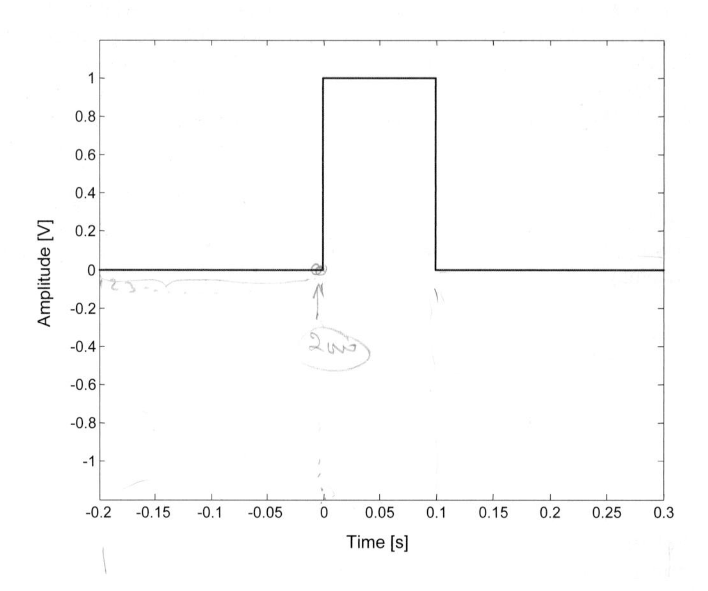

Figure 1–2 Unitary amplitude rectangular waveform with length t = 100 ms, rect_A.

To generate `rect_B`, one must introduce the following parameters in Function 1.1: `width=1e-1; points=1000; f0=1e3`. The command line for the second waveform is:

```
[rect_B,dt_B] = cp0101_genrect;
```

The graphical result of this operation is shown in Figure 1–3.

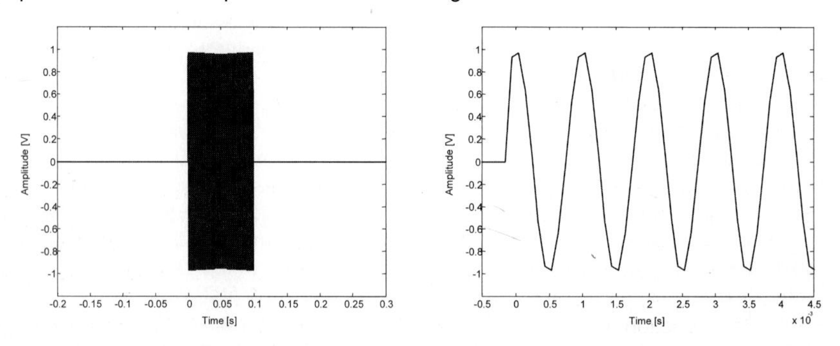

Figure 1–3 Unitary amplitude waveform with length t=100 ms, rect_B, modulated by a sinusoid at frequency f_0 = 1 kHz (left) and a detail of the same waveform in the range [0, 4.5 ms] (right).

The analysis of both `rect_A` and `rect_B` in the frequency domain requires the introduction of Function 1.2.

Function 1.2 (see Appendix 1.A) evaluates and represents the bandwidth of any input signal. The function receives three inputs: the array of samples representing the waveform to be analyzed in the frequency domain, vector `signal`; the corresponding value of the sampling period `dt`; and the threshold for evaluating bandwidth occupation `threshold`. Function 1.2 returns the single-sided ESD of the analyzed waveform `ss_E`, the value of the highest frequency `f_high`, the value of the lowest frequency `f_low`, and the resulting bandwidth `BW = f_high - f_low`.

In the following, we will use Function 1.2 for evaluating the bandwidth occupation for both signals `rect_A` and `rect_B`. With reference to `rect_A` (which is a base-band signal), we can evaluate both the bandwidth at -3 dB and the bandwidth at -10 dB. In the first case, threshold at -3 dB, we use the following command lines:

```
threshold=-3;
[ss_E,f_high,f_low,BW_A_minus3] = ...
   cp0101_bandwidth(rect_A, dt_A,threshold);
```

which provide the output:

```
>> Frequency Bandwidth = 6.000000 [Hz]
>> High Frequency = 6.000000 [Hz]
>> Low Frequency = 0.000000 [Hz]
```

In the second case, threshold at -10 dB, the commands are:

```
threshold=-10;
[ss_E,f_high,f_low,BW_A_minus10] = ...
   cp0101_bandwidth(rect_A,dt_A,threshold);
```

which return:

```
>> Frequency Bandwidth = 8.000000 [Hz]
>> High Frequency = 8.000000 [Hz]
>> Low Frequency = 0.000000 [Hz]
```

We can observe that the introduction of a different threshold has the effect of changing the width of the bandwidth that we assume as occupied by the same signal. In particular, we have a *computed* bandwidth $BW_{A,3}$ = 6 Hz with a threshold at -3 dB, and a *computed* bandwidth $BW_{A,10}$ = 8 Hz with a threshold at -10 dB. This result is shown in the graphical output (see Figure 1–4) provided by the function.

Note that Function 1.2 evaluates the bandwidth occupation of the input signal by first determining a vector representing the ESD, and then by identifying the first element of this vector that is below the given threshold. Because of the limited resolution in the representation of the ESD, the computed bandwidth can be greater than the theoretical one. This effect is represented in Figure 1–4 by the two lines that are generated by Function 1.2. The first line is horizontal and reproduces the threshold that is considered for evaluating the bandwidth; the intersection of this line with the plot identifies the theoretical bandwidth occupation of the ESD. The second line is vertical and crosses the plot in correspondence with the computed bandwidth. Obviously, the computed bandwidth approaches the theoretical bandwidth as the resolution of the ESD increases. Note that the upper and lower plots of Figure 1–4 have same resolution: The discrepancy between the computed and theoretical bandwidth, which appears to be reduced in the lower plot, is a matter of pure chance.

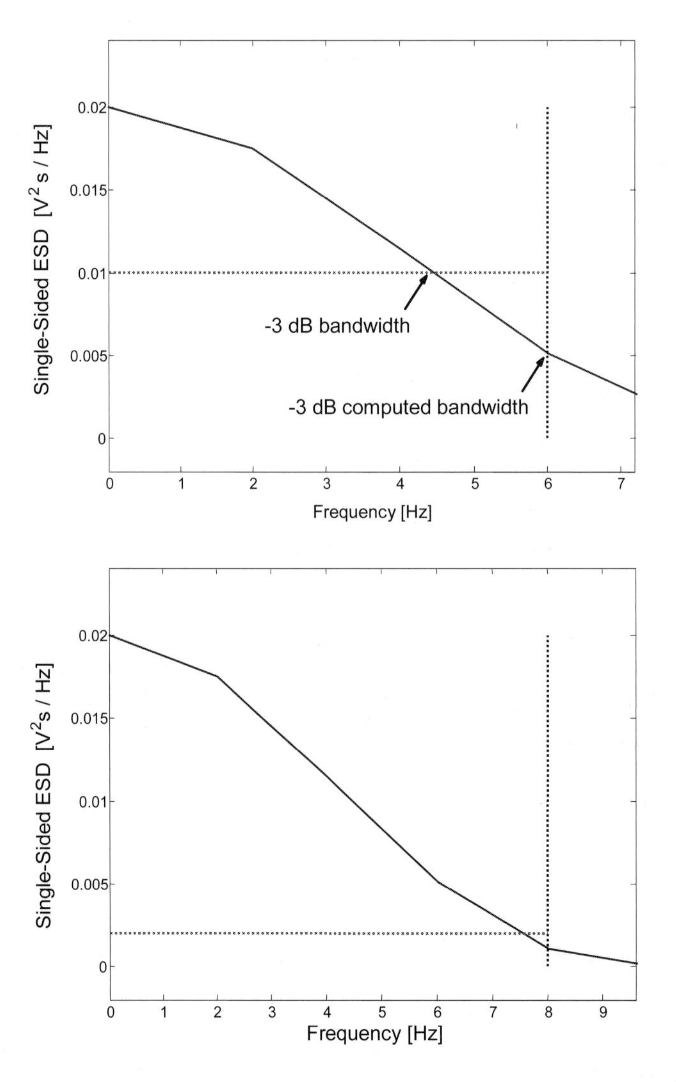

Figure 1–4 Energy Spectral Density (solid line) and occupied bandwidth (dashed line) in the case of signal rect_A (see Figure 1–2); (upper) bandwidth at -3 dB, (lower) bandwidth at -10 dB.

We can repeat the same operations for `rect_B` (which is modulated by a sinusoidal carrier with frequency f_0 = 1 kHz). In particular, we first evaluate the -3 dB bandwidth by typing the following command lines:

```
threshold=-3;
[ss_E,f_high,f_low,BW_B_minus3] = ...
   cp0101_bandwidth(rect_B,dt_B,threshold);
```

which provide the following output:

```
>> Frequency Bandwidth = 12.000000 [Hz]
>> High Frequency = 1006.000000 [Hz]
>> Low Frequency = 994.000000 [Hz]
```

In the case of signal `rect_B`, we have:

```
threshold=-10;
[ss_E,f_high,f_low,BW_B_minus10] = ...
   cp0101_bandwidth(rect_B,dt_B,threshold);
```

and the following output is observed:

```
>> Frequency Bandwidth = 16.000000 [Hz]
>> High Frequency = 1008.000000 [Hz]
>> Low Frequency = 992.000000 [Hz]
```

As expected, the single-sided bandwidth occupation of the modulated signal is twice the value of the bandwidth evaluated in the base-band case. In particular, we have a single-sided bandwidth occupation $BW_{B,3}$ = 12 Hz with the threshold at -3 dB and a single-sided bandwidth occupation $BW_{B,10}$ = 16 Hz with the threshold at -10 dB. A graphical output is provided by the function (see Figure 1–5).

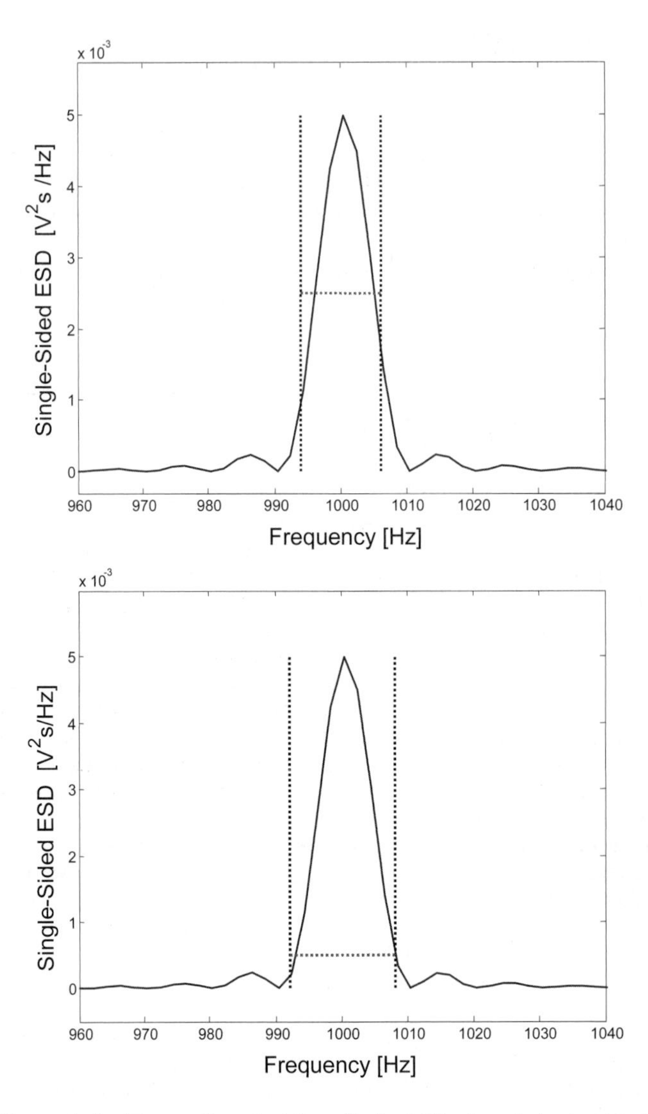

Figure 1–5 Energy Spectral Density (solid line) and single-sided bandwidth occupation (dashed line) in the special case of signal rect_B (see Figure 1–3); (upper) bandwidth at -3 dB, (lower) bandwidth at -10 dB.

CHECKPOINT 1–1

1.2 UWB vs. Non-UWB

The above UWB definition respects the traditional concept given in radio communications that a signal is wideband or broadband (vs. narrowband) when its bandwidth is large (vs. small) with respect to the modulating carrier frequency. In the case of UWB, due to the absence in general of a modulating carrier frequency, reference is made to the center frequency of the spectrum.

When it comes to using the UWB concept in communication systems, it is intuitive to note that signals formed by pulses with a duration on the order of fractions of nanoseconds will very likely be UWB.

CHECKPOINT 1–2

In this checkpoint, we will introduce computer simulation for generating different pulses. For each pulse, we will verify whether the pulse can be assumed to be UWB. Two MATLAB functions will be considered. The first, Function 1.3, generates pulses with fixed time duration T_P, composed of N_c complete cycles of a sinusoidal waveform. The N_c value determines the frequency of the sinusoid, that is, the higher N_c, the higher the frequency. The second function, Function 1.4, generates pulses composed of N_c complete cycles of a sinusoidal waveform with fixed frequency F_P. In this case, the N_c value determines the time duration T_P of the pulse, that is, the higher N_c, the longer the pulse. In both cases, sinusoidal pulses with finite length and constant envelope are considered. We can thus reasonably make use of the rule of thumb that states that the bandwidth is about $2/T_P$ around the frequency of the sinusoidal wave. By doing so we assume to discard the side lobes of the sin(f)/f function, which is the Fourier transform of a rectangular window of duration T_P. The first zeroes of this function are in fact located at $-1/T_P$ and $1/T_P$.

Function 1.3 (see Appendix 1.A) generates sinusoidal pulses with fixed time duration. The user must set the following parameters within the function: the time duration of the pulse `Tp`, the number of cycles of the sinusoidal waveform `Nc`, the amplitude of the pulse `A`, and the number of samples representing the pulse `smp`. Function 1.3 returns two outputs: the pulse waveform, vector `sinpulse`, and the value of the sampling period `dt`.

We will use Function 1.3 for generating two pulses: `sinpulse_A1` and `sinpulse_A2`. Both have time duration T_P = 10 ns, but `sinpulse_A2` contains a number of sinusoidal cycles that is twice the number in `sinpulse_A1`. In particular, we assume N_c = 8 for `sinpulse_A1` and N_c = 16 for `sinpulse_A2`. Pulse `sinpulse_A1` is generated by setting the following parameters in Function 1.3: `Tp=1e-8; Nc=8; A=1; smp=1000`. The command line for generating the waveform is the following:

```
[sinpulse_A1,dt] = cp0102_sinpulse_one;
```

This command stores in memory both the pulse waveform and the value of the sampling period. The numerical output provided by the function is:

```
>> Central Frequency = 0.800000 [GHz]
>> Bandwidth = 0.200000 [GHz]
>> Fractional Bandwidth = 0.250000
```

The corresponding plot is shown in Figure 1–6.

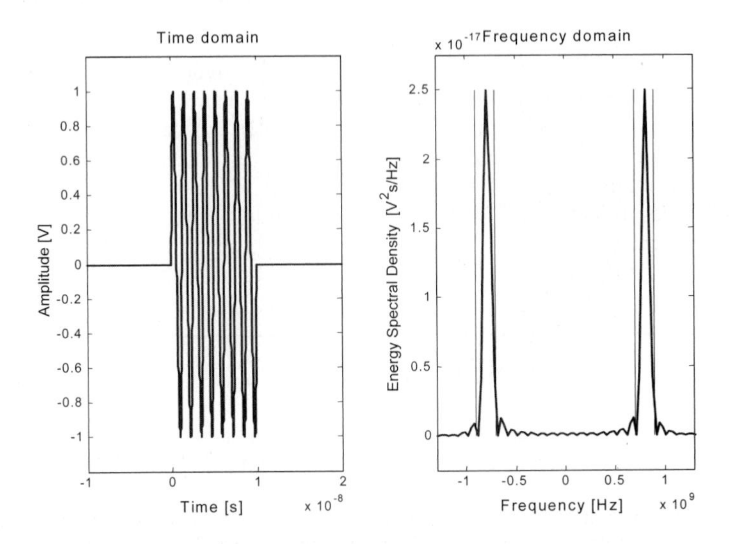

Figure 1–6 Representation of the pulse waveform sinpulse_A1 in the time domain (left) and the frequency domain (right). Vertical lines in the right plot delimit the null-to-null bandwidth.

Since the fractional bandwidth of the pulse is higher than 0.20, we conclude that `sinpulse_A1` is UWB.

To generate `sinpulse_A2`, we set the following parameters in Function 1.3: `Tp=1e-8; Nc=16; A=1; smp=1000`. The command line for generating the waveform is:

```
[sinpulse_A2,dt] = cp0102_sinpulse_one;
```

which returns the following output:

```
>> Central Frequency = 1.600000 [GHz]
>> Bandwidth = 0.200000 [GHz]
>> Fractional Bandwidth = 0.125000
```

Figure 1–7 shows the generated waveform in both the time and frequency domains.

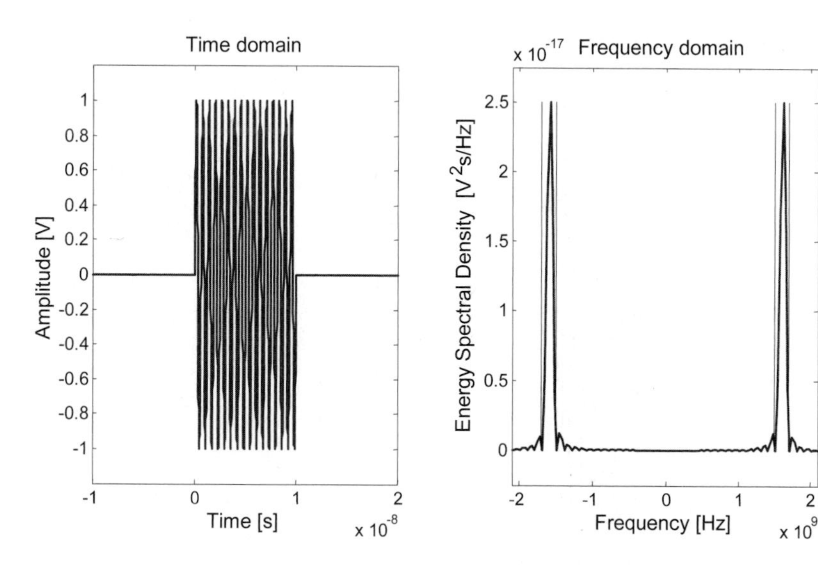

Figure 1–7 Representation of the pulse waveform sinpulse_A2 in the time domain (left) and the frequency domain (right). Vertical lines in the right plot delimit the null-to-null bandwidth.

The output of Function 1.3 in the case of `sinpulse_A2` shows that this pulse cannot be considered as UWB. Despite `sinpulse_A2` having same pulse duration of `sinpulse_A1`, that is, the same bandwidth occupation, the fractional bandwidth is decreased because of the shift of the signal spectrum to a higher frequency. This example shows that the definition of an UWB signal can be given only relatively to the central frequency of its spectrum. As the central frequency increases, the bandwidth must also increase to keep the fractional bandwidth at a constant value.

Function 1.4 (see Appendix 1.A) generates sinusoidal pulses with a fixed number of cycles. The user must set the following parameters within the function: the frequency of the sinusoidal waveform `Fp`, the number of cycles composing the waveform `Nc`, the amplitude of the pulse `A`, and the number of samples representing each cycle `smp`. Function 1.4 returns two outputs: the pulse waveform, vector `sinpulse`, and the value of the sampling period `dt`.

We will use Function 1.4 for generating two pulses: `sinpulse_B1` and `sinpulse_B2`. In both cases, we will adopt the same frequency F_P = 800 MHz for the sinusoidal waveform. Pulse `sinpulse_B2` contains a number of sinusoidal cycles that is twice the number in `sinpulse_B1`. In particular, we assume N_c = 8 for `sinpulse_B1`, and N_c = 16 for `sinpulse_B2`. Pulse `sinpulse_B1` is generated by setting the following parameters in Function 1.3: `Fp=8e8; Nc=8; A=1; smp=1000`. The command line for generating the waveform is:

```
[sinpulse_B1,dt] = cp0102_sinpulse_two;
```

This command stores both the pulse waveform and the value of the sampling period in memory. The output provided by the function (see Figure 1–8) is:

```
>> Central Frequency = 0.800000 [GHz]
>> Bandwidth = 0.200000 [GHz]
>> Fractional Bandwidth = 0.250000
```

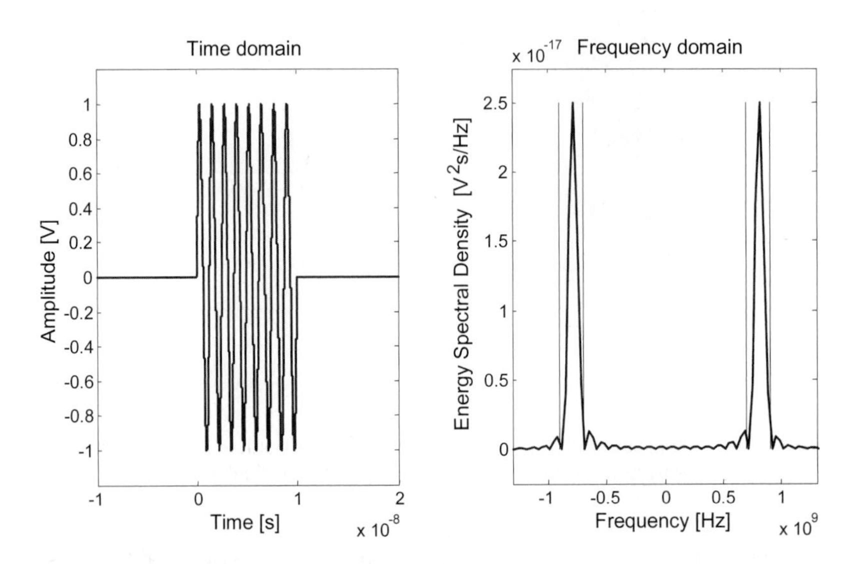

Figure 1–8 Representation of the pulse waveform sinpulse_B1 in the time domain (left) and the frequency domain (right). Vertical lines in the right plot delimit the null-to-null bandwidth.

Since the fractional bandwidth of the pulse is higher than 0.20, `sinpulse_B1` is UWB.

To generate `sinpulse_B2`, we set the following parameters in Function 1.4: `Fp=8e8; Nc=16; A=1; smp=1000`. The command line for generating the waveform is:

```
[sinpulse_B2,dt] = cp0102_sinpulse_two;
```

which returns the following output:

```
>> Central Frequency = 0.800000 [GHz]
>> Bandwidth = 0.100000 [GHz]
>> Fractional Bandwidth = 0.125000
```

Figure 1–9 shows the generated waveform in both the time and frequency domains.

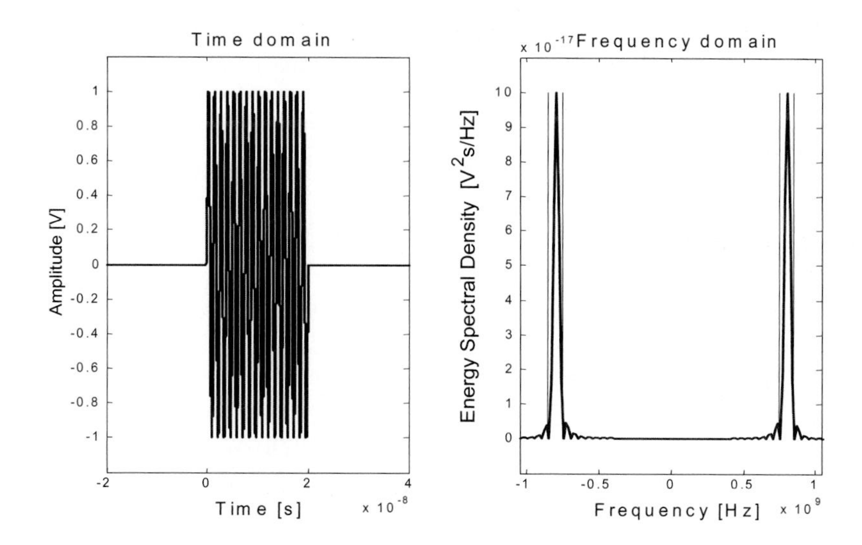

Figure 1–9 Representation of the pulse waveform sinpulse_B2 in the time domain (left) and the frequency domain (right). Vertical lines in the right plot delimit the null-to-null bandwidth.

The output of Function 1.4 in the case of `sinpulse_B2` shows that this pulse cannot be considered UWB. Despite the same central frequency of `sinpulse_B1`, the fractional bandwidth has decreased because of the longer duration of the pulse, that is because of the decreased width of the bandwidth occupied by the signal.

CHECKPOINT 1–2

Note that the signal used for transmission consists of a sequence of pulses of the type of Checkpoint 1–2, which are modulated by information data symbols. It remains to be verified that the bandwidth of the transmitted signal is large with respect to the center frequency. This verification requires the Power Spectral Density (PSD) of the radio signal to be expressed. Finding the PSD of UWB radio signals will be addressed in Chapters 3, "The PSD of TH-UWB Signals," 4, "The PSD of DS-UWB Signals," and 5, "The PSD of MB-UWB Signals."

Further Reading

There are currently no other published books on UWB radio in communications. The reader may find a general background on UWB radio when applied to radars in the two existing published books on UWB radar (Taylor, 1995; Taylor, 2001). The first reference covers

UWB radars, transmitters, antennas, radiation and propagation, radar cross-section and target scattering, receivers, signal processing, and performance prediction and modeling. The second reference is more recent and covers new semiconductor devices for pulse generation, micro-power impulse radar, vehicle sensing applications, and additional radar-specific topics.

REFERENCES

Defense Advanced Research Projects Agency, Office of the Secretary of Defense, "Assessment of Ultra-Wideband (UWB) Technology," Report R-6280, prepared by OSD/DARPA UWB Radar Review Panel (July 1990).

Di Benedetto, M.-G., and B.R. Vojcic, "Ultra Wide Band (UWB) Wireless Communications: A Tutorial," *Journal of Communication and Networks, Special Issue on Ultra-Wideband Communications*, Volume: 5, Issue: 4 (December 2003), 290–302.

Federal Communications Commission, "Revision of Part 15 of the Commission's rules Regarding Ultra-Wideband Transmission Systems: First report and order," *Technical Report FCC 02-48* (adopted February 14, 2002; released April 22, 2002).

Taylor, J.D. (Editor), *Introduction to Ultra-Wideband RADAR Systems*, Boca Raton, Florida: CRC Press (1995).

Taylor, J.D. (Editor), *Ultra-Wideband RADAR Technology*, Boca Raton, Florida: CRC Press (2001).

APPENDIX 1.A

Function 1.1 Generation of Rectangular Waveforms

Function 1.1 is composed of three steps. Step Zero contains all the parameters characterizing the waveform to be generated: the time duration of the rectangular waveform `width`, the number of samples representing the rectangular waveform `points`, and the frequency `f0` of the carrier, which can be eventually introduced to transmit the waveform at radio frequencies. Step One contains the code for generating the rectangular waveform and for modulating it at radio frequencies. Note that the generated signal has a time duration that is five times the length of the rectangle, that is, the total number of samples is `5*points`. Also note that we must set `f0=0` to generate a base-band signal. Step Two contains the code for the graphical representation of the generated waveform.

```
%
% FUNCTION 1.1 : "cp0101_genrect"
%
% Generates a unitary amplitude rectangular waveform
% that is modulated by a sinusoid at frequency 'f0'
% 'signal' is the output waveform
% 'dt' is the sampling period
%
% Programmed by Guerino Giancola
%

function [signal,dt] = cp0101_genrect

% ----------------------------
% Step Zero - Input parameters
% ----------------------------

width = 1e-1;        % width of the rectangle [s]
points = 1000;       % number of samples for representing the
                     % rectangle

f0 = 0;              % carrier frequency [Hz]

% ----------------------------
% Step One - Output evaluation
% ----------------------------
```

```
dt = width / points;        % sampling period
signal = zeros(1,5*points);
signal(2*points:3*points-1)=ones(1,points);

mod=cos(2.*pi.*f0.*linspace(1,5*width,5*points));
signal=signal.*mod;


% ----------------------------
% Step Two - Graphical output
% ----------------------------

figure(1)
time=linspace(-2*width,3*width,5*points);
P1=plot(time,signal);
set(P1,'LineWidth',[2]);
axis([-2*width 3*width -1.2 1.2]);
AX=gca;
set(AX,'FontSize',12);
X=xlabel('Time [s]');
set(X,'FontSize',14);
Y=ylabel('Amplitude [V]');
set(Y,'FontSize',14);
```

Function 1.2 Bandwidth evaluation

Function 1.2 is composed of three steps. Step One contains the code for evaluating the single-sided ESD of the input signal. The change of domain, for instance from time domain to frequency domain, is performed by means of the Fast Fourier Transform (FFT) algorithm. Different scaling factors are then introduced to operate the conversion from the discrete output of the FFT algorithm to the continuous spectrum required for our purposes. Step Two evaluates the bandwidth occupation of the signal under examination by applying an iterative algorithm that calculates the highest and lowest frequencies of the ESD according to a specified threshold. This algorithm works with both base-band and modulated signals. Finally, Step Three contains the code for the graphical output.

```
%
% FUNCTION 1.2 : "cp0101_bandwidth"
%
% Evaluates the bandwidth of the input 'signal' with
% sampling period 'dt'
%
% Bandwidth is evaluated according to the given 'threshold'
% (in dB)
% 'BW' is the bandwidth
% 'f_high' is the higher limit
% 'f_low' is the lower limit
%
% Programmed by Guerino Giancola
%

function [ss_E,f_high,f_low,BW] = ...
   cp0101_bandwidth(signal,dt,threshold)

% ---------------------------------------------
% Step One - Evaluation of the single-sided ESD
% ---------------------------------------------

fs = 1 / dt;           % sampling frequency
N = length(signal);    % number of samples (i.e., size of the
                       % FFT)
T = N * dt;            % time window
df = 1 / T;            % fundamental frequency

X = fft(signal); % double-sided MATLAB amplitude spectrum
X = X/N;         % conversion from MATLAB spectrum to
                 % fourier spectrum
ds_E = abs(X).^2/(df^2);      % double-sided ESD
```

```
ss_E = 2.*ds_E(1:floor(N/2)); % single-sided ESD

% ---------------------------------------------------
% Step Two - Evaluation of the frequency bandwidth
% ---------------------------------------------------

[Epeak,index] = max(ss_E);       % Epeak is the peak value
                                 % of the ESD

f_peak = index * df;             % peak frequency

Eth = Epeak*10^(threshold/10);   % Eth is the value of the
                                 % ESD corresponding to the
                                 % given threshold


% iterative algorithm for evaluating high and low
% frequencies

imax = index;
E0h = ss_E(index);

while (E0h>Eth)&(imax<=(N/2))

    imax = imax + 1;
    E0h = ss_E(imax);

end % while E0h > Eth

f_high = (imax-1) * df;              % high frequency

imin = index;
E0l = ss_E(index);

while (E0l>Eth)&(imin>1)&(index>1)

    imin = imin - 1;
    E0l = ss_E(imin);

end % while E0l > Eth

f_low = (min(index,imin)-1) * df;    % low frequency

% end of iterative algorithm
```

```
BW = f_high - f_low;    % signal frequency bandwidth


fprintf('\nFrequency Bandwidth = %f [Hz]\nHigh Frequency =
%f [Hz]\nLow Frequency =  %f [Hz]\n',BW,f_high,f_low);

% -----------------------------
% Step Three - Graphical output
% -----------------------------

figure(2)

frequency=linspace(0,fs/2,length(ss_E));
PF=plot(frequency,ss_E);
set(PF,'LineWidth',[2]);
L1=line([f_high f_high],[min(ss_E) max(ss_E)]);
set(L1,'Color',[0 0 0],'LineStyle',':')
L1=line([f_low f_low],[min(ss_E) max(ss_E)]);
set(L1,'Color',[0 0 0],'LineStyle',':')
L1=line([f_low f_high],[Eth Eth]);
set(L1,'LineWidth',[2],'Color','red','LineStyle',':')
axis([0.8*f_low 1.2*f_high -0.1*Epeak 1.2*Epeak]);
AX = gca;
set(AX,'FontSize',12);
T=title('Frequency domain');
set(T,'FontSize',14);
X=xlabel('Frequency [Hz]');
set(X,'FontSize',14);
Y=ylabel('Single-Sided ESD  [V^2s/Hz]');
set(Y,'FontSize',14);
```

Function 1.3 Generation of Sinusoidal Pulses with Fixed Time Duration

Function 1.3 is composed of five steps. Step Zero contains all the parameters characterizing the pulse to be generated: the time duration of the pulse `Tp`, the number of cycles of the sinusoidal waveform `Nc`, the amplitude of the pulse `A`, and the number of samples representing the pulse `smp`. Step One contains the code for generating the pulse. Note that the generated signal has a time duration that is three times the length of the rectangular envelope of the pulse, meaning that the total number of samples is `3*smp`. Step Two contains the code for analyzing the pulse in the frequency domain, following a procedure for calculating the ESD that is similar to the one of Function 1.2. Step Three evaluates the fractional bandwidth of the pulse. Note that in this case, the code does not require the introduction of a threshold since the bandwidth is evaluated as $2/T_P$. Finally, Step Four contains the code for the graphical output.

```
%
% FUNCTION 1.3 : "cp0102_sinpulse_one"
%
% Generates a pulse with length 'Tp',
% which is composed of 'Nc' cycles of a sinusoidal waveform
% 'sinpulse' is the waveform representing the pulse
%
% Programmed by Guerino Giancola
%

function [sinpulse,dt]=cp0102_sinpulse_one

% ----------------------------
% Step Zero - Input parameters
% ----------------------------

Tp = 1e-8;              % pulse duration [s]
Nc = 8;                 % number of cycles
A = 1;                  % pulse amplitude [V]

smp = 1000;             % number of samples for representing
                        % the pulse

% -----------------------------------------
% Step One - Generation of the reference pulse
% -----------------------------------------

f = Nc / Tp;
p = sin(2.*pi.*f.*linspace(0,Tp,smp));
```

```
sinpulse = zeros(1,3*smp);  % The pulse is represented in
sinpulse(1+smp:2*smp) = p;  % the center of a time window
                            % with length 3*Tp

% ------------------------------------------
% Step Two - Analysis in the frequency domain
% ------------------------------------------

fs = smp / Tp;          % sampling frequency
dt = 1 / fs;            % sampling period

N = length(sinpulse);   % number of samples (i.e. size of the
                        % FFT)

T = N * dt;             % time window
df = 1 / T;             % fundamental frequency

X=fft(sinpulse);  % double-sided MATLAB amplitude spectrum
X=X/N;            % conversion from MATLAB spectrum to
                  % Fourier spectrum

E = fftshift(abs(X).^2/(df^2));     % double-sided ESD

% ------------------------------
% Step Three - Output computation
% ------------------------------

fc = f;
fh = f + 1/Tp;
fl = f -1/Tp;
BW = 2/Tp;
FBW = 2*(fh-fl)/(fh+fl);

fprintf('\nCentral Frequency = %f [GHz]\nBandwidth = %f
[GHz]\nFractional Bandwidth = %f\n\n',fc*1e-9,BW*1e-9,FBW);


% -----------------------------
% Step Four - Graphical output
% -----------------------------

F=figure(3);
set(F,'Position',[100 190 850 450]);

subplot(1,2,1);
```

```
time=linspace(-Tp,2*Tp,3*smp);
PT=plot(time,sinpulse);
set(PT,'LineWidth',[2]);
axis([-Tp 2*Tp -1.2*A 1.2*A]);
AX=gca;
set(AX,'FontSize',12);
T=title('Time domain');
set(T,'FontSize',14);
X=xlabel('Time [s]');
set(X,'FontSize',14);
Y=ylabel('Amplitude [V]');
set(Y,'FontSize',14);

subplot(1,2,2)
frequency=linspace(-(fs/2),(fs/2),N);
PF=plot(frequency,E);
set(PF,'LineWidth',[2]);
L1=line([fh fh],[0 max(E)]);
set(L1,'Color',[0 0 0],'LineStyle',':')
L1=line([fl fl],[0 max(E)]);
set(L1,'Color',[0 0 0],'LineStyle',':')
L1=line([-fh -fh],[0 max(E)]);
set(L1,'Color',[0 0 0],'LineStyle',':')
L1=line([-fl -fl],[0 max(E)]);
set(L1,'Color',[0 0 0],'LineStyle',':')
fref=f+(5/Tp);
axis([-fref fref -(0.1*max(E)) 1.1*max(E)]);
AX=gca;
set(AX,'FontSize',12);
T=title('Frequency domain');
set(T,'FontSize',14);
X=xlabel('Frequency [Hz]');
set(X,'FontSize',14);
Y=ylabel('ESD  [V^2s/Hz]');
set(Y,'FontSize',14);
```

Function 1.4 Generation of Sinusoidal Pulses with Fixed Number of Cycles

Function 1.4 is composed of five steps. Step Zero contains all the parameters characterizing the pulse to be generated: the frequency of the sinusoidal waveform `Fp`, the number of cycles composing the waveform `Nc`, the amplitude of the pulse `A`, and the number of samples representing each cycle `smp`. Step One contains the code for generating the pulse. Note that the generated signal has a time duration that is three times the length of the pulse, meaning that the total number of samples is `3*Nc*smp`. Step Two contains the code for analyzing the pulse in the frequency domain, while Step Three evaluates the fractional bandwidth. Finally, Step Four contains the code for the graphical output.

```
%
% FUNCTION 1.4 : "cp0102_sinpulse_two"
%
% Generates a pulse composed of 'Nc' cycles of a
% sinusoidal waveform with fixed frequency 'Fp'
%
% Programmed by Guerino Giancola
%

function [sinpulse,dt]=cp0102_sinpulse_two

% ----------------------------
% Step Zero - Input parameters
% ----------------------------

Fp = 8e8;               % frequency of the sinusoid [Hz]
Nc = 8;                 % number of cycles composing the pulse
A = 1;                  % pulse amplitude [V]

smp = 1000;             % number of samples for representing
                        % each cycle

% -------------------------------------------
% Step One - Generation of the reference pulse
% -------------------------------------------

Tp = 1 / Fp;
p = sin(2.*pi.*Fp.*linspace(0,Nc*Tp,Nc*smp));

sinpulse=zeros(1,3*Nc*smp);      % the pulse is represented
sinpulse(1+Nc*smp:2*Nc*smp)=p;   % in the center of a time
                                 % window with length 3*Nc*Tp
```

```
% ------------------------------------------
% Step Two - Analysis in the frequency domain
% ------------------------------------------

fs = smp / Tp;              % sampling frequency
dt = 1 / fs;                % sampling period
N = length(sinpulse);       % number of samples (i.e., size
                            % of the FFT)

T = N * dt;                 % time window
df = 1 / T;                 % fundamental frequency

X=fft(sinpulse); % double-sided MATLAB amplitude spectrum
X=X/N;           % conversion from MATLAB spectrum to
                 % Fourier spectrum

E = fftshift(abs(X).^2/(df^2));    % double-sided ESD

% ------------------------------
% Step Three - Output computation
% ------------------------------

fc = Fp;
fh = Fp + 1/(Nc*Tp);
fl = Fp - 1/(Nc*Tp);
BW = 2/(Nc*Tp);
FBW = 2*(fh-fl)/(fh+fl);

fprintf('\nCentral Frequency = %f [GHz]\nBandwidth = %f
[GHz]\nFractional Bandwidth = %f\n\n',fc*1e-9,BW*1e-9,FBW);

% ----------------------------
% Step Four - Graphical output
% ----------------------------

F=figure(4);
set(F,'Position',[100 190 850 450]);

subplot(1,2,1);
time=linspace(-Tp*Nc,2*Tp*Nc,3*Nc*smp);
PT=plot(time,sinpulse);
set(PT,'LineWidth',[2]);
axis([-Tp*Nc 2*Tp*Nc -1.2*A 1.2*A]);
AX=gca;
set(AX,'FontSize',12);
T=title('Time domain');
```

```
set(T,'FontSize',14);
X=xlabel('Time [s]');
set(X,'FontSize',14);
Y=ylabel('Amplitude [V]');
set(Y,'FontSize',14);

subplot(1,2,2)
frequency=linspace(-(fs/2),(fs/2),N);
PF=plot(frequency,E);
set(PF,'LineWidth',[2]);
L1=line([fh fh],[0 max(E)]);
set(L1,'Color',[0 0 0],'LineStyle',':')
L1=line([fl fl],[0 max(E)]);
set(L1,'Color',[0 0 0],'LineStyle',':')
L1=line([-fh -fh],[0 max(E)]);
set(L1,'Color',[0 0 0],'LineStyle',':')
L1=line([-fl -fl],[0 max(E)]);
set(L1,'Color',[0 0 0],'LineStyle',':')
fref=Fp+(5/(Nc*Tp));
axis([-fref fref -(0.1*max(E)) 1.1*max(E)]);
AX=gca;
set(AX,'FontSize',12);
T=title('Frequency domain');
set(T,'FontSize',14);
X=xlabel('Frequency [Hz]');
set(X,'FontSize',14);
Y=ylabel('ESD  [V^2s/Hz]');
set(Y,'FontSize',14);
```

CHAPTER 2

The UWB Radio Signal

The most common and traditional way of emitting an UWB signal is by radiating pulses that are very short in time. This transmission technique goes under the name of *Impulse Radio* (IR). The way by which the information data symbols modulate the pulses may vary; Pulse Position Modulation (PPM) and Pulse Amplitude Modulation (PAM) are commonly adopted modulation schemes (Welborn, 2001; Guvenc and Arslan, 2003). In addition to modulation and in order to shape the spectrum of the generated signal, the data symbols are encoded using pseudorandom or pseudonoise (PN) codes. In a common approach, the encoded data symbols introduce a time dither on generated pulses leading to the so-called Time-Hopping UWB (TH-UWB). Direct-Sequence Spread Spectrum (DS-SS), that is, amplitude modulation of basic pulses by encoded data symbols, in the IR version indicated as Direct-Sequence UWB (DS-UWB), also seems particularly attractive (Huang and Li, 2001; Foerster, 2002; Vojcic and Pickholtz, 2003; Roberts, 2003). As is well-known, DS-SS has been adopted as the basic radio access technology for third-generation wireless communication systems (UMTS/IMT2000 in both Europe and Japan), and applying consolidated concepts is definitely appealing.

The UWB definition released by the FCC (FCC, 2002), as mentioned in the previous chapter, does not limit, however, the generation of UWB signals to IR and opens the way, at least in the United States, for alternative (nonimpulsive) schemes. An ultra wide bandwidth, say 500 MHz, might be produced by a very high data rate, independently of the characteristics of the pulses. The pulses might, for example, satisfy the Nyquist criterion at an operating pulse rate $1/T$, which would require a minimum bandwidth of $B = 1/(2T)$ and thus be limited in frequency, but unlimited in time having the classical raised-cosine infinity-bouncing shape with nulls at multiples of $1/T$. Systems with an ultra wide bandwidth of emission due to high-speed data rate rather than pulse width, provided that the fractional bandwidth or minimum bandwidth requirements are verified at all times of the transmission, are not precluded. Methods such as Orthogonal Frequency Division Multiplexing (OFDM) and Multi-Carrier Code Division Multiple Access (MC-CDMA) are capable of generating UWB signals at appropriate data rates.

Recent proposals in the United States, and in particular in the IEEE 802.15.TG3a Working Group, refer to a multi-band (MB) alternative to DS-UWB (Roberts, 2003) in which the overall available bandwidth is divided into sub-bands of at least 500 MHz (Foester et al., 2003; Yeh et al., 2003; Batra et al., 2003). Frequency-Hopping Spread Spectrum (FH-SS) might also be a viable path. This approach is currently less in vogue due to the difficulty in measuring the bandwidth at all instances of time, that is, stopping the frequency sweep and therefore complying with minimum bandwidth requirements. All of these methods are well-known in the wireless communications world, and are generously described in scientific journals and books (Proakis and Salehi, 1994; Webb and Hanzo, 1995; Hanzo et al., 2003). A detailed analysis of all these methods, which would represent a mere duplication of a well-consolidated literature, is definitely out of the scope of this book. We will, however, briefly describe and review the current MB proposal in Section 2.3.

The main focus of this chapter will be given to IR-UWB, and specifically to TH-UWB (Section 2.1) and DS-UWB (Section 2.2), in which spectrum expansion is obtained by using very short pulses in addition to the spreading introduced by coding. Also note that DS-SS, as well as OFDM and MC-CDMA, usually operate at Radio Frequencies (RFs), that is the base-band signal amplitude or phase modulates a carrier frequency. RF modulation is rarely mentioned in conjunction with UWB systems, which typically operate in the base-band. While RF modulation is obviously applicable to UWB as well, a shift in operating frequency can be potentially obtained here by pulse shaping.

TH-UWB and DS-UWB may adopt in principle either PPM or PAM for data modulation. A specific modulation method might, however, be more appropriate for one or the other as a function of the resulting spectrum shape and characteristics.

We will now focus on the generation process for TH-UWB, DS-UWB, and MB signals. Derivation of related Power Spectral Densities (PSDs) will be presented in Chapters 3, 4, and 5.

2.1 Generation of TH-UWB Signals

In TH-UWB combined with binary PPM (binary PPM-TH-UWB or 2PPM-TH-UWB), the UWB signal can be schematized to be generated as follows (see Figure 2–1 which represents the transmission chain).

Given the binary sequence to be transmitted $\mathbf{b} = (\ldots, b_0, b_1, \ldots, b_k, b_{k+1}, \ldots)$, generated at a rate of $R_b = 1/T_b$ bits/s, a first system repeats each bit N_s times and generates a binary sequence $(\ldots, b_0, b_0, \ldots, b_0, b_1, b_1, \ldots, b_1, \ldots, b_k, b_k, \ldots, b_k, b_{k+1}, b_{k+1}, \ldots, b_{k+1}, \ldots) = (\ldots, a_0, a_1, \ldots, a_j, a_{j+1}, \ldots) = \mathbf{a}$ at a rate of $R_{cb} = N_s/T_b = 1/T_s$ bits/s. This system introduces redundancy and is a $(N_s,1)$ block coder indicated as a code repetition coder. In the classical terminology this is a channel coder.

A second block called a transmission coder applies an integer-valued code $\mathbf{c} = (\ldots, c_0, c_1, \ldots, c_j, c_{j+1}, \ldots)$ to the binary sequence $\mathbf{a} = (\ldots, a_0, a_1, \ldots, a_j, a_{j+1}, \ldots)$ and generates a new sequence $\mathbf{d}$. The generic element of the sequence $\mathbf{d}$ is expressed as follows:

$$d_j = c_j T_c + a_j \varepsilon \tag{2–1}$$

where T_c and ε are constant terms that satisfy the condition $c_jT_c+\varepsilon < T_s$ for all c_j. One also has, in general, $\varepsilon < T_c$.

Note that **d** is a real-valued sequence as opposed to **a**, which is binary and to **c**, which is integer-valued. For now, we shall follow the most common trend and assume that **c** is a pseudorandom code, its generic element c_j being an integer verifying $0 \leq c_j \leq N_h - 1$. The code **c** might be periodic, and in that case, its period is indicated by N_p. Two particular cases are worth discussing. The first corresponds to the absence of periodicity in the code, that is, $N_p \rightarrow \infty$, and the second to $N_p = N_s$. In the second case, which is the most commonly adopted, the periodicity of the code coincides with the length of the repetition code. The effect of transmission coding will be made clear in Chapter 3, "The PSD of TH-UWB Signals," in which the PSD of the TH-UWB signal will be computed. However, we must keep in mind that the transmission coder plays a double role of code division multiple access coder and of spectrum shaper of the transmitted signal.

The coded real-valued sequence **d** enters a third system, the PPM modulator, which generates a sequence of unit pulses (Dirac pulses $\delta(t)$) at a rate of $R_p = N_s / T_b = 1/T_s$ pulses/s. These pulses are located at times jT_s+d_j, and are therefore shifted in time from nominal positions jT_s by d_j. Pulses occur at times $(jT_s+c_jT_c+a_j\varepsilon)$. Note that code **c** introduces a TH shift on the generated signal, and it is for this reason that it is indicated as TH code. Note that the shift introduced by the PPM modulator, $a_j\varepsilon$, is usually much smaller than the shift introduced by the TH code, c_jT_c, that is, $a_j\varepsilon < c_jT_c$, except for $c_j = 0$. T_c is called chip time.

The last system is the pulse shaper filter with impulse response $p(t)$. The impulse response $p(t)$ must be such that the signal at the output of the pulse shaper filter is a sequence of strictly non-overlapping pulses. The most commonly adopted pulse shapes will be analyzed in Chapter 7, "The Pulse Shaper."

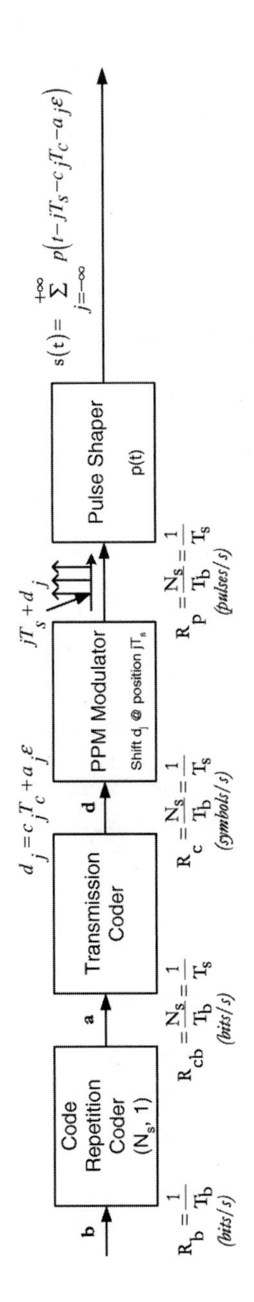

Figure 2–1 Transmission scheme for a PPM-TH-UWB signal.

The signal $s(t)$ at the output of the cascade of the above systems can be expressed as follows:

$$s(t) = \sum_{j=-\infty}^{+\infty} p\left(t - jT_s - c_j T_c - a_j \varepsilon\right) \quad (2\text{–}2)$$

Note that the bit interval, or bit duration, that is, the time used to transmit one bit T_b is: $T_b = N_s T_s$. Also note that in Eq. (2–2), the term $c_j T_c$ defines pulse randomization or dithering with respect to the nominal instances of time occurring at multiples of T_s. If we represent the time shift introduced by the TH code $c_j T_c$ by a random TH dither η_j, which can be assumed to be distributed between 0 and $T_\eta < T_s$, we obtain:

$$s(t) = \sum_{j=-\infty}^{+\infty} p\left(t - jT_s - \eta_j - a_j \varepsilon\right) \quad (2\text{–}3)$$

As noticed above, η_j is usually much larger than ε. The global effect of these two terms is to introduce a random time shift, distributed between 0 and $T_\eta + \varepsilon < T_s$, which will be indicated by θ_j leading to the following expression for the transmitted signal:

$$s(t) = \sum_{j=-\infty}^{+\infty} p\left(t - jT_s - \theta_j\right) \quad (2\text{–}4)$$

The concept leading to the signal of Eq. (2–2) can be generalized, the idea being that two different pulse shapes $p_0(t)$ and $p_1(t)$ can be transmitted in correspondence to the information bits "0" and "1". Note that the above analyzed case of a PPM modulator, which introduces a time shift ε depending on the bit to be represented, is a particular case in which $p_1(t)$ is a shifted version of $p_0(t)$. A more general expression is:

$$s(t) = \sum_{j=-\infty}^{+\infty} p_{a_j}\left(t - jT_s - c_j T_c\right) = \sum_{j=-\infty}^{+\infty} p_{a_j}\left(t - jT_s - \eta_j\right) \quad (2\text{–}5)$$

Equation (2–5) also represents the case of TH-UWB in combination with PAM (PAM-TH-UWB), when $p_1(t)$ is set to be $-p_0(t)$.

CHECKPOINT 2–1

This checkpoint is dedicated to the simulation of the whole transmission chain for the PPM-TH-UWB. The system model under examination is shown in Figure 2–2.

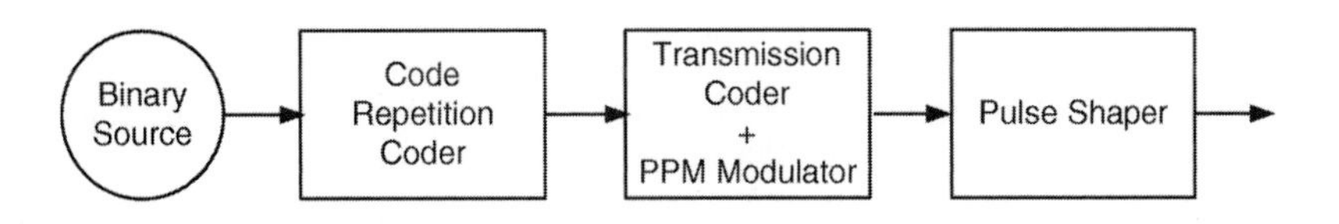

Figure 2–2 System model for the PPM-TH-UWB transmitter.

The first block in Figure 2–2 represents a binary source. The output of this block is a stream of bits to be transmitted on the physical channel. The activity of the binary source is simulated by introducing the MATLAB Function 2.1.

Function 2.1 (see Appendix 2.A) receives as input the number of bits to be generated, `numbits`, and returns as output vector `bits`, composed of `numbits` equiprobable binary values (0 and 1). The command line for generating `N` bits with Function 2.1 is:

```
bits = cp0201_bits(N);
```

This command stores in memory vector `bits`, composed of `N` equally probable binary values.

The second block of Figure 2–2 represents the code repetition coder. Each bit of the binary stream is repeated `Ns` times. The activity of the code repetition coder is simulated with Function 2.2.

Function 2.2 (see Appendix 2.A) receives two inputs: the original stream of bits `bits` and the cardinality of the code repetition coder `Ns`. Function 2.2 returns vector `repbits`. The command line for introducing redundancy with cardinality `Ns` into the input stream `bits` is the following:

```
repbits = cp0201_repcode(bits,Ns);
```

The third block of Figure 2–2 simulates both TH coding and binary PPM. Pseudorandom TH codes are considered. Each TH code is a sequence of `Np` integers randomly selected with uniform distribution on the interval `(0, Nh-1)`. Two MATLAB functions simulate the activity of this block. The first, Function 2.3, generates the pseudorandom TH code, and the second, Function 2.4, implements TH coding and PPM modulation.

Function 2.3 (see Appendix 2.A) generates a pseudorandom TH code with periodicity `Np` and cardinality `Nh`. Both `Np` and `Nh` are given as input to the function. Function 2.3 returns vector `THcode`, which contains the generated code with length `Np` and cardinality `Nh`. Given `Nh` and `Np`, the command line for generating the TH code is the following:

```
[THcode] = cp0201_TH(Nh,Np);
```

Function 2.4 (see Appendix 2.A) receives as input a binary stream and generates a train of Dirac pulses. The position in time of each output pulse is determined according to the pulse repetition period `Ts`, the TH code, and the binary PPM scheme. Function 2.4 receives the following inputs: the binary sequence to be modulated `seq`, the sampling frequency for the output signal `fc`, the value in seconds of the chip time `Tc`, the value in seconds of the average pulse repetition period `Ts`, the value in seconds of the PPM time shift `dPPM`, and the TH code to be applied on generated pulses `THcode`. Function 2.4 returns two output sequences. The first, `PPMTHseq`, is the train of pulses resulting from TH coding and 2PPM. The second, `THseq`, is a reference train of pulses corresponding to the same binary input but without PPM (TH coding only). Such a sequence will be fundamental for implementing the receiver blocks since it contains all the required information about the code without any indication of modulation. The command line for simulating the activity of the block implementing TH and 2PPM is the following:

```
[PPMTHseq,THseq] = ...
   cp0201_2PPM_TH(seq,fc,Tc,Ts,dPPM,THcode);
```

The last block of Figure 2–2 is the pulse shaper. The impulse response of this block represents the waveform of the basic pulse of the UWB signal to be transmitted. In this checkpoint, we adopt a pulse shaper filter with an impulse response equal to the second derivative of a Gaussian waveform as introduced in (Scholtz, 1993; Win and Scholtz, 2000). (See Chapter 7 for an extensive discussion regarding pulse shaping.) The impulse response of the pulse shaper is generated in Function 2.5. Note that different pulse waveforms can be introduced by simply modifying this block.

Function 2.5 (see Appendix 2.A) receives three inputs: the sampling frequency for representing the output signal `fc`, the time duration in seconds of the impulse response `Tm`, and the shape factor in seconds for the pulse waveform `tau`. The presence of a parameter for the time duration of the pulse is justified by the fact that the second derivative Gaussian pulse has nominally infinite duration. Consequently, it is reasonable to assume non-zero values only within a time window with length `Tm`. The choice of `Tm` strictly depends on the value of the shape factor `tau`. In particular, it has been shown in (Giancola, 2002) that `Tm=2.2*tau` leads to a truncation error energy that is 50 dB smaller than the original energy. Function 2.5 returns signal `w0`, which is the energy-normalized version of the second derivative Gaussian waveform having `tau` as the shape factor. The command line for generating the impulse response is the following:

```
[w0] = cp0201_waveform(fc,Tm,tau);
```

We can use Function 2.5 for evaluating the effect of changing the shape factor on the second derivative Gaussian pulse. We define the following variables:

```
fc = 10e10;
Tm = 0.9e-9;
tau1 = 0.2e-9;
tau2 = 0.3e-9;
tau3 = 0.4e-9;
```

Then we create the following waveforms:

```
[a1] = cp0201_waveform(fc,Tm,tau1);
[a2] = cp0201_waveform(fc,Tm,tau2);
[a3] = cp0201_waveform(fc,Tm,tau3);
```

The time axis for representing these signals can be generated by means of MATLAB function `linspace(x1,x2,N)`, which generates a row vector with `N` equally spaced points between `x1` and `x2`:

```
time = linspace(-Tm/2,Tm/2,length(a1));
```

We can now compare the three waveforms by generating the following plot:

```
P = plot(time,a1,time,a2,time,a3);
```

Figure 2–3 shows the plot `P` generated by running the above command, completed with axes labels and change of style for the three lines. As expected, the width of the pulses increases with the shaping factor.

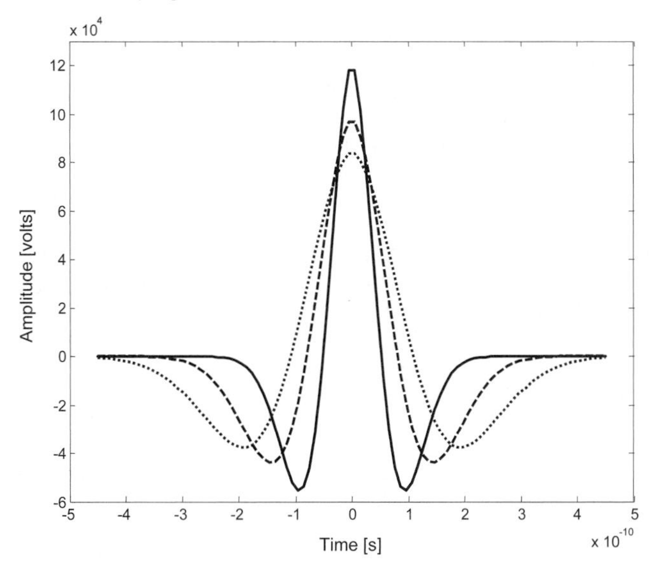

Figure 2–3 Second derivative Gaussian waveform in the special case of tau = 2 ns (solid line), tau = 3 ns (dashed line), and tau = 4 ns (dotted line).

The transmitted signal can be obtained by convolution of the train of pulses generated by Function 2.4 with the pulse shaper impulse response generated by Function 2.5. Convolution can be simulated by using MATLAB function `conv(A,B)`, which computes the convolution of vectors `A` and `B`.

With the five functions introduced above, we are capable of simulating the single blocks of the transmission system of Figure 2–2. We can simulate the whole transmission chain by introducing a new MATLAB function, **Function 2.6**, which executes in sequence all the command lines that have previously been explained. Within Function 2.6, the user can set the following parameters characterizing the signal: the average transmitted power in dBm `Pow`, the sampling frequency for representing the signal `fc`, the number of bits generated by the binary source `numbits`, the average pulse repetition time in seconds `Ts`, the number of pulses to be produced for each bit `Ns`, the chip time in seconds `Tc`, the cardinality and periodicity of the TH code `Nh` and `Np`, the impulse response duration `Tm`, the shaping factor for the pulse waveform in seconds `tau`, and the time shift of the PPM in seconds `dPPM`. Moreover, a graphical output is obtained by setting `G=1`.

Function 2.6 produces four outputs: the sequence of bits generated by the source `bits`, the TH sequence used by the TH coder `THcode`, the transmitted signal `Stx`, and a reference signal `ref` obtained by excluding the PPM modulator in the transmission chain. When implementing the receiver, we will use signal `ref` for implementing the correlator mask and vector `bits` for evaluating how many errors occurred during transmission over the channel.

The command line for simulating the activity of the UWB transmitter with Function 2.6 is:

```
[bits,THcode,Stx,ref] = cp0201_transmitter_2PPM_TH;
```

Figure 2–4 shows the UWB signal `Stx` generated by Function 2.6 in the special case of `Pow=-30, fc=50e9, numbits=2, Ts=3e-9, Ns=5, Tc=1e-9, Nh=3, Np=5, Tm = 0.5e-9, tau=0.25e-9`, and `dPPM=0.5e-9`.

In the case of Figure 2–4, Function 2.1 generated the sequence `bits=[1 0]`. We observe in fact that the first five pulses of the output sequence are located in the middle of the corresponding slots, while the last five pulses are located at the beginning of the corresponding slots. The effect of PPM on the pulse position is also shown in Figures 2–5 and 2–6. Figure 2–5, in particular, represents signal `Stx` in the time window corresponding to the first frame of the first bit. Here we observe that the pulse waveform starts `dPPM` seconds after the beginning of the second slot. In Figure 2–6, which represents `Stx` in the time window corresponding to the first frame of the second bit, the same pulse starts just at the beginning of the same slot.

With reference to the TH code, we can easily verify from Figure 2–4 that Function 2.3 generated the TH sequence `THcode = [1 0 1 0 2]`.

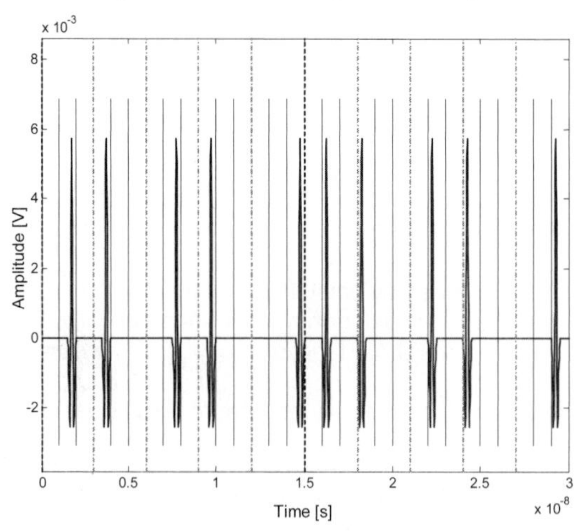

Figure 2–4 Signal generated by the PPM-TH-UWB transmitter in the case of Pow=-30, fc=50e9, numbits=2, Ts=3e-9, Ns=5, Tc=1e-9, Nh=3, Np=5, Tm = 0.5e-9, tau=0.25e-9, and dPPM=0.5e-9.

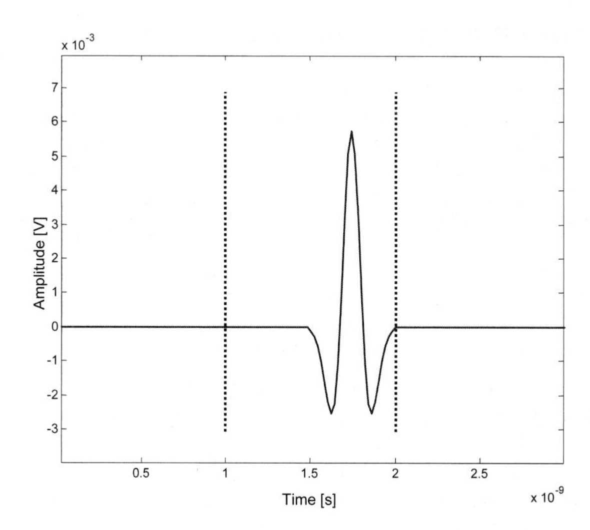

Figure 2–5 Detail of one pulse of the signal generated by the PPM-TH-UWB transmitter (see Figure 2–4). The plot shows the first frame of the first bit.

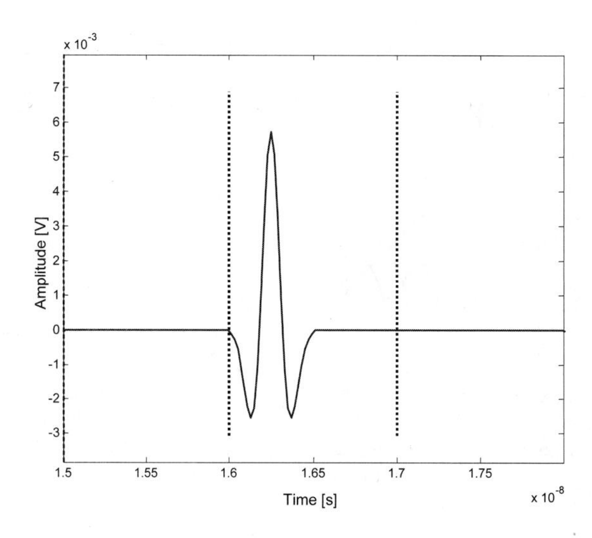

Figure 2–6 Detail of one pulse of the signal generated by the PPM-TH-UWB transmitter (see Figure 2–4). The plot shows the first frame of the second bit.

CHECKPOINT 2–1

2.2 GENERATION OF DS-UWB SIGNALS

Direct-Sequence Spread Spectrum (DS-SS) is a well-known digital modulation method and extensive literature is available on the topic. We will review its basic principles here, with particular focus on its extension to UWB.

Signals with an ultra wide bandwidth can be generated by first coding the binary sequence to be transmitted with a pseudorandom or PN binary-valued sequence, and then amplitude modulating a train of short pulses. This operation can be seen as an extreme case of the DS-SS systems currently in use, in which pulses are of the Nyquist type or rectangular in time with a typical time T_c indicated as chip time. The analytical expressions for this DS-SS-UWB signal can be easily derived by adopting a pulse width that is much smaller than the chip interval. In traditional DS-SS systems, the signal obtained is used to amplitude-modulate a carrier to bring the resulting signal at RF, usually with Binary Phase Shift Keying (BPSK) modulation. This operation can be avoided in DS-UWB unless it is specifically requested.

In more detail, the above signal can be generated as follows (see Figure 2–7, which shows the transmission chain).

Given the binary sequence to be transmitted $\mathbf{b} = (\ldots, b_0, b_1, \ldots, b_k, b_{k+1}, \ldots)$, generated at a rate of $R_b = 1/T_b$ bits/s, a first system repeats each bit N_s times and generates a binary sequence $(\ldots, b_0, b_0, \ldots, b_0, b_1, b_1, \ldots, b_1, \ldots, b_k, b_k, \ldots, b_k, b_{k+1}, b_{k+1}, \ldots, b_{k+1}, \ldots) = \mathbf{a}^*$ at a rate of $R_{cb} = N_s/T_b = 1/T_s$ bits/s. As in the TH scheme, this system introduces redundancy and is an $(N_s,1)$ code repetition coder.

A second system transforms the $\mathbf{a}^*$ sequence into a positive- and negative-valued sequence $\mathbf{a} = (\ldots, a_0, a_1, \ldots, a_j, a_{j+1}, \ldots)$, i.e.: $(a_j = 2a_j^* \text{-} 1, -\infty < j < +\infty)$.

The transmission coder applies a binary code $\mathbf{c} = (\ldots, c_0, c_1, \ldots, c_j, c_{j+1}, \ldots)$ composed of ± 1's and period N_p to the sequence $\mathbf{a} = (\ldots, a_0, a_1, \ldots, a_j, a_{j+1}, \ldots)$, and generates a new sequence $\mathbf{d} = \mathbf{a} \cdot \mathbf{c}$ composed of elements $d_j = a_j c_j$. N_p is commonly assumed to be equal to N_s. A more general assumption is to set N_p as a multiple of N_s. Note that $\mathbf{d}$ is a ± 1's sequence as is $\mathbf{a}$ and is generated at a rate $R_c = N_s/T_b = 1/T_s$ bits/s .

Sequence $\mathbf{d}$ enters a third system, the PAM modulator, which generates a sequence of unit pulses (Dirac pulses $\delta(t)$) at a rate of $R_p = N_s/T_b = 1/T_s$ pulses/s. These pulses are located at times jT_s.

The output of the modulator enters the pulse shaper filter with impulse response $p(t)$. In traditional DS-SS systems, the impulse response $p(t)$ is rectangular with duration T_s. In the DS-UWB case, $p(t)$ is a pulse with a duration much smaller than T_s, as analyzed in the TH case. Pulse shape will be analyzed in detail in Chapter 7.

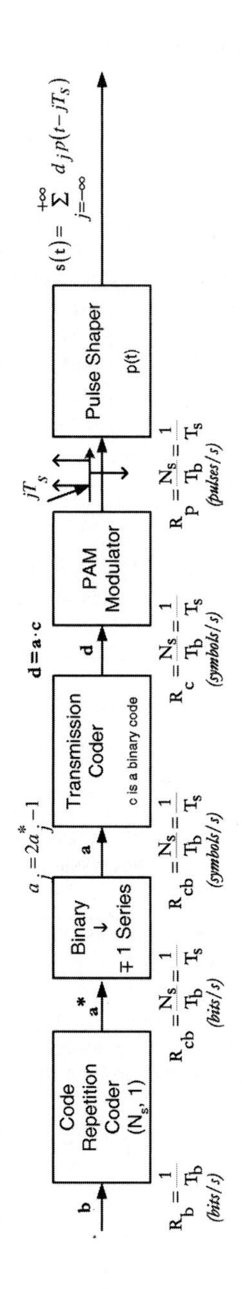

Figure 2–7 Transmission scheme for a PAM-DS-UWB signal.

The signal $s(t)$ at the output of the cascade of the above systems can be expressed as follows:

$$s(t) = \sum_{j=-\infty}^{+\infty} d_j p\left(t - jT_s\right) \tag{2-6}$$

Note that as in the TH case, the bit interval, or bit duration, that is, the time used to transmit one bit, T_b, is $T_b = N_s T_s$.

The resulting waveform is a straightforward PAM waveform. One can expect that evaluating the PSD of the signal of Eq. (2–6) is simpler than for Eq. (2–2) due to the absence of time shift and given the occurrence of pulses at regular time intervals.

A variant of the above method using a PPM modulator instead of a PAM modulator results in a signal expressed by:

$$s(t) = \sum_{j=-\infty}^{+\infty} p\left(t - jT_s - \varepsilon \frac{d_j + 1}{2}\right) \tag{2-7}$$

Note that in Eq. (2–7) the coding operation has the effect of whitening the spectrum due to the pseudorandom characteristics of the code.

CHECKPOINT 2–2

This checkpoint is dedicated to the simulation of the whole transmission chain for PAM-DS-UWB. The system model under examination is shown in Figure 2–8.

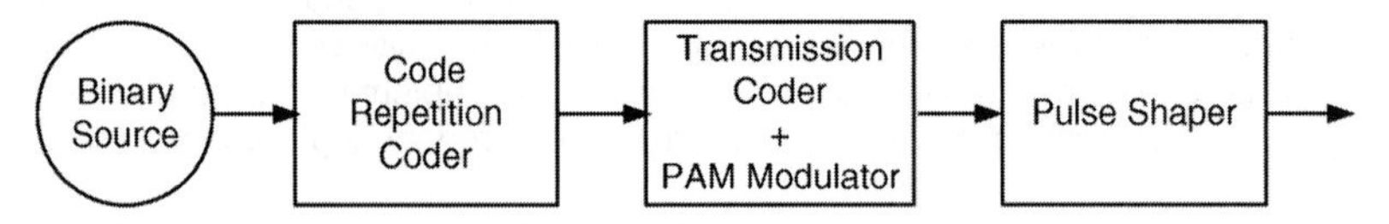

Figure 2–8 System model for the transmitter.

The first two blocks in Figure 2–8 correspond to the binary source and code repetition coder. The activity in these blocks can be simulated with Functions 2.1 and 2.2, introduced in the previous checkpoint.

The third block of Figure 2–8 implements both DS coding and binary PAM on the output of the code repetition coder. We consider pseudorandom DS codes, that is, the DS code that is assigned to a generic user is a sequence of N_p binary values. To simulate the activity of the block, we introduce two MATLAB functions: Function 2.7, which generates the pseudorandom DS code, and Function 2.8, which implements DS coding and PAM.

Function 2.7 (see Appendix 2.A) generates a pseudorandom DS code with periodicity `Np`. The `Np` value is given as input to the function. Given `Np`, the command line for generating the DS code is the following:

```
[DScode] = cp0202_DS(Np);
```

Function 2.8 (see Appendix 2.A) receives in input a stream of bipolar values and generates a train of Dirac pulses. All pulses are equally spaced in time with a fixed pulse repetition period `Ts`. The amplitude of each pulse is determined according to both the DS code and the PAM scheme. Function 2.8 receives the following inputs: the binary sequence to be modulated `seq`, the sampling frequency for the output signal `fc`, the value in seconds of the average pulse repetition period `Ts`, and the TH code to be applied on the generated pulses `DScode`. Function 2.8 returns two output sequences. The first, `PAMDSseq`, is the train of pulses resulting from DS coding and binary PAM. The second, `DSseq`, is a reference train of pulses corresponding to the same binary input but without PAM (DS coding only). Such a sequence will be fundamental for implementing the receiver blocks since it contains all the required information about the code without any indication on the modulation. The command line for simulating the activity of the block implementing DS and binary PAM is as follows:

```
[PAMDSseq,DSseq] = cp0202_2PAM_DS(seq,fc,Ts,DScode);
```

The last block in Figure 2–8 is the pulse shaper. The impulse response of the pulse shaper is generated by using Function 2.5 of Checkpoint 2–1.

The whole transmission chain in Figure 2–8 can now be simulated by introducing a new MATLAB function, **Function 2.9**, which executes in sequence all the command lines that have been previously described. Within Function 2.9, the user can set the following parameters characterizing the signal: the average transmitted power in dBm `Pow`, the sampling frequency for representing the signal `fc`, the number of bits generating from the binary source `numbits`, the average pulse repetition time in seconds `Ts`, the number of pulses to be produced for each bit `Ns`, the periodicity of the DS code `Np`, the impulse response duration `Tm`, and the shaping factor for the pulse waveform in seconds `tau`. A final parameter `G` determines whether the graphical output is required, `G = 1`, or not, `G = 0`. Function 2.9 generates four outputs: the sequence of bits generated by the source `bits`, the DS sequence `DScode` used by the DS coder, the transmitted signal `Stx`, and a reference signal obtained by excluding the PAM modulator in the transmission chain `ref`. When the receiver is to be implemented, we will use signal `ref` for implementing the correlator mask and vector `bits` for evaluating how many errors occurred during transmission over the channel.

The command line for simulating the activity of the UWB transmitter with Function 2.9 is:

```
[bits,DScode,Stx,ref] = cp0202_transmitter_2PAM_DS;
```

Figure 2–9 shows the UWB signal `Stx` generated by Function 2.9 in the special case of `Pow=-30`, `fc=50e9`, `numbits=2`, `Ts=2e-9`, `Ns=10`, `Np=10`, `Tm=0.5e-9`, and `tau=0.25e-9`. This signal consists of two groups of ten pulses. Each group is associated with one single bit generated by the source. In the case represented in Figure 2–9, in particular, Function 2.1 generated the sequence `bits=[1 0]`. We observe, in fact, that the second group of ten pulses is inverted in amplitude with respect to the first group of ten pulses.

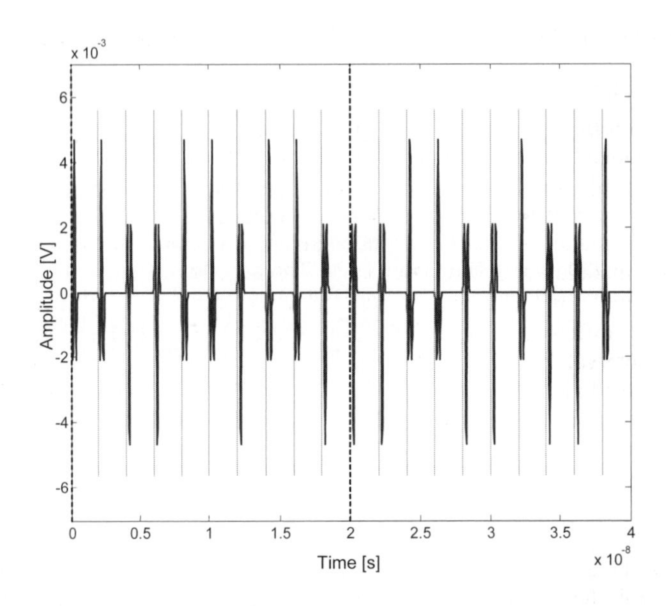

Figure 2–9 Signal generated by the PAM-DS-UWB transmitter in the case of Pow=-30, fc=50e9, numbits=2, Ts=2e-9, Ns=10, Np=10, Tm = 0.5e-9, and tau=0.25e-9.

CHECKPOINT 2–2

2.3 GENERATION OF MULTI-BAND UWB SIGNALS

The MB approach moves away from the IR principle analyzed in the previous sections of this chapter. In particular, based on the UWB definition released by the FCC (FCC, 2002) that a signal is UWB if its bandwidth exceeds 500 MHz, the overall 7.5 GHz bandwidth, that is, frequencies in the range 3.1 GHz to 10.6 GHz as based on the FCC ruling, is split into smaller frequency bands of at least 500 MHz each. The transmission of data for a given user occurs on different sub-bands in subsequent periods of time, leading to a system that can avoid nonintentional interference in certain bands without RF notch filters.

Different types of modulation can be adopted for data modulation within each sub-band. IR is not a must, and the correct spectrum occupancy can be obtained at appropriate bit rates. The most popular criterion, which is currently under analysis in the IEEE 802.15.TG3a (Batra et al., 2003), is based on the well-known Orthogonal Frequency

Division Multiplexing (OFDM) concept (Hanzo et al., 2003) and will be briefly summarized here.

An OFDM-modulated signal consists of the parallel transmission of several signals that are modulated at different carrier frequencies f_m. These carriers are equally spaced by Δf in the frequency domain. The binary sequence, which forms the input of the OFDM modulator, is subdivided into groups of K bits used to generate blocks of N symbols $\{d_0, \ldots, d_m, \ldots, d_{N-1}\}$, where the generic d_m assumes one of L possible values, with $K = N \log_2 L$. Finally, each symbol modulates a different carrier. To transmit the N symbols of the block in parallel, the signals modulating different carriers must be orthogonal in frequency. If T_0 is the time that is used for transmitting each symbol on the corresponding carrier, orthogonality among different transmissions can be achieved by adopting $\Delta f = 1/T_0$. In addition, a guard interval T_G is introduced between transmission of subsequent blocks mainly to prevent Inter-Symbol Interference (ISI). The total OFDM symbol duration is thus $T = T_0 + T_G$, leading to a maximum symbol rate of:

$$R_S = \frac{N}{T} = \frac{N}{T_0 + T_G} \tag{2–8}$$

The length of the guard interval is usually on the order of 20–30% of the total symbol duration T. In general, the guard interval is used for transmitting a copy of the final border section of the OFDM symbol, called the *cyclic prefix*, and is mainly introduced to maintain carrier synchronization at the receiver in the presence of time-dispersive channels. The length of the cyclic prefix is obviously limited by the duration of the guard interval. At the receiver, the cyclic prefix is discarded.

All modulators use the same rectangular shape $g_T(t)$ of finite duration T such that:

$$g_T(t) = \begin{cases} \sqrt{1/T} & for \quad -T_G = T_0 - T \le t \le T_0 \\ 0 & elsewhere \end{cases} \tag{2–9}$$

If $c_m = a_m + jb_m$ indicates the point in the constellation associated with symbol d_m, the OFDM signal corresponding to a block of N symbols is given by:

$$x(t) = g_T(t) \sum_{m=0}^{N-1} \left(a_m \cos\left(2\pi\left(f_p + f_m\right)t + \phi\right) - b_m \sin\left(2\pi\left(f_p + f_m\right)t + \phi\right)\right) \tag{2–10}$$

while the corresponding complex envelope is:

$$\underline{x}(t) = g_T(t) \sum_{m=0}^{N-1} c_m e^{j2\pi f_m t} \equiv \sum_{m=0}^{N-1} c_m \varphi_m(t) \equiv g_T(t) S(t) \tag{2–11}$$

in which $\varphi_m(t) = g_T(t) e^{j2\pi f_m t}$, and $S(t)$ is a periodic function of period T_0.

The simplest way of implementing an OFDM modulator is to adopt a digital structure by using the Discrete Fourier Transform (DFT), as first suggested by (Weinstein and Ebert, 1971).

The digital transmission of the OFDM signal in Eq. (2–10) corresponds to the transmission of a sampled version of the complex envelope in Eq. (2–11), that is, the transmission of the following sequence:

$$\underline{x}[n] = \underline{x}(nt_c) = g_T(nt_c)\sum_{m=0}^{N-1} c_m e^{j2\pi f_m t_c} \quad (2\text{–}12)$$

where t_c is the sampling period. According to Eq. (2–10), $x(t)$ consists of the parallel transmission of N signals with frequency occupancy $2\Delta f$, which modulate carriers spaced by Δf. As a consequence, it is reasonable to assume that the complex envelope in Eq. (2–11) occupies frequencies in the range $-B \div B$, with $B = N\Delta f/2$. The complex envelope of the OFDM signal can therefore be represented by samples taken at multiples of $t_c = T_0/N$:

$$\underline{x}[n] = g_T(nt_c)\sum_{m=0}^{N-1} c_m e^{j\frac{2\pi f_m n T_0}{N}} \quad (2\text{–}13)$$

Moreover, we can assume $f_m = m\,\Delta f\text{-}(N/2)$ and $\Delta f = m\,/T_0\text{-}N\,/(2T_0)$, leading to:

$$\underline{x}[n] = g_T(nt_c)\sum_{m=0}^{N-1} c_m e^{j\frac{2\pi mn}{N}} e^{-j\pi n} = g_T(nt_c)(-1)^n \sum_{m=0}^{N-1} c_m e^{j\frac{2\pi mn}{N}} \quad (2\text{–}14)$$

The summation in Eq. (2–14) corresponds to the n-th element of vector C representing the Inverse Discrete Fourier Transform (IDFT) of the vector $\{c_0, \ldots, c_n, \ldots, c_{N-1}\}$:

$$\underline{x}[n] = g_T(nt_c)(-1)^n C_n \quad (2\text{–}15)$$

Equation (2–15) indicates that the samples of the complex envelope of Eq. (2–10) can be obtained by computing the IDFT of the set of points of the coefficients $\{c_0, \ldots, c_m, \ldots, c_{N-1}\}$, which produces the IDFT sequence $\{C_0, \ldots, C_m, \ldots, C_{N-1}\}$. As suggested in (Van Nee and Prasad, 1999), however, the serial transmission of the sequence in Eq. (2–15) does not allow the reproduction of a real OFDM signal since the absence of oversampling would introduce intolerable aliasing when passing this sequence to the digital-to-analog converter. A solution to the above problem is to introduce zero padding within the input sequence $\{c_0, \ldots, c_m, \ldots, c_{N-1}\}$ before computing the IDFT. Note that the introduction of zeroes in the input sequence corresponds to the introduction of additional sub-carriers with zero amplitude. As a consequence, the zeroes of the oversampling should be added in the middle of the input vector rather than appending them at the end. In this way, the additional sub-carriers will be located at frequencies close to plus and minus half of the sampling frequency, and they will not interfere with the original sub-carriers of the OFDM signal.

Since the IDFT is a periodic sequence with period N, the introduction of the cyclic prefix in the digital domain is performed by simply appending the last $N_G = T_G/t_c$ elements of the original sequence in Eq. (2–15) to the beginning of the sequence itself.

Note that the symbol rate of Eq. (2–8) is a raw bit rate, that is, it corresponds to an encoded user data rate. If the raw bit rate is fixed, the redundancy introduced by a transmission coder depends on the user data rate, as further explained in the next checkpoint.

CHECKPOINT 2–3

In this checkpoint, we will use computer simulation to generate MB-UWB signals. The special case where Quadrature Phase Shift Keying (QPSK) symbols modulate the different carriers of the OFDM system is taken into account. Two new MATLAB functions are required: Function 2.10, which receives a stream of binary values and returns a sequence of QPSK symbols, and Function 2.11, which produces the multi-carrier signal in Eq. (2–10).

Function 2.10 (see Appendix 2.A), receives in input a stream of bipolar values `bits` and returns the corresponding sequence of QPSK symbols `S`, plus the two sequences `Sc` and `Ss` containing the real and imaginary parts of `S`. We can apply Function 2.10 by first generating a stream of binary values `bits` with Function 2.1 and by then executing the following command line:

```
[S,Sc,Ss] = cp0203_qpsk_mod(bits);
```

Function 2.11 simulates the whole transmission chain that produces the signal of Eq. (2–10). Within Function 2.11, the user can set the following parameters characterizing the signal: the total number of bits to be transmitted `numbits`, the central frequency of the modulated signal `fp`, the value of the sampling frequency `fc`, the time used for transmitting each symbol on the corresponding carrier `T0`, the time duration of the cyclic prefix `TP`, the time duration of the guard interval `TG`, the amplitude `A` of the rectangular impulse response `gT(t)`, and the number `N` of different carriers of the OFDM system. The command line for executing Function 2.11 is:

```
[bits,S,SI,SQ,Stx,fc,fp,T0,TP,TG,N] = cp0203_OFDM_qpsk;
```

We will use Function 2.11 for simulating the physical layer of the MB-OFDM scheme, which was proposed to the IEEE 802.15.TG3a in July 2003 (Batra et al., 2003). This proposal was submitted in response to the P802.15 Alternate PHY Call for Proposal by several industrial and academic contributors. The basic idea of the proposed scheme consists of dividing the spectrum into several sub-bands and then using OFDM modulation for transmission on each sub-band. The bandwidth of each sub-band is greater than 500 MHz, so the resulting signal is UWB according to FCC rules (FCC, 2002). Information bits are interleaved across all bands to better exploit frequency diversity over the whole frequency spectrum. Inside each band, however, transmitter architecture is similar to that of a conventional OFDM system, that is, the signal transmitted in each sub-band is derived according to the analysis that was previously presented in Section 2.3.

Different bit rates at the physical layer can be sustained as follows: 55 Mbits/s, 80 Mbits/s, 110 Mbits/s, 160 Mbits/s, 200 Mbits/s, 320 Mbits/s, and 480 Mbits/s (see Table 5–1). In addition, different coding rates and different spreading factors are applied to the original binary stream, to obtain a fixed 640 Mbits/s value at the input of the OFDM modulator. As an example, a coding rate of 11/32 and a spreading rate of 4 are applied in the 55 Mbits/s case, while a coding rate of 5/8 and a spreading rate of 2 are applied in the 200 Mbits/s case. In the first case, the coding operation produces 32 coded bits every 11 input bits, and the spreading operation associates 4 binary values for each coded bit, leading to a raw bit rate at the modulator input equal to:

$$R_b = 55 \times 10^6 \times \left(\frac{11}{32}\right)^{-1} \times 4 = 640 \times 10^6 \quad bits/s \qquad (2–16)$$

In the second case, the coding operation produces 8 coded bits every 5 input bits, and the spreading operation associates 2 binary values for each coded bit, leading to a raw bit rate at the modulator input of:

$$R_b = 200 \times 10^6 \times \left(\frac{5}{8}\right)^{-1} \times 2 = 640 \times 10^6 \quad bits/s \qquad (2–17)$$

All bits at the receiver input are mapped onto QPSK symbols. The resulting symbol rate R_S is therefore:

$$R_s = \frac{R_b}{\log_2 4} = 320 \times 10^6 \quad symbols/s \qquad (2–18)$$

On each sub-band, the transmitter generates OFDM symbols with a time duration of 312.5 ns. The number of symbols that must be transmitted within each OFDM symbol is thus:

$$N_s = R_s \times 312.5 \times 10^{-9} = 100 \qquad (2–19)$$

Each OFDM symbol consists of a guard time of 70.1 ns and an information period of 242.4 ns. The guard time contains a cyclic prefix of 60.6 ns, plus an initial interval of 9.5 ns, which is necessary for providing a sufficient time for switching between different sub-bands. The information period is used for transmitting the 100 information symbols (see Eq. (2–19)) plus 28 pilot symbols. 128 different carriers are thus necessary for transmitting the symbols within each block. In other words, each symbol of the block (both data and pilot) modulates the corresponding carrier for a period of T_0=242.4 ns, leading to a frequency separation of 4.1254 MHz between adjacent carriers. The bandwidth of each OFDM signal is thus 128x4.1254 = 528.5 MHz.

To simulate the generation of the OFDM signal inside one of the sub-bands, we can apply Function 2.11 with the following parameters: `numbits = 4096; fp = 1e9; fc = 1e11; T0 = 242.4e-9; TP = 60.6e-9; TG = 70.1e-9; A = 1; N = 128`. The command line for generating the signal is:

```
[bits,S,SI,SQ,Sofdm,fc,fp,T0,TP,TG,N]=cp0203_OFDM_qpsk;
```

The above command line stores in memory vector `Sofdm`, representing the output of the OFDM modulator. The first OFDM symbol of signal `Sofdm` is shown in Figure 2–10. Here, we can observe the nonimpulsive nature of the transmitted signal, although still UWB, given the bandwidth occupation greater than 500 MHz. Figure 2–10 also highlights the presence of the cyclic prefix at the beginning of the transmitted symbol.

The spectral analysis of OFDM-UWB signals will be performed in Chapter 5, "The PSD of MB-UWB Signals."

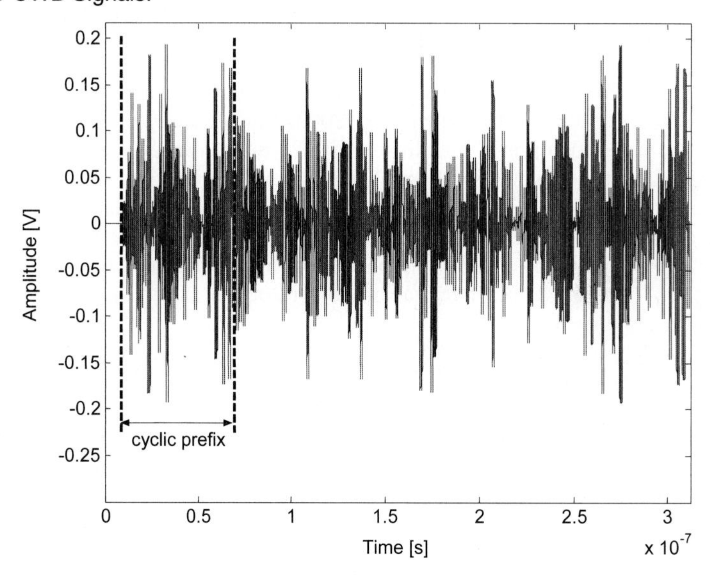

Figure 2–10 First OFDM symbol of signal Sofdm.

CHECKPOINT 2–3

A variant of OFDM called MC-CDMA consists of applying the above scheme to a code-repeated sequence, similar to the concept of using multiple pulses for conveying one bit. In this case, each sub-carrier of the OFDM symbol is modulated by the same source symbol, which eventually is complex, as in OFDM (Prasad and Hara, 1996). This procedure provokes a spreading of the information over the available bandwidth, and can thus be used for granting access to multiple users based on a one-to-one user code assignment: Each user has a unique code $Q^{(k)} = \{q^{(k)}(0), q^{(k)}(1), \ldots, q^{(k)}(N-1)\}$, which multiplies the code-repeated symbol $c^{(k)}$ before modulating the sub-carriers. The complex envelope of Eq. (2–11) for one MC-CDMA symbol of user k can be written as follows:

$$\underline{x}^{(k)}(t) = g_T(t)\sum_{m=0}^{N-1} c^{(k)} q^{(k)}(m) e^{j2\pi f_m t} = g_T(t) c^{(k)} \sum_{m=0}^{N-1} q^{(k)}(m) e^{j2\pi f_m t} \qquad (2\text{–}20)$$

FURTHER READING

Concerning PPM, it is particularly interesting to understand the case of analogue PPM, which was used in the old days. This analysis will be carried out in the next chapter, in which we will make frequent reference to the work by (Bennett, 1933, 1944, and 1947) and (Rowe, 1965). General PPM principles are reported in (Zhao and Haimovich, 2002; Souilmi and Knopp, 2003; Zhang et al., 2003; August et al. 2003; Jung et al. 2003; Chu and Murch, 2003). Recent work by (Hoctor and Tomlinson, 2002) and (Van Stralen et al., 2002) describes the experimental hardware design and implementation for a delay hopped transmitted reference consisting of two versions, one modulated by the data and the other unmodulated, of a wideband carrier. (Trindade et al., 2003) introduce a signal processing model for the Hoctor and Tomlinson delay hopped, transmitted reference. Finally, the work by (Buchegger et al., 2003) describes two transmitter prototypes based on a step recovery diode and an avalanche transistor.

For more information concerning PAM modulation, DS-CDMA, and OFDM, we cite works that address general concepts, such as (Hengstler et al., 2002; Venkatesan et al., 2003; Runkle et al., 2003).

Several papers were published in the past year proposing the MB approach. Most of these papers were published in the *Proceedings of the IEEE Conference on Vehicular Technology 2003*, the *IEEE Conferences on Ultra Wideband Systems and Technologies 2002 & 2003*, the *IEEE GLOBECOM 2002 & 2003*, the *IEEE MILCOM 2003*, and the *IEEE International Conference on Communications 2003*. Among these papers, we will mention the modulation schemes described by (Gerakoulis and Salmi, 2002; Saberinia et al., 2003; Saberinia and Tewfik, 2003a, 2003b; Tewfik and Saberinia, 2002; Askar et al., 2003; Balakrishnan et al., 2003; Nassar et al., 2003). The use of Hadamard codes is investigated in (Gao et al., 2003). (Lu et al., 2003) report the first published architecture for a MB transceiver. The work by (Okamoto et al., 2003) proposes a wavelet approach.

A different approach based on modulation of the pulse spectrum is presented by (D'Souza and Postula, 2003). Different modulation schemes, including pulse shape modulation, are compared in recent papers by (Ney da Silva and de Campos, 2003) and by (Wilson and Scholtz, 2003). Pulse polarity modulation is investigated in (Xu et al., 2003).

Pulse interval modulation, that is, differential PPM, is addressed by (Cariolaro et al., 2000, 2001). In particular, the authors discuss the interesting property of gain in efficiency of differential PPM vs. PPM in terms of capacity and bandwidth requirements.

REFERENCES

Askar, N.K., S.C. Lin, H.D. Pfister, G.E. Rogerson, and D.S. Furuno, "Spectral Keying™: A Novel Modulation Scheme for UWB Systems," *IEEE Conference on Ultra Wideband Systems and Technologies* (November 2003), 418–422.

August, N.J., W.C. Chung, and D.S. Ha, "Energy Efficient Methods of Increasing Data Rate for Ultra Wideband (UWB) Communications Systems," *IEEE Conference on Ultra Wideband Systems and Technologies* (November 2003), 280–284.

Balakrishnan, J., A. Batra, and A. Dabak, "A Multi-Band OFDM System for UWB Communication," *IEEE Conference on Ultra Wideband Systems and Technologies* (November 2003), 354–358.

Batra, A., et al., *Multi-band OFDM Physical Layer Proposal for IEEE 802.15 Task Group 3a*, Available at *www.multibandofdm.org/papers/15-03-0268-01-003a-Multi-band-CFP-Document.pdf* (September 2003).

Bennett, W.R., "New results in the calculation of modulation products," *Bell System Tech. J.*, no. 12 (1933), 228–243.

Bennett, W.R., "Response of a linear rectifier to signal and noise," *J. Acoust. Soc. Am.*, no. 15 (1944), 164–172.

Bennett, W.R., "The biased ideal rectifier," *Bell System Tech. J.*, no. 26 (1947), 139–169.

Buchegger, T., G. Ossberger, A. Reisenzahn, A. Stelzer, and A. Springer, "Pulse Delay Techniques for PPM Impulse Radio Transmitters," *IEEE Conference on Ultra Wideband Systems and Technologies* (November 2003), 37–41.

Cariolaro, G., T. Erseghe, and L. Vangelista, "Stationary Model of Pulse Interval Modulation and Exact Spectral Evaluation," *IEEE International Conference on Communications*, Volume: 2 (June 2000), 660–664.

Cariolaro, G., T. Erseghe, and L. Vangelista, "Exact Spectral Evaluation of the Family of Digital Pulse Interval Modulated Signals," *IEEE Transactions on Information Theory*, Volume: 47, Issue: 7 (November 2001), 2983–2992.

Chu, X., and R.D. Murch, "Quadrature Modulation for UWB Wireless Multipath Channels," *IEEE Global Telecommunications Conference* (December 2003), 431–435.

D'Souza, M., and A. Postula, "Novel Ultra-Wideband Pulse Spectrum Modulation Scheme," *IEEE Conference on Ultra Wideband Systems and Technologies* (November 2003), 240–244.

Da Silva, J.A.N., and M.L.R. De Campos, "Performance Comparison of Binary and Quaternary UWB Modulation Schemes," *IEEE Global Telecommunications Conference* (December 2003), 789–793.

Federal Communications Commission, "Revision of Part 15 of the Commission's rules Regarding Ultra-Wideband Transmission Systems: First report and order," *Technical Report FCC 02-48* (adopted February, 14 2002; released April 22, 2002).

Foerster, J.R., "The performance of a Direct-Sequence Spread Ultra-Wideband system in the presence of Multipath, Narrowband Interference, and Multiuser Interference," *IEEE Conference on Ultra Wideband Systems and Technologies* (May 2002), 87–91.

Foerster, J.R., V. Somayazulu, and S. Roy, "A Multi-Banded System Architecture for Ultra-Wideband Communications," *IEEE Military Communications Conference* (October 2003), 903–908.

Gao, X., R. Yao, and Z. Feng, "Multi-band UWB System with Hadamard Coding," *IEEE Conference on Vehicular Technology* (October 2003), 1288–1292.

Gerakoulis, D., and P. Salmi, "An Interference Suppressing OFDM System for Ultra Wide Bandwidth Radio Channels," *IEEE Conference on Ultra Wideband Systems and Technologies* (May 2002), 259–264.

Giancola, G., "Comparative Analysis of UWB, CDMA, and OFDM transmission techniques," in *Networking with Ultra Wide Band*, M.-G. Di Benedetto editor, Ingegneria2000 (2002), 1.1–1.28.

Guvenc, I., and H. Arslan, "On the modulation options for UWB Systems," *IEEE Military Communications Conference* (October 2003), 892–897.

Hanzo, L., M. Münster, B.J. Choi, and T. Keller, *OFDM and MC-CDMA for Broadband Multi-User Communications, WLANs and Broadcasting*, Chichester, West Sussex, England: John Wiley and Sons, Inc. (2003).

Hengstler, S., D.P. Kasilingam, and A.H. Costa, "A Novel Chirp Modulation Spread Spectrum Technique for Multiple Access," *IEEE Seventh International Symposium on Spread Spectrum Techniques and Applications*, Volume: 1 (September 2002), 73–77.

Hoctor, R., and H. Tomlinson, "Delay-Hopped Transmitted-Reference RF Communications," *IEEE Conference on Ultra Wideband Systems and Technologies* (May 2002), 265–269.

Huang, X., and Y. Li, "Generating Near-White Ultra-Wideband Signals with Period Extended PN Sequences," *IEEE Conference Vehicular Technology Conference* Volume: 2 (May 2001), 1184–1188.

Jung, S.Y., D.J. Park, Y.H. Kwon, and S.M. Lee, "Design and Performance Analysis of UWB TH-MA Scheme Using Multi-code Based PPM," *IEEE Conference on Ultra Wideband Systems and Technologies* (November 2003), 463–467.

Lu, I.S.C., N. Weste, and S. Parameswaran, "A Digital Ultra-Wideband Multiband Transceiver Architecture with Fast Frequency Hopping Capabilities," *IEEE Conference on Ultra Wideband Systems and Technologies* (November 2003), 448–452.

Nassar, C.R., F. Zhu, and Z. Wu, "Direct Sequence Spreading UWB Systems: Frequency Domain Processing for Enhanced Performance and Throughput," *IEEE International Conference on Communications*, Volume: 3 (May 2003), 2180–2186.

Okamoto, E., Y. Iwanami, and T. Ikegami, "Multimode Transmission Using Wavelet Packet Modulation and OFDM," *IEEE Conference on Vehicular Technology* (October 2003), 1458–1462.

Prasad, R., and S. Hara, "An overview of multi-carrier CDMA," *IEEE International Symposium on Spread Spectrum Techniques and Applications*, Volume: 1 (September 1996), 107–114.

Proakis, J.G., *Digital Communications*, 3rd Edition, New York: McGraw-Hill International Editions (1995).

Proakis, J.G., and M. Salehi, *Communication Systems Engineering*, Englewood Cliffs, New Jersey: Prentice-Hall International (1994).

Roberts, R., *XtremeSpectrum CFP Document*, Available at *grouper.ieee.org/groups/802/15/pub/2003/Jul03/03154r3P802-15_TG3a-XtremeSpectrum-CFP-Documentation.pdf* (July 2003).

Rowe, H.E., *Signals and Noise in Communications Systems*, Princeton, New Jersey: D. Van Nostrand Company, Inc (1965).

Runkle, P., J. McCorkle, T. Miller, and M. Welborn, "DS-CDMA: The Modulation Technology of Choice for UWB Communications," *IEEE Conference on Ultra Wideband Systems and Technologies* (November 2003), 364–368.

Saberinia, E., A.H. Tewfik, and R. Gupta, "Pilot Assisted Multi-User UWB Communications," *IEEE Conference on Vehicular Technology Conference* (October 2003), 1885–1889.

Saberinia, E., and A.H. Tewfik, "Single and Multi-Carrier UWB Communications," *IEEE Seventh International Symposium on Signal Processing and its Applications*, Volume: 2 (July 2003a), 343–346.

Saberinia, E., and A.H. Tewfik, "Pulsed and Non-Pulsed OFDM Ultra Wideband Wireless Personal Area Networks," *IEEE Conference on Ultra Wideband Systems and Technologies* (November 2003b), 275–279.

Scholtz, R.A., "Multiple Access with Time-Hopping Impulse Modulation," *IEEE Military Communications Conference*, Volume: 2 (October 1993), 447–450.

Souilmi, Y., and R. Knopp, "On the Achievable Rates of Ultra-Wideband PPM with Non-Coherent Detection in Multipath Environments," *IEEE International Conference on Communications*, Volume: 5 (May 2003), 3530–3534.

Tewfik, A.H., and E. Saberinia, "High Bit Rate Ultra-Wideband OFDM," *IEEE Global Telecommunications Conference*, Volume: 3 (November 2002), 2260–2264.

Trindade, A., Q.H. Dang, and A.J. Van der Veen, "Signal Processing Model for a Transmit-Reference UWB Wireless Communication System," *IEEE Conference on Ultra Wideband Systems and Technologies* (November 2003), 270–274.

Van Nee, R., and R. Prasad, *OFDM for Wireless Multimedia Communications*, Norwood, Massachusetts: Arthec House Publishers (2000).

Van Stralen, N., A. Dentinger, K. Welles II, R. Gaus Jr., R. Hoctor, and H. Tomlinson, "Delay Hopped Transmitted Reference Experimental Results," *IEEE Conference on Ultra Wideband Systems and Technologies* (May 2002), 93–98.

Venkatesan, V., H. Liu, C. Nielsen, R. Kyker, and M.E. Magana, "Performance of an Optimally Spaced PPM Ultra-Wideband System with Direct Sequence Spreading for Multiple Access," *IEEE Conference on Vehicular Technology* (October 2003), 602–606.

Vojcic, B.R., and R.L. Pickholtz, "Direct-Sequence Code Division Multiple Access for Ultra-Wide Bandwidth Impulse Radio," *IEEE Military Communications Conference* (October 2003), 898–902.

Webb, W.T., and L. Hanzo, *Modern Quadrature Amplitude Modulation - Principles and Applications for Fixed and Wireless Channels*, London: Co-publication of the IEEE Press and Pentech Press (1995).

Weinstein, S.B., and P.M. Ebert, "Data transmission by frequency-division multiplexing using the discrete Fourier transform," *IEEE Transactions on Communications*, Volume: 19, Issue: 5 (October 1971), 628–634.

Welborn, M.L., "System considerations for ultrawideband wireless networks," *IEEE Radio and Wireless Conference* (August. 2001), 5–8.

Wilson, R.D., and R.A. Scholtz, "Comparison of CDMA and Modulation Schemes for UWB Radio in a Multipath Environment," *IEEE Global Telecommunications Conference* (December 2003), 754–758.

Win, M.Z., and R.A. Scholtz, "Ultra-Wide Bandwidth Time-Hopping Spread-Spectrum Impulse Radio for Wireless Multiple-Access Communications," *IEEE Transactions on Communications*, Volume: 48, Issue: 4 (April 2000), 679–691.

Xu, W., R. Yao, Z. Guo, W. Zhu, and Z. Zhou, "A Power Efficient M-ary Orthogonal Pulse Polarity Modulation for TH-UWB System Using Modified OVSF Codes," *IEEE Global Telecommunications Conference* (December 2003), 436–440.

Yeh, P.-C., J.D. Choi, S. Zummo, M.D. Casciato, and W.E. Stark, "Performance Analysis of Coded Multi-Carrier Wideband Systems over Multipath Fading Channels," *IEEE Military Communications Conference* (October 2003), 909–914.

Zhang, J., R.A. Kennedy, and T.D. Abhayapala, "New Results on the Capacity of M-ary PPM Ultra WideBand Systems," *IEEE International Conference on Communications*, Volume: 4 (May 2003), 2867–2871.

Zhao, L., and A.M. Haimovich, "Multi-User Capacity of M-ary PPM Ultra-Wideband Communications", *IEEE Conference on Ultra Wideband Systems and Technologies* (May 2002), 175–179.

APPENDIX 2.A

Function 2.1 Generation of Equiprobable Binary Values

Function 2.1 generates binary values by means of the MATLAB function `rand(N,M)`, which generates an `N·M` matrix with random entries chosen from a uniform distribution on the interval (0.0, 1.0).

```
%
% FUNCTION 2.1 : "cp0201_bits"
%
% Generates a stream of equiprobable binary values ('bits')
% The number of bits ('numbits') is an input parameter
%
% Programmed by Guerino Giancola
%

function [bits]=cp0201_bits(numbits)

% ----------------------------------------------
% Step One - Generation of the reference pulse
% ----------------------------------------------

bits=rand(1,numbits)>0.5;
```

Function 2.2 Code Repetition Coding

Redundancy within a code repetition coder is introduced by first transforming the input stream into a train of discrete type Dirac pulses spaced by N_s samples, and then by operating a discrete convolution between this train of pulses and a discrete-type rectangular shape filter of length N_s samples.

```
%
% FUNCTION 2.2 : "cp0201_repcode"
%
% Introduces a repetition code for the stream of 'bits' in
%  input:
% 'Ns' identical binary values are generated from the same
%  bit
% 'repbits' represents the output binary sequence
%
% Programmed by Guerino Giancola
%

function [repbits]=cp0201_repcode(bits,Ns)

% ----------------------------------------------
% Step One - Introduction of the repetition code
% ----------------------------------------------

numbits = length(bits);

temprect=ones(1,Ns);
temp1=zeros(1,numbits*Ns);
temp1(1:Ns:1+Ns*(numbits-1))=bits;
temp2=conv(temp1,temprect);
repbits=temp2(1:Ns*numbits);
```

Function 2.3 TH Coding

The generation of the TH code within Function 2.3 is similar to the generation of the binary stream described in Function 2.1. First, we generate a vector of size N_p, composed of random entries chosen from a uniform distribution on the interval (0, N_h-1), then we apply the MATLAB function `floor(x)`, which rounds the elements of `x` to the closest lower integers.

```
%
% FUNCTION 2.3 : "cp0201_TH"
%
% Generates a pseudorandom TH code
% with periodicity 'Np' and cardinality 'Nh'
%
% Programmed by Guerino Giancola
%

function [THcode]=cp0201_TH(Nh,Np);

% ------------------------------------
% Step One - Generation of the TH code
% ------------------------------------

THcode = floor(rand(1,Np).*Nh);
```

Function 2.4 PPM-TH Modulation

Function 2.4 is composed of two steps. Step One operates the conversion from the continuous time domain of the input quantities to the discrete time domain of the simulation. Moreover, it allocates in memory the resource for representing the output signals to be generated. Step Two consists of a main loop to be repeated for all bits of the input sequence. For each input bit, the function evaluates the position of the corresponding pulse according to both the TH code and the PPM shift. The loop itself is composed of three parts. In the first part, the position of the pulse is fixed according to the uniform pulse repetition period. Then, the additional time shift due to the TH code is introduced. Finally, depending on the value of the input bit, the PPM time shift is eventually introduced.

```
%
% FUNCTION 2.4 : "cp0201_2PPM_TH"
%
% Introduces the TH code given by 'THcode'
% and implements binary PPM modulation
% 'seq' is the input binary stream
% 'fc' is the sampling frequency for the generated signal
% 'Tc' is the chip time
% 'Ts' is the average pulse repetition time
% 'dPPM' is the PPM delta shift
% 'THcode' is the TH code
%
% The function generates two output streams
% '2PPMTHseq' is the output with both TH and 2PPM
% 'THseq' is the output with TH only
%
% Programmed by Guerino Giancola
%

function [PPMTHseq,THseq] = ...
  cp0201_2PPM_TH(seq,fc,Tc,Ts,dPPM,THcode)


% ----------------------------------------------------
% Step One - Implementation of the 2PPM-TH modulator
% ----------------------------------------------------

dt = 1 ./ fc;                     % sampling period
framesamples = floor(Ts./dt);     % no. of samples between
                                  % pulses
chipsamples = floor (Tc./dt);     % no. of samples for the
                                  % chip duration
```

```
PPMsamples = floor (dPPM./dt);  % no. of samples for the
                                % PPM shift

THp = length(THcode);           % TH-code periodicity

totlength = framesamples*length(seq);
PPMTHseq=zeros(1,totlength);
THseq=zeros(1,totlength);

% -----------------------------------------------
% Step Two - Main loop for introducing TH and 2PPM
% -----------------------------------------------

for k = 1 : length(seq)

    % uniform pulse position
    index = 1 + (k-1)*framesamples;

    % introduction of TH
    kTH = THcode(1+mod(k-1,THp));
    index = index + kTH*chipsamples;

    THseq(index) = 1;

    % introduction of 2PPM
    index = index + PPMsamples*seq(k);
    PPMTHseq(index) = 1;

end % for k = 1 : length(seq)
```

Function 2.5 Pulse Shaper

The code for generating the pulse waveform evaluates the number of samples OVER of the resulting signal and separates the case where OVER is odd from the case where OVER is even.

```
%
% FUNCTION 2.5 : "cp0201_waveform"
%
% Generates the energy-normalized pulse waveform
%
% Special case of the second derivative Gaussian pulse:
% SCHOLTZ'S MONOCYCLE
%
% ************************************************************
% ref:
% Scholtz R.A. "Multiple Access with Time-Hopping Impulse
%               Modulation"
% in Proceedings of MILCOM'93, 1993, pp. 679-691
% ************************************************************
%
% 'fc' is the sampling frequency
% 'Tm' is the pulse duration
% 'tau' is the shaping parameter
%
% Programmed by Guerino Giancola
%

function [w0]= cp0201_waveform(fc,Tm,tau);

% -------------------------------------
% Step One - Pulse waveform generation
% -------------------------------------

dt = 1 / fc;               % reference sampling period
OVER = floor(Tm/dt);       % number of samples representing
                           % the pulse

e = mod(OVER,2);
kbk = floor(OVER/2);
tmp = linspace(dt,Tm/2,kbk);
s = (1-4.*pi.*((tmp./tau).^2)).* ...
     exp(-2.*pi.*((tmp./tau).^2));

if e                       % OVER is odd
```

```
    for k=1:length(s)
        y(kbk+1)=1;
        y(kbk+1+k)=s(k);
        y(kbk+1-k)=s(k);
    end
else                            % OVER is even
    for k=1:length(s)
        y(kbk+k)=s(k);
        y(kbk+1-k)=s(k);
    end
end

E = sum((y.^2).*dt);            % pulse energy
w0 = y ./ (E^0.5);              % energy normalization
```

Function 2.6 PPM-TH Transmitter

Function 2.6 is composed of three steps. Step Zero contains all the parameters characterizing the UWB signal to be generated: the average transmitted power in dBm `Pow`; the sampling frequency for representing the signal `fc`; the number of bits generated by the binary source `numbits`; the average pulse repetition time in seconds `Ts`; the number of pulses to be produced for each bit `Ns`; the chip time in seconds `Tc`; the cardinality and the periodicity of the TH code `Nh` and `Np`; the impulse response duration `Tm`; the shaping factor for the pulse waveform in seconds `tau`; and the time shift of the PPM in seconds `dPPM`. A graphical output is obtained by setting `G=1`. Step One contains the command lines for simulating the single blocks of the transmission chain. Note that the bits in excess originating from the convolution are trimmed off from the final vector to obtain a signal with a length equal to `Ts·Ns·numbits`. Step Two contains the code for generating the graphical output. This graphical representation allows the user to verify the subdivision of the time axis into bit intervals, frames, and slots.

```
%
% FUNCTION 2.6 : "cp0201_transmitter_2PPM_TH"
%
% Simulation of a UWB transmitter implementing 2PPM with TH
%
% Transmitted Power is fixed to 'Pow'
% The signal is sampled with frequency 'fc'
% 'numbits' is the number of bits generated by the source
% 'Ns' pulses are generated for each bit, and these pulses
% are spaced in time by an average pulse repetition period
% 'Ts'
% The TH code has periodicity 'Np', and cardinality 'Nh'
% The chip time has time duration 'Tc'
% Each pulse has time duration 'Tm' and shaping factor
% 'tau'
% The PPM introduces a time shift of 'dPPM'
%
% The function returns:
% 1) the generated stream of bits ('bits')
% 2) the generated TH code ('THcode')
% 3) the generated signal ('Stx')
% 4) a reference signal without data modulation ('ref')
%
% Programmed by Guerino Giancola
%

function [bits,THcode,Stx,ref]=cp0201_transmitter_2PPM_TH
```

```
% -----------------------------
% Step Zero - Input parameters
% -----------------------------

Pow = -30;          % average transmitted power (dBm)

fc = 50e9;          % sampling frequency

numbits = 2;        % number of bits generated by the source

Ts = 3e-9;          % frame time, i.e., average pulse
                    %  repetition period [s]
Ns = 5;             % number of pulses per bit

Tc = 1e-9;          % chip time [s]
Nh = 3;             % cardinality of the TH code
Np = 5;             % periodicity of the TH code

Tm = 0.5e-9;        % pulse duration [s]
tau = 0.25e-9;      % shaping factor for the pulse [s]
dPPM = 0.5e-9;      % time shift introduced by the PPM [s]

G = 1;
% G=0 -> no graphical output
% G=1 -> graphical output

% ------------------------------------------
% Step One - Simulating transmission chain
% ------------------------------------------

% binary source
bits = cp0201_bits(numbits);

% repetition coder
repbits = cp0201_repcode(bits,Ns);

% TH code
THcode = cp0201_TH(Nh,Np);

% PPM + TH
[PPMTHseq,THseq] = ...
   cp0201_2PPM_TH(repbits,fc,Tc,Ts,dPPM,THcode);

% shaping filter
power = (10^(Pow/10))/1000;      % average transmitted power
```

```
                                        % (watt)
Ex = power * Ts;                        % energy per pulse
w0 = cp0201_waveform(fc,Tm,tau);% energy normalized pulse
                                        % waveform
wtx = w0 .* sqrt(Ex);                   % pulse waveform
Sa = conv(PPMTHseq,wtx);                % output of the filter
                                        % (with modulation)
Sb = conv(THseq,wtx);                   % output of the filter
                                        % (without modulation)

% Output generation

L = (floor(Ts*fc))*Ns*numbits;
Stx = Sa(1:L);
ref = Sb(1:L);

% ---------------------------
% Step Two - Graphical output
% ---------------------------

if G

F = figure(1);
set(F,'Position',[32 223 951 420]);

tmax = numbits*Ns*Ts;
time = linspace(0,tmax,length(Stx));
P = plot(time,Stx);
set(P,'LineWidth',[2]);
ylow=-1.5*abs(min(wtx));
yhigh=1.5*max(wtx);
axis([0 tmax ylow yhigh]);
AX=gca;
set(AX,'FontSize',12);
X=xlabel('Time [s]');
set(X,'FontSize',14);
Y=ylabel('Amplitude [V]');
set(Y,'FontSize',14);
for j = 1 : numbits
    tj = (j-1)*Ns*Ts;
    L1=line([tj tj],[ylow yhigh]);
    set(L1,'Color',[0 0 0],'LineStyle', ...
       '--','LineWidth',[2]);
    for k = 0 : Ns-1
        if k > 0
            tn = tj + k*Nh*Tc;
```

```
            L2=line([tn tn],[ylow yhigh]);
            set(L2,'Color',[0.5 0.5 0.5],'LineStyle', ...
               '-.','LineWidth',[2]);
        end
        for q = 1 : Nh-1
            th = tj + k*Nh*Tc + q*Tc;
            L3=line([th th],[0.8*ylow 0.8*yhigh]);
            set(L3,'Color',[0 0 0],'LineStyle', ...
               ':','LineWidth',[1]);
        end
    end
end

end % end of graphical output
```

Function 2.7 DS Coding

The generation of DS code within Function 2.7 is similar to the generation of TH code described in Function 2.3, that is, we generate a vector with size `Np` composed of bipolar (+1 and -1) and equiprobable random entries. Function 2.7 returns vector `Dscode`, which contains the generated code with length `Np`.

```
%
% FUNCTION 2.7 : "cp0202_DS"
%
% Generates a random DS code
% with periodicity 'Np'
%
% Programmed by Guerino Giancola
%

function [DScode]=cp0202_DS(Np);

% -------------------------------------
% Step One - Generation of the DS code
% -------------------------------------

DScode = ((rand(1,Np)>0.5).*2)-ones(1,Np);
```

Function 2.8 PAM-DS Modulation

Function 2.8 is composed of two steps. Step One operates the conversion from the continuous time to the discrete time domain. Moreover, it allocates in memory the resource for representing the output signals to be generated. Step Two consists of a main loop to be repeated for all bits originating from the input sequence. For each input bit, the function evaluates the position and amplitude of the corresponding pulses, according to both DS code and PAM. The loop itself is composed of three parts. In the first part, the position of the pulse is determined according to the uniform pulse repetition period. Then, the amplitude of each pulse is determined according to the DS code. Finally, the PAM scheme is applied, that is, the pulse is inverted in amplitude if the corresponding bit is 0.

```
%
% FUNCTION 2.8 : "cp0202_2PAM_DS"
%
% Introduces the DS code given by 'DScode'
% and implements binary PAM modulation
% 'seq' is the input binary stream
% 'fc' is the sampling frequency for the generated signal
% 'Ts' is the average pulse repetition time
% 'DScode' is the DS code
%
% The function generates two output streams
% '2PPMDSseq' is the output with both TH and 2PPM
% 'DSseq' is the output with the TH only
%
% Programmed by Guerino Giancola
%

function [PAMDSseq,DSseq] = ...
   cp0201_2PAM_DS(seq,fc,Ts,DScode)


% ---------------------------------------------------
% Step One - Implementation of the 2PPM+DS modulator
% ---------------------------------------------------

dt = 1 ./ fc;                      % sampling period
framesamples = floor(Ts ./ dt);    % number of samples
                                   % between pulses

DSp = length(DScode);              % DS code periodicity
```

```
totlength = framesamples*length(seq);
PAMDSseq = zeros(1,totlength);
DSseq = zeros(1,totlength);

% -------------------------------------------------
% Step Two - Main loop for introducing DS and 2PAM
% -------------------------------------------------

for k = 1 : length(seq)

    % uniform pulse position
    index = 1 + (k-1)*framesamples;

    % introduction of DS
    kDS = DScode(1+mod(k-1,DSp));

    DSseq(index)=kDS;

    % introduction of 2PAM
            PAMDSseq(index) = kDS*((seq(k)*2)-1);

end % for k = 1 : length(seq)
```

Function 2.9 PAM-DS Transmitter

Function 2.9 is composed of three steps. Step Zero contains all the parameters characterizing the UWB signal to be generated: the average transmitted power in dBm `Pow`; the sampling frequency for representing the signal `fc`; the number of bits generating from the binary source `numbits`; the average pulse repetition time in seconds `Ts`; the number of pulses to be produced for each bit `Ns`; the periodicity of the DS code `Np`; the impulse response duration `Tm`; and the shaping factor for the pulse waveform in seconds `tau`. A final parameter, `G`, determines whether the graphical output is required, `G = 1`, or not, `G = 0`. Step Two contains the command lines for simulating the single blocks of the transmission chain. Note that the bits in excess originating from the convolution are trimmed off from the final vector, to obtain a signal with a length equal to `Ts·Ns·numbits`. Step Three contains the code for generating the graphical output. This graphical representation allows the user to verify the subdivision of the time axis into bit intervals and frames.

```
%
% FUNCTION 2.9 : "cp0202_transmitter_2PAM_DS"
%
% Simulation of a UWB transmitter implementing 2PAM with DS
%
% Transmitted power is fixed to 'Pow'
% The signal is sampled with frequency 'fc'
% 'numbits' is the number of bits generated by the source.
% 'Ns' pulses are generated for each bit, and these pulses
% are spaced in time by an average pulse repetition period
% 'Ts'
% The DS code has periodicity 'Np'
% Each pulse has time duration 'Tm' and shaping factor
% 'tau'
%
% The function returns:
% 1) the generated stream of bits ('bits')
% 2) the generated DS code ('DScode')
% 3) the generated signal ('Stx')
% 4) a reference signal without data modulation ('ref')
%
% Programmed by Guerino Giancola
%

function [bits,DScode,Stx,ref]=cp0202_transmitter_2PAM_DS

% ----------------------------
% Step Zero - Input parameters
```

```
% ----------------------------

Pow = -30;        % average transmitted power (dBm)

fc = 50e9;        % sampling frequency

numbits = 2;      % number of bits generated by the source

Ts = 2e-9;        % frame time, i.e., average pulse
                  % repetition period [s]

Ns = 10;          % number of pulses per bit

Np = 10;          % periodicity of the DS code

Tm = 0.5e-9;      % pulse duration [s]
tau = 0.25e-9;    % shaping factor for the pulse [s]

G = 1;
% G=0 -> no graphical output
% G=1 -> graphical output

% ----------------------------------------
% Step One - Simulating transmission chain
% ----------------------------------------

% binary source
bits = cp0201_bits(numbits);

% repetition coder
repbits = cp0201_repcode(bits,Ns);

% Direct Sequence code
DScode = cp0202_DS(Np);

% Pulse Amplitude Modulation + DS
[PAMDSseq,DSseq] = cp0202_2PAM_DS(repbits,fc,Ts,DScode);

% Shaping filter
power = (10^(Pow/10))/1000;      % average transmitted power
                                 % (watt)
Ex = power * Ts;                 % energy per pulse
w0 = cp0201_waveform(fc,Tm,tau);% energy normalized pulse
                                 % waveform
wtx = w0 .* sqrt(Ex);            % pulse waveform
Sa = conv(PAMDSseq,wtx);         % output of the filter
```

```
                                    % (with modulation)
Sb = conv(DSseq,wtx);               % output of the filter (no
                                    % modulation)

L = (floor(Ts*fc))*Ns*numbits;
Stx = Sa(1:L);
ref = Sb(1:L);

% --------------------------
% Step Two - Graphical output
% --------------------------

if G

F = figure(1);
set(F,'Position',[32 223 951 420]);

tmax = numbits*Ns*Ts;
time = linspace(0,tmax,length(Stx));
P = plot(time,Stx);
set(P,'LineWidth',[2]);
ylow=-1.5*max(wtx);
yhigh=1.5*max(wtx);
axis([0 tmax ylow yhigh]);
AX=gca;
set(AX,'FontSize',12);
X=xlabel('Time [s]');
set(X,'FontSize',14);
Y=ylabel('Amplitude [V]');
set(Y,'FontSize',14);
for j = 1 : numbits
    tj = (j-1)*Ns*Ts;
    L1=line([tj tj],[ylow yhigh]);
    set(L1,'Color',[0 0 0],'LineStyle','--', ...
       'LineWidth',[2]);
    for k = 0 : Ns-1
     if k > 0
       tn = tj + k*Ts;
       L2=line([tn tn],[0.8*ylow 0.8*yhigh]);
       set(L2,'Color',[0.5 0.5 0.5], ...
          'LineStyle','-.','LineWidth',[1]);
     end
    end
end

end % end of graphical output
```

Function 2.10 QPSK Modulation

Function 2.10 generates the QPSK sequence generating from the input binary stream `bits`. Note that vector `bits` is zero-padded to have a number of bits that is a multiple of 2.

```
%
% FUNCTION 2.10 : "cp0203_qpsk_mod"
%
% This function receives a binary stream in input ('bits')
% and returns the corresponding sequence of QPSK symbols
%  ('S'),
% plus the two sequences 'Sc' and 'Ss' containing the real
% and imaginary part of each symbol
%
% Programmed by Guerino Giancola
%

function [S,Sc,Ss] = cp0203_qpsk_mod(bits)

nb = length(bits);       % number of bits
ns = ceil(nb/2);         % number of symbols

b0 = zeros(1,ns*2);      % zero padding
b0(1:nb) = bits;

j = sqrt(-1);

for s = 1 : ns

    ba = b0(((s-1)*2)+1);
    bb = b0(((s-1)*2)+2);
    k = bb + ba*2;
    p = ((pi/4)*(2*k-1))-pi;

    Sc(s) = cos(p);
    Ss(s) = sin(p);
    S(s) = Sc(s) + j*Ss(s);

end
```

Function 2.11 OFDM Modulation

Function 2.11 is composed of two steps. Step Zero contains all parameters characterizing the signal to be transmitted: the total number of bits to be transmitted `numbits`; the central frequency of the modulated signal `fp`; the value of the sampling frequency `fc`; the time used for transmitting each symbol on the corresponding carrier `T0`; the time duration of the cyclic prefix `TP`; the time duration of the guard interval `TG`; the amplitude `A` of the rectangular impulse response `gT(t)`; and the number `N` of different carriers of the OFDM system. Step One contains the code for generating the OFDM signal. The generation of the original stream of bits is performed by means of Function 2.1; the conversion of this stream into the sequence of QPSK symbols is performed by Function 2.10. Note that the code is composed of a main loop that operates on a block basis. For each block, the zero padding is first introduced in the middle of the vector to achieve oversampling on the digital sequence. Zero padding has the effect of doubling the size of the input vector. The Inverse FFT (IFFT) is applied to the zero-padded vector by introducing the corresponding MATLAB function `ifft(x)`, then the cyclic prefix is appended at the beginning of the resulting sequence `c`. Note that the zero padding, which is operated on each vector of QPSK symbols, has the effect of shifting the sub-carriers of the OFDM systems just at the center of the 0 Hz value. As a consequence, the term $(-1)^n$ in Eq. (2-15) should not be applied at the output of the IFFT block in the presence of zero padding since it would produce an additional and useless shift. After the introduction of the cyclic prefix, we can generate the shape of the complex envelope of the OFDM symbol by operating the convolution of the IFFT sequence with a rectangular signal of duration `T0/2*NT`, where `NT` is the total number of QPSK symbols transmitted with an OFDM symbol (the useful symbols plus the symbols of the cyclic prefix). At the end of the main loop, the complete complex envelope is obtained. We then extract the in-phase signal `I` and the quadrature signal `Q` and modulate the carrier at frequency `fc`.

```
%
% FUNCTION 2.11 : "cp0203_OFDM_qpsk"
%
% Simulation of a transmitter implementing
% the OFDM transmission chain with QPSK modulation
% on each sub-carrier
%
% 'numbits' is the number of bits generated by the source
% 'fp' is the carrier frequency of the generated signal
% 'fc' is the sampling frequency
% 'T0' is the block length in [s], i.e., 1/T0 is the
% carrier separation
%
% 'TP' is the length of the cyclic prefix [s]
% 'TG' is guard time
% 'A'  is the amplitude of the rectangular impulse response
```

```
%  [V]
% 'N' is the number of carriers (tones) used in the OFDM
%  system
%
% The function returns:
% 1) the generated stream of bits ('bits')
% 2) the corresponding stream of QPSK symbols ('S')
% 3) the I component of the generated signal ('SI')
% 4) the Q component of the generated signal ('SQ')
% 5) the generated OFDM signal ('Stx')
% 6) the value of the sampling frequency ('fc')
% 7) the value of the carrier frequency ('fp')
% 8)9)10) the values of T0, TP, and TG
% 11) the number of tones used for transmission
%
% Programmed by Guerino Giancola
%

function [bits,S,SI,SQ,Stx,fc,fp,T0,TP,TG,N] = ...
   cp0203_OFDM_qpsk;

% ----------------------------
% Step Zero - Input parameters
% ----------------------------

numbits = 1024;      % number of bits to be transmitted

fp = 1e9;            % central frequency

fc = 50e9;           % sampling frequency

T0 = 242.4e-9;       % information length
TP = 60.6e-9;        % cyclic prefix
TG = 70.1e-9;        % total guard time

A = 1;               % amplitude of the rectangular impulse
                     % response

N = 128;             % number of carriers of the OFDM system


% -------------------------
% Step One - OFDM modulator
% -------------------------

tc = T0 / N;         % chip time
```

```
ntcp = floor(TP/tc);        % number of tones of the cyclic
                            % prefix
n = (-ntcp+1:1:N);          % tone counter
NT = length(n);             % total number of tones per symbol

% Bit generation
[bits] = cp0201_bits(numbits);


% QPSK modulator
[S,Sc,Ss] = cp0203_qpsk_mod(bits);

% OFDM modulator

nb = ceil(length(S)/N);          % number of OFDM blocks to be
                                 % transmitted
S0 = zeros(1,nb*N);              % zero padding
S0(1:length(S))=S;

dt = 1 / fc;                     % sampling period

if ntcp>0
    tc = (T0+TP)/NT;             % tone duration
end
tonesamples = floor(tc/dt);      % samples per tone
toneres = floor((TG-TP)/dt);     % samples for the residual
                                 % part

symsamp = (tonesamples*NT)+toneres;
% number of samples representing one OFDM symbol

totsamp = symsamp * nb;
% number of samples representing the transmitted signal

X = [zeros(1,totsamp)'];

for b = 1 : nb

    c = S0((1+(b-1)*N):(N+(b-1)*N));      % block extraction

    % Serial to Parallel conversion and zero padding
    A = length(c);
    a1 = floor(A/2);
    a2 = A - a1;
    FS = 2*A;
    Czp=zeros(FS,1);
```

```
        Czp(1:a1)=[c(1:a1).'];
        Czp(FS-a2+1:FS)=[c(A-a2+1:A).'];

        C = ifft(Czp);  % IFFT of the zero-padded input


        if ntcp>0 % insertion of the cyclic prefix
            C1=zeros(length(C)+2*ntcp,1);
            C1(1:(2*ntcp))=C(2*N+1-(2*ntcp):2*N);
            C1(2*ntcp+1:length(C1))=C;
        else
            C1=C;
        end
        %

        zp = floor(tonesamples/2);
        C2 = [C1.';zeros((zp-1),length(C1))];
        C3 = C2(:);
        g = ones(1,zp);
        C4 = conv(g,C3);

        C4 = C4(1:(zp*NT*2));

        ics = 1 + (b-1)*symsamp + toneres;
        X(ics:ics+length(C4)-1)=C4;


    end % for b = 1 : nb

    XM = X';                    % Parallel to Serial conversion
    XM = XM(1:totsamp);

    I = real(XM);
    Q = imag(XM);

    % carrier modulation
    time = linspace(0,totsamp*dt,length(I));
    SI = I.*(cos((2*pi*fp).*time));
    SQ = Q.*(sin((2*pi*fp).*time));

    Stx = SI - SQ;
```

CHAPTER 3

The PSD of TH-UWB Signals

Power Spectral Density (PSD) for TH-UWB signals using PPM is derived in this chapter. The adopted approach (Di Benedetto and Vojcic, 2003) follows the analog PPM theory of the old days — practical pulse communication equipment is described in Black, Beyer, Grieser, and Polkinghorn as early as 1946 (Black et al., 1946) — and reconciles this well-known modulation method with its digital variant currently in vogue in wireless communications.

3.1 Borrowing from PPM

The signal of Eq. (2–4) has strong similarities with the output of a PPM modulator in its analog form. Given the modulating signal *m(t)*, the PPM wave *x(t)* in its analog form consists of a train of identically shaped and strictly non-overlapping pulses that are shifted from nominal instances of time T_s by the signal samples $m(kT_s)$. The expression of the PPM analog wave is given as:

$$x_{PPM}(t) = \sum_{j=-\infty}^{+\infty} p(t - jT_s - m(jT_s)) = p(t) * \sum_{j=-\infty}^{+\infty} \delta(t - jT_s - m(jT_s)) \quad (3\text{–}1)$$

Equation (3–1) is commonly known as the uniform sampling representation. A slightly different version of analog PPM, called natural sampling, is obtained without explicit sampling of the modulating wave *m(t)*. The two forms, however, differ very little if the maximum shift is small compared to the pulse period T_s. We should also note that the idea back in time behind natural sampling was to avoid sampling to simplify the equipment used at that time. This is the last of our worries nowadays. We can, therefore, consider Eq. (3–1) as a valid expression for the PPM analog wave.

A sufficient condition for the pulses of Eq. (3–1) to be strictly non-overlapping is as follows:

$$p(t) = 0 \quad for \quad |t| \geq \frac{T_s}{2} - |m(t)|_{\max} \Rightarrow |m(t)|_{\max} \leq \frac{T_s}{2} \tag{3–2}$$

If the condition expressed by Eq. (3–2) is verified, then the pulses are non-overlapping and the order of the pulses unaltered. For special modulating signals, for example, a sine wave at a frequency much lower than $1/T_s$, weaker conditions than Eq. (3–2) can be found. For the purpose of generality we will suppose that Eq. (3–2) is verified in all the cases we will examine.

The PSD of a PPM signal is difficult to evaluate due to the non-linear nature of PPM modulation. The complete derivation of this spectrum can be found in (Rowe, 1964), and is based on various articles by Bennett published from 1933 to 1947 (Bennett, 1933, 1944, and 1947). We shall report here only the principal results for three particular cases that are of interest for understanding the spectrum of a TH-UWB signal. The three relevant cases are: a) sinusoidal modulating signals; b) generic periodic modulating signals; c) random modulating signals.

3.1.1 Sinusoidal Modulating Signals

Consider the sinusoidal modulating signal $m(t)$ at frequency f_0:

$$m(t) = A\cos\left(2\pi f_0 t\right) \tag{3–3}$$

Note that according to the sampling theorem, to be able to reconstruct $m(t)$ from the modulated waveform the sampling frequency, $1/T_s$, must be at least equal to $2f_0$. Common practice sets $f_0 << 1/T_s$.

The PPM wave of Eq. (3–1) in this case becomes:

$$x_{PPM}(t) = p(t) * \sum_{j=-\infty}^{+\infty} \delta\left(t - jT_s - A\cos\left(2\pi f_0 jT_s\right)\right) \tag{3–4}$$

If one indicates by $P(f)$ the Fourier transform of $p(t)$:

$$P(f) = \int_{-\infty}^{+\infty} p(t)e^{-j2\pi ft}dt \tag{3–5}$$

an expansion of $x_{PPM}(t)$ into sinusoidal components as shown by (Rowe, 1965) can be found:

$$
\begin{aligned}
x_{PPM}(t) &= p(t) * \sum_{j=-\infty}^{+\infty} \delta\left(t - jT_s - A\cos\left(2\pi f_0 jT_s\right)\right) \\
&= \frac{1}{T_s} \sum_{m=-\infty}^{+\infty} \sum_{n=-\infty}^{+\infty} (-j)^n J_n\left(2\pi A\left(m\frac{1}{T_s} + nf_0\right)\right) \cdot \\
&\quad \cdot P\left(m\frac{1}{T_s} + nf_0\right) e^{j2\pi(m\frac{1}{T_s}+nf_0)t}
\end{aligned} \tag{3–6}
$$

where $J_n(\cdot)$ are the Bessel functions of the first kind. Properties and curves for the Bessel functions can be found in many communication systems books since these functions appear in the computation of the spectrum of angle modulated signals (see, for example, Proakis and Salehi, 1994). The general expression and a few of the properties of the Bessel functions should be recalled here:

$$
\begin{aligned}
&J_n(x) = \frac{1}{2\pi} \int_{-\pi}^{+\pi} e^{jx\sin\psi} e^{-jn\psi} d\psi \\
&J_{-n}(x) = (-1)^n J_n(x), \quad J_n(-x) = (-1)^n J_n(x) \\
&\text{and} \\
&J_n(x) \cong 0 \quad for \quad |n| > |x|
\end{aligned} \tag{3–7}
$$

From Eq. (3–6), one can derive the PSD of $x_{PPM}(t)$:

$$
\begin{aligned}
P_{x_{PPM}}(f) &= \frac{1}{T_s^2} \sum_{m=-\infty}^{+\infty} \sum_{n=-\infty}^{+\infty} \left| J_n\left(2\pi A\left(m\frac{1}{T_s} + nf_0\right)\right)\right|^2 \cdot \\
&\quad \cdot \left| P\left(m\frac{1}{T_s} + nf_0\right)\right|^2 \delta\left(f - \left(m\frac{1}{T_s} + nf_0\right)\right)
\end{aligned} \tag{3–8}
$$

Equation (3–8) shows that, for sinusoidal modulating signals, the PPM signal has discrete frequency components located at the sinewave and pulse repetition frequencies and their harmonics, and at their sum and difference frequencies and harmonics. The amplitude of the pulses in frequency is governed by two terms, $P(f)$ and $J_n(x)$. We will now analyze the effect of each of these two terms separately.

Effect of the $|P(f)|^2$ Term

The values assumed by $|P(f)|^2$ in correspondence to the frequencies of the Dirac pulses contribute to determine the amplitudes of the Dirac pulses. If $P(f)$ has a limited bandwidth, the bandwidth of the PPM signal is limited as well.

Effect of the $J_n(x)$ Term

First, observe that for $n = 0$, $J_n(x) = 1$ if $x = 0$. Since $x = 2\pi Am/T_s$, one has $x = 0$ when $m = 0$. Recall that since m is the index referring to the harmonics of $1/T_s$ for all $m \neq 0$, discrete frequency components at frequencies m/T_s are present with amplitude given by $|J_n(2\pi Am/T_s)|^2$. Since m spans the infinite interval of summation, we can therefore assert that the term $J_n(x)$ does not introduce a limitation in the bandwidth of the signal of Eq. (3–8). However, it regulates the presence or absence of Dirac pulses located at multiples of f_0 and of linear combinations of f_0 and $1/T_s$.

From Eq. (3–7) we know that $J_n(x)$ approaches zero for $|n|>|x|$. Since $J_{-n}(x) = (-1)^n J_n(x)$ then $|J_{-n}(x)|^2 = |J_n(x)|^2$; therefore, we can limit the analysis to the case, $J_n(x) \geq 0$. In this case one has:

$$J_n(x) \cong 0 \quad for \quad |n| > |x|$$
$$i.e., for\ n\ positive, n > \left| 2\pi A \left(m\frac{1}{T_s} + nf_0 \right) \right| \tag{3–9}$$

We now suppose that $m > 0$ since the case $m < 0$ can be obtained by symmetry. In this case, note that the condition of Eq. (3–9) becomes $n > 2\pi A(m/T_s + nf_0)$, which implies:

$$n > \left(\frac{2\pi A / T_s}{1 - 2\pi Af_0} \right) m \tag{3–10}$$

Observe that $A < T_s/2$ and put $A = \beta T_s/2$. Equation (3–10) becomes:

$$n > \left(\frac{\pi\beta}{1 - \pi\beta T_s f_0} \right) m \tag{3–11}$$

Note that since $f_0 << 1/T_s$, the quantity $(1\text{-}\pi f_0 T_s)$ approaches 1 and the condition of Eq. (3–11) becomes:

$$n > \pi\beta m \tag{3–12}$$

CHECKPOINT 3–1

In this checkpoint we will use computer simulation to analyze analog PPM in the case of a sinusoidal modulating signal.

Two MATLAB functions are introduced. The first, Function 3.1, generates a train of rectangular pulses that are analog PPM modulated by a sinusoidal wave. The second, Function 3.2, evaluates the spectrum of the signal generated by Function 3.1. We choose a rectangular waveform for the transmitted pulses to simplify the analysis in the frequency domain. The Fourier transform of a rectangular waveform has the well-known sin(x)/x shape.

Function 3.1 (see Appendix 3.A) generates a PPM-UWB signal in the case of a sinusoidal modulating signal and rectangular pulses. Within the function, the user must set the following parameters: the average transmitted power in dBm `Pow`, the sampling frequency for representing the signal `fc`, the number of pulses to be generated `np`, the time duration of each rectangular pulse `Tr`, the average pulse repetition period in seconds `Ts`, the amplitude and frequency of the sinusoidal modulating signal `A` and `f0`. Function 3.1 returns two outputs: the generated train of pulses `Stx`, and the corresponding sampling frequency `fc`. The command line for generating the signal is:

```
[Stx,fc]=cp0301_PPM_sin;
```

We will use Function 3.1 to generate two signals. The first, signal `S1`, represents the output of the transmitter in the absence of modulation. The second, signal `S2`, represents the output of the transmitter when the PPM scheme is applied in the special case of a sinusoidal modulating signal.

With reference to signal `S1`, we set the following parameters within Function 3.1: `Pow=-30; fc=1e11; np=10000; Tr=0.5e-9; Ts=2e-9; A=0; f0=0`. The output signal is composed of 10,000 equally spaced rectangular pulses. The average pulse repetition period is 2 ns, that is, four times the length of each pulse. The command line for generating signal `S1` is:

```
[S1,fc]=cp0301_PPM_sin;
```

In the case of signal `S2`, the PPM block is introduced within the transmission chain. The following parameters characterize the generated waveform: `Pow=-30; fc=1e11; np=10000; Tr=0.5e-9; Ts=2e-9; A=1e-9; f0=5e7`. Note that the amplitude of the modulating signal `A` is half the value of the average pulse repetition period `Ts`, while the frequency `f0` is ten times smaller than the average pulse repetition frequency `1/Ts`. The command line for generating signal `S2` is:

```
[S2,fc]=cp0301_PPM_sin;
```

To better understand the effect of PPM on the PSD of UWB signals, we compare `S1` and `S2` in the frequency domain. This comparison is carried out by Function 3.2.

In **Function 3.2** (see Appendix 3.A), the PSD of an input signal `x(t)` is computed by dividing the ESD of `x(t)` derived in Checkpoint 1–1 by the length `T` of the time window in which `x(t)` is represented. Function 3.1 receives in input vector `x` representing the signal in the time domain, and the value of the sampling frequency `fc`. Function 3.1 returns two outputs: vector `PSD` containing the PSD of the input signal and the value `df` of frequency separation between the samples of the PSD. This value is useful to derive the amount of power P which is associated with the input signal from the PSD,

```
P = sum(PSD.*df)
```

For a given signal `x(t)`, one can thus verify the exactness of the PSD provided by Function 3.2 by comparing the above `P` value with the one that can be evaluated in the time domain:

```
P = (1/T) * sum((x.^2).*(1/fc))
```

The command line for evaluating the PSD of signal `S1` is the following;

```
[PSD1,df] = cp0301_PSD(S1,fc);
```

The above command line stores vector `PSD1` in memory and produces the graphical output shown in Figures 3–1 and 3–2.

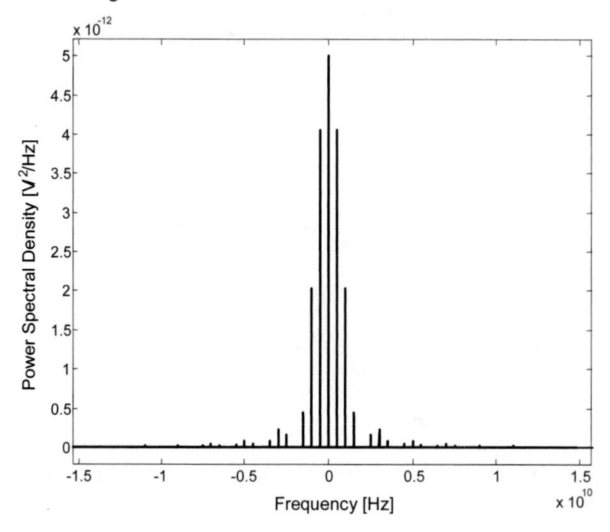

Figure 3–1 PSD of a train of equally spaced rectangular pulses (signal S1), in the absence of modulation.

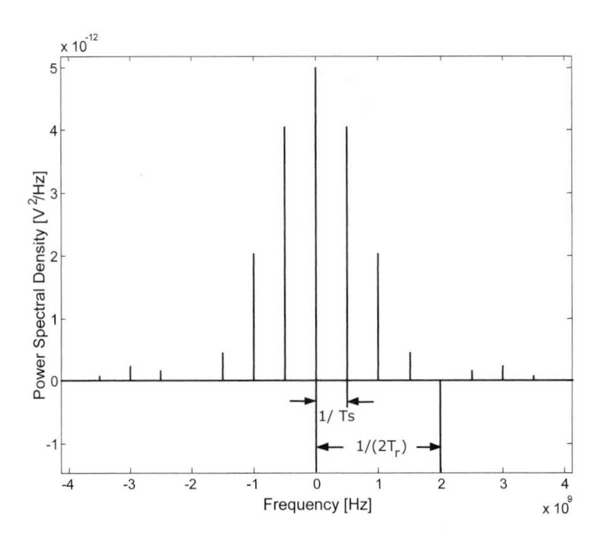

Figure 3–2 Detail of Figure 3–1: PSD of a train of equally spaced rectangular pulses (signal S1).

As expected, the PSD of signal `S1` is characterized by the presence of equally spaced spectral lines. This result is due to the absence of modulation in the time domain. The output of the transmitter is a periodic signal with its period equal to the average pulse repetition period T_s. As a consequence, the Fourier transform of `S1` is non-zero only for those frequencies that are integer multiples of the average pulse repetition frequency $f_s = 1/T_s$ = 500 MHz. The envelope of the PSD in Figures 3–1 and 3–2 follows the sin(x) / x shape of the Fourier transform of the rectangular pulse. As shown in Figure 3–2, the PSD presents in fact zero values for all frequency multiples of $1/(2T_r)$ = 2 GHz, where `Tr` is the time duration of the rectangular pulse.

We conclude the analysis of signal `S1` with the verification of the amount of transmitted power. When considering `S1` in the time domain, we obtain:

```
P_time=(fc/length(S1))*sum((S1.^2).*(1/fc))
>> P_time = 1.0000e-006
```

where the term `(fc/length(S1))` is the inverse of the time duration of the signal. In the frequency domain, one has:

```
P_freq=sum(PSD1.*df)
>> P_freq = 1.0000e-006
```

As expected, `P_time` and `P_freq` are identical. Moreover, these values confirm the input parameter `Pow` in Function 3.1, which was set equal to -30 dBm.

The spectral analysis of signal `S1` can be repeated in the case of the modulated signal `S2`. The following command must be executed:

```
[PSD2,df] = cp0301_PSD(S2,fc);
```

which stores vector `PSD2` in memory and produces the graphical output of Figures 3–3 and 3–4:

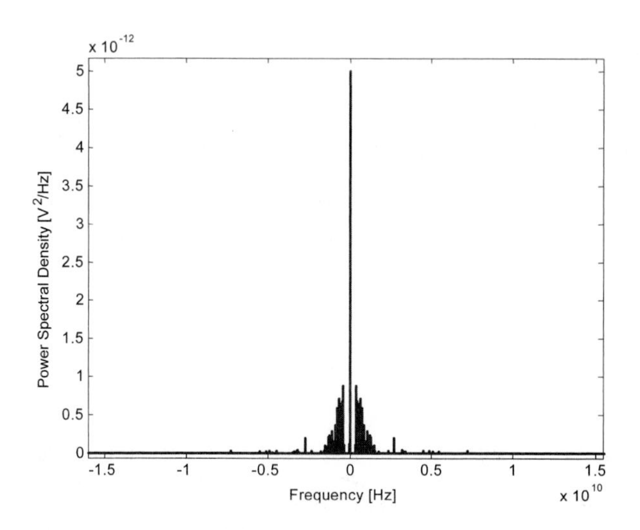

Figure 3–3 PSD of a train of rectangular pulses that is PPM-modulated by a sinusoidal modulating signal (signal S2).

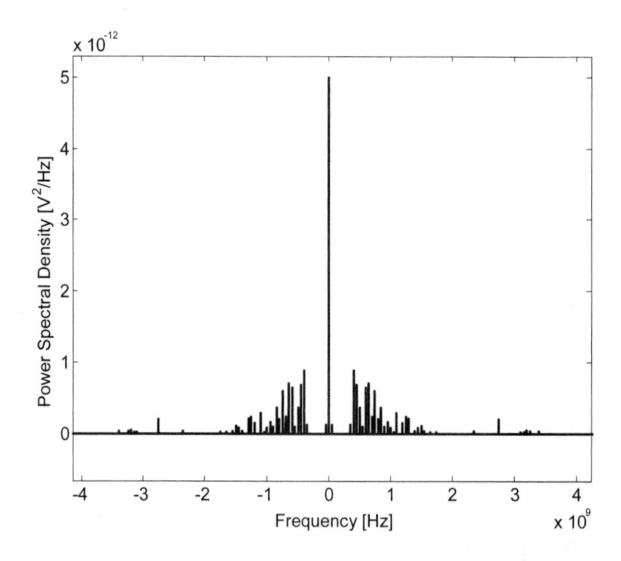

Figure 3–4 Detail of Figure 3–3: PSD of a train of PPM-modulated rectangular pulses in the case of a sinusoidal modulating signal (signal S2).

Figures 3–3 and 3–4 represent the `PSD` of signal `S2` in the same frequency range of Figures 3–1 and 3–2. We can observe that the presence of the sinusoidal modulating signal has the effect of altering the sin(x)/x shape of the signal spectrum. In addition, a higher number of spectral lines is present in the PSD than in the preceding case. According to Eq. (3–8), these new lines are located at frequencies $f(m,n) = (m/T_s+nf_0)$, that is, at the sum and difference frequencies of the modulating signal frequency f_0 and the pulse repetition frequency $1/T_s$ and their harmonics. The amplitude of the spectral line at frequency f(m,n) depends on two terms: the modulus of the Fourier transform of the rectangular pulse evaluated in f(m,n) and the modulus of the Bessel function $J_n(x)$ of order n and argument $x = (2\pi Af(m,n))$. In the case under examination, that is, $A = T_s/2$ and $f_0 = 1/(10T_s)$, the spectral lines are equally spaced with frequency separation $df = 0.1/T_s$. In other words, the spectral lines of `PSD2` are ten times closer compared to the PSD of signal `S1`. Each spectral line has an amplitude that depends on the Fourier transform of the pulse waveform and on the modulus of the following function:

$$J_n\left(2\pi\frac{T_S}{2}\left(\frac{m}{T_S}+\frac{n}{10T_S}\right)\right)=J_n\left(m\pi+\frac{n\pi}{10}\right) \quad (3\text{–}13)$$

According to Eq. (3–7), the amplitude of the Bessel function of order n and argument x tends to zero when $|n| > |x|$. As a consequence, we derive from Eq. (3–13) that the PSD of signal `S2` is composed of "clusters" of spectral lines, each cluster being located in correspondence of a multiple of the average pulse repetition frequency $f_s = 1/T_s$. This result is shown in Figure 3–5, which compares the PSD of signals `S1` and `S2`.

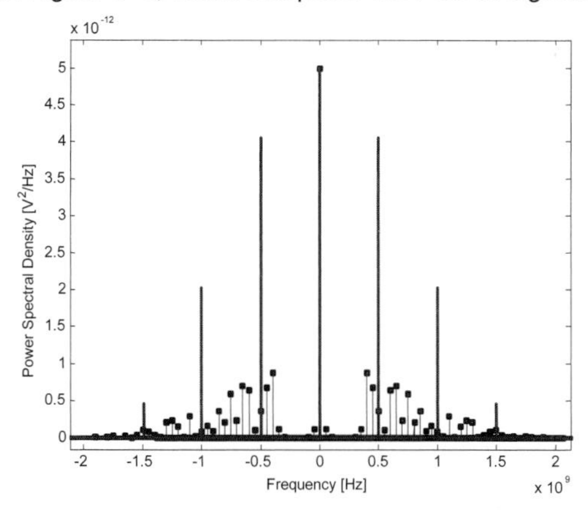

Figure 3–5 Comparison of the PSD of signals S1 (solid lines) and S2 (dashed lines).

The analysis of each cluster of spectral lines can be performed by introducing the MATLAB function `besselj(nu,z)`, which computes the value of the Bessel function of the first kind with order `nu` and argument `z`.

We start by considering the cluster of spectral lines located around frequency zero. According to Eqs. (3–8) and (3–13), this cluster consists of spectral lines at frequencies f(n) =

$n/(10T_s)$. For the single cluster, we may neglect the shaping effect of the Fourier transform of the pulse waveform. The amplitude of the spectral line at frequency f(n) is therefore proportional to:

$$A_0(n) = \left| J_n\left(\frac{n\pi}{10}\right)\right|^2 \quad (3–14)$$

Equation (3–14) can be visualized by executing the following MATLAB code:

```
n=(-20:1:20);
A0=abs(besselj(n,(pi/10).*n)).^2;
figure(1)
stem(n,A0)
```

The above code stores vector `A0` in memory and generates the plot of Figure 3–6, in which we can recognize the cluster of spectral lines located at the center of the PSD of Figure 3–5.

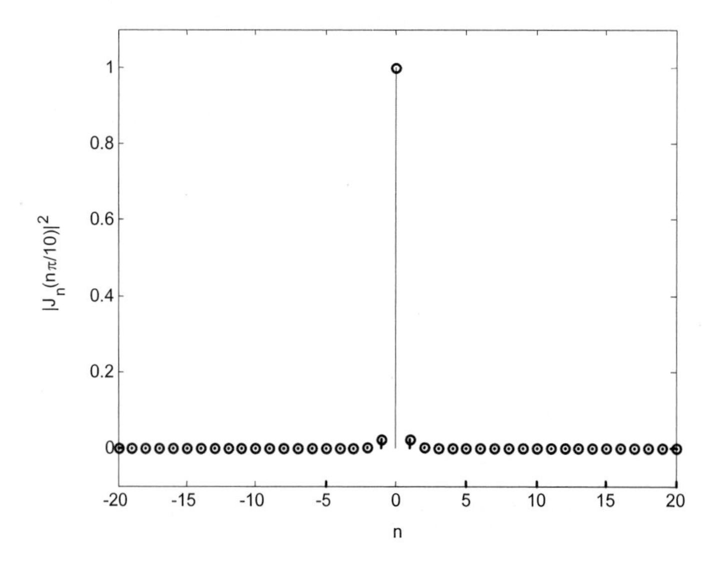

Figure 3–6 Cluster of spectral lines located around frequency zero — $|J_n(n\pi/10)|^2$ for different values of n.

The pair of clusters located around $\pm 1/T_s = \pm 500$ MHz give:

$$A_{+1}(n) = \left| J_n\left(\pi + \frac{n\pi}{10}\right) \right|^2 \quad (3\text{–}15)$$

$$A_{-1}(n) = \left| J_n\left(-\pi + \frac{n\pi}{10}\right) \right|^2 \quad (3\text{–}16)$$

where $A_{+1}(n)$ and $A_{-1}(n)$ represent approximated amplitude values of the spectral lines located around $1/T_s$ and $-1/T_s$, respectively. The code for evaluating the expressions in Eqs. (3–15) and (3–16) is:

```
n=(-20:1:20);
Ap1=abs(besselj(n,pi+(pi/10).*n)).^2;
Am1=abs(besselj(n,((pi/10).*n)-pi)).^2;
figure(2)
stem(n,Ap1)
figure(3)
stem(n,Am1)
```

The graphical output generated by the above code is shown in Figures 3–7 and 3–8.

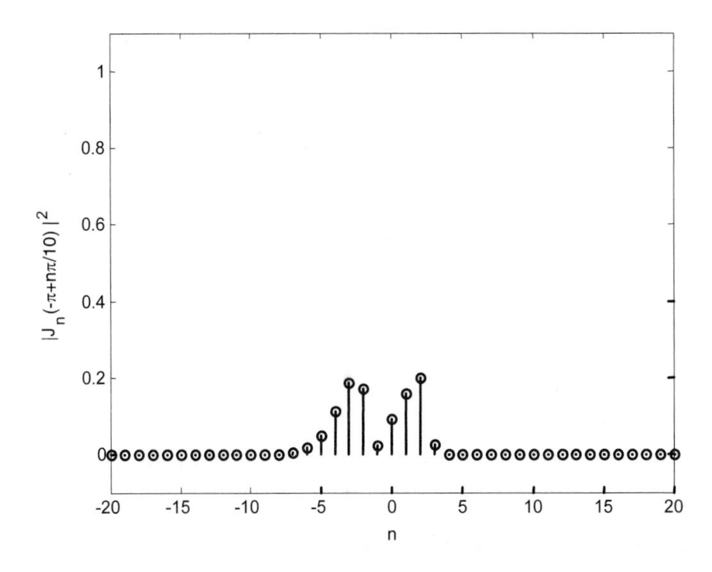

Figure 3–7 Cluster of spectral lines located around frequency -500 MHz — $|J_n(-\pi+n\pi/10)|^2$ for different values of n.

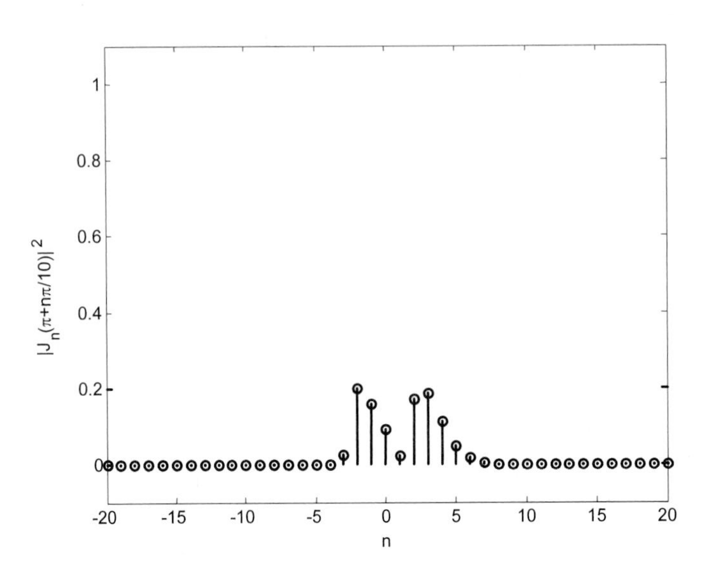

Figure 3–8 Cluster of spectral lines located around frequency +500 MHz — $|J_n(\pi+n\pi/10)|^2$ for different values of n.

Figure 3–7 represents the cluster of spectral lines located around frequency $-1/T_s = -500$ MHz, while Figure 3–8 represents the cluster of spectral lines located around frequency 1/Ts = 500 MHz. When comparing the plots in Figures 3–7 and 3–8 with the PSD in Figure 3–5, we notice the agreement between theoretical analysis and simulation results. In addition, if we compare the plots in Figures 3–7 and 3–8 vs. Figure 3–6 , we observe that the clusters at $\pm 1/T_s$ are composed of more spectral lines with respect to the cluster at the center of the PSD. The same result is verified when analyzing the pair of clusters located at $\pm 2/T_s = \pm 1$ GHz. In this case, the following expressions must be considered:

$$A_{+2}(n) = \left| J_n\left(2\pi + \frac{n\pi}{10}\right)\right|^2 \quad (3\text{–}17)$$

$$A_{-2}(n) = \left| J_n\left(-2\pi + \frac{n\pi}{10}\right)\right|^2 \quad (3\text{–}18)$$

which can be reproduced with the following code lines:

```
n=(-20:1:20);
Ap2=abs(besselj(n,2*pi+(pi/10).*n)).^2;
```

```
Am2=abs(besselj(n,((pi/10).*n)-2*pi)).^2;
figure(4)
stem(n,Ap2)
figure(5)
stem(n,Am2)
```

The graphical output generated by the above code is shown in Figures 3–9 and 3–10.

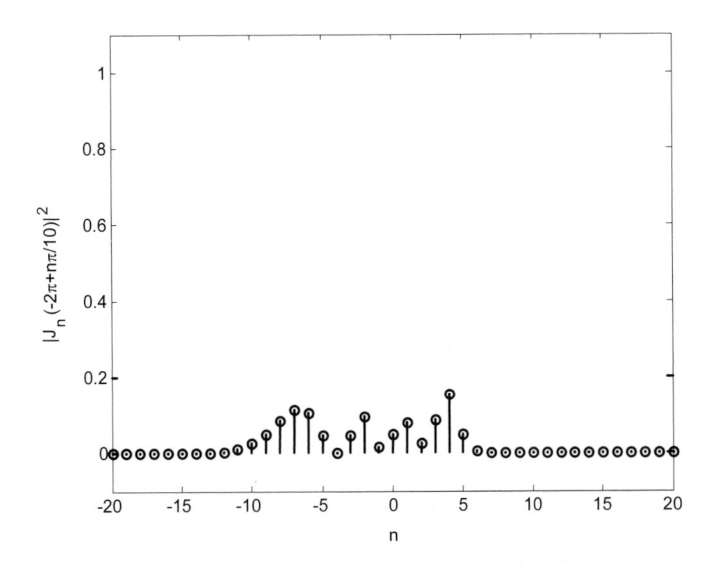

Figure 3–9 Cluster of spectral lines located around frequency -1GHz — $|J_n(-2\pi+n\pi/10)|^2$ for different values of n.

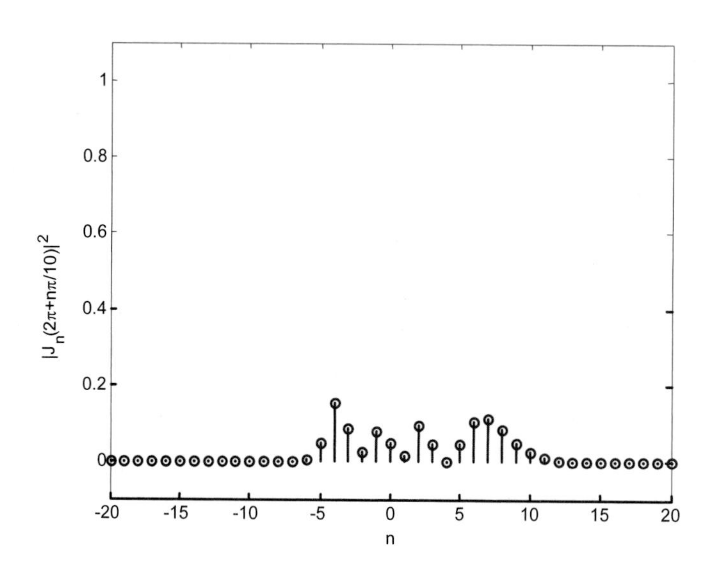

Figure 3–10 Cluster of spectral lines located around frequency +1GHz — $|J_n(2\pi+n\pi/10)|^2$ for different values of n.

Figure 3–9 represents the cluster of spectral lines located around frequency $-2/T_s = -1$ GHz, while Figure 3–10 represents the cluster of spectral lines located around frequency $2/T_s = 1$ GHz. Once again, we observe that the number of spectral lines composing these clusters is increased with respect to the previous cases, that is, the number of spectral lines composing one cluster increases when the central frequency of the cluster increases.

CHECKPOINT 3–1

3.1.2 Generic Periodic Modulating Signals

When $m(t)$ is periodic of period T_p, that is, $m(t+T_p) = m(t)$ for all t, the Fourier series representation is valid for all t and is expressed as follows:

$$m(t) = \sum_{n=-\infty}^{+\infty} m_n e^{jn2\pi t/T_p} \tag{3–19}$$

where m_n is the n-th Fourier coefficient given by:

$$m_n = \frac{1}{T_p}\int_{\alpha}^{\alpha+T_p} m(t)e^{-jn2\pi t/T_p}dt \tag{3–20}$$

If $m(t)$ is real, then one obtains:

$$m_n = m^*_{-n} \tag{3–21}$$

The Fourier series expansion shows that a periodic signal $m(t)$ can be represented for all t as a sum of components of different frequencies, all multiples of the fundamental frequency $1/T_p$. The n-th term of the summation corresponds to frequency n/T_p.

The analysis of sinusoidal modulating waves can be expanded to periodic modulating signals by observing that in this last case, the modulating signal is composed of the sum of sinusoidal waves at frequencies that are multiples of the fundamental. Note that in this case, the condition $A < T_s/2$ becomes:

$$\sum_{n=-\infty}^{+\infty} m_n < \frac{T_s}{2} \tag{3–22}$$

We put:

$$M = \sum_{n=-\infty}^{+\infty} m_n \tag{3–23}$$

By applying the multiple Fourier series method as proposed by (Bennett, 1933, 1944, and 1947) and as further suggested by (Rowe, 1965), we find:

$$x_{PPM}(t) = \frac{1}{T_s}\sum_{m=-\infty}^{+\infty}\sum_{n=-\infty}^{+\infty}\sum_{l=-\infty}^{+\infty}(-j)^n J_n\left(2\pi M\left(m\frac{1}{T_s}+nl\frac{1}{T_p}\right)\right)\cdot P\left(m\frac{1}{T_s}+nl\frac{1}{T_p}\right)e^{j2\pi\left(m\frac{1}{T_s}+nl\frac{1}{T_p}\right)t} \tag{3–24}$$

Equation (3–24) easily leads to the PSD of a periodic modulating PPM signal $x_{PPM}(t)$ which is expressed by:

$$P_{x_{PPM}}(f) = \frac{1}{T_s^2}\sum_{m=-\infty}^{+\infty}\sum_{n=-\infty}^{+\infty}\sum_{l=-\infty}^{+\infty}\left|J_n\left(2\pi M\left(m\frac{1}{T_s}+nl\frac{1}{T_p}\right)\right)\right|^2\cdot\left|P\left(m\frac{1}{T_s}+nl\frac{1}{T_p}\right)\right|^2\delta\left(f-\left(m\frac{1}{T_s}+nl\frac{1}{T_p}\right)\right) \tag{3–25}$$

Similar to the sinusoidal case, when the modulating signal is a periodic waveform, the PPM signal contains discrete frequency components located at the fundamental and pulse repetition frequencies and their harmonics, as well as at the sum and difference frequencies of the modulating signal and the pulse repetition frequency and harmonics. The amplitude of the pulses in frequency is governed by two terms: $P(f)$ and $J_n(x)$. The analysis of the effect of these two terms closely follows the analysis of the sinusoidal case.

CHECKPOINT 3–2

In this checkpoint, we will use computer simulation to analyze the spectral occupation of an UWB signal implementing PPM in the case of a periodic modulating signal. The modulating signal m(t) is chosen to have a negative exponential amplitude decay within a period T_p:

$$m(t) = \sum_{k=-\infty}^{+\infty} Ae^{B(kT_p - t)} rect_{T_p}\left(t - \frac{3}{2}kT_p\right) \quad (3\text{–}26)$$

where A and B are two real constant terms.

Assume, for example, T_p = 20 ns, A = $1 \cdot 10^{-9}$ V, and B = 10. We can generate the waveform in Eq. (3–26) by using the following MATLAB code:

```
A = 1e-9;
B = 10;
Tp = 20e-9;
fc = 1e11;
dt = 1 / fc;
T = 1e-6;
time = (0:dt:T);
m = A.*exp(-(B/Tp).*mod(time,Tp));
plot(time,m);
```

The above code stores in memory vector `m`, which contains the samples of signal m(t). Figure 3–11 represents m(t) in the time domain.

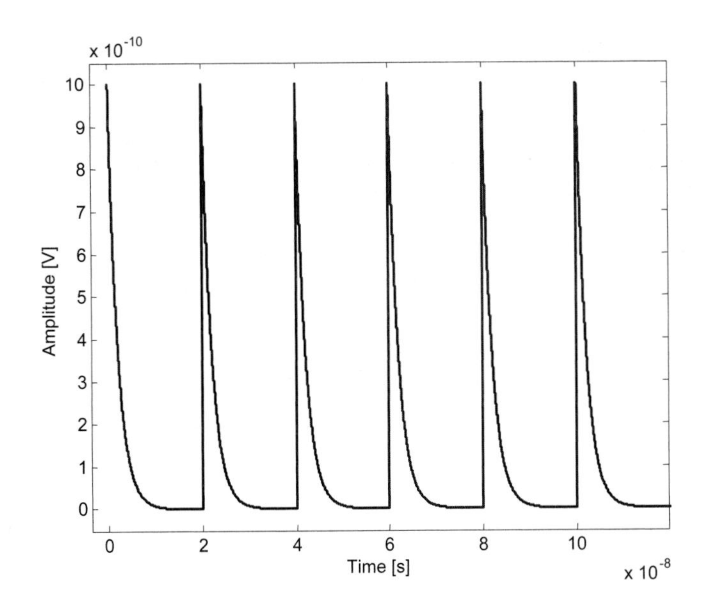

Figure 3–11 Periodic modulating signal m(t).

Given vector `m`, we can isolate a single period of the periodic signal. The following code line extracts the first period of m(t) and stores it in vector `x`:

```
x = m(1:floor(Tp/dt));
```

Given vector `x`, we can use the MATLAB function `fft(x)` to evaluate the coefficients of the Fourier series representing the periodic signal m(t):

```
X = fftshift((1/length(x)).*fft(x));
```

Figure 3–12 shows the modulus of the coefficients of the Fourier series of m(t).

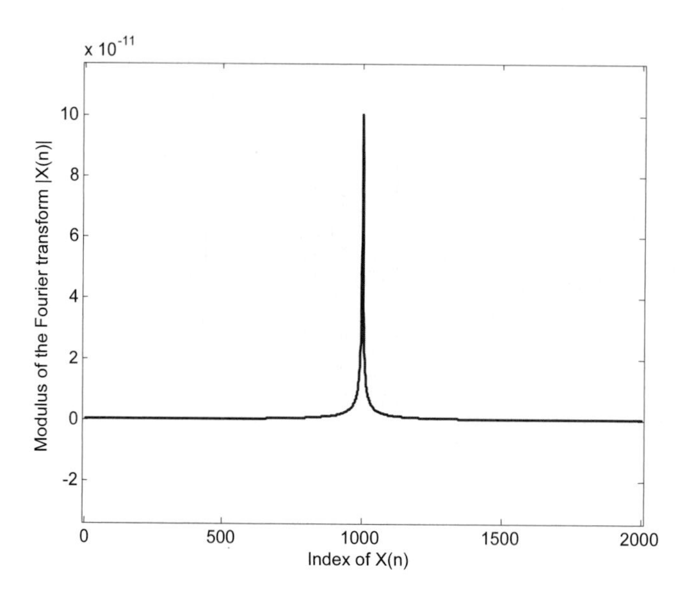

Figure 3–12 Modulus of the coefficients of the Fourier series of m(t).

Finally, we can evaluate the value in Eq. (3–23) as follows:

```
M = real(sum(X))
>> M = 1.0000e-009
```

In the above command line, the MATLAB function `real()` is necessary to take into account the approximation errors due to the sampling of the original waveform m(t). In the case of no approximation errors, the summation in Eq. (3–23) is always real if m(t) is real.

To analyze the spectral characteristics of PPM-UWB signals with periodic modulating signals, we must introduce a new MATLAB function.

Function 3.3 (see Appendix 3.A) generates a train of rectangular pulses that are modulated in position by the signal m(t) in Eq. (3–26). Within the function, the user must set the following parameters: the average transmitted power in dBm `Pow`, the sampling frequency `fc`, the number of pulses to be generated `np`, the time duration of each rectangular pulse `Tr`, the average pulse repetition period in seconds `Ts`, and finally, parameters `A`, `B`, and `Tp`, which characterize the modulating signal. Function 3.3 returns two outputs: the generated train of pulses `Stx` and the corresponding sampling frequency `fc`. The command line is:

```
[Stx,fc]=cp0302_PPM_periodic;
```

We will use Function 3.3 for generating an UWB signal with a periodic modulating signal m(t) as in Figure 3–11. The following parameters are set within the function: `Pow=-30; fc=1e11; np=10000; Tr=0.5e-9; Ts=2e-9; A=1e-9; B=10; Tp=20e-9`. Figure 3–13 represents a portion of the generated signal `Stx` in the time domain.

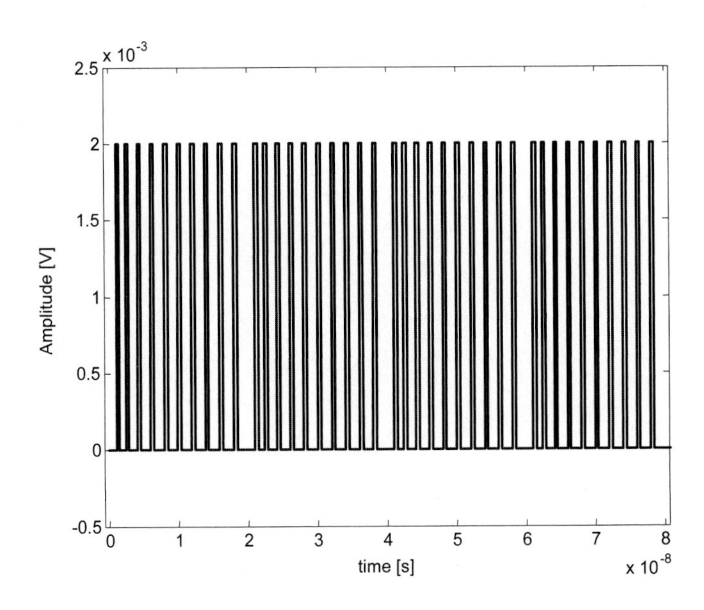

Figure 3–13 A portion of signal Stx generated with Function 3.3. The modulating signal is the periodic waveform represented in Figure 3–11.

The spectral analysis of signal `Stx` can be carried out using Function 3.2. The following command must be executed:

```
[PSD,df] = cp0301_PSD(Stx,fc);
```

which stores vector `PSD` in memory and produces the graphical output in Figures 3–14 and 3–15.

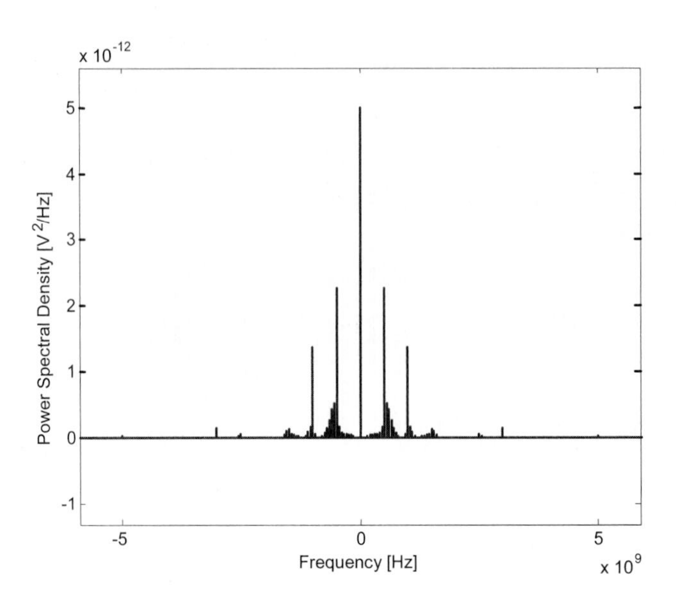

Figure 3–14 PSD of signal Stx (see Figure 3–13).

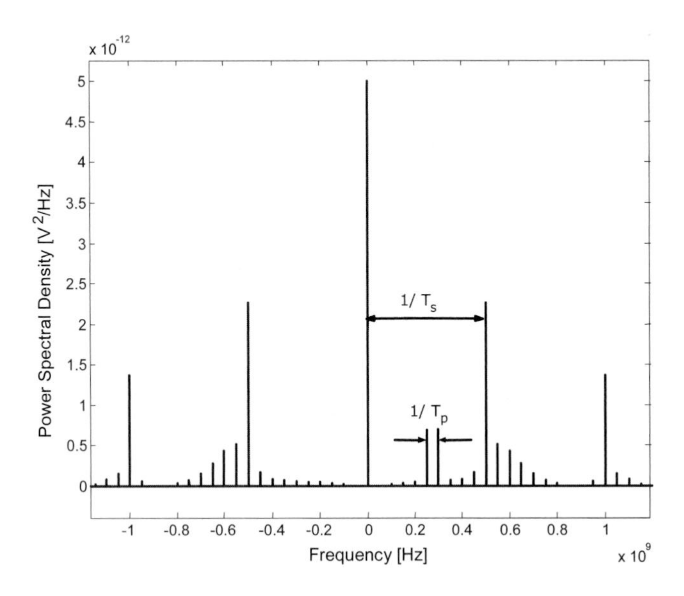

Figure 3–15 Detail of Figure 3–14: PSD of signal Stx.

Figures 3–14 and 3–15 show that the PSD of signal `stx` is composed of spectral lines located at each integer multiple of the pulse repetition frequency $1/T_s$, and at all the sum and difference frequencies of the modulating signal frequency $1/T_p$ and the pulse repetition frequency $1/T_s$. According to Eq. (3–25), the amplitude of the spectral line at frequency $f_x = (m/T_s + nl/T_p)$ depends on both the value in f_x of the Fourier transform of the rectangular pulse and the value of the Bessel function $J_n(2\pi M f_x)$, where M is the constant term in Eq. (3–23). In the case under examination, we found that $M = 1 \cdot 10^{-9} = T_s/2$. Moreover, since $T_p = 10T_s$, we can simplify the argument of the Bessel function as follows:

$$J_n\left(2\pi\frac{T_S}{2}\left(\frac{m}{T_S}+\frac{nl}{10T_S}\right)\right) = J_n\left(m\pi+\frac{nl\pi}{10}\right) \qquad (3\text{–}27)$$

As shown in Checkpoint 3–1, we can use the MATLAB function `besselj(nu,z)` to analyze each cluster of spectral lines separately. The code for generating the cluster located at the zero frequency, for example, is:

```
Jm0=zeros(1,51);
for n = -5 : 5
for l = -5 : 5
i = n * l;
index = i + 26;
Jm0(index)=Jm0(index)+abs(besselj(n,n*l*pi/10))^2;
end
end
abscissa = (-25:1:25);
figure(1)
stem(abscissa,Jm0)
```

The above code lines store vector `Jm0` in memory and produce the plot of Figure 3–16.

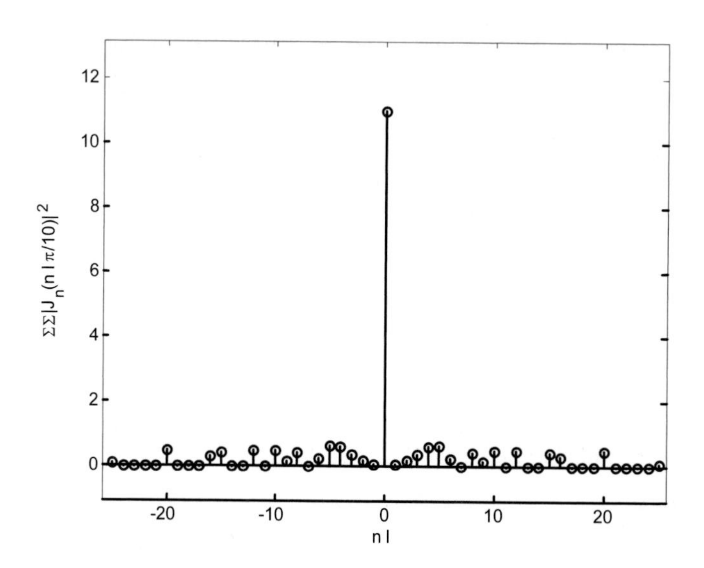

Figure 3–16 Cluster of spectral lines located around frequency zero

The two clusters at $\pm 1/T_s$ can be generated as follows:

```
Jmp1 = zeros(1,51);
Jmn1 = zeros(1,51);
for n = -5 : 5
for l = -5 : 5
i = n * l;
index = i + 26;
Jmp1(index) = Jmp1(index) + abs(besselj(n,pi + ...
   (n*l*pi/10))) ^2;
Jmn1(index) = Jmn1(index) + abs(besselj(n,(n*l*pi/10)...
   - pi)) ^2;
end
end
abscissa = (-25:1:25);
figure(2)
stem(abscissa,Jmn1);
figure(3)
stem(abscissa,Jmp1);
```

The graphical output resulting from the above code lines is shown in Figures 3–17 and 3–18.

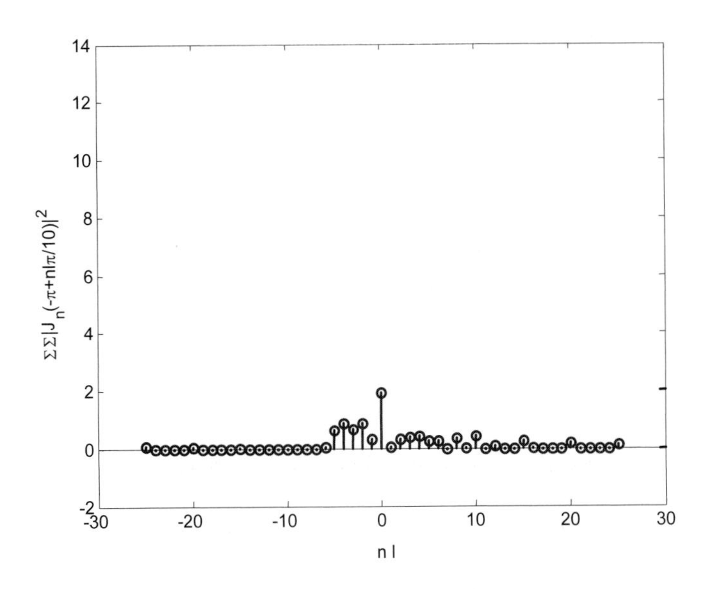

Figure 3–17 Cluster of spectral lines located around frequency -500 MHz

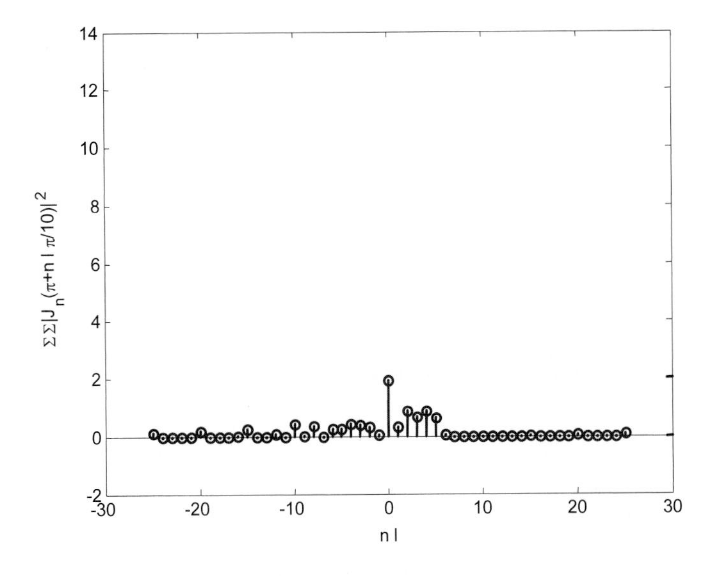

Figure 3–18 Cluster of spectral lines located around frequency +500 MHz

Finally, we can generate the clusters at $\pm 2/T_s$ with the following code lines:

```
Jmp2 = zeros(1,51);
Jmn2 = zeros(1,51);
for n = -5 : 5
for l = -5 : 5
i = n * l;
index = i + 26;
Jmp2(index) = Jmp2(index) + abs(besselj(n,2*pi + ...
   (n*l*pi/10))) ^2;
Jmn2(index) = Jmn2(index) + ...
   abs(besselj(n,(n*l*pi/10)- pi*2)) ^2;
end
end
abscissa=(-25:1:25);
figure(4)
stem(abscissa,Jmn2)
figure(5)
stem(abscissa,Jmp2)
```

The output provided by the above MATLAB code is shown in Figures 3–19 and 3–20.

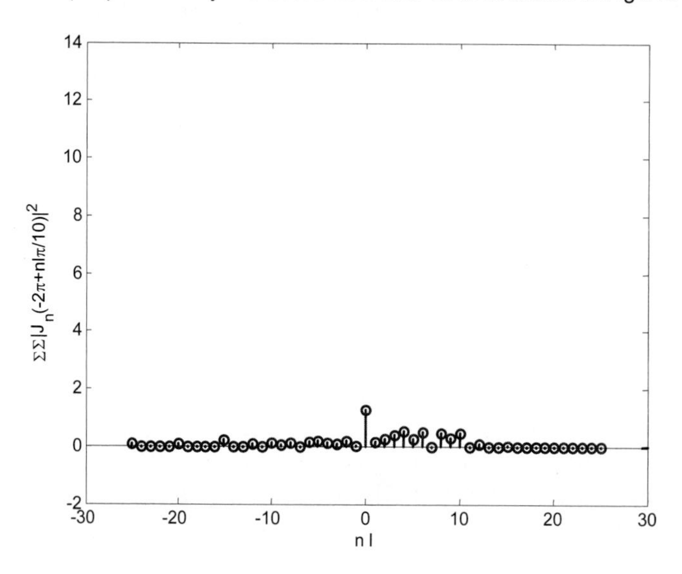

Figure 3–19 Cluster of spectral lines located around frequency -1 GHz

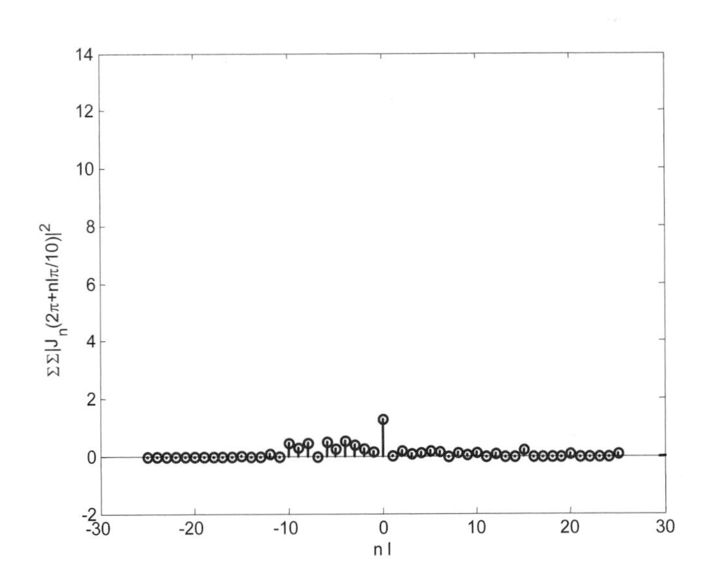

Figure 3–20 Cluster of spectral lines located around frequency +1 GHz

Remember that the clusters represented in Figures 3–16 to 3–20 only take into account the effect on the amplitude of the Bessel function. The exact values of the amplitude of the spectral lines should also take into account the effect of the Fourier transform of the rectangular pulse waveform.

CHECKPOINT 3–2

3.1.3 Random Modulating Signals

The derivation of the PSD of Eq. (3–1) can be performed under the hypothesis that $m(kT_s)$ is a strict-sense stationary discrete random process, where $m(kT_s)$ are the samples of a strict-sense stationary continuous process $m(t)$ and the different $m(kT_s)$ are statistically independent with a common probability density function $w(m(kT_s))$.

Since the signal of Eq. (3–1) is not wide-sense stationary, the PSD $P_{x_{PPM}}(f)$ for this signal can be found by applying the following steps:

1. Compute the autocorrelation function of a particular $x_{PPM}(t)$.
2. Average over the ensemble to find the ensemble average.

3. Obtain $P_{x_{PPM}}(f)$ by taking the Fourier transform of the ensemble average.

As shown by (Rowe, 1965), $P_{x_{PPM}}(f)$ can be expressed as follows:

$$P_{x_{PPM}}(f) = \frac{|P(f)|^2}{T_s}\left[1 - |W(f)|^2 + \frac{|W(f)|^2}{T_s}\sum_{n=-\infty}^{+\infty}\delta(f - \frac{n}{T_s})\right] \quad (3\text{–}28)$$

where $W(f)$ is the Fourier transform of the probability density w and coincides with the characteristic function of w computed in $-2\pi f$:

$$W(f) = \int_{-\infty}^{+\infty} w(s)e^{-j2\pi fs}ds = \left\langle e^{-j2\pi fs}\right\rangle = C(-2\pi f) \quad (3\text{–}29)$$

Equation (3–28) shows that the spectrum of a random modulating PPM signal is composed of a continuous part controlled by the term $1-|W(f)|^2$, and of a discrete part formed by line components at frequency $1/T_s$, that is, the pulse repetition frequency and harmonics. The discrete part corresponds to a periodic component of the PPM signal of Eq. (3–1).

Since $W(f)$ is the Fourier transform of a probability density function, its value at 0 is 1, that is, $W(0) = 1$; therefore, the continuous component of the spectrum is zero at frequency zero and it necessarily rises at higher frequencies. The discrete components that are weighted by $|W(f)|^2$ are larger at low frequencies, then decrease at high frequencies. The predominance of the continuous term over the discrete term depends on the values of $m(kT_s)$. If these are small, then the PPM signal of Eq. (3–1) resembles a periodic signal and the discrete components dominate the low frequencies. If, however, the $m(kT_s)$ values are large, then the PPM signal of Eq. (3–1) loses its resemblance with a periodic signal and the continuous component dominates the low frequency as well as the high frequency range of the spectrum.

Finally, as in the case of sinusoidal modulating signals, the term $|P(f)|^2$ shapes the overall spectrum and limits bandwidth occupation to finite values.

When the different $m(kT_s)$ are not independent, the PSD is found to be (Rowe, 1965):

$$P_{x_{PPM}}(f) = \frac{|P(f)|^2}{T_s}\sum_{n=-\infty}^{+\infty}\left\langle e^{-j2\pi f\left(m((l+n)T_s)-m(lT_s)\right)}\right\rangle e^{-j2\pi fnT_s} \quad (3\text{–}30)$$

CHECKPOINT 3–3

In this checkpoint, we will use computer simulation to analyze the spectral occupation of a PPM-UWB signal in the presence of a random modulating signal. In particular, we will assume that the samples $m(kT_s)$ are statistically independent and Gaussian distributed random variables. The transmitted UWB signal is characterized by rectangular pulses with duration T_r and an average pulse repetition period T_s. To simulate the generation of the UWB signal under examination, we will introduce Function 3.4.

Function 3.4 (see Appendix 3.A) generates a PPM-UWB signal in the case of a random modulating signal. Within the function, the user must set the following parameters: the average transmitted power in dBm `Pow`, the sampling frequency for representing the signal `fc`, the number of pulses to be generated `np`, the time duration of each rectangular pulse `Tr`, the average pulse repetition period in seconds `Ts`, and the standard deviation of the Gaussian distributed modulating signal `sigma`. Function 3.4 returns three outputs: the generated train of pulses `Stx`, the corresponding sampling frequency `fc`, and vector `M0` containing all the time shifts applied to the transmitted pulses due to the presence of the modulating signal. The command line for generating the signal is:

```
[Stx,fc,M0]=cp0303_PPM_random;
```

We will use Function 3.4 for generating the UWB signal `RS0`, which is characterized by the following parameters: `Pow=-30`; `fc=1e11`; `np=10000`; `Tr=0.5e-9`; `Ts=2e-9`; `sigma=0.1e-9`. The command line for generating signal `RS0` is:

```
[RS0,fc,M0]=cp0303_PPM_random;
```

Figure 3–21 represents a fragment of the generated signal `RS0` in the time domain.

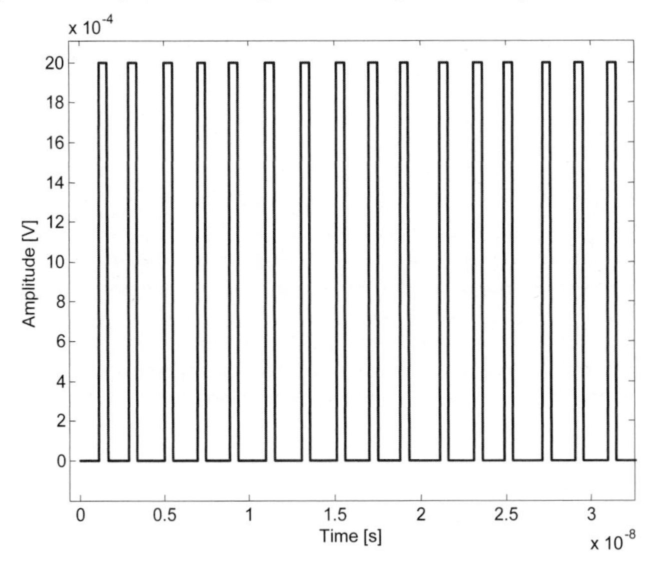

Figure 3–21 Signal RS0 in the time domain — PPM-UWB signal in the case of a random modulating signal.

Figure 3–21 shows that the effect of PPM on the position of the transmitted pulses is barely appreciable, due to the small value chosen for the standard deviation, `sigma=0.1·10-9`. We can analyze the spectral characteristics of signal `RS0` by using Function 3.2, i.e.:

```
[PSD0,df]=cp0301_PSD(RS0,fc);
```

The above command line stores in memory the PSD `PSD0` and produces the graphical output in Figure 3–22.

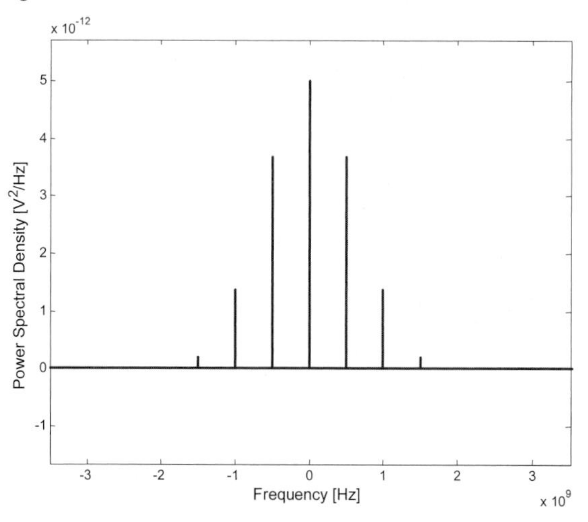

Figure 3–22 PSD of signal RS0 (see Figure 3–21).

The PSD in Figure 3–22 is composed of discrete contributions only. The discrete terms in Eq. (3–28) are thus predominant over the continuous term. This result is due to the presence of very small values for the randomly generated PPM shifts. We also observe that the envelope of the PSD is considerably different from the sin(x)/x shape since no side lobes are visible in the plot of Figure 3–22. This result can be analyzed by taking into account the statistical characteristics of the modulating signal. Given vector `M0`, resulting from the execution of Function 3.4, we can evaluate the probability density function of the time shifts, introduced within the signal by executing the following code lines:

```
dt = 1/fc;
NI = ((2e-9)-(0.5e-9))/dt;
h = hist(M0,NI);
h0 = (1/sum(h)).*h;
time=linspace(0,2e-9,length(h))-(2e-9/2);
stem(time,h0)
```

The above code makes use of the MATLAB function `H=hist(Y,N)`, which groups the elements of vector `Y` into `N` equally spaced containers and returns vector `H`, which contains the number of elements in each container. The above set of commands also produces the plot in Figure 3–23, which represents the probability density function of the PPM shifts for signal `RS0`. The shape of this function resembles the well-known bell shape of a Gaussian.

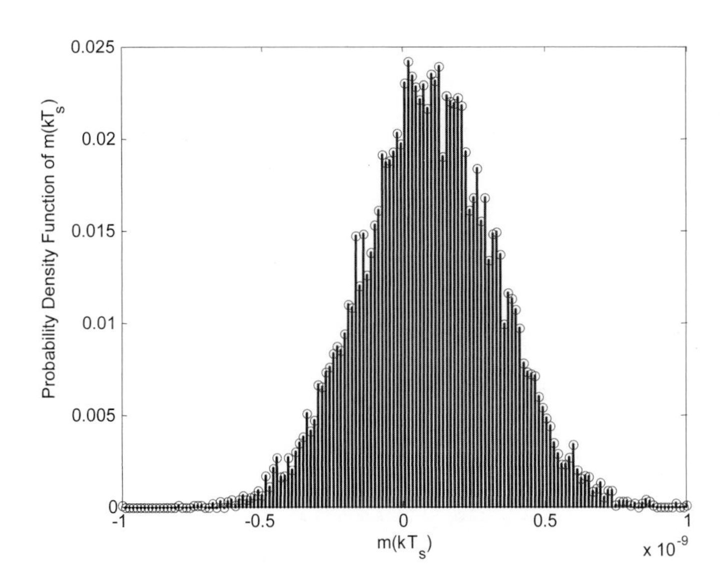

Figure 3–23 Probability density function of the PPM shifts for signal RS0.

We can now apply Function 3.2 to vector `h0`:

```
fcx=1/((2e-9)/length(h));
[W,df]=cp0301_PSD(h0,fcx);
```

The above commands store in memory vector `W`, which represents a function in the frequency domain that is proportional to the term $|W(f)|^2$ of Eq. (3–28). The graphical output provided by the above commands is represented in Figure 3–24.

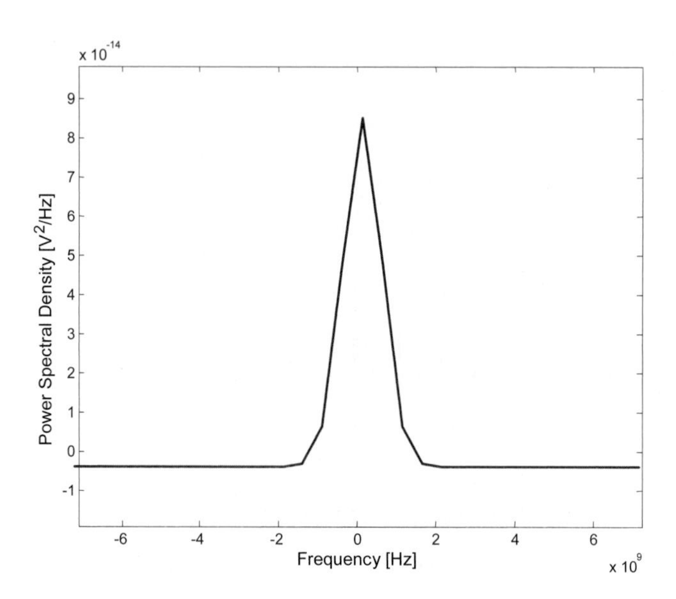

Figure 3–24 PSD of vector h0 — squared modulus of the Fourier transform of the probability density function of the PPM shifts.

The result shown in Figure 3–24 confirms the effect of the Fourier transform W(f) on the PSD of the generated UWB signal. When comparing Figures 3–24 and 3–22, we conclude that in the present case, the bandwidth of the transmitted signal is limited by the Fourier transform W(f) of the probability density of the PPM shifts, not by the Fourier transform P(f) of the pulse waveform.

To verify the presence of a continuous part in the PSD of PPM-UWB signals modulated with a random signal, we can consider the case of signal `RS1` with an increased variability in the PPM shifts. To better exploit the variability of the PPM shift, we choose higher values for both standard deviation `sigma` and average pulse repetition period `Ts`. The following parameters are set within Function 3.4: `Pow=-30; fc=1e11; np=10000; Tr=0.5e-9; Ts=10e-9; sigma=4e-9`. The command line for generating signal `RS1` is:

```
[RS1,fc,M1]=cp0303_PPM_random;
```

Figure 3–25 represents a section of the generated signal `RS1` in the time domain.

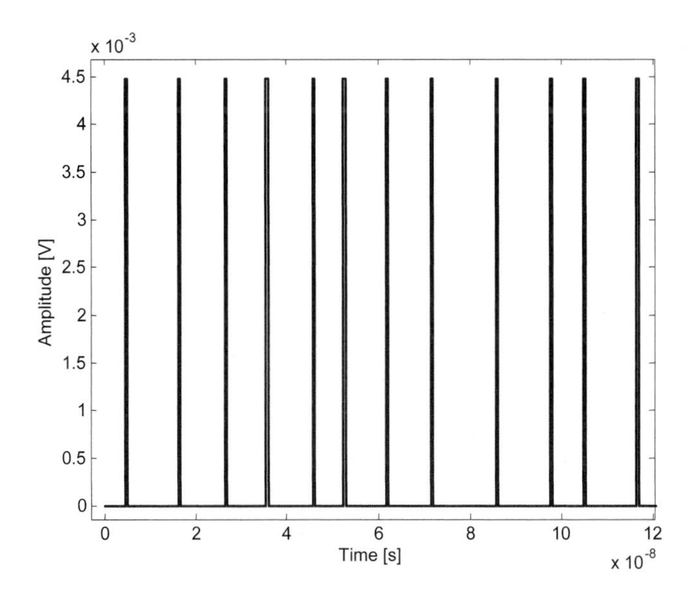

Figure 3–25 Signal RS1 in the time domain.

Figure 3–25 shows that the effect of PPM is considerable in the case of signal `RS1`. We can analyze the spectral characteristics of signal `RS1` by executing the following command line:

```
[PSD1,df]=cp0301_PSD(RS1,fc);
```

The graphical output that results from the above command is shown in Figure 3–26.

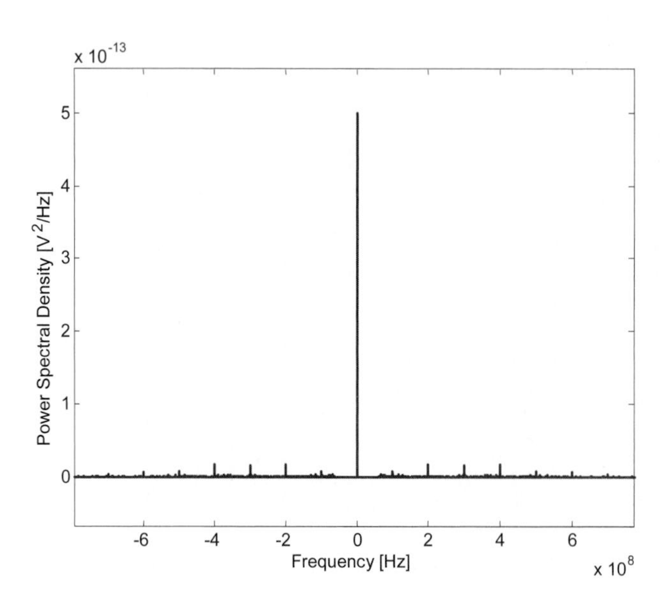

Figure 3–26 PSD of signal RS1 (see Figure 3–25).

Figure 3–26 shows that the PSD of signal `RS1` is dominated by a strong peak at zero frequency. We also observe that the PSD is not composed of spectral lines only, since we notice a few spurious contributions between these lines. This observation is confirmed by zooming in on Figure 3–26 in the region of frequencies included between ±200 MHz (see Figure 3–27).

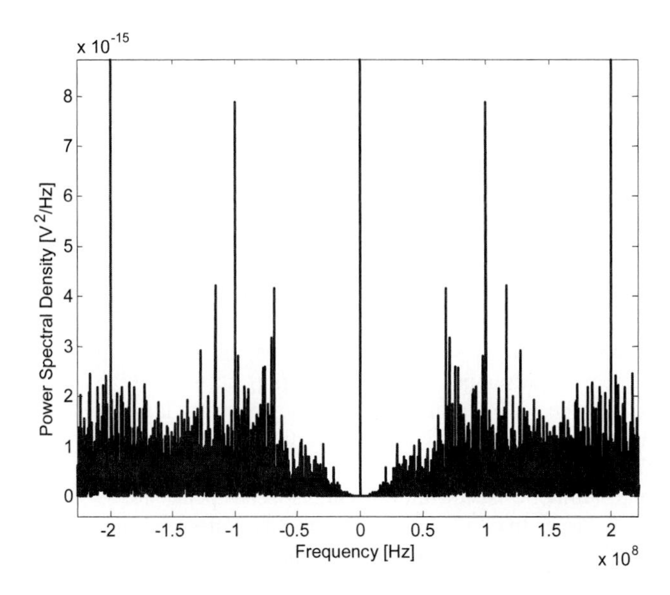

Figure 3–27 Details of the PSD of signal RS1 .

In the case of signal `RS1`, we can recognize the presence of a continuous part in the PSD, which is controlled by the function W(f).

Note that a similar effect is not observed if one tries to zooming in on Figure 3–22, as can be easily checked by the interested reader.

CHECKPOINT 3–3

3.2 The PPM-TH-UWB Case

We can now refer back to the PPM-TH-UWB signal, as expressed by Eq. (2–4), and establish a correspondence with the PPM signal of Eq. (3–1), that is, between the $m(kT_s)$ process and the θ time dither process, which, as previously defined, incorporates the time shift introduced by the TH code η and the time shift introduced by the PPM modulator ε.

Since ε is much smaller than η, θ is quasi-periodic and closely follows the periodicity of the TH code. We can, as a reasonable first approximation, make the hypothesis that the effect of the ε shift on the PSD is not significant with respect to η. Therefore, the signal of Eq. (2–4) is modulated by a periodic signal and its PSD follows Eq. (3–25). In other words, the PSD is discrete and contains discrete frequency components located at the fundamental. In addition, it contains components at pulse repetition frequencies and their harmonics, and at linear combinations. In the present case, the period of the modulating periodic waveform corresponds to the period of the code N_p multiplied by the pulse interval T_s, that is, $T_p = N_p T_s$; therefore, the fundamental frequency of the modulating waveform is $f_p = 1/T_p$.

We shall first note that the pulse repetition frequency $1/T_s = N_p/T_p$ is a multiple of the fundamental frequency of the periodic waveform; therefore, the PSD is composed of lines occurring at $1/T_p$ and its harmonics (see Checkpoint 3–2).

The case $N_p = 1$ corresponds to the actual absence of coding and generates a signal with a PSD composed of lines at $1/T_s$ and harmonics. Power concentrates on spectrum lines with the undesirable effect of presenting spectral line peaks. This is not a surprise since having neglected the effect of ε, Eq. (2–4) forms a periodic train of pulses occurring at multiples of T_s.

If, as common practice, N_p is set equal to N_s, that is, the periodicity of the code coincides with the number of pulses per bit, spectrum lines occur at $1/T_b$ and its harmonics, where $T_b = N_s T_s = N_p T_s$ is the bit interval. Although the spectrum is still discrete, spectrum lines occur at frequencies that are more numerous than in the previous case for equal bandwidth since $1/T_b < 1/T_s$. The whitening effect of the code is visible in that power distributes over a larger number of spectrum lines and spectral peaks are less accentuated.

When we make N_p larger than N_s, the above effect is more prominent, and if N_p is not a multiple of N_s, several spectral lines generated by linear combination of $1/T_p$ and $1/T_s$ fill up the power spectrum with a beneficial smoothing effect.

The extreme case $N_p \rightarrow \infty$ corresponds to a lack of periodicity in the signal of Eq. (2–4). In this case, the time dither process can be assimilated to the random modulating signal $m(kT_s)$ and its power spectrum is given by Eq. (3–28). All comments made on Eq. (3–28) are valid here, that is, the spectrum has two components, one continuous and one discrete. The discrete term corresponds to a periodic component of the signal, which reduces with increasing variance in the position of the pulses. In addition, note that in the present case, the TH code generates time shifts that span over the entire T_s interval. Therefore, the θ values cannot be considered as small, and we can expect a reduction of the periodic component in the signal, that is, of the discrete component in the spectrum.

In the presence of many such signals, or in the case of a multi-user system, we can expect that the resulting signal shows little periodicity, and the comments corresponding to the random modulating case apply.

The case of a system composed of a few users using the same value for N_p could possibly be considered as of a periodic type, with the resulting cumulative signal having a discrete spectrum if all users were synchronized. Note, however, that under the realistic hypothesis of asynchronous users we can expect that, as is the case for several users, the multi-user signal looses its periodicity and its spectrum is well represented by Eq. (3–28).

A more detailed analysis requires relaxing the hypothesis of an inconsequential effect of the PPM time shift ε. A straightforward solution to this problem corresponding to the common case $N_p = N_s$ is to consider Eq. (2–4) in which we first neglect the effect of ε, that is, we define a signal $v(t)$ given by:

$$v(t) = \sum_{j=1}^{N_s} p\left(t - jT_s - \eta_j\right) \tag{3–31}$$

The Fourier transform of the above signal is:

$$P_v(f) = P(f)\sum_{m=1}^{N_s} e^{-j(2\pi f(mT_s+\eta_m))} \tag{3–32}$$

If we now consider *v(t)* as the basic multi-pulse used for transmission and apply the ε PPM shift, we obtain the following expression for the transmitted signal:

$$s(t) = \sum_{j=-\infty}^{+\infty} v\left(t - jT_b - \varepsilon b_j\right) \tag{3–33}$$

which is a PPM modulated waveform in which the shift is ruled by the sequence of data symbols **b**, that is, the **b** process emitted by the source. Note that the repetition code is now incorporated in the multi-pulse. If we can assume that **b** is a strict-sense stationary discrete random process, and the different extracted random variables b_k are statistically independent with a common probability density function *w*, then the signal of Eq. (3–33) has the PSD of Eq. (3–28) in which the Fourier transform of the pulse waveform *P(f)* is substituted by the PSD of the multi-pulse given by Eq. (3–32).

Given that the multi-pulse repetition rate is T_b, one obtains the following spectrum of a PPM-TH-UWB signal:

$$P_s(f) = \frac{\left|P_v(f)\right|^2}{T_b}\left[1 - \left|W(f)\right|^2 + \frac{\left|W(f)\right|^2}{T_b}\sum_{n=-\infty}^{+\infty}\delta\left(f - \frac{n}{T_b}\right)\right] \tag{3–34}$$

Equation (3–34) shows the double effect on one side of the TH code through $P_v(f)$, and on the other side of the time shift introduced by the PPM modulator which has characteristics following the statistical properties of the source. Note that the discrete component of the spectrum has lines at $1/T_b$. The amplitude of the lines is weighted by the statistical properties of the source represented by $|W(f)|^2$. If p indicates the probability of emitting a 0 bit (no shift) and 1-p the probability of emitting a '1' bit (ε shift), one can write:

$$\left|W(f)\right|^2 = 1 + 2p^2\left(1 - \cos\left(2\pi f\varepsilon\right)\right) - 2p\left(1 - \cos\left(2\pi f\varepsilon\right)\right) \tag{3–35}$$

If the source emits equiprobable symbols 0 and 1, then Eq. (3–35) simplifies as follows:

$$\left|W(f)\right|^2 = \frac{1}{2}\left(1 + \cos\left(2\pi f\varepsilon\right)\right) \tag{3–36}$$

Note here that the time shift is small and therefore the discrete components dominate the spectrum. In the simplifying hypothesis made in the beginning of this paragraph that ε is negligible, Eq. (3–34) is periodic with period $1/T_b$. Note that Eq. (3–34) can also be applied to any type of source, not necessarily binary.

CHECKPOINT 3–4

In this checkpoint, we will use computer simulation to analyze the spectral occupation of a PPM-TH-UWB signal. The MATLAB functions that are required for such an analysis have already been introduced in previous checkpoints: Function 2.6 in Checkpoint 2–1 for generating the PPM-TH-UWB signal and Function 3.2 in Checkpoint 3–1 for representing the PSD. Different simulations will be performed to analyze the effect of the main parameters of the UWB signal under examination on the PSD.

In the first simulation, we consider the case of an UWB signal with no PPM and no TH coding, denoted as signal `u0`. To generate signal `u0`, we execute Function 2.6 with the following parameters: `Pow=-30; fc=50e9; numbits=1000; Ts=10e-9; Ns=5; Tc=1e-9; Nh=10; Np=1; Tm=0.5e-9; tau=0.25e-9; dPPM=0; G=0`. The command line for generating signal `u0` is:

```
[bits,THcode,u0,ref]=cp0201_transmitter_2PPM_TH;
```

The above command line stores vector `u0` in memory, to represent the signal under examination. This signal is characterized by the transmission of five pulses per bit. All pulses are equally spaced in time with a pulse repetition period T_s. Because of the value `Np=1`, all pulses occupy the same position inside each T_s interval. Each pulse has the second derivative Gaussian shape with a maximum length of 0.5 ns.

We can analyze signal `u0` in the frequency domain by executing the command:

```
[PSDu0,df]=cp0301_PSD(u0,50e9);
```

The above command stores vector `PSDu0` in memory, to represent the PSD of signal `u0` and produce Figure 3–28. An enlarged version of Figure 3–28 is shown in Figure 3–29.

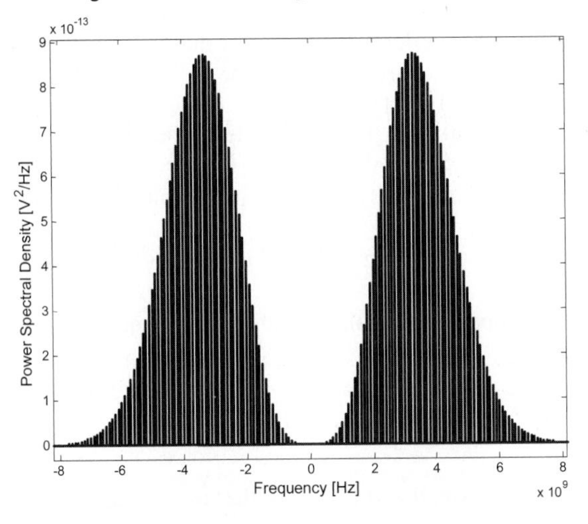

Figure 3–28 PSD of signal u0 — no PPM and no TH coding, and $N_p = 1$.

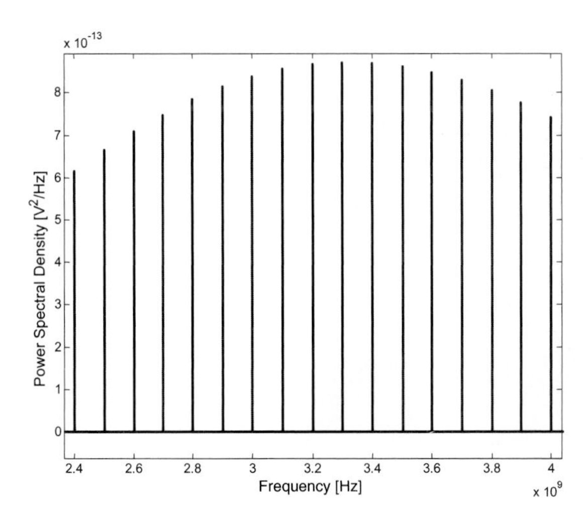

Figure 3–29 Detail of Figure 3–28 — PSD of signal u0.

As expected, Figures 3–28 and 3–29 show that the PSD of signal `u0` is composed of spectral lines occurring at $1/T_s$ = 0.1 GHz and harmonics, that is, the transmitted power is concentrated at multiples of the pulse repetition frequency. The envelope of the PSD has the shape of the Fourier transform of the second derivative Gaussian waveform.

In the second simulation, we consider the same parameters characterizing signal `u0`, but with an increased periodicity `Np` of the TH code. In particular, we set `Np=Ns`, that is, `Np=5`. The resulting signal `u1` can be generated as follows:

```
[bits,THcode,u1,ref] = cp0201_transmitter_2PPM_TH;
```

The above command line stores vector `u1` in memory, to represent the signal under examination. This signal is characterized by the transmission of five pulses per bit. The position of the pulse within each T_s interval depends on the corresponding coefficient of the TH code.

We can analyze signal `u1` in the frequency domain by executing the following command:

```
[PSDu1,df] = cp0301_PSD(u1,50e9);
```

The above command stores vector `PSDu1` in memory, to represent the PSD of signal `u1` and produce the plot in Figure 3–30 (see the detailed plot in Figure 3–31).

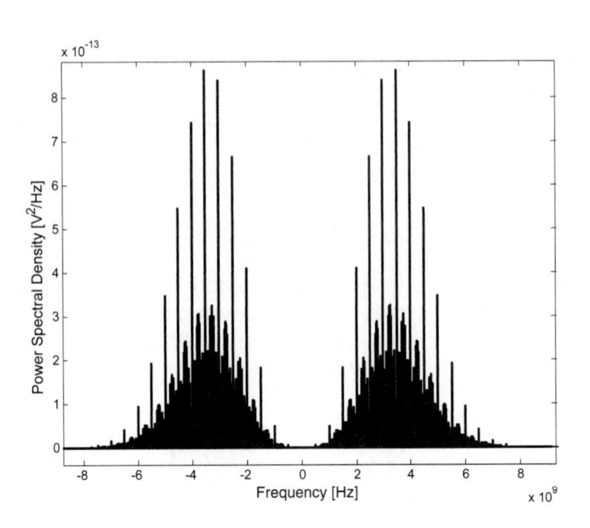

Figure 3–30 PSD of signal u1 — no PPM and with TH coding, and $N_p = 5$.

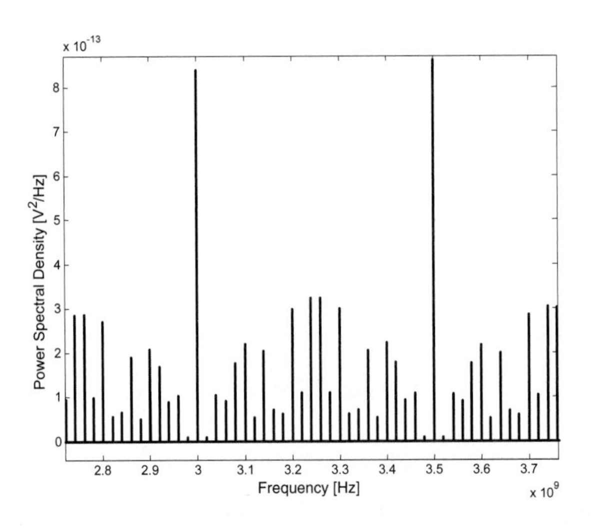

Figure 3–31 Detail of Figure 3–30 — PSD of signal u1.

Figures 3–30 and 3–31 show that the PSD of signal `u1` is composed of spectral lines at distances of $1/(N_sT_s) = 20$ MHz, that is, the transmitted power is concentrated at multiples of the bit repetition frequency (the bit rate). This result is justified by the periodicity of the TH code, which coincides with N_s. Signal `u1` is, therefore, periodic with a period equal to the bit period. The envelope of the PSD still resembles the Fourier transform of the second derivative Gaussian waveform. When comparing the PSD of signal `u1` with the PSD of signal u0, we can verify that the TH code has the effect of diminishing the number of peaks with the highest power contribution since the same power is distributed over a larger number of spectral lines. This effect should be more prominent when one further increases the N_p value. This analysis can be performed by generating a new signal u2, with N_p equal to the total number of transmitted pulses. In the case being examined, we run Function 2.6 with the same parameters of signal u1, but we set N_p = 5000. The command line for generating the signal is:

```
[bits,THcode,u2,ref]=cp0201_transmitter_2PPM_TH;
```

which stores vector u2 in memory, representing the signal under examination. This signal is still characterized by five pulses per bit. Each pulse, however, occupies a position that is given by a discrete random variable uniformly distributed between 0 and N_h-1. We can analyze signal u2 in the frequency domain by executing:

```
[PSDu2,df]=cp0301_PSD(u2,50e9);
```

The above command stores vector PSDu2 in memory, to represent the PSD of signal u2 and produce the plot of Figure 3–32 .

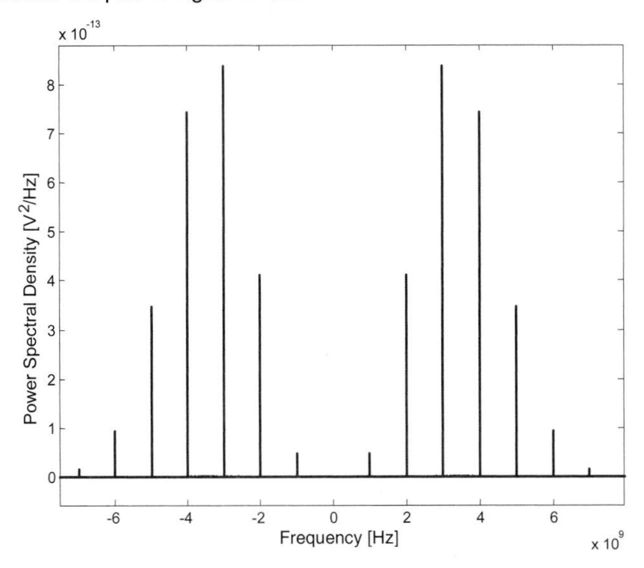

Figure 3–32 PSD of signal u2 — no PPM and with TH coding, and N_p = 5000.

Figure 3–32 shows that the PSD is still composed of a discrete part. The number of peaks, however, is smaller with respect to signals u0 and u1. These peaks are located at multiples of frequency $1/T_c$ = 1 GHz. Although the TH code is not periodic, the positions of the pulses inside each T_s interval are not random. Each T_s interval is, in fact, divided into N_h slots with length T_c, and the pulses are forced to locations at the beginning of these intervals. In the case under examination, N_h = 10, there are only ten possible positions for each pulse within one T_s interval. One can conclude, therefore, that the loss of periodicity of the TH code does not guarantee by itself the possibility of removing all peaks in the PSD. The number of peaks, however, is definitely reduced. Figure 3–33 shows the detail of the PSD of signal `u2` in the range between 2 and 3 GHz. We can verify the presence of a continuous part with smaller peaks at distances $1/(N_s T_s)$ = 20 MHz.

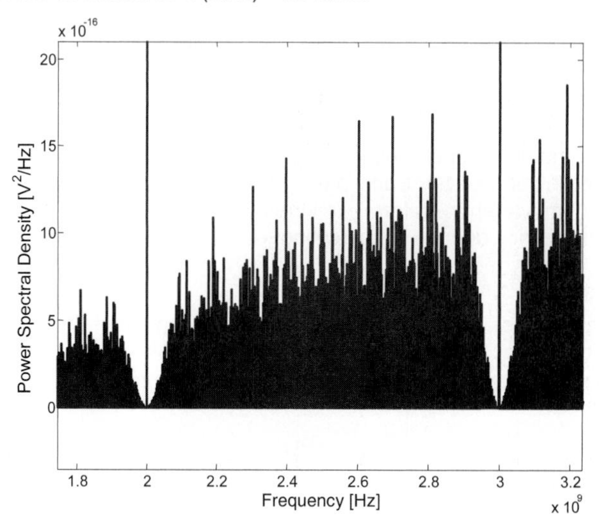

Figure 3–33 Detail of Figure 3–32 — PSD of signal u2.

The spectral analysis of signal `u2` showed that it is not possible to remove all the peaks of the PSD by only increasing the periodicity of the TH code. To decrease the peaks, we should allow each pulse to assume random positions inside each T_s interval. In the following simulation, we generate an UWB signal with the same parameters of signal `u2`, but with an increased cardinality of the TH code, that is, we divide the T_s interval into a higher number of T_c intervals. In particular, we consider the following parameters: `Pow=-30`; `fc=50e9`; `numbits=1000`; `Ts=10e-9`; `Ns=5`; `Tc=0.1e-9`; `Nh=100`; `Np=5000`; `Tm=0.5e-9`; `tau=0.25e-9`; `dPPM=0`; `G=0`. The number of slots per frame is increased tenfold, with a higher variance in the position of the pulses. We generate the resulting signal, signal `u3`, by executing the following command:

```
[bits,THcode,u3,ref] = cp0201_transmitter_2PPM_TH;
```

which stores vector `u3` in memory, representing the signal under examination. The PSD of signal `u3` is obtained as follows:

```
[PSDu3,df] = cp0301_PSD(u3,50e9);
```

The above command stores vector `PSDu3` in memory and provides the plot of Figure 3–34.

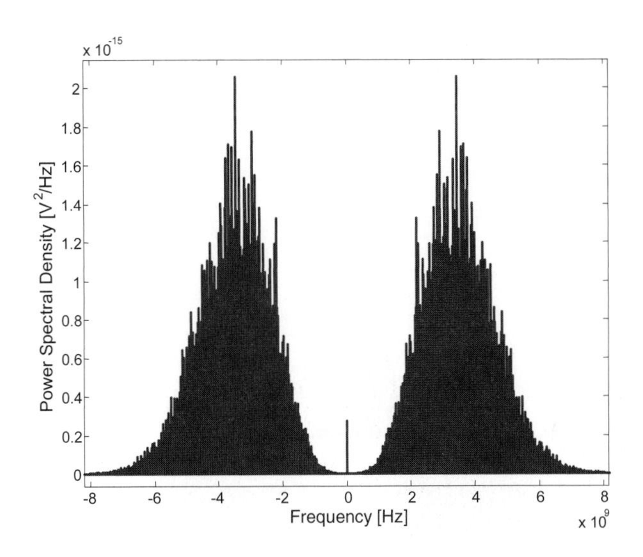

Figure 3–34 PSD of signal u3 — no PPM and with TH coding. Effect of an increased cardinality of the TH code, N_h.

When comparing the PSD in Figure 3–34 with the PSD in Figure 3–32, we conclude that the increased N_h value has the effect of removing the strong peaks in the PSD. As a matter of fact, these lines have been simply moved out of the bandwidth of the Fourier transform of the pulse waveform, that is, $1/T_c$= 10 GHz.

A final simulation can be performed to take into account the effect of the PPM. The following parameters are set within Function 2.6: `Pow=-30`; `fc=50e9`; `numbits=1000`; `Ts=10e-9`; `Ns=5`; `Tc=0.1e-9`; `Nh=100`; `Np=5000`; `Tm=0.5e-9`; `tau=0.25e-9`; `dPPM=0.25e-9`; `G=0`. The resulting signal, signal `u4`, is generated as follows:

```
[bits,THcode,u4,ref] = cp0201_transmitter_2PPM_TH;
```

The above command stores vector `u4` in memory. The PPM block is included within the transmission chain with a PPM shift of 0.25 ns for representing 1 bits. The spectral analysis of signal `u4` is performed as follows:

```
[PSDu4,df] = cp0301_PSD(u4,50e9);
```

The above command stores vector `PSDu4` in memory and provides the plot of Figure 3–35, which shows the PSD of a PPM-TH-UWB signal. This PSD is composed of a continuous part plus spectral lines located at multiples of $1/T_b$.

Figure 3–36 compares this PSD with that of signal `u0`. We can verify that the introduction of both TH coding and PPM has the effect of distorting the original Gaussian shape of the PSD. Figure 3–36 also shows that the PSD of a PPM-TH-UWB signal is fully contained within the envelope of the PSD, which results from the transmission of equally spaced pulses with the same shape and same average repetition frequency, that is, with the same average power.

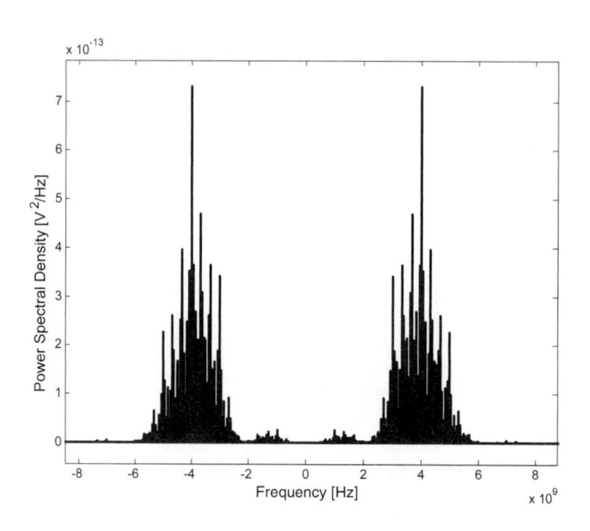

Figure 3–35 PSD of signal u4 — with PPM and TH coding.

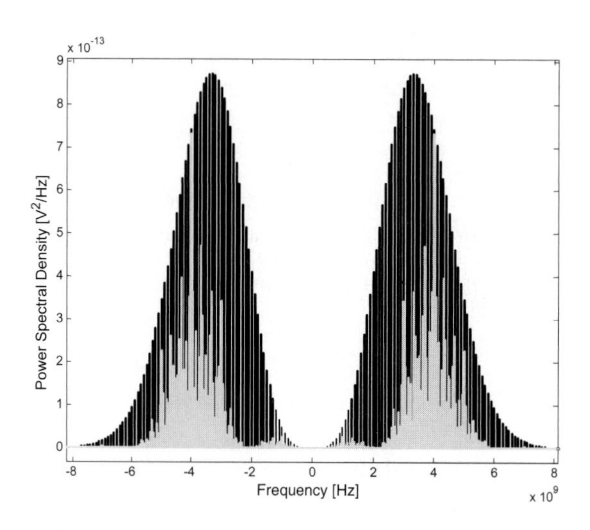

Figure 3–36 Comparison between the PSD of signal u0 (black), that is, no PPM and no TH coding, and the PSD of signal u4 (gray), that is, with PPM and TH coding.

CHECKPOINT 3–4

FURTHER READING

A PSD for PPM-TH-UWB following a different approach of the one adopted here can be found in (Kissik, 2001) and (Win, 2002), in which a unified spectral analysis for TH-SS in the presence of timing jitter is introduced. A recent paper by (Lehman and Haimovich, 2003) includes several PSD analytical expressions for TH-IR signals. The work by (Padgett, 2003) assimilates dithering to a modulation of the pulse repetition frequency.

REFERENCES

Bennett, W.R., "New results in the calculation of modulation products," *Bell System Tech. J.*, no. 12 (1933), 228–243.

Bennett, W.R., "Response of a linear rectifier to signal and noise," *J. Acoust. Soc. Am.*, no. 15 (1944), 164–172.

Bennett, W.R., "The biased ideal rectifier," *Bell System Tech. J.*, no. 26 (1947), 139–169.

Black, H.S., J.W. Beyer, T.J. Grieser, and F.A. Polkinghorn, "A multichannel microwave radio relay system," *AIEE Transaction Electrical Engineering*, Volume: 65 (1946), 798–805.

Di Benedetto, M.-G., and B.R. Vojcic, "Ultra Wide Band (UWB) Wireless Communications: A Tutorial," *Journal of Communication and Networks, Special Issue on Ultra-Wideband Communications*, Volume: 5, Issue: 4 (December 2003), 290–302.

Kissik, A.W., U.S. Department of Commerce, *The Temporal and Spectral Characteristics of Ultra-wideband Signals*, Available at *www.its.bldrdoc.gov/pub/ntia-rpt/01-383/01-383.pdf* (January 2001).

Lehmann, N.H., and A.M. Haimovich, "The Power Spectral Density of a Time Hopping UWB Signal: A Survey," *IEEE Conference on Ultra Wideband Systems and Technologies* (November 2003), 234–239.

Padgett, J.E., "The Power Spectral Density of a UWB Signal with Pulse Repetition Frequency (PRF) Modulation," *IEEE Conference on Ultra Wideband Systems and Technologies* (November 2003), 15–20.

Proakis, J.G., and M. Salehi, *Communication Systems Engineering*, Englewood Cliffs, New Jersey: Prentice-Hall International (1994).

Rowe, H.E., *Signals and Noise in Communications Systems*, Princeton, New Jersey: D. Van Nostrand Company, Inc (1965).

Win, M.Z., "A Unified Analysis of Generalized Time-Hopping Spread-Spectrum Signals in the Presence of Timing Jitter," *IEEE Journal on Selected Areas in Communications*, Volume: 20, Issue: 9 (December 2002), 1664–1676.

APPENDIX 3.A

Function 3.1 Analogue PPM with Sinusoidal Modulating Signals

Function 3.1 is composed of two steps. Step Zero contains all the parameters characterizing the signal to be generated: the average transmitted power in dBm `Pow`, the sampling frequency for representing the signal `fc`, the number of pulses to be generated `np`, the time duration of each rectangular pulse `Tr`, the average pulse repetition period in seconds `Ts`, the amplitude and frequency of the sinusoidal modulating signal, `A` and `f0`. Step One contains the code for generating the PPM signal. PPM is implemented in Function 3.1 by introducing vector `Mtot`, which collects all the time shifts that must be applied to the pulses.

```
%
% FUNCTION 3.1 : "cp0301_PPM_sin"
%
% Generation of a PPM-UWB signal in the case of a
% sinusoidal modulating signal and rectangular pulses
%
% Transmitted power is fixed at 'Pow'
% The signal is sampled with frequency 'fc'
% 'np' is the number of generated pulses
% 'Ts' is the average pulse repetition period
% Each rectangular pulse has time duration 'Tr'
% The modulating signal is a sinusoid with
% amplitude 'A' and frequency 'f0'
%
% The function returns the generated signal 'Stx'
% and the corresponding sampling frequency 'fc'
%
% Programmed by Guerino Giancola
%

function [Stx,fc]=cp0301_PPM_sin;

% ----------------------------
% Step Zero - Input parameters
% ----------------------------

Pow = -30;  % average transmitted power (dBm)
```

```
fc = 1e11;  % sampling frequency

np = 10000; % number of pulses

Tr = 0.5e-9;% time duration of the rectangular pulse [s]

Ts = 2e-9;  % average pulse repetition period [s]
A = Ts/2;   % maximum time shift provided by the
            % modulation [s]
f0 = 5e7;   % frequency of the modulating signal [Hz]

% ----------------------------------------
% Step One - Simulating transmission chain
% ----------------------------------------

dt = 1 / fc;           % sampling period
sTs = floor(Ts/dt);    % number of samples per frame
sTot = sTs * np;       % total number of samples
Stx = zeros(1,sTot);   % output vector

% pulse position modulation
j = (0:1:np-1);
M0 = A.*cos((2*pi*f0).*(j.*Ts));
M1 = j.*Ts;
Mtot = M0 + M1;
for k = 1 : np
  Stx(1+floor(Mtot(k)/dt))=1;
end

% shaping filter
sP = floor(Tr/dt);     % number of samples per pulse

p0 = (1/sqrt(Tr)).*ones(1,sP); % energy normalized rect
power = (10^(Pow/10))/1000;    % average transmitted power
                               % (watt)
Ex = power * Ts;          % energy per pulse
ptx= p0 .* sqrt(Ex);      % pulse waveform

Stx = conv(Stx,ptx);
Stx = Stx(1:sTot);
```

Function 3.2 PSD

Function 3.2 is composed of two steps. Step One contains the code for evaluating the PSD. Step Two contains the code for the graphical representation of the PSD. Note that Function 3.2 makes use of the MATLAB function `fftshift(x)`, which shifts zero frequency components of `x` to the center of the spectrum. The command `fftshift(x)` is useful for visualizing double-sided Fourier transforms.

```
%
% FUNCTION 3.2 : "cp0301_PSD"
%
% Evaluates the PSD of the
% signal represented by the input vector 'x'
% The input signal is sampled with frequency 'fc'
%
% This function returns the PSD ('PSD')
% and the corresponding frequency resolution ('df')
%
% Programmed by Guerino Giancola
%

function [PSD,df]=cp0301_PSD(x,fc)

% --------------------------------
% Step One - Evaluation of the PSD
% --------------------------------

dt=1/fc;
N=length(x);
T=N*dt;
df=1/T;
X = fft(x);
X = X / N;
mPSD=abs(X).^2/(df^2);
PSD = fftshift(mPSD);
PSD = (1/T).*PSD;

% ------------------------------------
% Step Two - Graphical representation
% ------------------------------------

frequency = linspace(-fc/2,fc/2,length(PSD));
PF=plot(frequency,PSD);
set(PF,'LineWidth',[2]);
```

```
AX=gca;
set(AX,'FontSize',12);
X=xlabel('Frequency [Hz]');
set(X,'FontSize',14);
Y=ylabel('Power Spectral Density [V^2/Hz]');
set(Y,'FontSize',14);
```

Function 3.3 Analog PPM with Generic Periodic Modulating Signals

Function 3.3 is composed of two steps. Step Zero contains all the parameters characterizing the signal to be generated, that is, the average transmitted power in dBm `Pow`, the sampling frequency `fc`, the number of pulses to be generated `np`, the time duration of each rectangular pulse, `Tr`, the average pulse repetition period in seconds `Ts`, and finally, parameters `A`, `B`, and `Tp`, characterizing the modulating signal (see Checkpoint 3–2). Step One contains the code for generating the PPM signal. Similarly to Function 3.1, the PPM scheme is implemented by introducing vector `Mtot` which collects all the time-shifts that must be applied on the transmitted pulses (see also Function 3.1).

```
%
% FUNCTION 3.3 : "cp0302_PPM_periodic"
%
% Generation of a PPM-UWB signal in the case of a generic
% periodic modulating signal and rectangular pulses
% Modulating signal is chosen to be characterized by
% an exponential decay exp(-t)
%
% Transmitted power is fixed at 'Pow'
% The signal is sampled with frequency 'fc'
% 'np' is the number of generated pulses
% 'Ts' is the average pulse repetition period
% Each rectangular pulse has time duration 'Tr'
% The periodic signal is characterized by
% shape parameters 'A' and 'B', and period 'Tp'
%
% The function returns the generated signal 'Stx'
% and the corresponding sampling frequency 'fc'
%
% Programmed by Guerino Giancola
%

function [Stx,fc]=cp0302_PPM_periodic;

% ----------------------------
% Step Zero - Input parameters
% ----------------------------

Pow = -30;      % average transmitted power (dBm)

fc = 1e11;      % sampling frequency
```

```
np = 10000;     % number of pulses

Tr = 0.5e-9;    % time duration of the rectangular pulse [s]

Ts = 2e-9;      % average pulse repetition period [s]

A = 1e-9;       % first shape parameter
B = 10;         % second shape parameter
Tp = 20e-9;     % period of the modulating signal [s]

% ----------------------------------------
% Step One - Simulating transmission chain
% ----------------------------------------

dt = 1 / fc;          % sampling period
sTs = floor(Ts/dt); % number of samples per frame
sTot = sTs * np;    % total number of samples
Stx = zeros(1,sTot);% output vector

% PPM
j = (0:1:np-1);
M0 = A.*exp(-(B/Tp).*mod(j*Ts,Tp));
M1 = j.*Ts;
Mtot = M0 + M1;
for k = 1 : np
  Stx(1+floor(Mtot(k)/dt))=1;
end

% shaping filter
sP = floor(Tr/dt);          % number of samples per pulse

p0 = (1/sqrt(Tr)).*ones(1,sP); % energy normalized rect
power = (10^(Pow/10))/1000;     % average transmitted power
                                % (watt)
Ex = power * Ts;          % energy per pulse
ptx= p0 .* sqrt(Ex);      % pulse waveform

Stx = conv(Stx,ptx);
Stx = Stx(1:sTot);
```

Function 3.4 Analog PPM with Random Modulating Signals

Function 3.4 is composed of two steps. Step Zero contains all the parameters characterizing the signal to be generated, that is, the average transmitted power in dBm `Pow`, the sampling frequency for representing the signal `fc`, the number of pulses to be generated `np`, the time duration of each rectangular pulse `Tr`, the average pulse repetition period in seconds `Ts`, and the standard deviation of the Gaussian distributed modulating signal `sigma`. Step One contains the code for generating the PPM signal. The PPM scheme is implemented by introducing vector `Mtot`, which collects all time shifts that must be applied on the transmitted pulses (see also Functions 3.1 and 3.3). The generation of the random shifts is performed by means of the MATLAB function `randn(1,N)`, which generates a vector of `N` random entries, chosen from a normal distribution with mean zero, variance one, and standard deviation one. To avoid pulse overlapping, the values of the random shifts are limited within the range $[0,T_s-T_r]$.

```
%
% FUNCTION 3.4 : "cp0303_PPM_random"
%
% Generation of a PPM-UWB signal in the case of
% a random modulating signal and rectangular pulses
% The modulating signal is characterized by
% a normal distribution
%
% Transmitted Power is fixed at 'Pow'
% The signal is sampled with frequency 'fc'
% 'np' is the number of generated pulses
% 'Ts' is the average pulse repetition period
% Each rectangular pulse has time duration 'Tr'
% The random modulating signal is characterized
% by standard deviation 'sigma'
%
% The function returns the generated signal 'Stx',
% the corresponding sampling frequency 'fc',
% and vector 'M0' of all the PPM time shifts
%
% Programmed by Guerino Giancola
%

function [Stx,fc,M0]=cp0303_PPM_random;

% ----------------------------
% Step Zero - Input parameters
% ----------------------------
```

```
Pow = -30;      % average transmitted power (dBm)

fc = 1e11;      % sampling frequency

np = 10000;     % number of pulses

Tr = 0.5e-9;    % time duration of the rectangular pulse [s]

Ts = 2e-9;      % average pulse repetition period [s]

sigma = 0.1e-9;  % standard deviation of the modulating
signal

% -----------------------------------------
% Step One - Simulating transmission chain
% -----------------------------------------

dt = 1 / fc;                % sampling period
sTs = floor(Ts/dt);         % number of samples per frame
sTot = sTs * np;            % total number of samples
Stx = zeros(1,sTot);        % output vector

% PPM
j = (0:1:np-1);
M0 = max(zeros(1,np), min((Ts -Tr).*...
   ones(1,np),((Ts/2)+sigma.*randn(1,np))));
M1 = j.*Ts;
Mtot = M0 + M1;
for k = 1 : np
  Stx(1+floor(Mtot(k)/dt))=1;
end

% shaping filter
sP = floor(Tr/dt);                  % number of samples per
                                    % pulse
p0 = (1/sqrt(Tr)).*ones(1,sP);  % energy normalized rect
power = (10^(Pow/10))/1000;         % average transmitted power
                                    % (Watt)
Ex = power * Ts;            % energy per pulse
ptx= p0 .* sqrt(Ex);        % pulse waveform

Stx = conv(Stx,ptx);
Stx = Stx(1:sTot);
```

CHAPTER 4

The PSD of DS-UWB Signals

The PSD of a DS-UWB signal is more easily derived with respect to the TH-UWB case since pulses occur at multiples of T_s (Di Benedetto and Vojcic, 2003). In this chapter, we will focus on the derivation of the PSD of DS-UWB signals. Spectral properties of DS-UWB signals are of interest with respect to the current DS proposal submitted to the IEEE 802.15.TG3a (Roberts, 2003).

At the end of this chapter, we will include the PAM-TH case and show how it can be obtained by combining previously examined system blocks.

4.1 DS-UWB

As is well-known (see, for example, Proakis, 1995) the signal of Eq. (2–6) is not wide-sense stationary, but it can be made this way by introducing a random phase epoch Θ, which is uniformly distributed on $[0, T_s]$ and independent of **d**, and which further reflects uncertainty about the phase of the signal. The DS-UWB random process modifies into:

$$s(t+\Theta)=\sum_{j=-\infty}^{+\infty} d_j p(t-jT_s+\Theta) \tag{4–1}$$

The PSD of Eq. (4–1) can be determined in a straightforward manner by first computing the autocorrelation function of *s(t)*, and then taking the Fourier transform of the autocorrelation function.

The autocorrelation function of Eq. (4–1) is given by:

$$\left\langle s(t+\Theta+\tau)s^*(t+\Theta)\right\rangle=\left\langle \sum_{k=-\infty}^{+\infty}\sum_{h=-\infty}^{+\infty} d_k d_h^* p(t-kT_s+\Theta+\tau)p^*(t-hT_s+\Theta)\right\rangle \tag{4–2}$$

Since Θ is independent of **d**, one gets:

$$\left\langle s(t+\Theta+\tau)s^{*}(t+\Theta)\right\rangle=\sum_{m=-\infty}^{+\infty}\left\langle d_{k}d_{k+m}^{*}\right\rangle\cdot$$
$$\cdot\sum_{k=-\infty}^{+\infty}\left\langle p\left(t-kT_{s}+\Theta+\tau\right)p^{*}\left(t-(k+m)T_{s}+\Theta\right)\right\rangle \tag{4–3}$$

where the first expected value is the autocorrelation of the sequence **d**, $R_d(m)$, while the second summation, since Θ is uniformly distributed, is:

$$\begin{aligned}
&\sum_{k=-\infty}^{+\infty}\left\langle p\left(t-kT_{s}+\Theta+\tau\right)p^{*}\left(t-kT_{s}-mT_{s}+\Theta\right)\right\rangle=\\
&=\sum_{k=-\infty}^{+\infty}\frac{1}{T_{s}}\int_{0}^{T_{s}}p\left(t-kT_{s}+\Theta+\tau\right)p^{*}\left(t-kT_{s}-mT_{s}+\Theta\right)d\Theta=\\
&=\sum_{h=-\infty}^{+\infty}\frac{1}{T_{s}}\int_{t+hT_{s}}^{t+(h+1)T_{s}}p\left(\xi+\tau\right)p^{*}\left(\xi-mT_{s}\right)d\xi=\\
&=\frac{1}{T_{s}}\int_{-\infty}^{+\infty}p\left(\xi+\tau\right)p^{*}\left(\xi-mT_{s}\right)d\xi
\end{aligned} \tag{4–4}$$

Equation (4–3) thus becomes:

$$\left\langle s\left(t+\Theta+\tau\right)s^{*}\left(t+\Theta\right)\right\rangle=\frac{1}{T_{S}}\sum_{m=-\infty}^{+\infty}R_{d}\left(m\right)\int_{-\infty}^{+\infty}p\left(\xi+\tau\right)p^{*}\left(\xi-mT_{S}\right)d\xi \tag{4–5}$$

in which the last term is simply the autocorrelation integral of $p(t)$. Note that Eq. (4–5) is independent of t, as expected, since the process of Eq. (4–1) is wide-sense stationary. The PSD is obtained by taking the Fourier transform of Eq. (4–5), and since the Fourier transform of the autocorrelation integral of $p(t)$ is $|P(f)|^2$, one obtains:

$$P_{x_{DS}}\left(f\right)=\frac{\left|P\left(f\right)\right|^{2}}{T_{s}}\sum_{m=-\infty}^{+\infty}R_{d}\left(m\right)e^{-j2\pi fmT_{s}}=\frac{\left|P\left(f\right)\right|^{2}}{T_{s}}P_{c}\left(f\right) \tag{4–6}$$

where $P_c(f)$ is the so-called code spectrum and is the discrete time Fourier transform of the autocorrelation function of **d**. Since $R_d(m)$ is an even function, it may be written as follows:

$$P_{c}\left(f\right)=\sum_{m=-\infty}^{+\infty}R_{d}\left(m\right)e^{-j2\pi fmT_{s}}=R_{d}\left(0\right)+2\sum_{m=1}^{\infty}R_{d}\left(m\right)\cos 2\pi fmT_{s} \tag{4–7}$$

Equation (4–6) shows that the spectrum of the DS-UWB signal is governed by two terms: the transfer function of the pulse shaper $P(f)$, as is the case for TH-UWB signals, and the code spectrum $P_c(f)$. Note that if sequence **d** was composed of independent symbols,

$R_d(m)$ would be different from 0 only for $m = 0$, and therefore, $P_c(f)$ would be independent of f. In this case, the spectrum of the DS signal would be entirely governed by the properties of the pulse $p(t)$.

CHECKPOINT 4–1

In this checkpoint, we will use computer simulation for evaluating the PSD of a DS-UWB signal. The MATLAB functions that are required for such an analysis have already been introduced in previous checkpoints (Function 2.9 in Checkpoint 2–2 for generating the DS-UWB signal, and Function 3.2 in Checkpoint 3–1 for representing the PSD of the generated signal).

In a first simulation, we will generate signal `u0` by applying Function 2.9 with the following parameters: `Pow=-30; fc=50e9; numbits=5000; Ts=2e-9; Ns=10; Np=10; Tm=0.5e-9; tau=0.25e-9`. The command line for generating `u0` is:

```
[bits,DScode,u0,ref]=cp0202_transmitter_2PAM_DS;
```

The above command line stores vector `u0` in memory, representing the signal under examination. This signal is characterized by the transmission of $N_p = N_s = 10$ pulses per bit. All pulses are equally spaced in time with a pulse repetition period T_s. Each pulse has the second derivative Gaussian shape with a maximum length of 0.5 ns.

We may analyze signal `u0` in the frequency domain by executing the command:

```
[PSDu0,df]=cp0301_PSD(u0,50e9);
```

The above command stores vector `PSDu0` in memory, representing the PSD of signal `u0`, and produces the plot in Figure 4–1.

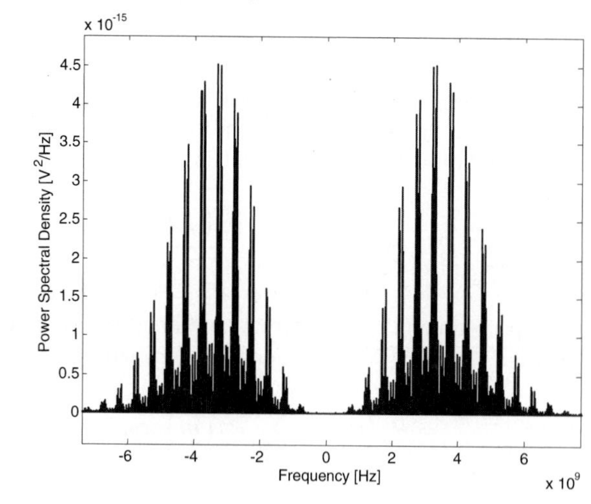

Figure 4–1 PSD of the PAM-DS-UWB signal u0 ($N_p = 10$).

Figure 4–1 shows the PSD of signal u0. We first observe that the envelope of the PSD has the Gaussian shape of the Fourier transform of the basic pulse. In addition, we note that transmitted power concentrates on peaks located at multiples of $1/T_s = 500$ MHz. This is due to the effect of code spectrum $P_c(f)$, which is defined in Eq. (4–7). The code spectrum is, in fact, composed by the superposition of several sinusoids in the frequency domain, and the period of the sinusoid with the highest amplitude is exactly $1/T_s$. We can analyze in detail the shape of the PSD by analytically reproducing function $P_c(f)$. According to Eq. (4–7), $P_c(f)$ is completely defined once we derive the autocorrelation function $R_d(m)$ of the sequence **d**, which contains all the binary antipodal values modulating the pulses. $R_d(m)$ can be evaluated since Function 2.9 returns both the sequence of transmitted bits and the code used for generating signal u0. First, we reconstruct sequence **d** as follows:

```
for nb = 1 : 5000
d(1+(nb-1)*10:10+(nb-1)*10)=bits(nb).*DScode;
end
```

The above set of code lines stores vector d in memory, representing the sequence of binary values that modulate the pulses. Given d, we can evaluate its normalized autocorrelation function Rd0:

```
Rd = xcorr(d);
Rd0 = Rd/max(Rd);
```

Note that the autocorrelation is evaluated through the MATLAB function xcorr(a), which computes the autocorrelation sequence of the input vector a. Given Rd0, we derive a function C(f), having an amplitude proportional to $P_c(f)$. C(f) can be evaluated as follows:

```
freq=linspace(-25e9,25e9,length(PSDu0));
[m,i]=max(Rd0);
C = Rd0(i).*ones(1,length(freq));
for j = 1 : 100
C = C + (2*Rd0(i+j)).*cos((j*2*pi*2e-9).*freq);
end
plot(freq,C)
```

The above set of command lines stores vector C in memory, representing function C(f). Note that vector C was evaluated by only taking into account the first 100 terms of the summation in Eq. (4–7). This approximation was sufficient for the evaluation of C(f) with good accuracy. The above code produces the plots in Figures 4–2 and 4–3.

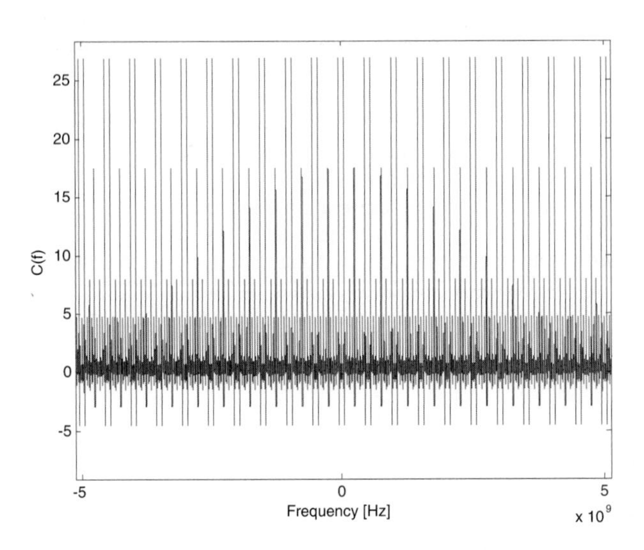

Figure 4–2 Graphical representation of function C(f). The amplitude of C(f) is proportional to the code spectrum $P_c(f)$.

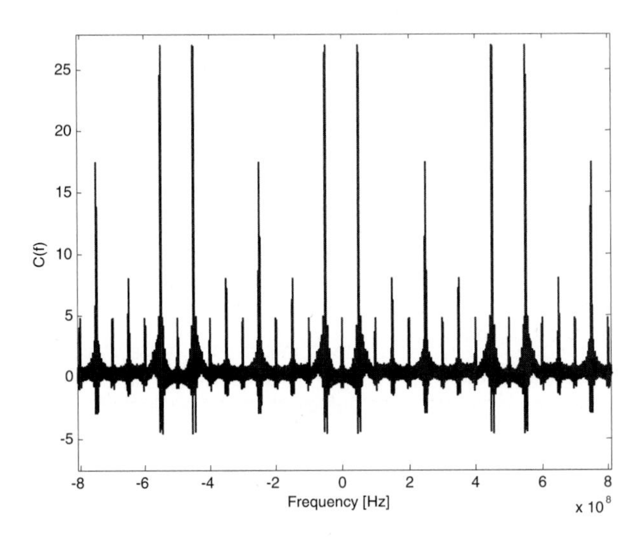

Figure 4–3 Detail of Figure 4–2: function C(f).

The plot in Figure 4–3 shows that the code spectrum $P_c(f)$ causes the presence of peaks of power in the PSD of signal `u0`. In particular, we can observe that $P_c(f)$ presents a regular pattern in the frequency domain, with a pair of peaks located at each multiple of $1/T_s$ = 0.5 GHz. In Figure 4–4, we compare the PSD of signal `u0` with a scaled version of function C(f). Here, it is possible to verify an agreement between theoretical results (in gray) and simulation output (in black).

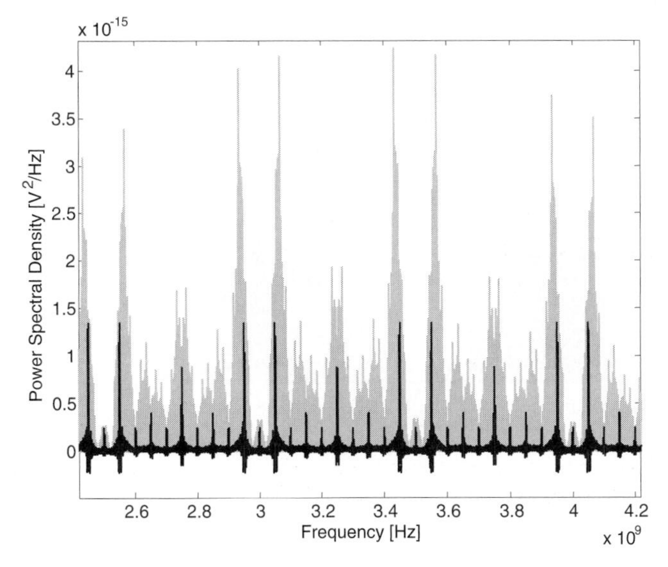

Figure 4–4 Comparison between PSDu0 (gray plot), that is, the PSD of u0, and a scaled version of C(f), kC(f) (black plot) with $k = 5 \cdot 10^{-16}$, that is, the code spectrum.

To decrease the effect of the peaks in Figure 4–4, we can follow the same approach described previously in the case of PPM-TH-UWB signals (see Checkpoint 3–4), that is, we can decrease the amount of power located at the peaks by lengthening the code period. To verify the effect of code periodicity on spectral characteristics of the transmitted signal, we run a second simulation with the same parameters of signal `u0`, but with an increased `Np` value, that is, `Np=50` instead of `Np=10`. The code lines for generating the new signal `u1` and for representing the corresponding PSD are:

```
[bits,DScode,u1,ref]=cp0202_transmitter_2PAM_DS;
[PSDu1,df]=cp0301_PSD(u1,50e9);
```

The above command lines store signals `u1` and `PSDu1` in memory, and produce the plot in Figure 4–5.

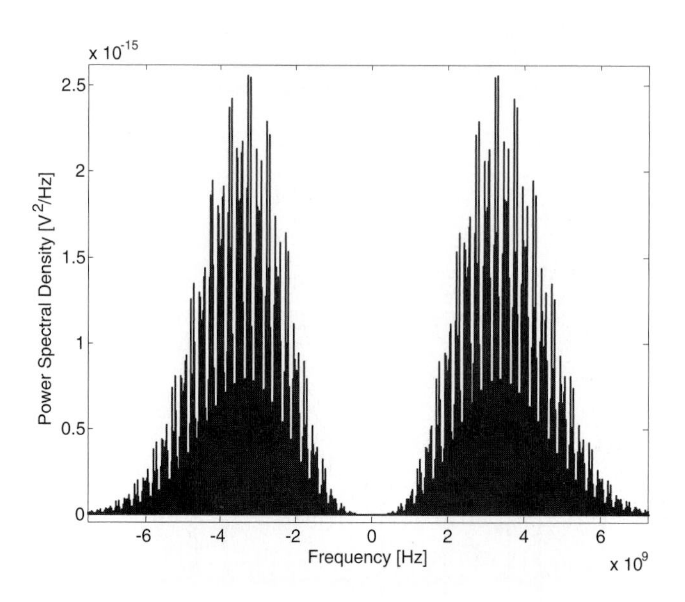

Figure 4–5 PSD of the PAM-DS-UWB signal u1 ($N_p = 50$).

Signal `u1` conveys the same power as signal `u0` and occupies the same bandwidth in the frequency domain. The power of signal `u1`, however, has a larger distribution over the spectrum, that is, the amplitude of the peaks in the PSD of signal `u1` is reduced with respect to the case of signal `u0`. This effect should be mostly evident when the `Np` value is further increased. A third simulation can be performed with `Np=50000`, that is, when the code period is equal to the total number of transmitted pulses:

```
[bits,DScode,u2,ref]=cp0202_transmitter_2PAM_DS;
[PSDu2,df]=cp0301_PSD(u2,50e9);
```

The above code lines store vectors `u2` and `PSDu2` in memory, and produce the plot in Figure 4–6.

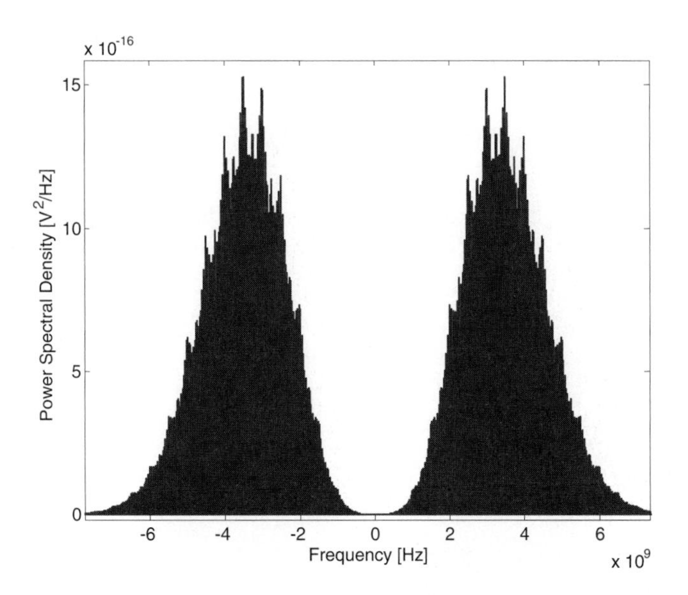

Figure 4–6 PSD of the PAM-DS-UWB signal u2 ($N_p = 50000$).

Figure 4–6 shows the PSD of signal `u2`. Here, we observe that the PSD approaches the Gaussian shape of the Fourier transform of the basic pulse. In this case, in fact, the code sequence is composed of independent symbols, and the autocorrelation $R_d(m)$ in Eq. (4–7) consists of the term $R_d(0)$ only.

CHECKPOINT 4–1

4.2 THE CASE OF PAM-TH-UWB

Consider as an additional case a sequence composed of ±1 values whose amplitude modulates a TH train of pulses. The generated signal is a PAM-TH-UWB signal, which is given by:

$$s(t) = \sum_{j=-\infty}^{+\infty} a_j p\left(t - jT_s - c_j T_c\right) \tag{4–8}$$

Equation (4–8) is Eq. (2–5) in the specific case of transmission of *p(t)* for a '1' value and of *–p(t)* for a '-1' value. The generation scheme for the PAM-TH-UWB signal is shown in Figure 4–7. Note that we may now consider a basic multi-pulse waveform *v(t)* for the common case of $N_p = N_s$, as in Eq. (3–31), which is amplitude-modulated by a source emitting a ±1 sequence. Thus, Eq. (4–8) becomes:

$$s(t) = \sum_{j=-\infty}^{+\infty} b_j v(t - jT_b) \qquad (4\text{–}9)$$

Equation (4–9) is a straightforward PAM waveform with basic multi-pulse $v(t)$ having the Fourier transform given by Eq. (3–32). For this type of waveform, the PSD can be easily derived (Proakis, 1995).

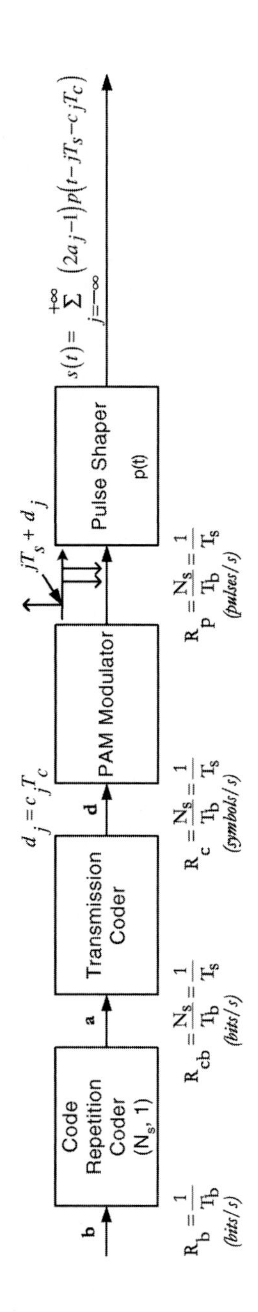

Figure 4–7 Transmission scheme for a PAM-TH-UWB signal.

CHECKPOINT 4–2

In this checkpoint, we will use computer simulation for generating a PAM-TH-UWB signal and for analyzing the main characteristics of the corresponding PSD.

To generate a PAM-TH-UWB signal, we combine the MATLAB functions introduced in Chapter 2, "The UWB Radio Signal," for generating DS-PAM and TH-PPM signals. We start by considering the system model introduced in Checkpoint 2–2 for simulating the transmission of a DS-PAM-UWB signal (see Figure 2–8). We observe, in fact, that the PAM-TH-UWB signal can be generated by simply introducing some small modifications in the part of the system that implements the transmission coder. We can then introduce a new MATLAB function, **Function 4.1** (see Appendix 4.A), which is based on the same algorithm of Function 2.8 (see Appendix 2.A), except for the part of the code that implements DS coding, which we substitute with the set of commands implementing TH coding in Function 2.4 (see Appendix 2.A).

The whole transmission chain, which generates a PAM-TH-UWB signal, is simulated in **Function 4.2** (see Appendix 4.A). Within the function, the user must set the following parameters: the average transmitted power `Pow`, the value of the sampling frequency `fc`, the number of bits to be generated `numbits`, the average pulse repetition period `Ts`, the number of pulses per bit `Ns`, the value of the chip time `Tc`, the cardinality of the TH code `Nh`, the periodicity of the TH code `Np`, the value of the time duration of the pulse `Tm`, the shaping factor `tau`, and a flag `G`, which must be set to 1 to produce a graphical output. Function 4.2 generates four outputs: the sequence of bits generated by the source `bits`, the TH sequence used by the TH coder `THcode`, the transmitted signal `Stx`, and a reference signal `ref` obtained by excluding the PAM modulator in the transmission chain.

The command line for generating PAM-TH-UWB signals is:

```
[bits,THcode,Stx,ref]=cp0402_transmitter_2PAM_TH;
```

The above command line stores the stream of bits generated by the source `bits`, in memory; a vector representing the TH code `THcode`; the transmitted signal `Stx`; and a reference signal with no modulation `ref`. Figure 4–8 shows an example of the output provided by Function 4.2 in the case of `Pow=-30; fc=50e9; numbits=3; Ts=5e-9; Ns=5; Tc=1e-9; Nh=5; Np=5; Tm=0.5e-9; tau=0.25e-9; G=1`. Here we recognize the transmission of the binary sequence `[0 1 1]`, and the presence of the TH code `[2 3 1 1 3]`.

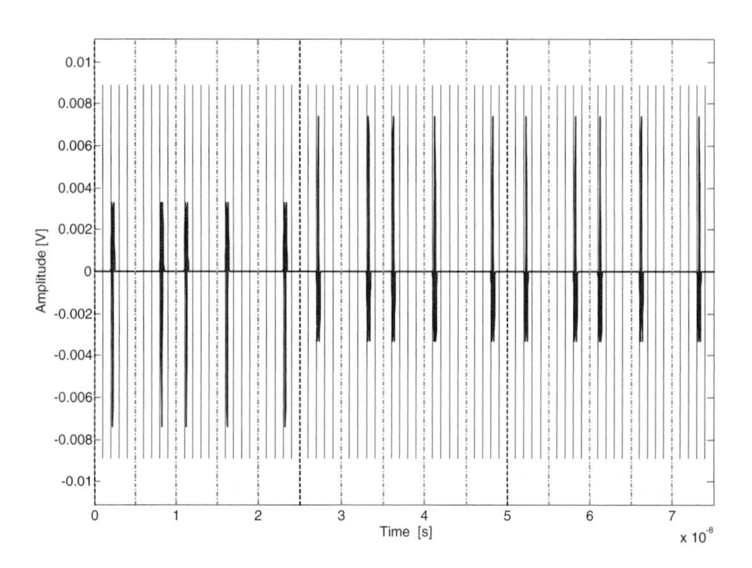

Figure 4–8 Example of a PAM-TH-UWB signal (Pow=-30; fc=50e9; numbits=3; Ts=5e-9; Ns=5; Tc=1e-9; Nh=5; Np=5; Tm=0.5e-9; tau=0.25e-9; G=1).

We use Function 4.2 for analyzing the spectral characteristics of PAM-TH-UWB signals. In particular, we use Function 4.2 for generating different signals with increasing values of the parameter N_p. The first signal, signal `v0`, is characterized by the following parameters: `Pow=-30`; `fc=50e9`; `numbits=5000`; `Ts=5e-9`; `Ns=5`; `Tc=1e-9`; `Nh=5`; `Np=5`; `Tm=0.5e-9`; `tau=0.25e-9`; `G=0`. The second signal, signal `v1`, is characterized by the same parameters of signal `v0`, but with an increased `Np` value, `Np=50`. Finally, we generate a third signal, signal `v2`, with `Np=25000`. The PSD for these signals can be evaluated through Function 3.2, which provides for the three cases under examination the plots in Figures 4–9, 4–10, and 4–11, respectively.

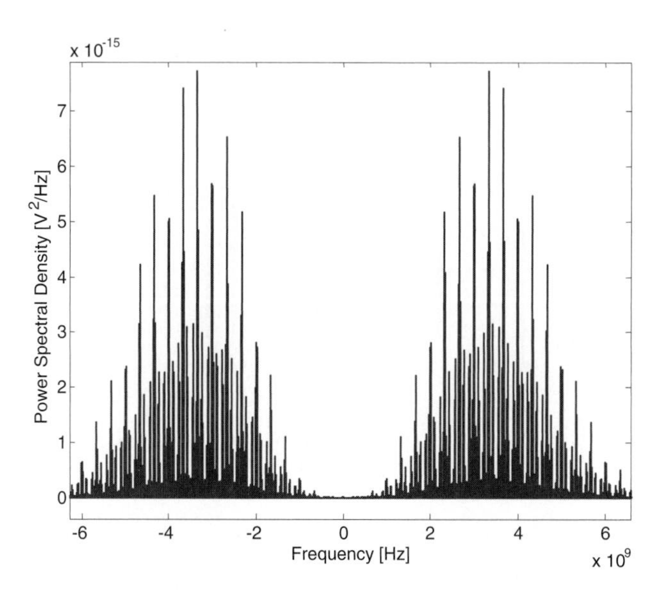

Figure 4–9 PSD of the PAM-TH-UWB signal v0 ($N_p = 5$).

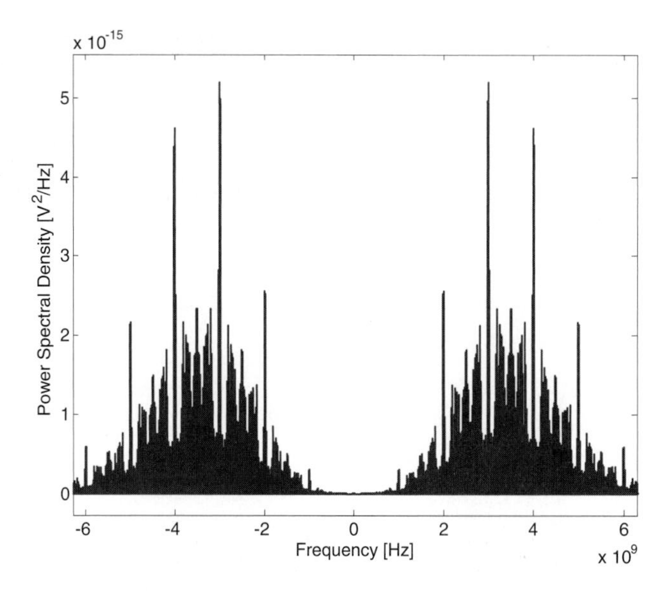

Figure 4–10 PSD of the PAM-TH-UWB signal v1 ($N_p = 50$).

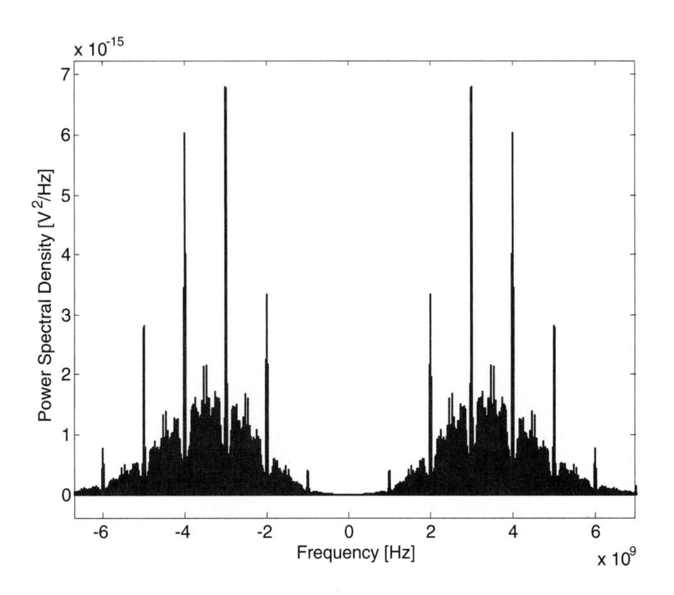

Figure 4–11 PSD of the PAM-TH-UWB signal v2 (N_p = 25000).

Figures 4–9, 4–10 and 4–11 show that the PSD of a PAM-TH-UWB is characterized by the presence of spectral lines located at multiples of the average pulse repetition frequency, $1/T_s$, and at multiples of $1/T_c$. Some of these lines, that is, those lines located at multiples of $1/T_s$, are strongly reduced when increasing the N_p value. The other lines, that is, those lines located at multiples of $1/T_c$, still maintain a high concentration of power even when N_p reaches the total number of transmitted pulses (see Figure 4–11). Similarly to the case of PPM-TH-UWB systems, these lines can be removed only by reducing the chip duration T_c, that is, by increasing the cardinality of the TH code.

CHECKPOINT 4–2

Further Reading

As analyzed above, the DS-UWB case can be derived in a straightforward fashion from DS-CDMA (Viterbi, 1995). DS-CDMA is a well-established concept, which has been fully analyzed in the traditional literature of digital communication theory. Among the vast bibliography, reference is made, in particular, to a few classical textbooks (Benedetto et al., 1987; Lee and Messerschmitt, 1994; Proakis, 1995; Benedetto and Biglieri, 1999).

Performance analyses of DS-UWB systems can be found in (Vojcic and Pickholtz, 2003). In particular, comparisons with TH-PPM are derived in (Sadler and Swami, 2002), and (Ziemer et al., 2003).

REFERENCES

Benedetto, S., E. Biglieri, and V. Castellani, *Digital Transmission Theory*, Englewood Cliffs, New Jersey: Prentice Hall, Inc. (1987).

Benedetto, S., and E. Biglieri, *Principles of Digital Transmission*, New York: Kluwer Academic/Plenum Publishers (1999).

Di Benedetto, M.-G., and B.R. Vojcic, "Ultra Wide Band (UWB) Wireless Communications: A Tutorial," *Journal of Communication and Networks, Special Issue on Ultra-Wideband Communications*, Volume: 5, Issue: 4 (December 2003), 290–302.

Lee, E.A., and D.G. Messerschmitt, *Digital Communication*, 2nd Edition, Boston, Massachusetts: Kluwer Academic Publishers (1994).

Proakis, J.G., *Digital Communications*, 3rd Edition, New York: McGraw-Hill International Editions (1995).

Roberts, R., *XtremeSpectrum CFP Document*, Available at *grouper.ieee.org/groups/802/15/pub/2003/Jul03/03154r3P802-15_TG3a-XtremeSpectrum-CFP-Documentation.pdf* (July 2003).

Sadler, B.M., and A. Swami, "On the performance of UWB and DS-spread spectrum communication systems," *IEEE Conference on Ultra Wideband Systems and Technologies* (May 2002), 289–292.

Viterbi, A. J., *CDMA — Principles of Spread Spectrum Communication*, Reading, Massachusetts: Addison-Wesley Publishing Company (1995).

Vojcic, B.R., and R.L. Pickholtz, "Direct-Sequence Code Division Multiple Access for Ultra-Wide Bandwidth Impulse Radio," *IEEE Military Communications Conference* (October 2003), 898–902.

Ziemer, R., M. Wickert, and T. Williams, "A Comparison Between UWB and DSSS for use in a Multiple Access Secure Wireless Sensor Network," *IEEE Conference on Ultra Wideband Systems and Technologies* (November 2003), 428–432.

APPENDIX 4.A

Function 4.1 PAM-TH Modulation

Function 4.1 is based on the same algorithm as Function 2.8, except for the part of the code that implements DS coding, which we substitute with the set of commands implementing TH coding in Function 2.4.

```
%
% FUNCTION 4.1 : "cp0402_2PAM_TH"
%
% Introduces the DS code given by 'DScode'
% and implements binary PAM modulation
% 'seq' is the input binary stream
% 'fc' is the sampling frequency for the generated signal
% 'Ts' is the average pulse repetition period
% 'DScode' is the DS code
%
% The function generates two output streams:
% 'PAMTHseq' is the output with both TH and PAM
% 'THseq' is the output with TH only
%
% Programmed by Guerino Giancola
%

function [PAMTHseq,THseq] = ...
   cp0402_2PAM_TH(seq,fc,Tc,Ts,THcode)


% --------------------------------------------------
% Step One - Implementation of the PAM+TH modulator
% --------------------------------------------------

dt = 1 ./ fc;                  % sampling period
framesamples=floor(Ts./dt);% number of samples between
                           % pulses
chipsamples = floor (Tc ./ dt);  % number of samples for
                                 % the chip duration

THp = length(THcode);            % TH-code length
```

```
totlength = framesamples*length(seq);
PAMTHseq=zeros(1,totlength);
THseq=zeros(1,totlength);

% ----------------------------------------
% Step Two - Main loop for introducing TH
% ----------------------------------------

for k = 1 : length(seq)

    % uniform pulse position
    index = 1 + (k-1)*framesamples;


    % introduction of TH
    kTH = THcode(1+mod(k-1,THp));
    index = index + kTH*chipsamples;

    THseq(index)=1;
    PAMTHseq(index)=((seq(k)*2)-1);

end % for k = 1 : length(seq)
```

Function 4.2 PAM-TH Transmitter

Function 4.2 is composed of three steps. Step Zero contains all the parameters characterizing the signal to be generated: the average transmitted power `Pow`, the value of the sampling frequency `fc`, the number of bits to be generated `numbits`, the average pulse repetition period `Ts`, the number of pulses per bit `Ns`, the value of the chip time `Tc`, the cardinality of the TH code `Nh`, the periodicity of the TH code `Np`, the value of the time duration of the pulse `Tm`, the shaping factor `tau`, and a flag `G`, which must be set to 1 to produce a graphical output. Step One is composed of the command lines for sequentially activating the different blocks of the transmission chain. Step Two produces the graphical output. Except for the presence of Function 4.1, all the blocks were already described in Appendix 2.A.

```
%
% FUNCTION 4.2 : "cp0402_transmitter_2PAM_TH"
%
% Simulation of a UWB transmitter implementing PAM with TH
%
% Transmitted power is fixed at 'Pow'
% The signal is sampled with frequency 'fc'
% 'numbits' is the number of bits generated by the source
% 'Ns' pulses are generated for each bit, and these pulses
% are spaced in time by an average pulse repetition period
% 'Ts'
% The TH code has periodicity 'Np' and cardinality 'Nh'
% Each pulse has time duration 'Tm' and shaping factor
% 'tau'
%
% The function returns:
% 1) the generated stream of bits ('bits')
% 2) the generated TH code ('THcode')
% 3) the generated signal ('Stx')
% 4) a reference signal without data modulation ('ref')
%
% Programmed by Guerino Giancola
%

function [bits,THcode,Stx,ref]=cp0402_transmitter_2PAM_TH

% ----------------------------
% Step Zero - Input parameters
% ----------------------------

Pow = -30;       % average transmitted power (dBm)
```

```
fc = 50e9;          % sampling frequency

numbits = 5;        % number of bits generated by the source

Ts = 10e-9;         % frame time, or average pulse repetition
                    % period [s]
Ns = 5;             % number of pulses per bit

Tc = 1e-9;          % chip time [s]
Nh = 10;            % cardinality of the TH code
Np = 5;             % period of the TH code

Tm = 0.5e-9;        % pulse duration [s]
tau = 0.25e-9;      % shaping factor for the pulse [s]

G = 1;
% G=0 -> no graphical output
% G=1 -> graphical output

% ----------------------------------------
% Step One - Simulating transmission chain
% ----------------------------------------

% binary source
bits = cp0201_bits(numbits);

% repetition coder
repbits = cp0201_repcode(bits,Ns);

% TH code
[THcode]=cp0201_TH(Nh,Np);

% PAM + TH
[PAMTHseq,THseq] = cp0402_2PAM_TH(repbits,fc,Tc,Ts,THcode);

% shaping filter
power = (10^(Pow/10))/1000;        % average transmitted
                                   % power (watts)
Ex = power * Ts;                   % energy per pulse
w0 = cp0201_waveform(fc,Tm,tau);   % energy normalized pulse
                                   % waveform
wtx = w0 .* sqrt(Ex);              % pulse waveform
Sa = conv(PAMTHseq,wtx);           % output of the filter
                                   % (with modulation)
Sb = conv(THseq,wtx);              % output of the filter
                                   % (without modulation)
```

```
% Output generation

L = (floor(Ts*fc))*Ns*numbits;
Stx = Sa(1:L);
ref = Sb(1:L);

% ---------------------------
% Step Two - Graphical output
% ---------------------------

if G

F = figure(1);
set(F,'Position',[32 223 951 420]);

tmax = numbits*Ns*Ts;
time = linspace(0,tmax,length(Stx));
P = plot(time,Stx);
set(P,'LineWidth',[2]);
ylow=-1.5*max(wtx);
yhigh=1.5*max(wtx);axis([0 tmax ylow yhigh]);
AX=gca;
set(AX,'FontSize',12);
X=xlabel('Time [s]');
set(X,'FontSize',14);
Y=ylabel('Amplitude [V]');
set(Y,'FontSize',14);
for j = 1 : numbits
    tj = (j-1)*Ns*Ts;
    L1=line([tj tj],[ylow yhigh]);
    set(L1,'Color',[0 0 0],'LineStyle','--',...
       'LineWidth', [2]);
    for k = 0 : Ns-1
        if k > 0
            tn = tj + k*Nh*Tc;
            L2=line([tn tn],[ylow yhigh]);
            set(L2,'Color',[0.5 0.5 0.5],...
               'LineStyle','-.', 'LineWidth', [2]);
        end
        for q = 1 : Nh-1
            th = tj + k*Nh*Tc + q*Tc;
            L3=line([th th],[0.8*ylow 0.8*yhigh]);
            set(L3,'Color',[0 0 0], 'LineStyle',...
               ':', 'LineWidth', [1]);
        end
    end
```

```
end
end % end of graphical output
```

CHAPTER 5

The PSD of MB-UWB Signals

In the MB approach, the overall bandwidth is split into smaller frequency bands of at least 500 MHz each, compliant with FCC rules (FCC, 2002). Transmission of data for a given user occurs on different sub-bands in subsequent periods of time, using OFDM, and leading to a system that can avoid nonintentional interference in certain bands without the need for RF notch filters.

In this chapter, spectral properties of OFDM signals will be investigated with the purpose of further analyzing the MB proposal submitted to the IEEE 802.15.TG3a (Batra et al., 2003).

5.1 Spectral Characteristics of OFDM Signals

As discussed in Section 2.3 the MB approach moves away from the IR principle, and current proposals favoring MB within the IEEE 802.15.TG3a (Batra et al., 2003) are based on the OFDM concept (Hanzo et al., 2003). Checkpoint 2–3 analyzed the proposal by (Batra et al., 2003), and provided the tools for generating an OFDM signal having specific features in agreement with system specifications.

In this chapter, we want to complete the analysis of the OFDM alternative by briefly describing the spectral characteristics and PSD of an OFDM signal. Spectral characterization of OFDM is widely described in related literature (Hanzo et al., 2003). We will review here the basic features of OFDM spectral properties.

Recall that as previously illustrated in Section 2.3, an OFDM symbol consists of the parallel transmission of N signals, which are modulated at different carrier frequencies spaced by Δf. The binary source sequence is subdivided into groups of K bits, used to generate blocks of N symbols defined over a set of L possible values, with $K = N\log_2 L$. Each of these symbols modulates a different carrier. To discriminate the N signals of the OFDM symbol at the receiver, the signals must be orthogonal in frequency. If T_0 is the time used for transmitting each signal on the corresponding carrier, orthogonality among different transmissions can be achieved by adopting $\Delta f = 1/T_0$. A guard interval T_G is then introduced between transmission of subsequent blocks, mainly to prevent ISI, leading to a total OFDM symbol duration of $T = T_0 + T_G$. The length of the guard interval is commonly on the order of 20–30% of total symbol duration T. The final border section of the OFDM symbol is usually replicated within this guard interval and is referred to as the cyclic prefix. The cyclic prefix is mainly introduced to maintain carrier synchronization at the receiver in the presence of time-dispersive channels. The length of the cyclic prefix is obviously limited by the duration of the guard interval. At the receiver, the cyclic prefix is discarded.

It was shown in Section 2.3 that a sampled version of the complex envelope of the OFDM signal $\underline{x}_k(t)$ corresponding to the k-th OFDM symbol can be obtained by computing the Inverse Discrete Fourier Transform (IDFT) of the sequence of complex symbols $\{c_0, \ldots, c_j, \ldots, c_{N-1}\}$, which modulate the in-phase and in-quadrature carriers (Weinstein and Ebert, 1971). We found in particular that the n-th element C_n of vector C representing the IDFT of $\{c_0, \ldots, c_j, \ldots, c_{N-1}\}$ is used to generate the n-th sample of $\underline{x}_k(t)$ (see Eq. (2–15)). The generic coefficient C_n can be expressed as follows:

$$C_n \equiv \sum_{m=0}^{N-1} c_m e^{j2\pi\frac{nm}{N}} \quad (5–1)$$

The complex envelope of the signal corresponding to a set of OFDM symbols can therefore be expressed as:

$$\underline{s}(t) \equiv \sum_{n=-\infty}^{+\infty} \underline{x}_n(t - nT) \quad (5–2)$$

Let us first compute the total power $P_{\underline{s}}$ of the complex envelope $\underline{s}(t)$ of Eq. (5–2). Using the central limit theorem, based on Eq. (5–1), we can make the hypothesis that the quantities C_n have a complex Gaussian distribution. This hypothesis is reasonable if the number of sub-carriers N is high. We further assume that the expected value of each C_n is zero; that all C_n are uncorrelated, and since Gaussian, are also independent; and that all C_n have same variance σ_C^2. By also taking into account Eq. (2–15), we can write:

$$P_{\underline{s}} = \sigma_C^2 = \sum_{j=0}^{N-1} \sigma_{c_m}^2 \quad (5–3)$$

The PSD of the OFDM signal of Eq. (5–2) can be computed by first introducing a random phase epoch Θ, uniformly distributed in $[0,T]$. Equation (5–2) becomes:

$$\underline{s}(t) \equiv \sum_{n=-\infty}^{+\infty} \underline{x}_n(t - nT + \Theta) \quad (5–4)$$

The PSD of Eq. (5–4) can be determined by computing the autocorrelation function of $\underline{s}(t)$, and by taking the Fourier transform of the autocorrelation function. Note, however, that within each symbol, it is reasonable to assume statistical independence of the complex values c_m modulating the sub-carriers, and that furthermore, the values modulating a generic sub-carrier in different symbol periods can also be assumed to be statistically independent. The PSD of Eq. (5–4) can therefore be found by adding up the PSDs of individual sub-carriers, for a generic OFDM symbol $\underline{x}_k(t) = \underline{x}(t)$.

The complex envelope of an OFDM symbol as defined in Eq. (2–11) can be rewritten here by introducing a rectangular window of duration T, where T is the OFDM symbol duration:

$$\underline{x}(t) = rect\left(\frac{t}{T}\right)\sum_{m=0}^{N-1} c_m e^{j2\pi f_m t} \tag{5–5}$$

Equation (5–5) indicates that the spectrum of an OFDM symbol is a sin(x)/x function corresponding to the Fourier transform of the rectangular window, convolved with a sequence of Dirac pulses located at the sub-carrier frequencies f_m. After removal at the receiver of the cyclic prefix, the spectrum centered on the m-th sub-carrier f_m can be expressed as:

$$P_{f_m}(f) = \frac{\sin\left(\pi T_0 (f - f_m)\right)}{\pi T_0 (f - f_m)} = \frac{\sin\left(\dfrac{\pi (f - f_m)}{\Delta f}\right)}{\dfrac{\pi (f - f_m)}{\Delta f}} \tag{5–6}$$

The spectrum of an OFDM symbol, and by extension of Eq. (5–4), is therefore:

$$P_{\underline{s}}(f) = P_{\underline{s}} \sum_{m=0}^{N-1} \frac{\sin\left(\dfrac{\pi (f - f_m)}{\Delta f}\right)}{\dfrac{\pi (f - f_m)}{\Delta f}} \tag{5–7}$$

5.2 THE MB IEEE 802.15.TG3A PROPOSAL

As anticipated in Chapter 2, Checkpoint 2–3, in 2003, an OFDM scheme in which the OFDM signal complies with the FCC rules set for UWB was submitted to Task Group 3a of the IEEE 802.15, as a candidate standard for a high-rate WPAN physical layer. This approach goes under the name of the *Multi-Band (MB) approach* (Batra et al., 2003).

The proposed scheme is based on the idea of dividing the available spectrum, which is typically considered between 3.1 GHz and 10.6 GHz according to FCC rules (see Chapter 6), into several sub-bands, characterized by a width greater than the FCC limit of 500 MHz.

In the next checkpoint (Checkpoint 5–1), the spectral characteristics of a signal complying with the MB description will be analyzed. For a description of the main features of MB, reference is made to Chapter 2, Checkpoint 2–3. Table 5–1 reports the main parameters of the proposal.

Table 5–1 MB OFDM System Parameters

Bit Rate (Mbits/s)	Modulation - Constellation	FFT Size	Coding Rate	Spreading Rate	T_O (ns)	T_P (ns)	T_G (ns)
55	OFDM - QPSK	128	11/32	4	242.4	60.6	70.1
80	OFDM - QPSK	128	1/2	4	242.4	60.6	70.1
110	OFDM - QPSK	128	11/32	2	242.4	60.6	70.1
160	OFDM - QPSK	128	1/2	2	242.4	60.6	70.1
200	OFDM - QPSK	128	5/8	2	242.4	60.6	70.1
320	OFDM - QPSK	128	1/2	1	242.4	60.6	70.1
480	OFDM - QPSK	128	3/4	1	242.4	60.6	70.1

CHECKPOINT 5–1

In this checkpoint, we will evaluate by simulation the spectral characteristics of OFDM signals. The MATLAB functions that are required for such an analysis were introduced in previous checkpoints: Function 2.11 (Appendix 2.A) for the generation of OFDM signals, and Function 3.2 (Appendix 3.A) for the evaluation of PSD.

As a first example, we generate an OFDM signal `ofdm01`, characterized by four sub-carriers and symbol duration T_O = 100 ns. The PSD of such a signal should be characterized by the superimposition of four different spectra, one for each sub-carrier. The spacing between adjacent sub-carriers should be $1/T_O$ = 10 MHz, for a total occupied bandwidth of about 40 MHz around the central frequency (plus another 40 MHz at negative frequencies). The parameters for generating `ofdm01` with Function 2.11 are: `numbits=128; fp=1e9; fc=50e9; T0=100e-9; TP=0; TG=50e-9; A=1; N=4.` The resulting signal is composed of 16 OFDM symbols of duration 100 ns, separated by a guard interval of 50 ns. No cyclic prefix is introduced. The complex symbols resulting from the IFFT modulate the in-phase and in-quadrature components of a carrier at frequency 1 GHz. The commands for generating signal `ofdm01` and for evaluating the corresponding PSD are:

```
[bits,S,SI,SQ,ofdm01,fc,fp,T0,TP,TG,N]=cp0203_OFDM_qpsk;
[PSD01,df]=cp0301_PSD(ofdm01,fc);
```

The PSD of signal `ofdm01` is shown in Figure 5–1 and, in logarithmic scale, in Figure 5–2. By observing these figures, we can verify that the PSD of signal `ofdm01` occupies about 40 MHz around the central frequency of 1 GHz.

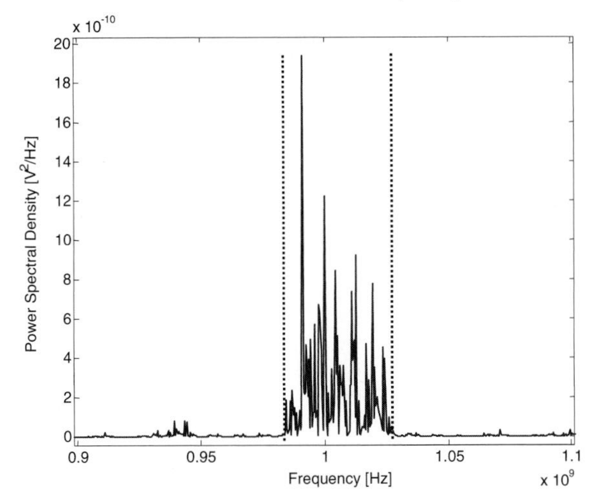

Figure 5–1 PSD of the OFDM signal ofdm01 (four sub-carriers).

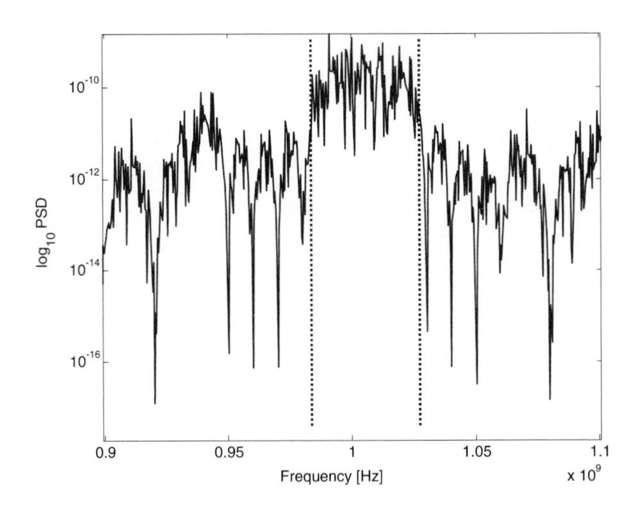

Figure 5–2 PSD of the OFDM signal ofdm01 (four sub-carriers) in logarithmic units.

The same analysis can be repeated with a second signal, `ofdm02`, which is characterized by the same parameters as signal `ofdm01`, except for an increased number of sub-carriers `N=32`, and a correspondingly increased number of transmitted bits `numbits=1024`. Note that we are increasing the number of sub-carriers without changing symbol duration. In other words, a higher number of symbols can be transmitted during the same period of time, at the price of an increased bandwidth occupation. The PSD of signal `ofdm02`, in particular, should consist of the superimposition of 32 sub-carrier spectra, with a total occupied bandwidth of 320 MHz (for positive frequencies). The commands for generating signal `ofdm02` and for evaluating the corresponding PSD are:

```
[bits,S,SI,SQ,ofdm02,fc,fp,T0,TP,TG,N]=cp0203_OFDM_qpsk;
[PSD02,df]=cp0301_PSD(ofdm02,fc);
```

The PSD of signal `ofdm02` is shown in Figures 5–3 and 5–4 (logarithmic scale), where we verify a bandwidth occupation of 320 MHz around the central frequency of 1 GHz.

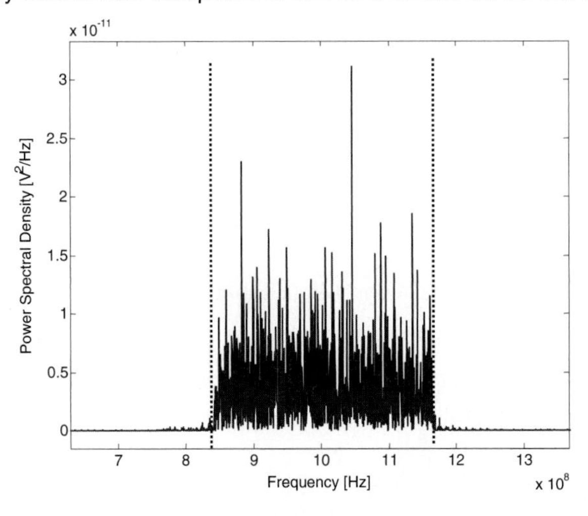

Figure 5–3 PSD of the OFDM signal ofdm02 (32 sub-carriers).

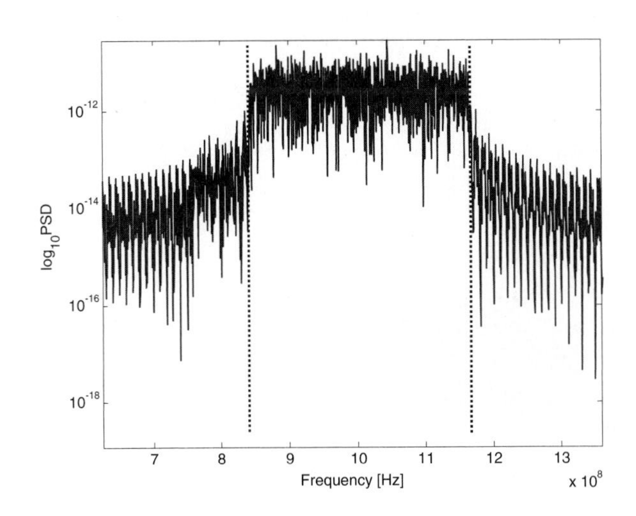

Figure 5–4 PSD of the OFDM signal ofdm02 (32 sub-carriers) in logarithmic units.

We can repeat the same analysis with a third signal, `ofdm03`, which is characterized by the same parameters of signals `ofdm01` and `ofdm02`, except for an increased number of sub-carriers `N=64`, and a corresponding increase in the number of transmitted bits `numbits=2048`. The PSD of signal `ofdm03` should occupy a bandwidth of 640 MHz (at positive frequencies), and can therefore be assumed to be an OFDM-UWB signal according to FCC rules (FCC, 2002). The commands for generating signal `ofdm03` and for evaluating the corresponding PSD are:

```
[bits,S,SI,SQ,ofdm03,fc,fp,T0,TP,TG,N]=cp0203_OFDM_qpsk;
[PSD03,df]=cp0301_PSD(ofdm03,fc);
```

The PSD of signal `ofdm03` is shown in Figures 5–5 and 5–6 (logarithmic scale), where we can verify a bandwidth occupation of 640 MHz around the central frequency of 1 GHz.

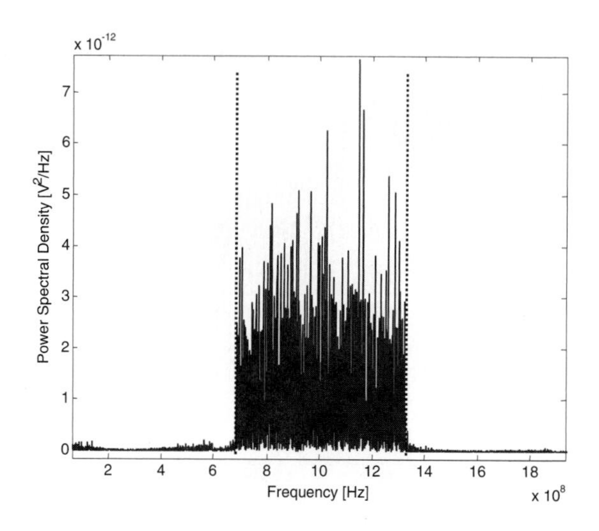

Figure 5–5 PSD of the OFDM signal ofdm03 (64 sub-carriers).

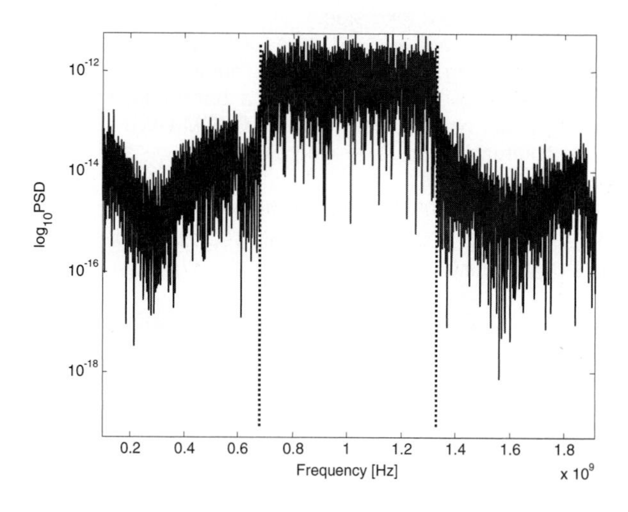

Figure 5–6 PSD of the OFDM signal ofdm03 (64 sub-carriers) in logarithmic units.

We can now evaluate the spectral characteristics of the UWB signal format proposed to the IEEE 802.15.TG3a by the MB coalition (see Checkpoint 2–3). The following parameters are introduced within Function 2.11: `numbits=4096; fp=3.432e9; fc=50e9; T0=242.4e-9; TP=60.6e-9; TG=70.1e-9; A=1; N=128.` The resulting waveform represents an OFDM signal that is composed of 128 sub-carriers equally spaced by 4.1254 MHz. These sub-carriers are located around a central frequency f_c = 3.432 GHz, that is, in the first slot of the band plan, which is described in the MB proposal (Batra et al., 2003). The commands for generating an OFDM-UWB signal `mb01` and for evaluating the corresponding PSD are:

```
[bits,S,SI,SQ,mb01,fc,fp,T0,TP,TG,N]=cp0203_OFDM_qpsk;
[PSDmb01,df]=cp0301_PSD(mb01,fc);
```

The PSD of signal `mb01` is shown in Figures 5–7 and 5–8 (logarithmic scale), where we can verify that it is contained within a frequency band of 528 MHz.

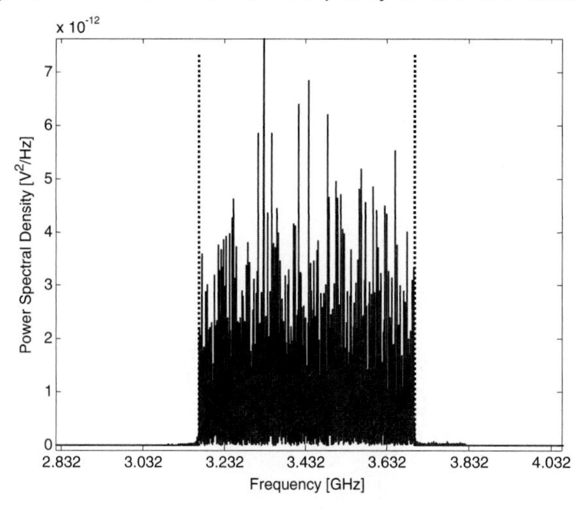

Figure 5–7 PSD of the OFDM signal mb01 (128 sub-carriers).

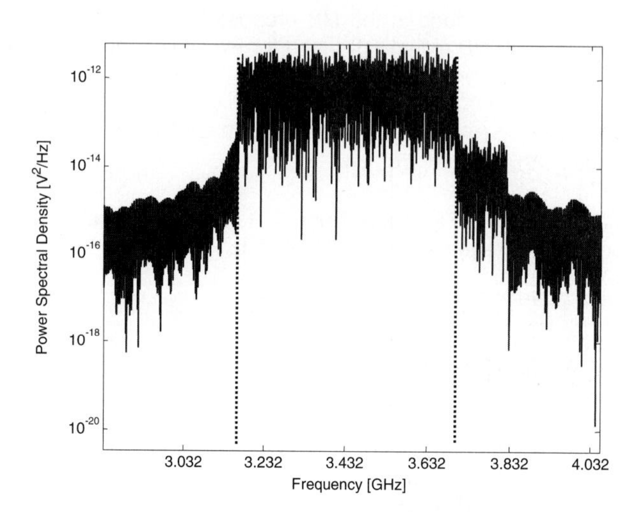

Figure 5–8 PSD of the OFDM signal mb01 (128 sub-carriers) in logarithmic units.

It is interesting to verify the possibility of using different bands at the same time without causing interference between signals. In particular, we can use Function 2.11 for generating other two OFDM-UWB signals, `mb02` and `mb03`, which are characterized by the same parameters as signal `mb01`, except for the central frequency of the signal. We can use `fp=3.960e9` for `mb02` and `fp=4.488e9` for `mb03`. In this case, the resulting signals occupy the second and third slots of a band plan defined in (Batra et al., 2003). Figure 5–9 shows the superimposition of the PSD of the three signals `mb01`, `mb02`, and `mb03`, where we can verify that each signal does not interfere in frequency with the other two. As a consequence, the parallel transmission of three UWB signals in different bands is allowed since the receiver is always capable of isolating the useful signal by filtering in frequency the contributions that are outside the band of interest. The possibility of applying a frequency-hopping scheme for a single signal is also available, that is, the transmission of a user can use different bands in different periods of time. A "Mode 1 device," in particular, should be capable of using all three bands n_b 1,n_b 2, and n_b 3, which are represented in Figure 5–9.

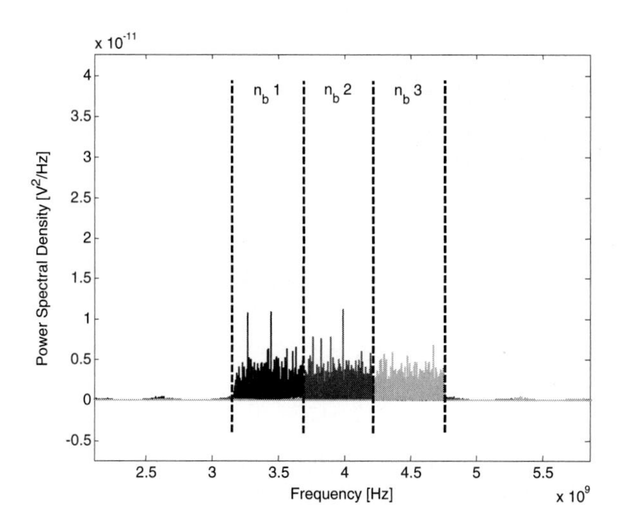

Figure 5–9 PSD of the OFDM signals mb01, mb02, and mb03, which correspond to the transmission in the first three bands of the band plan proposed by the MB coalition (see Batra et al., 2003)

CHECKPOINT 5–1

Further Reading

A recent book by (Hanzo et al., 2003) contains over 500 references to papers on OFDM! This gives a flavor for the tremendous interest and effort that has been devoted to this topic, previously and after its adoption in a variety of industrial standards. To cite a few: IEEE 802.11a (IEEE, 1999), HIPERLAN/2 (Khun-Jush et al., 2000), and Digital Audio Broadcasting (Thibault and Le, 1997).

References

Batra, A., et al., *Multi-band OFDM Physical Layer Proposal for IEEE 802.15 Task Group 3a*, Available at *www.Multi-Bandofdm.org/papers/15-03-0268-01-003a-Multi-band-CFP-Document.pdf* (September 2003).

Federal Communications Commission, "Revision of Part 15 of the Commission's rules Regarding Ultra-Wideband Transmission Systems: First report and order," *Technical Report FCC 02-48* (adopted February 14, 2002; released April 22, 2002).

Hanzo, L., M. Münster, B.J. Choi, and T. Keller, *OFDM and MC-CDMA for Broadband Multi-User Communications, WLANs and Broadcasting*, Chichester, West Sussex, England: John Wiley and Sons, Inc. (2003).

IEEE Std 802.11a-1999, *Supplement to IEEE standard for information technology telecommunications and information exchange between systems - local and metropolitan area networks - specific requirements, Part 11: wireless LAN Medium Access Control (MAC) and Physical Layer (PHY)* (December 1999).

Khun-Jush, J., G. Malmgren, P. Schramm, and J. Torsner, "Overview and performance of HIPERLAN type 2-a standard for broadband wireless communications," *Proceedings of IEEE 51st Vehicular Technology Conference*, Volume: 1 (May 2000), 112–117.

Thibault, L., and M.T. Le, "Performance evaluation of COFDM for digital audio broadcasting. I. Parametric study," *IEEE Transactions on Broadcasting*, Volume: 43, Issue: 1 (March 1997), 64–75.

Weinstein, S., and P. Ebert, "Data Transmission by Frequency-Division Multiplexing using the Discrete Fourier Transform," *IEEE Transactions on Communications*, Volume: 19, Issue: 5 (October 1971), 628–634.

CHAPTER 6

Performance Analysis for the UWB Radio Link

UWB radio signals must, in principle, coexist with other radio signals. Possible interference from and onto other communication systems must be contained within regulated values that indicate the maximum tolerable power to be present in the air interface at any given frequency, as set by emission masks. In this chapter, we will first analyze how to read and apply an emission mask, and second, we will introduce the methodology for performing a link budget, that is, we will determine the maximum distance of propagation at a given data rate under a maximum probability of error constraint for the UWB point-to-point link.

6.1 Power Limits and Emission Masks

The power limitation set by emission masks is on the effective radiated power, that is the Effective Isotropic Radiated Power (*EIRP*) for a given range of operating frequencies, and is given by the product of the available power of the transmitter P_{TX}, which is the maximum power that the transmitter can transfer to the transmitter antenna and the gain of the transmitter antenna G_{AT}.

$$EIRP = P_{TX} G_{AT} \tag{6–1}$$

The quantity *EIRP* is usually measured in dBm, that is as $10\log_{10} EIRP_{\text{mWatts}}$.

The available power of the transmitter P_{TX} is effectively transferred from the transmitter to the antenna when the condition for maximum power transfer between the output impedance of the transmitter Z_{oTX} and input impedance of the antenna Z_{AT} is verified, $Z_{AT} = Z_{oTX}^{*}$.

An equivalent way of ruling emitted radiations is to impose limits on the field strength V_s. The field strength represents the voltage one should apply to an impedance equal to the characteristic impedance of free-space Z_{FS} to obtain an available power P_{TX} after propagation over a distance D. Z_{FS} is related to permeability and permittivity of free space and is equal to approximately 377 ohms (the exact value being 120π). In principle, Z_{FS} is independent of frequency. The relation between field strength (expressed in V/m) and available power (expressed in W) is thus:

$$EIRP \cong \frac{V_s^2}{377} 4\pi D^2 \tag{6–2}$$

The power defined by Eq. (6–2) is an average power. In a binary IR scheme, the average power is computed by averaging over the bit interval T_b. Given the energy of a single pulse, E_p, and the total energy of the pulses representing one bit, $N_s E_p$, the average power P_{av} under the hypothesis $T_b = N_s T_s$ is thus expressed by:

$$P_{av} = \frac{N_s E_p}{T_b} = \frac{N_s E_p}{N_s T_s} = \frac{E_p}{T_s} \tag{6–3}$$

where $1/T_s$ is the pulse repetition rate.

As shown by Eq. (6–3), different signals can exhibit the same P_{av} with different pulse energy E_p, depending on pulse repetition rate. At equal average power, signals with low repetition rate may have higher E_p. For equal pulse duration, this is equivalent to saying that the maximum instantaneous power may be markedly different among signals with similar average power.

Emission masks impose limits on the PSD of emitted signals, that is, on *EIRP* spectral density, expressed, for example, in dBm/Hz or dBm/MHz. Emission masks are, however, commonly provided in terms of power values, typically dBm, at a given frequency, rather than power density values. The value of the emission mask at a given frequency f_c indicates, in fact, the maximum allowed *EIRP* within a measured bandwidth (*mb*) centered around f_c. Such a power value, denoted as $EIRP_{mb}$, coincides with the total allowed *EIRP* only when the bandwidth B of the transmitted signal is equal to the measured bandwidth *mb*. For a signal occupying a bandwidth greater than *mb*, the maximum allowed *EIRP* is equal to the sum of the $EIRP_{mb}$ values that are provided by the mask corresponding to the frequency range occupied by the signal. In the particular case where the $EIRP_{mb}$ value indicated by the mask is constant over B, one has:

$$EIRP_{mb} \frac{B}{mb} = EIRP \simeq \frac{V^2}{377} 4\pi D^2 \tag{6–4}$$

Note that Eq. (6–4) provides a formal definition of $EIRP_{mb}$ as the maximum allowed *EIRP* for a signal having $B = mb$.

Currently, the only available emission masks for UWB radio communications are those issued by the FCC in the United States (FCC, 2002). Regarding indoor UWB systems, the mask limits operation to a -10 dB bandwidth lying between 3.1 and 10.6 GHz, and sets very stringent limits on out-of-band emission masks. The emission mask values are shown in Table 6–1 and Figure 6–1. The rule also specifies a limit on the peak level of emission within a 50 MHz bandwidth centered on the frequency f_M, at which the highest radiated emission occurs. The limit is set to 0 dBm/50 MHz, that is, the power computed over a frequency range of 50 MHz around f_M is limited to 0 dBm. Outdoor FCC UWB masks are provided in Appendix 6.B.

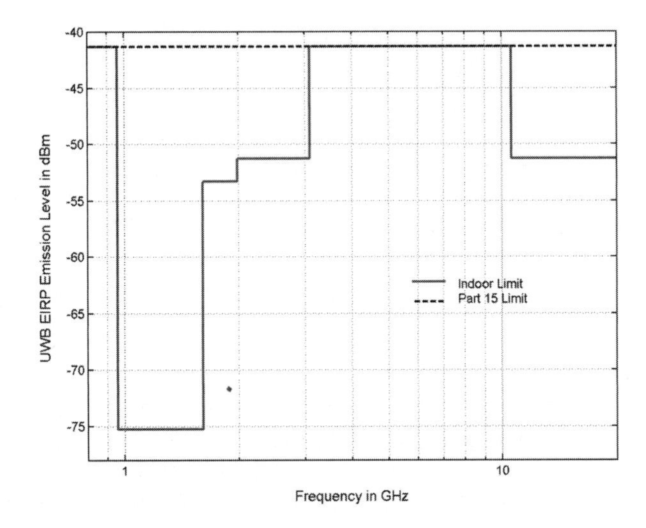

Figure 6–1 FCC indoor emission mask for UWB devices (FCC, 2002).

Table 6–1 Average Power Limits Set by the FCC in the U.S. for Indoor UWB Devices

Frequency in MHz	$EIRP_{mb}$ in dBm
0–960	-41.3
960–1610	-75.3
1610–1990	-53.3
1990–3100	-51.3
3100–10600	-41.3
Above 10600	-51.3

It is important to note that the FCC definition refers to a unilateral PSD $P_M^+(f)$; therefore, the maximum allowed total power $P_{M\max}$ for a signal occupying frequencies between f_L and f_H is:

$$P_{M\max} = \int_{f_L}^{f_H} P_M^+(f)\, df \tag{6–5}$$

When, for example, f_L =3.1 GHz and f_H =10.6 GHz, we have:

$$
\begin{aligned}
P_{M\max}\Big|_{dBm} &= 10\log_{10}\left(\int_{3.1\cdot 10^3}^{10.6\cdot 10^3} P_M^+(f)\,df\right) \\
&= 10\log_{10}\left(EIRP_{mb}\frac{\left(10.6\cdot 10^3 - 3.1\cdot 10^3\right)\Big|_{MHz}}{mb}\right) \\
&= -41.3 + 10\log_{10}\left(7.5\cdot 10^3/1\right) \\
&\cong -2.8\ dBm \\
&\quad i.e. \\
P_{M\max} &\cong 0.55\ mW
\end{aligned}
\qquad (6\text{–}6)
$$

CHECKPOINT 6–1

Table 6–1 shows the limits set by the FCC for $EIRP_{mb}$ expressed in dBm. These values indicate the maximum allowed EIRP for a signal with bandwidth equal to the measured bandwidth (mb). The average field strength limits can be derived from the average EIRP limits using Eq. (6–2). These limits, expressed in μV/m, are shown in Table 6–2.

Table 6–2 Average Field Strength Limits at D = 3 Meters Set by the FCC for Indoor UWB Communications

Frequency in MHz	Field Strength in μV/m
0–960	500
960–1610	10
1610–1990	125
1990–3100	157
3100–10600	500
Above 10600	157

For a signal with B = 500 MHz, the $EIRP_{mb}$ limitation of Table 6–1 limits $P_{M\max}$ to:

$$
P_{M\max} \cong \frac{V^2}{377}4\pi D^2 \frac{500}{1} = 500\ EIRP_{mb} \qquad (6\text{–}7)
$$

If B = 500 MHz is located within the frequency range 3.1–10.6 GHz, one has:

$$P_{M\max} \cong \frac{\left(5 \cdot 10^{-4}\right)^2}{377} 4\pi 3^2 \frac{500}{1} = 3.75 \cdot 10^{-5}\, W \quad (6\text{–}8)$$

which corresponds to:

$$P_{M\max}\Big|_{\mathrm{dBm}} \cong -41.3 + 10\log_{10}\left(\frac{500}{1}\right) = -14.3\ \mathrm{dBm} \quad (6\text{–}9)$$

FCC rules also limit peak emission to 0 dBm = 10^{-3} W for mb = 50 MHz around frequency f_M of maximum radiation. The field strength, corresponding to peak emission over a bandwidth B = mb, is thus:

$$V \cong \sqrt{EIRP \frac{377}{4\pi 3^2}} = \sqrt{10^{-3} \cdot \frac{377}{36\pi}} \cong 58\ mV/m \quad (6\text{–}10)$$

Emission power limits impose a limitation to pulse energy, as indicated in Eq. (6–3), showing that pulse energy is a function of pulse repetition period T_s for a fixed emitted power. One can thus write:

$$E_{p\max} = T_s P_{M\max} = T_s \int_{f_L}^{f_H} P_M^+(f)\, df \quad (6\text{–}11)$$

where $E_{p\max}$ is the maximum pulse energy given the limit $P_{M\max}$. Figure 6–2 shows $E_{p\max}$ as a function of T_s, for a signal with B = 500 MHz under the FCC average emission limit -41.3 dBm.

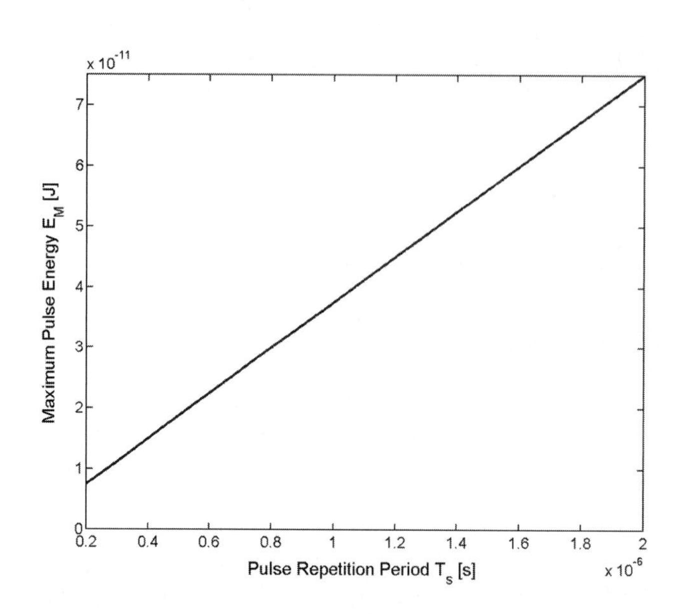

Figure 6–2 Maximum allowed pulse energy E_{pmax} as a function of T_s for a signal with B = 500 MHz under average emission limit $EIRP_{mb}$=-41.3 dBm, leading to $P_{Mmax} = 3.75 \cdot 10^{-5}$ W.

CHECKPOINT 6–1

Regulation processes are underway in several other areas of the world. At present, *draft* emission masks, which closely follow the FCC emission masks within the 3.1–10.6 GHz band, have been tentatively proposed in Europe (Sørensen, 2002). Other countries, such as Canada, Japan, Korea, and Russia, are monitoring developments and are close to releasing, but have not yet issued emission masks for UWB radiations (Hanna, 2003).

6.2 Link Budget

The PSD limitation defined by emission masks determines the maximum allowed transmitted power. Given the allowed power, we now will evaluate, under rather simplified hypotheses, the maximum distance over which propagation can occur when a predetermined probability of error must be guaranteed at the receiver, at a given data rate.

Let us first consider the case of IR transmission. We will then close this section with a reference to the OFDM hypothesis.

Decision at the receiver is based on the observation of a received energy E over a finite time interval, which is composed of mainly two terms: a signal term E_r and a noise term E_{noise}. The noise term may include several independent noise sources such as thermal noise, multi-user interference, external interference, and so on, that is:

$$E = E_r + \sum_{i=1}^{N} E_i = E_r + E_{noise} \tag{6–12}$$

and we can define the Signal to Noise Ratio (*SNR*) as follows:

$$SNR = \left(\sum_{i=1}^{N} \left(SNR_i \right)^{-1} \right)^{-1} \tag{6–13}$$

where SNR_i for i = 1, …, N are the SNRs corresponding to different noise sources, that is:

$$SNR_i = E_r / E_i \tag{6–14}$$

In the following analysis, let us assume that the only noise source at the receiver is Additive White Gaussian Noise (AWGN). This noise is typically thermal, introduced by the circuitry of the receiving antenna and the receiver. We do not take into account for the time being the presence of multi-user interference. This analysis will be carried out later in Chapter 9, "Multi-User Wireless Communications," in which the Gaussian model for multi-user interference will be discussed in detail. The bilateral thermal noise PSD $N_0/2$ expressed in Joule, that is, W/Hz is given by:

$$\frac{1}{2} N_0 = \frac{1}{2} kTemp_s(f) = \frac{1}{2} k\left(Temp_A + \left(F(f) - 1\right)Temp_0\right) \tag{6–15}$$

where $Temp_s$ is the so-called spot noise temperature at the receiver, that is, referring to a particular spot or frequency in the spectrum; $Temp_A$ is the receiving antenna temperature; $Temp_0$ = 290 °K (room temperature of 17 °C or 62.3 °F) is a standard temperature; k = $1.38 \cdot 10^{-23}$ Joules/°K is the Boltzman constant; and finally, $F(f)$ is the so-called spot noise figure of the receiving device. In the case of narrow-band transmissions, $F(f)$ is usually averaged over the frequencies of interest, that is, over the bandwidth B around the carrier frequency f_c, and the average noise figure F is used instead of $F(f)$ in Eq. (6–15) (see, for example, Couch, 1997):

$$\frac{1}{2} N_0 = \frac{1}{2} kTemp_s = \frac{1}{2} k \frac{1}{B} \int_{f_c - \frac{B}{2}}^{f_c + \frac{B}{2}} \left(Temp_A + \left(F(f) - 1\right)Temp_0\right) df = \\ = \frac{1}{2} k\left(Temp_A + \left(F - 1\right)Temp_0\right) \tag{6–16}$$

For UWB systems, averaging $F(f)$ over the frequencies of interest might not be significant due to the very large bandwidth of the signal. We will, therefore, for the time being, refer to Eq. (6–15) in the case of UWB receivers.

Let us also suppose that the signal propagates over a free-space propagation channel. The multi-path-affected UWB channel will be considered in Chapter 8, "Propagation over a Multi-Path Affected UWB Radio Channel." The free-space attenuation A_{FS} is expressed by:

$$A_{FS} = \frac{4\pi D^2}{G_T S_R} \tag{6–17}$$

where D is the distance of propagation, G_T is the transmitter antenna gain with respect to an isotropic antenna, and S_R is the receiving antenna effective area that represents the capability of the receiving antenna to intercept energy. The effective area and gain of an antenna, which is a reciprocal element, are related by:

$$G = \frac{4\pi S}{\lambda^2} \tag{6–18}$$

where $\lambda = c/f$ is the wavelength at which the system is operating, f is the frequency of operation, and c is the speed of light ($c = 3 \cdot 10^8$ m/s).

Equation (6–18) indicates that the behavior of the receiving antenna changes with the frequency at which the system operates. In the case of narrow-band systems, the operating frequency is the carrier frequency and we can compute the gain of the receiving antenna by giving the value of the carrier frequency f_c to the operating frequency. For UWB, moreover, where the involved frequencies span over an interval of several hundreds of MHz or even a few GHz, the concept of carrier frequency ceases to be valid, and the computation of the receiving antenna gain of Eq. (6–18) is not straightforward. We will leave in for the time being the frequency dependency in the free-space attenuation, which is expressed by:

$$A_{FS}(f) = \frac{(4\pi)^2 D^2 f^2}{G_T G_R c^2} \tag{6–19}$$

Suppose now that the transmitted waveform is characterized by a bilateral PSD $P_s(f)$, as derived in Chapters 3, "The PSD of TH-UWB Signals," 4, "The PSD of DS-UWB Signals," and 5, "The PSD of MB-UWB Signals," for UWB signals under various modulation hypotheses. The emission mask sets a limit to the value assumed by $P_s(f)$. Consider the case of the FCC masks illustrated in the previous section. We must be aware, however, that FCC masks are unilateral. Compliance with the mask limit must be verified for all frequencies of 2 $P_s^+(f)$ of the transmitted waveform.

The total transmitted power P_s can be computed by integrating $P_s(f)$ over the range of frequencies characterizing transmission. If we indicate by f_L and f_H the lowest and highest frequencies of interest, respectively, one has:

$$P_s = 2\int_{f_L}^{f_H} P_s(f)\,df \tag{6–20}$$

Note that the mask also imposes an upper limit on the total transmitted power as a consequence of the limitation over $P_s(f)$. Recalling that $P_M^+(f)$ indicates the unilateral PSD mask profile as for the FCC mask of Figure 6–1, one has:

$$P_s \leq \int_{f_L}^{f_H} P_M^+(f)\,df = P_{\max} \tag{6–21}$$

Following the approach proposed by (Sheng et al, 2003) and given the free-space attenuation expression of Eq. (6–19), we can express the power of the signal at the receiver P_r:

$$\begin{aligned} P_r &= 2\int_{f_L}^{f_H} \frac{P_s(f)}{A_{FS}(f)}\,df = 2\int_{f_L}^{f_H} P_s(f)\frac{G_T G_R c^2}{(4\pi)^2 D^2 f^2}\,df = \\ &= \frac{G_T G_R c^2}{(4\pi)^2 D^2}\,2\int_{f_L}^{f_H} \frac{P_s(f)}{f^2}\,df \end{aligned} \tag{6–22}$$

To meet the requirements of a given average SNR_{spec} at the receiver with a system margin set to M_S, the received power P_r must be such that:

$$P_r = M_s SNR_{spec} P_N \tag{6–23}$$

where P_N is the received thermal noise power at the receiver. We can now rewrite Eq. (6–23) by referring to the received energy E_r and the noise energy E_N:

$$E_r = M_s SNR_{spec} E_N = M_s SNR_{spec} \frac{1}{2} kTemp_s \tag{6–24}$$

Regarding SNR_{spec}, we can refer back to Eq. (6–14) and write:

$$M_s SNR_{spec} = \frac{E_r}{E_N} = \frac{P_r T_b}{\frac{1}{2}kTemp_s} = \frac{\frac{G_T G_R c^2}{(4\pi)^2 D^2}\,2T_b \int_{f_L}^{f_H} \frac{P_s(f)}{f^2}\,df}{\frac{1}{2}k\left(Temp_A + \left(F(f)-1\right)Temp_0\right)} \tag{6–25}$$

We can therefore deduce that for a given SNR_{spec} and system margin M_s, the maximum distance D, which can be covered by transmission, has a squared value equal to:

$$D^2 = \frac{\frac{G_T G_R c^2}{(4\pi)^2}\,2T_b \int_{f_L}^{f_H} \frac{P_s(f)}{f^2}\,df}{M_s SNR_{spec}\frac{1}{2}k\left(Temp_A + \left(F(f)-1\right)Temp_0\right)} \tag{6–26}$$

Equation (6–26) can be simplified if we introduce the hypothesis that $Temp_A = Temp_0$, that is, the noise temperature of the receiving antenna is the standard temperature. This hypothesis is reasonable for terrestrial links. Furthermore, given the current absence of UWB noise figure models, we assume a noise figure narrow-band model (IEEE 802.15.SG3a, 2003). Equation (6–26) becomes:

$$D^2 = \frac{\dfrac{G_T G_R c^2}{(4\pi)^2} 2T_b \displaystyle\int_{f_L}^{f_H} \frac{P_s(f)}{f^2} df}{M_s SNR_{spec} \dfrac{1}{2} FkTemp_0} \tag{6–27}$$

If we now assume for $P_s(f)$ a worst-case hypothesis, that is, we substitute it by its lower value P_{smin} within bandwidth, Eq. (6–27) becomes:

$$D^2 = \frac{\dfrac{G_T G_R c^2}{(4\pi)^2} 2T_b P_{s\min} \displaystyle\int_{f_L}^{f_H} \frac{1}{f^2} df}{M_s SNR_{spec} \dfrac{1}{2} FkTemp_0} = \frac{\dfrac{G_T G_R c^2}{(4\pi)^2} 2T_b P_{s\min} \dfrac{f_H - f_L}{f_H f_L}}{M_s SNR_{spec} \dfrac{1}{2} FkTemp_0} \tag{6–28}$$

System specification is usually defined in terms of probability of symbol error $\Pr_e$ rather than SNR. The relation between SNR and $\Pr_e$ depends on the modulation scheme and can be relatively easily expressed if the noise term is AWGN. In this case, the optimum receiver can be easily derived (Proakis, 1995). The optimum receiver consists of an optimum demodulator, which is typically a coherent demodulator or matched filter followed by an optimum detector, which in the case of equally probable signals, can be either a Maximum Likelihood (ML) or a Maximum A Posteriori (MAP) detector, as will be examined in detail in Chapter 8. For the time being, let us suppose that the receiver is optimum and let us evaluate its performance for various modulation schemes. We suppose for simplicity of notation that the system margin is zero ($M_s = 0$); given a target $\Pr_e$, we want to evaluate the required SNR_{spec}. These relations can be found in books dealing with digital communications [see Proakis (1995) and Lee and Messerschmitt (1994) for several modulation schemes and a summary in Guvenc and Arslan (2003)]. We will report here on the cases of binary PAM and PPM for which $\Pr_e$ reduces to the probability of bit error $\Pr_b$, and M-ary PAM and M-ary PPM.

In the case of binary PAM modulation with $p_1(t) = -p_0(t)$, that is, binary antipodal signals, it can be shown that the relation between $\Pr_b$ and SNR_{spec} is (*binary antipodal PAM*):

$$\Pr_b = \frac{1}{2} erfc(y)$$

$$\text{where} \quad y^2 = \frac{1}{2} SNR_{spec} = \frac{1}{2} \frac{P_r}{\frac{1}{2} FkTemp_0 2 \frac{f_b}{2}} = \frac{P_r T_b}{FkTemp_0} = \frac{E_b}{N_0} \tag{6–29}$$

$$\text{and} \quad erfc(y) = \frac{2}{\sqrt{\pi}} \int_y^{+\infty} e^{-\xi^2} d\xi$$

Binary orthogonal PPM corresponds to the transmission of binary orthogonal signals. In this case, as is well-known and will be demonstrated in Chapter 8, there is a 3 dB loss due to an increased effect of the noise that corresponds to considering in Eq. (6–29) either half of the useful power or a doubled noise power. One can write (*binary orthogonal PPM*):

$$\Pr_b = \frac{1}{2} erfc(y)$$

$$\text{where} \quad y^2 = \frac{1}{2} SNR_{spec} = \frac{1}{2} \frac{P_r}{2 \frac{1}{2} FkTemp_0 2 \frac{f_b}{2}} = \frac{E_b}{2N_0} \tag{6–30}$$

When the choice of the PPM shift ε is optimized (see Chapter 8), that is, for optimum PPM, performance is slightly improved over the orthogonal case. In other words, a target $\Pr_b$ value can be obtained at a slightly lower *SNR*.

Note that for a given $\Pr_b$, hence a given y^2, Eqs. (6–29) and (6–30) indicate that SNR_{spec} must be twice y^2. This SNR_{spec} is obtained for an E_b value, which, in the PPM case, must be twice the value needed for PAM.

The case of M-ary PAM can be described by the following relation (note that in this case, $\log_2 M$ bits are transmitted during an equivalent T_b interval *M-ary PAM*)):

$$\Pr_e = \left(1 - \frac{1}{M}\right) erfc(y)$$

$$\text{where} \quad y^2 = \frac{SNR_{spec}}{\frac{2}{3}\left(M^2 - 1\right)} = \frac{P_r T_b \log_2 M}{\frac{1}{2} FkTemp_0 \frac{2}{3}\left(M^2 - 1\right)} = \frac{E_b}{N_0} \frac{3 \log_2 M}{\left(M^2 - 1\right)} \tag{6–31}$$

The case of M-ary PPM can be described by the following equation (*M-ary orthogonal PPM*):

$$\Pr_e = \frac{1}{\sqrt{2\pi}} \int_{-\infty}^{+\infty} \left(1 - \left(\frac{1}{\sqrt{2\pi}} \int_{-\infty}^{y} e^{-x^2/2} dx \right)^{M-1} \right) \mathrm{e}^{-\frac{1}{2}\left(y - \sqrt{SNR_{spec}}\right)^2} dy$$

$$= \frac{1}{\sqrt{2\pi}} \int_{-\infty}^{+\infty} \left(1 - \left(\frac{1}{\sqrt{2\pi}} \int_{-\infty}^{y} e^{-x^2/2} dx \right)^{M-1} \right) \mathrm{e}^{-\frac{1}{2}\left(y - \sqrt{\frac{P_r T_b \log_2 M}{\frac{1}{2} FkTemp_0}}\right)^2} dy \qquad (6\text{–}32)$$

$$= \frac{1}{\sqrt{2\pi}} \int_{-\infty}^{+\infty} \left(1 - \left(\frac{1}{\sqrt{2\pi}} \int_{-\infty}^{y} e^{-x^2/2} dx \right)^{M-1} \right) \mathrm{e}^{-\frac{1}{2}\left(y - \sqrt{\frac{2 E_b \log_2 M}{N_0}}\right)^2} dy$$

It is possible to derive an upper bound on the probability of error for orthogonal signals, such as the M-ary orthogonal case (see, for example, Proakis, 1995, for a complete derivation), which provides a simpler relation than Eq. (6–32).

Provided that the *SNR* is sufficiently high, and specifically E_b/N_0 in dB is greater than $10\log_{10}(4\log_e 2) = 4.43$ dB, the following upper bound for Eq. (6–32) is sufficiently restringent (*M-ary orthogonal PPM — Upper bound for* $E_b/N_0 > 4.43$ *dB*):

$$\Pr_e < e^{-\log_2 M\left(E_b/N_0 - 2\log_e 2\right)/2} \qquad (6\text{–}33)$$

CHECKPOINT 6–2

This checkpoint shows a MATLAB implementation of Eqs. (6–31) and (6–33), which express the $\Pr_e$ for an M-ary PAM and the upper bound of the above probability for an M-ary PPM. We then proceed by working out an example of a link budget based on Eq. (6–27).

Function 6.1 evaluates the $\Pr_e$ for an M-ary PAM signal (Eq. (6–31)) and the upper bound of the symbol error probability $\Pr_e$ for an M-ary PPM (Eq. (6–33)). The command to invoke the function is:

```
cp0602_symbol_error_probability;
```

The function plots the symbol error probability for M equal to 2, 4, and 8 as a function of E_b/N_0. The result is plotted in Figure 6–3.

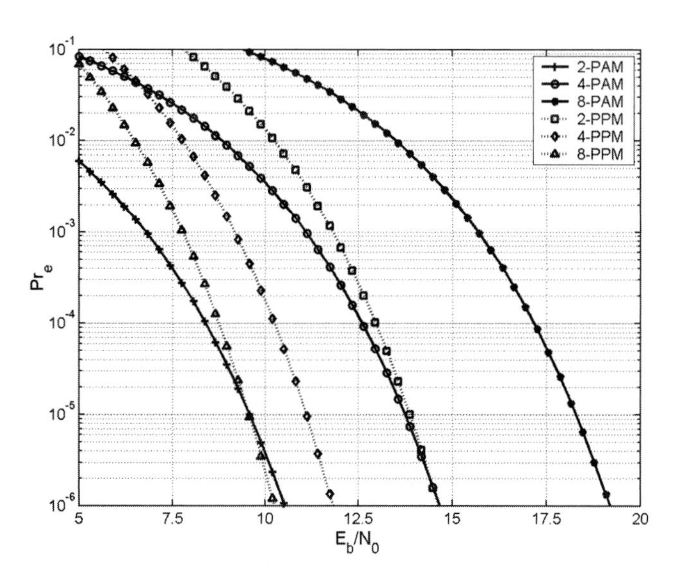

Figure 6–3 Probability of symbol error for M-ary PAM and PPM based on Eqs. (6–31) and (6–33).

Note that for M-PPM signals, performance improves as M increases, while the opposite is true for M-PAM signals. This indicates that in PPM, by increasing M, we can arbitrarily decrease Pr_e. The minimum required SNR for achieving $Pr_e \rightarrow 0$ can be computed by finding a sufficiently tight upper bound for Eq. (6–32) when the SNR is low, since Eq. (6–33) is valid only for higher SNR values (larger than 4.43 dB). For SNR values lower than 4.43 dB, a tighter bound can be found to be:

$$Pr_e < 2e^{-\log_2 M\left(\sqrt{E_b/N_0}-\sqrt{\log_e 2}\right)^2} \qquad (6–34)$$

which tends to 0 when $E_b/N_0\big|_{dB} > 10\log_{10}\left(\log_e 2\right) = -1.59\ dB$.

We will now use Eqs. (6–31) and (6–33) to perform a link budget for M-PAM and M-PPM UWB signals. We derive, for each number of levels M, the minimum value of y^2, which guarantees the target Pr_e. From y^2, we then compute the corresponding E_b/N_0 and determine the maximum distance D between transmitter and receiver by applying Eq. (6–27).

Note that D depends on the signal PSD. In our example, the pulse waveform is the 5th derivative of the Gaussian pulse, as adopted by (Sheng et al., 2003), thanks to the good fit of this derivative in the 3.1–10.6 GHz FCC interval. The PSD of the 5th derivative of the Gaussian pulse, after amplitude normalization to meet the FCC indoor emission mask, is given by (Sheng et al., 2003):

$$P_s(f) = A_{max} \frac{(2\pi f \sigma)^{10} e^{-(2\pi f \sigma)^2}}{5^5 e^{-5}} \quad (6\text{–}35)$$

where $A_{max} = 10^{-13.125}$ W/Hz and $\sigma = 51$ ps.

To evaluate the maximum distance D for M-PAM and M-PPM systems as a function of both the number of levels M and of data rate R_b, we introduce four new MATLAB functions: Functions 6.2, 6.3, 6.4, and 6.5.

Function 6.2 performs the link budget for M-PAM and M-PPM systems, and plots the maximum distance between transmitter and receiver as a function of the data rate. Three values of M are considered: M = 2, M = 4, and M = 8. The function receives as input all the parameters required to perform the link budget: antenna gains of transmitter and receiver `Gt` and `Gr`, link margin in dB `MargindB`, receiver noise figure in dB `FdB`, unilateral signal PSD `PSD` over a frequency range determined by vector `f`, and the required probability of symbol error `Pre`. The function returns matrices `dist_PAM` and `dist_PPM`, containing the maximum distance D for M = {2, 4, 8}, and R_b in the interval [20 Mbits/s, 200 Mbits/s] taken at steps of 10 Mbits/s for PAM and PPM signals.

The command to invoke Function 6.2 is:

```
[dist_PAM,dist_PPM] = ...
   cp0602_link_budget(Gt,Gr,MargindB,FdB,f,PSD,Pre);
```

Function 6.3 evaluates the maximum distance by implementing Eq. (6–27) in a straightforward manner. The function receives as input the antenna gains `Gt` and `Gr`, a vector `f` representing the frequency range of interest, the PSD of the transmitted waveform `PSD`, the link margin in linear units `Margin`, the bit period `Tb`, the N_0 value `N0`, and the E_b/N_0 value `Eb_N0`. Function 6.3 returns the maximum distance `D` between transmitter and receiver. The command to execute Function 6.3 is:

```
D = cp0602_max_distance(Gt, Gr, f, PSD,...
    Margin, Tb, N0, Eb_N0);
```

Function 6.4 generates the normalized PSD of the n-th derivative of the Gaussian pulse. The function receives as input vector `frequency` representing the frequency axis, the order of the derivative `n`, the standard deviation of the Gaussian pulse `sigma`, and the peak value `Amax` of the PSD. Function 6.4 vector `PSD` represents the normalized PSD evaluated over the frequencies of interest. Function 6.4 is executed as follows:

```
[PSD] = cp0602_Gaussian_PSD_nth(frequency, n,...
    sigma, Amax);
```

Function 6.5 evaluates the bandwidth of an input PSD according to a given threshold. The input PSD is limited within a specified frequency range [f_L,f_H]. The function receives as input a vector `f` representing the frequency range of interest, the PSD of the transmitted waveform `PSD`, and the threshold value in dB `threshold`. Function 6.5 returns vector `freq_th_dB`, representing the frequency range of the evaluated bandwidth, and vector

`PSD_th_dB`, containing the components of the original PSD inside the bandwidth of interest. The command to execute Function 6.5 is:

```
[f_th_dB, PSD_th_dB] = cp0602_thr_dB_vectors(f,...
    PSD, threshold);
```

We will start with an example of a link budget, with the definition of a positive frequency axis in the range (0, 30 GHz], with the command:

```
f = linspace (1, 30e9, 2048);
```

Next, we will generate the unilateral signal PSD in Eq. (6–35) with Function 6.4:

```
sigma = 51e-12;
n = 5;
Amax=10^-13.125;
PSD = cp0602_Gaussian_PSD_nth(f, n, sigma, Amax);
```

We must limit vectors `f` and `PSD` to the interval $[f_L, f_H]$ of interest before passing them to Function 6.2. A typical example of a reasonable range $[f_L, f_H]$ is given by the -3 dB bandwidth of the transmitted signal. We use Function 6.5 to determine the frequency interval of interest and truncate, accordingly, vectors `f` and `PSD`. Let us adopt a -3 dB bandwidth. The command to invoke Function 6.5 is:

```
[f_th_dB,PSD_th_dB] = cp0602_thr_dB_vectors(f,PSD,-3);
```

We now invoke Function 6.2 to perform the link budget and assume the following values for the system parameters: $G_t = G_r = 1$; link margin = 5 dB; noise figure = 7 dB; $Pr_e = 10^{-3}$. The command to invoke Function 6.2 is:

```
[distance_matrix_PAM,distance_matrix_PPM] = ...
    cp0602_link_budget(1,1,5,7,f_th_dB,PSD_th_dB,1e-3);
```

The resulting plot is shown in Figure 6–4.

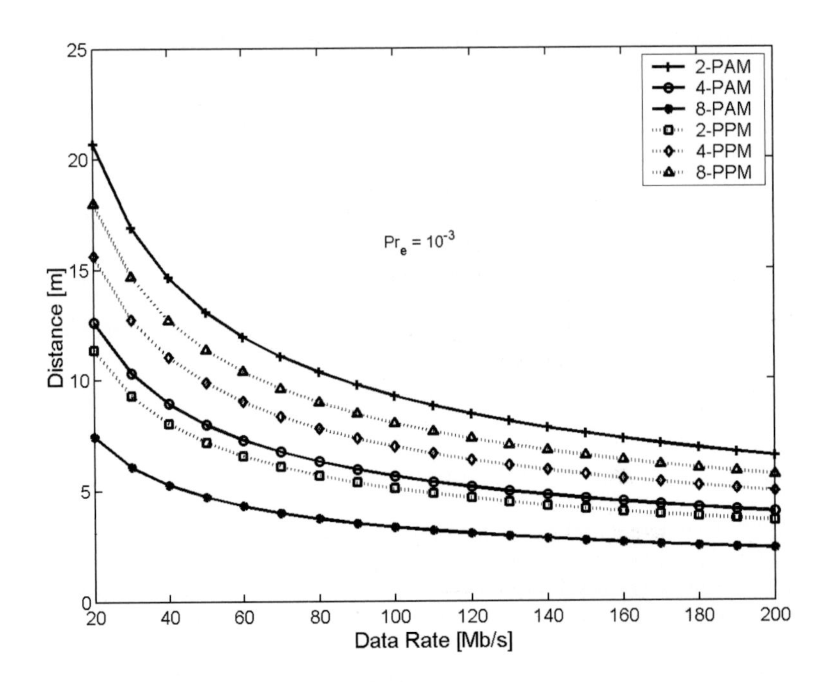

Figure 6–4 Maximum value of distance between transmitter and receiver, as a function of the data rate for M-PAM and M-PPM signals.

The above procedure is flexible regarding the input PSD. Similar curves of Figure 6–4 can be obtained by applying the same sequence of steps and providing Function 6.5 with a PSD vector containing the adopted power spectrum. Let us consider, for example, an ideal PSD that fully exploits the 3.1–10.6 GHz FCC interval. This case is of particular interest because it provides the upper limit for the distance reachable on an UWB communication link. Figures 6–5, 6–6, and 6–7 show such a limit for R_b in the intervals [1–100 Kbits/s], [1–20 Mbits/s], and [20–200 Mbits/s]. Figure 6–5, in particular, shows that, for low data rates, distances on the order of several hundreds of meters can be achieved, while for the highest data rates, as shown in Figure 6–7, communication distances are reduced to a few meters.

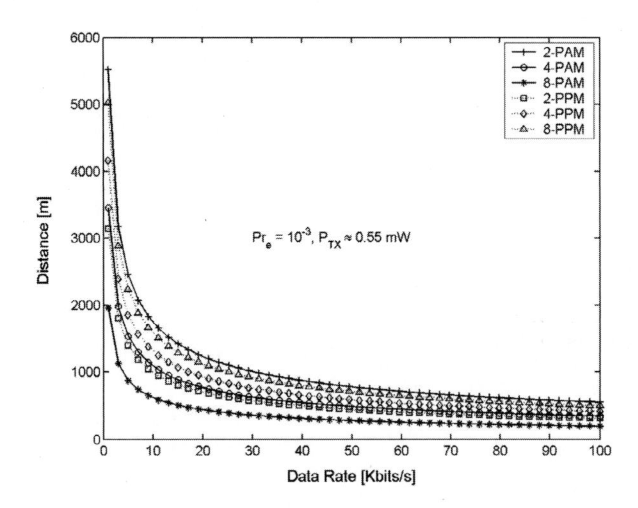

Figure 6–5 Maximum link distance as a function of the data rate for M-PAM and M-PPM signals when available power in the 3.1–10.6 GHz bandwidth is fully exploited for R_b in the range [1–100 Kbits/s].

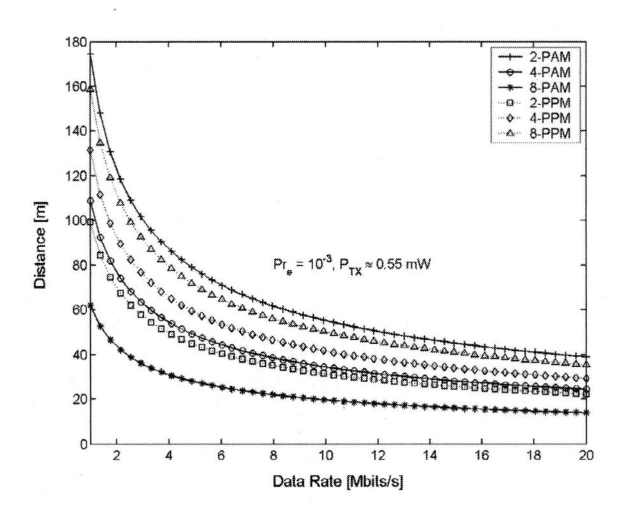

Figure 6–6 Maximum link distance, as a function of the data rate, for M-PAM and M-PPM signals, when available power in the 3.1–10.6 GHz bandwidth is fully exploited, for R_b in the range [1–20 Mbits/s].

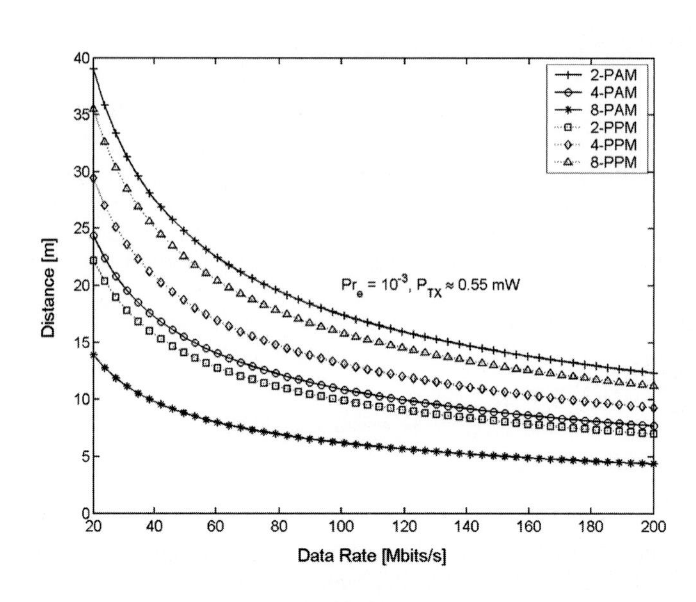

Figure 6–7 Maximum link distance, as a function of the data rate, for M-PAM and M-PPM signals, when available power in the 3.1–10.6 GHz bandwidth is fully exploited, for R_b in the range [20–200 Mbits/s].

CHECKPOINT 6–2

We finally conclude by including the case of OFDM with QAM modulation. It is possible to show that the probability of error for each sub-carrier is as follows (Lucky, Salz, and Weldon, 1968) (*OFDM with QAM modulation*):

$$\Pr_e \cong 2\left(1-\frac{1}{\sqrt{M}}\right) erfc\left\{\frac{1}{\sigma_{noise}\left(\sqrt{M}-1\right)}\right\}$$
$$\text{with} \quad \sigma^2_{noise} = \frac{\lambda^2}{2}\left(1+\frac{T_G}{T_0}\right)\frac{2N_0}{E_M} \tag{6–36}$$

where $\lambda^2 = 2\left(\sqrt{M}+1\right)/3\left(\sqrt{M}-1\right)$ is the constellation power and $E_m = (\lambda^2 T)/(2T_0)$ is the energy per symbol.

FURTHER READING

Several recently published papers focus on performance analysis for UWB systems. Most of these papers provide methods for determining the probability of error in the presence of multi-user interference and also external interference (Durisi et al., 2003; Durisi and Benedetto, 2003; Durisi et al., 2003; Hu and Beaulieu, 2003). We will refer to these papers again in Chapter 9, when we introduce the notion of multiple access and multi-user systems. Performance evaluation in the presence of timing jitter is analysed for example, in (Guvenc and Arslan, 2003) and (Lin and Chiueh, 2003). Performance for PPM modulation using biorthogonal signaling is investigated in (Clabaugh et al., 2003) and (Rouault et al. (2003).

REFERENCES

Clabaugh, D.J., M.A. Temple, R.A. Raines, and C.M. Canadeo, "UWB Multiple Access Performance Using Time Hopped Pulse Position Modulation with Biorthogonal Signaling," *IEEE Conference on Ultra Wideband Systems and Technologies* (November 2003), 330–333.

Couch II, L.W., *Digital and Analog Communication Systems*, 5th Edition, Upper Saddle River, New Jersey: Prentice-Hall International (1997).

Durisi, G., and S. Benedetto, "Performance Evaluation and Comparison of Different Modulation Schemes for UWB Multiaccess Systems," *IEEE International Conference on Communications*, Volume: 3 (May 2003), 2187–2191.

Durisi, G., A. Tarable, J. Romme, and S. Benedetto, "A General Method for Error Probability Computation of UWB Systems for Indoor Multiuser Communications," *Special Issue on Ultra-Wideband Communications of the Journal of Communications and Networks*, Volume: 5, Issue: 4 (December 2003), 354–364.

Durisi, G., J. Romme, and S. Benedetto, "A General Method for SER Computation of M-PAM and M-PPM UWB Systems for Indoor Multiuser Communications," *IEEE Global Telecommunications Conference* (December 2003), 734–738.

Federal Communications Commission, "Revision of Part 15 of the Commission's rules Regarding Ultra-Wideband Transmission Systems: First report and order," *Technical Report FCC 02-48* (adopted February 14, 2002; released April 22, 2002).

Guvenc, I., and H. Arslan, "On the Modulation Options for UWB Systems," *IEEE Military Communications Conference* (October 2003), 892–897.

Guvenc, I., and H. Arslan, "Performance Evaluation of UWB Systems in the Presence of Timing Jitter," *IEEE Conference on Ultra Wideband Systems and Technologies* (November 2003), 136–141.

Hanna, S.A., "Spectrum Management Issues Relevant to the Introduction of Ultra Wideband Technology," *IEEE Conference on Ultra Wideband Systems and Technologies* (November 2003), 468–472.

Hu, B., and N.C. Beaulieu, "Precise Bit Error Rate of TH-PPM UWB Systems in the Presence of Multiple Access Interference," *IEEE Conference on Ultra Wideband Systems and Technologies* (November 2003), 106–110.

IEEE 802.15.SG3a, "Channel modeling Sub-committee Report Final," IEEE P802.15-02/490r1-SG3a (February 2003).

Lee, E.A., and D.G. Messerschmitt, *Digital Communication*, 2nd Edition, Boston, Massachusetts: Kluwer Academic Publishers (1994).

Lin, S.C., and T.D. Chiueh, "Performance Analysis of Impulse Radio Under Timing Jitter Using M-ary Bipolar Pulse Waveform and Position Modulation," *IEEE Conference on Ultra Wideband Systems and Technologies* (November 2003), 121–125.

Lucky, R.W., J. Salz, and E.J. Weldon Jr., *Principles of Data Communication*, New York: McGraw-Hill (1968).

Proakis, J.G., *Digital Communications*, 3rd Edition, New York: McGraw-Hill International Editions (1995).

Rouault, L., S. Chaillou, and D. Helal, "A Priori Probability Calculation for UWB Systems using Correlated PPM and Polarity Modulation," *IEEE Conference on Ultra Wideband Systems and Technologies* (November 2003), 408–412.

Sheng, H., P. Orlik, A.M. Haimovich, L.J. Cimini Jr., and J. Zhang, "On the Spectral and Power Requirements for Ultra Wideband Transmission," *IEEE International Conference on Communications*, Volume: 1 (May 2003), 738–742.

Sørensen, S. B., *ETSI UWB Activities*, UWB Colloquium, London, UK, Available at *http://www.radio.gov.uk/topics/uwb/etsi-uwb_activities.pdf* (July 2002).

APPENDIX 6.A

Function 6.1 Pr_e for M-PAM and M-PPM

```
%
% FUNCTION 6.1 : "cp0602_symbol_error_probability"
%
% Analysis of the symbol error probability for M-PAM and
% M-PPM signals
%
% The function computes and plots the symbol error
% probability for M-PAM and M-PPM signals
% for M = {2, 4, 8} as a function of Eb/N0
%
% Programmed by Luca De Nardis

function cp0602_symbol_error_probability

% --------------------------
% Step Zero - Initialization
% --------------------------

Eb_N0 = logspace(0.5,2);        % Range of Eb/N0: 5-20 dB
M = [2 4 8];                    % Values of M

% ---------------------------------------------------
% Step One - Evaluation of symbol error probability
% ---------------------------------------------------

for i=1:3
    M_PAM(i,:)=(1-1/M(i))*erfc(sqrt(Eb_N0 * 3 *...
       log2(M(i))/(M(i)^2-1)));
    M_PPM(i,:)= exp(-0.5 * log2(M(i))*(Eb_N0 - 2 *...
       log(2)));
end

% --------------------------
% Step Two - Graphical output
% --------------------------
```

```
M_ary_BEP = [M_PAM' M_PPM']'  % Building a single matrix
                              % for all vectors
F=figure(1);
set(F,'Position',[100 190 650 450]);
set(gcf,'DefaultAxesColorOrder',[0 0 0],...
   'DefaultAxesLineStyleOrder','-+|-o|-*|:s|:d|:^');
PT=semilogy(10*log10(Eb_N0),M_ary_BEP);
set(PT,'LineWidth',[2]);
hold on
L= legend('2-PAM','4-PAM','8-PAM','2-PPM','4-PPM','8-PPM');
set(L,'FontSize',12);
set(PT,'LineWidth',[2]);
X=xlabel('E_b/N_0');
set(X,'FontSize',14);
Y=ylabel('Pr_e');
set(Y,'FontSize',14);
AX=gca;
set(AX,'FontSize',12);
axis([5 20 1e-6 1e-1]);
set(AX,'XTick', [5 7.5 10 12.5 15 17.5 20 ]);
grid on
```

Function 6.2 Link Budget

Function 6.2 first evaluates the required E_b/N_0 for each value of M, and then invokes Function 6.3 for each pair (M, R_b) .

```
%
% FUNCTION 6.2 : "cp0602_link_budget"
%
% Link budget for M-PAM and M-PPM UWB signals
%
% The function performs the link budget for an M-PAM and an
% M-PPM UWB signal for M = {2, 4, 8} and plots the maximum
% distance between transmitter and receiver as a function
% of the data rate
%
% The function receives as input the antenna gains 'Gt' and
% 'Gr', the link margin in dB 'MargindB', the noise factor
% in dB 'FdB', the frequency vector 'f' in the range of
% interest [fl, fh], the transmitted PSD in the same
% frequency range 'PSD', and the requested symbol error
% probability 'Pre'
%
% The function returns as output a matrix containing the
% maximum distance for each value of M as a function of
% rate Rb
%
% Programmed by Luca De Nardis

function [dist_PAM, dist_PPM] =cp0602_link_budget(Gt,...
   Gr, MargindB, FdB, f, PSD, Pre)

% -------------------------
% Step Zero - Initialization
% -------------------------

dist=zeros(3,19);
Rb=linspace(20e6,200e6,19);  % values of the bit rate
M = [2 4 8];                 % values of M
F = 10^(FdB/10);             % conversion of the noise
                             % factor to linear units
Margin = 10^(MargindB/10);   % conversion of the link
                             % margin to linear units
k = 1.38e-23;                % Boltzmann constant [J/K]
T0 = 300;                    % standard temperature [K]
```

```
N0 = F*k*T0;                    % noise unilateral PSD [W/Hz]

% ------------------------------------------------
% Step One - Evaluation of maximum distance
% ------------------------------------------------

Eb_N0 = logspace(0.5,2);        % range of Eb/N0: 5-20 dB

for i=1:3
    % Evaluation of the EB/N0 for the two cases, M-PAM and
    %  M-PPM
    % from Eqs.(6.31) and (6.33)
    Pr_M_PAM = 1;
    Pr_M_PPM = 1;
    j=1;
    while Pr_M_PAM > Pre
        Pr_M_PAM =(1-1/M(i))*erfc(sqrt(Eb_N0(j) * 3 *...
           log2(M(i))/(M(i)^2-1)));
        Eb_N0_PAM = Eb_N0(j);
        j=j+1;
    end
    j=1;
    while Pr_M_PPM > Pre
        Pr_M_PPM = exp(-0.5 * log2(M(i))*(Eb_N0(j) - 2 *...
           log(2)));
        Eb_N0_PPM = Eb_N0(j);
        j=j+1;
    end

    for k=1:19
        % Invoking Function 6.3
        dist_PAM(i,k) = cp0602_max_distance(Gt, Gr, f,...
           PSD,Margin, 1/Rb(k), N0, Eb_N0_PAM);
        % Invoking Function 6.3
        dist_PPM(i,k) = cp0602_max_distance(Gt, Gr, f,...
           PSD, Margin, 1/Rb(k), N0, Eb_N0_PPM);
    end
end

% --------------------------
% Step Two - Graphical output
% --------------------------

% Building a single matrix for all vectors
M_ary_Distance = [dist_PAM' dist_PPM']'
F=figure(2);
```

```
set(gcf,'DefaultAxesColorOrder',[0 0 0],...
   'DefaultAxesLineStyleOrder','-+|-o|-*|:s|:d|:^');
PT = plot(Rb/1e6,M_ary_Distance);
AX=gca;
set(AX,'FontSize',12);
X=xlabel('Data Rate [Mb/s]');
set(X,'FontSize',14);
Y=ylabel('Distance [m]');
set(Y,'FontSize',14);
axis([20 200 0 25]);
L= legend('2-PAM','4-PAM','8-PAM','2-PPM','4-PPM','8-PPM');
set(L,'FontSize',12);
text(100,15,'Pr_e = 10^{-3}','BackgroundColor',...
   [1 1 1],'FontSize',12);
```

Function 6.3 Evaluation of the Maximum Distance

Function 6.3 evaluates the integral in Eq. (6–27) over a frequency range determined by the input vector `f`. As a consequence, the input vector `f` must be limited to the range $[f_L,f_H]$ of interest, and the input vector `PSD` must take values only in this range.

```
%
% FUNCTION 6.3 : "cp0602_max_distance"
%
% Maximum distance between transmitter and receiver for a
% M-PAM UWB signal as a function of the signal PSD and the
% path loss, based on Eq. (6-27)
%
% The function receives as input the antenna gains 'Gt' and
% 'Gr', the frequency vector 'f' in the range of interest
% [fL, fH], the transmitted PSD in the same frequency range
% 'PSD', the link margin in linear units 'Margin', the bit
% period 'Tb', the noise unilateral PSD N0, and the Eb/N0
% value 'Eb_N0'
%
% The function returns as output the maximum distance
% between transmitter and receiver for the given rate and
% Eb/N0
%
% Programmed by Luca De Nardis

function D = cp0602_max_distance(Gt, Gr, f, PSD,...
   Margin, Tb, N0, Eb_N0)


c = 3e8;                                 % Light speed [m/s]
free_space_term =  (Gt * Gr * c^2) / (16 * pi^2);
frequency_term = sum(PSD./(f.^2))*(f(length(f))-...
   f(1))/length(f);
D = sqrt((free_space_term * frequency_term * Tb) /...
   (Margin * Eb_N0* N0));
```

Function 6.4 Normalized PSD for Gaussian Pulses

```
%
% FUNCTION 6.4 : "cp0602_Gaussian_PSD_nth"
%
% Generates the normalized PSD of the n-th derivative of
% the Gaussian pulse with variance sigma^2 as defined in:
%
%     Sheng, H., P. Orlik, A.M. Haimovich, L.J. Cimini Jr.,
%     and J. Zhang, "On the Spectral and Power
%     Requirements for Ultra Wideband Transmission,"
%     IEEE International Conference on Communications,
%     Volume: 1 (May 2003), 738-742.
%
% The function receives as input:
% a vector representing the frequency axis 'frequency',
% the order of the derivative 'n',
% the square root of the variance 'sigma',
% and the peak value of the PSD 'Amax'
%
% The function returns the PSD evaluated on the frequency
% range determined by 'frequency'
%
% Programmed by Luca De Nardis

function PSD = cp0602_Gaussian_PSD_nth(frequency, n,...
   sigma, Amax)
PSD = Amax * (1 / ((n^n) * exp(-n))) * [(2 * pi *...
   frequency * sigma) .^ (2 * n)] .* exp(-(2 * pi *...
   frequency * sigma) .^ 2);
```

Function 6.5 Bandwidth Evaluation and PSD Truncation

```
%
% FUNCTION 6.5 : "cp0602_thr_dB_vectors"
%
% Evaluates the bandwidth of the input 'PSD' with values in
% the frequency range given by the frequency vector 'f'
% Bandwidth is evaluated according to the given 'threshold'
% (in dB)
%
% The function returns:
% a truncated frequency vector 'freq_th_dB' corresponding
% to the bandwidth at -'threshold' dB of the input PSD,
% a truncated PSD vector 'PSD_th_dB' containing the
% corresponding components of the input PSD
% 'BW', and the -'threshold' dB bandwidth
% 'f_high' the higher limit of the bandwidth
% 'f_low' the lower limit of the bandwidth
%
%
% Programmed by Guerino Giancola / Luca De Nardis

function [f_th_dB, PSD_th_dB] = cp0602_thr_dB_vectors(f,...
   PSD, threshold)

N = length(f);
df = f(length(f))/length(f);

% -------------------------------------------------
% Step One - Evaluation of the frequency bandwidth
% -------------------------------------------------

[Ppeak,index] = max(PSD);  % Ppeak is the peak value of the
                           % PSD
f_peak = index * df;       % peak frequency

Pth = Ppeak*10^(threshold/10); % Pth is the value of the
                               % PSD corresponding to the
                               % given threshold

% iterative algorithm for evaluating high and low
% frequencies
```

```
imax = index;
P0h = PSD(index);

while (P0h>Pth)&(imax<=N)
    imax = imax + 1;
    P0h = PSD(imax);
end

f_high = (imax-1) * df;              % high frequency

imin = index;
P0l = PSD(index);

while (P0l>Pth)&(imin>1)&(index>1)
    imin = imin - 1;
    P0l = PSD(imin);
end

f_low = (min(index,imin)-1) * df;    % low frequency

% end of iterative algorithm

BW = f_high - f_low;            % signal frequency bandwidth

% -----------------------------------------------------
% Step Two - Cutting the vectors in the -'threshold' dB
%            bandwidth
% -----------------------------------------------------

f_th_dB = f(imin:imax);
PSD_th_dB = PSD(imin:imax);
```

APPENDIX 6.B

This appendix includes the outdoor UWB emission mask as set by (FCC, 2002). The outdoor FCC mask is reported in Figure 6B–1.

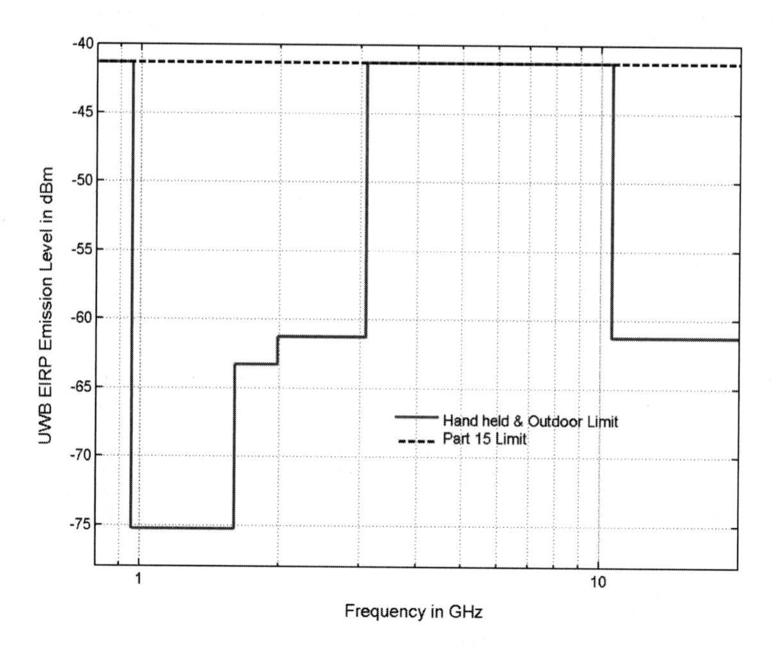

Figure 6B–1 FCC emission mask for UWB handheld and outdoor devices (FCC, 2002).

CHAPTER 7

The Pulse Shaper

The choice of the impulse response of the pulse shaper filter is crucial since it affects the PSD of the transmitted signal. In this chapter, we will first analyze the effect of pulse width variation and differentiation on pulse shape. We then investigate the effect of combining different waveforms to form a pulse that complies with power limitations as set by emission masks defined in Chapter 6, "Performance Analysis of the UWB Radio Link."

7.1 THE PULSE

The choice of the impulse response of the pulse shaper filter is crucial since it affects the PSD of the transmitted signal (Di Benedetto and Vojcic, 2003).

Often, the term "monocycle" is found instead of "pulse" because the pulse-modulated sinewave used in conventional radars is formed by one cycle of a sinewave, or several cycles, in which case it is called a polycycle.

Although often called a monocycle, the adopted pulse in UWB communication systems is rarely a cycle of a sine wave. It is in fact less difficult and less expensive to generate non-sinusoidal pulses than pulse-modulated sinewaves. Generating pulses of duration on the order of a nanosecond with an inexpensive technology (CMOS chips) has become possible after UWB Large Current Radiator (LCR) antennas were introduced by Harmuth (1990). The LCR antenna is driven by a current, and the antenna radiates a power that is proportional to the square of the derivative of current. When a step function current, for example, is applied to the antenna, a pulse is generated: The steeper the step function current, the narrower the generated pulse (Kardo-Sysoev, 2003).

The pulse shape that can be generated in the easiest way by a pulse generator actually has a bell shape such as a Gaussian. A Gaussian pulse *p(t)* can be described by the following expression:

$$p(t) = \pm \frac{1}{\sqrt{2\pi\sigma^2}} e^{-\left(\frac{t^2}{2\sigma^2}\right)} = \pm \frac{\sqrt{2}}{\alpha} e^{-\frac{2\pi t^2}{\alpha^2}} \tag{7–1}$$

where $\alpha^2 = 4\pi\sigma^2$ is the shape factor and σ^2 is the variance.

The waveform and corresponding ESD of the pulse of Eq. (7–1) with a minus sign are shown in Figure 7–1.

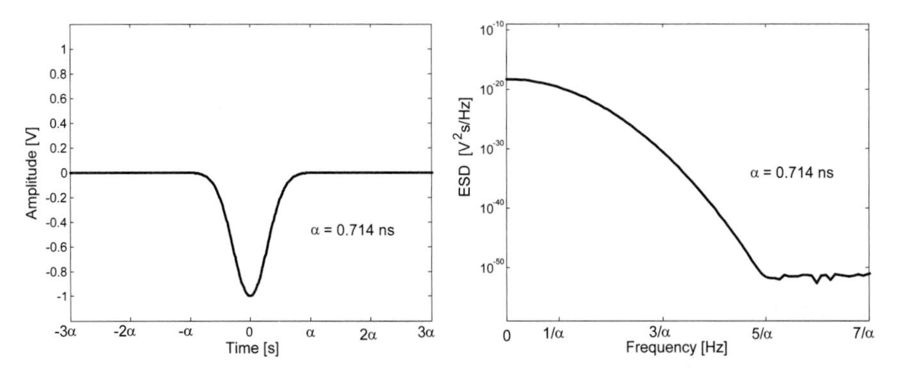

Figure 7–1 Example of a Gaussian pulse: (left) waveform, and (right) corresponding unilateral ESD (by permission from *Journal of Communications and Networks*, ISSN 1229-2370, © Korean Institute of Communication Sciences).

To be radiated in an efficient way, however, a basic feature of the pulse is to have a zero dc (direct current) offset. Several pulse waveforms might be considered, provided that this condition is verified. Gaussian derivatives are suitable. Actually, the most currently adopted pulse shape is modeled as the second derivative of a Gaussian function (Win and Scholtz, 2000), described by:

$$\frac{d^2 p(t)}{dt^2} = \left(1 - 4\pi \frac{t^2}{\alpha^2}\right) e^{-\frac{2\pi t^2}{\alpha^2}} \tag{7–2}$$

Note that this pulse has energy $3\alpha/8$ V^2s.

The second derivative Gaussian pulse of Eq. (7–2) is usually referred to as the pulse at the receiver, that is, after passing through the transmitter and receiver antennas. Ideally, a second derivative Gaussian pulse can be obtained at the output of the transmitting antenna if the antenna is fed with a current pulse shaped as the first derivative of a Gaussian waveform (and thus zero dc current), the radiating pulse being proportional to the derivative of the drive current in an ideal antenna (Immoreev and Sinyavin, 2002). One should note that for the reciprocity theorem, the behavior of the receiving antenna should not be assimilated to the transmitter one. In particular, when receiving, the antenna does not act as a derivator on the incoming signal, but rather presents a flat frequency response (Sheers et al., 2000).

Other pulse shapes have also been proposed such as the Laplacian (Conroy et al., 1999), compositions of Gaussian pulses having same length and reversed amplitudes with a fixed time gap between the pulses (Hämäläinen et al., 2001), and Hermite pulses (Ghavami et al., 2002).

Shaping the spectrum by changing the pulse waveform is an interesting feature of IR. Basically, the spectrum may be shaped in three different ways: pulse width variation, pulse differentiation, and a combination of base functions.

In the following, we will consider the Gaussian pulse as a case study. The Gaussian pulse is well suited for our analysis, since its shape can be modified in a straightforward way by modifying the shape factor α, and infinite waveforms can be obtained by differentiating the original pulse.

In Section 7.2, we will analyze the effect of pulse width variation and differentiation on pulse shape and corresponding ESD characteristics. In Section 7.3, we will analyze the possibility of combining different Gaussian waveforms, such as different derivatives and different pulse widths, for the purpose of approximating emission masks such as those defined in Chapter 6.

7.2 PULSE WIDTH VARIATION AND PULSE DIFFERENTIATION

Pulse width is tightly related to the shape factor α. Reducing the value of α shortens the pulse, and thus enlarges the bandwidth of the transmitted signal. As a consequence, the same waveform can be used to occupy different bandwidths by adjusting the value of the shape factor.

CHECKPOINT 7–1

In this checkpoint, we will evaluate the effect of the shape factor α on pulse waveform and corresponding ESD.

Function 7.1 (Appendix A.7) generates a vector representing the Gaussian pulse waveform of Eq. (7–1) for different α values and evaluates the corresponding ESD. Finally, Function 7.1 plots the waveform and ESD for each of the considered α values. The function receives the following as input: the minimum α value `alphamin`, the increment step `alphastep`, and the number of α values `N_alphavalues` to be analyzed. The function does not return any output.

The command line to invoke Function 7.1 is:

```
cp0701_shape_factor_variation(alphamin,...
   alphastep, N_alphavalues);
```

The output of Function 7.1 is presented in Figure 7–2 for the values `alphamin = 0.414e-9`, `alphastep = 0.1e-9`, and `N_alphavalues = 7` corresponding to α varying in the range 0.414 ns - 1.014 ns.

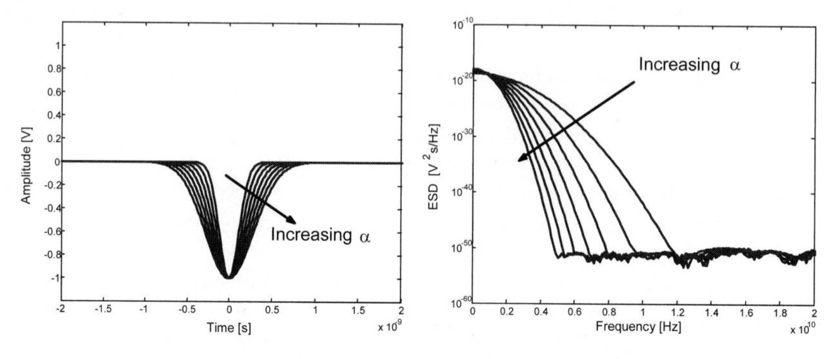

Figure 7–2 The Gaussian pulse case: effect of α on (left) pulse duration, and (right) ESD (by permission from *Journal of Communications and Networks*, ISSN 1229-2370, © Korean Institute of Communication Sciences).

CHECKPOINT 7–1

Note that the Gaussian pulse has infinite duration leading to unavoidable overlap between pulses and ISI. It is reasonable, however, to consider for the Gaussian pulse a limited duration T_M as defined by limiting the cutout energy below a given threshold. Under this assumption, an upper limit for α is given by pulse duration T_M, which cannot exceed chip duration T_c, while a lower limit is given by technological limitations in generating extremely short pulses.

Differentiation of the Gaussian pulse influences the ESD as well; both peak frequency and bandwidth of the pulse vary with increasing differentiation order. In particular, it is possible to find a general relationship between the peak frequency f_{peak}, the order of differentiation k, and the shape factor α by observing that the Fourier transform of the k-th derivative has the property:

$$X_k^{'}(f) \propto f^k e^{-\frac{\pi f^2 \alpha^2}{2}} \tag{7–3}$$

which leads to a peak frequency for the k-th derivative:

$$f_{peak,k} = \sqrt{k}\,\frac{1}{\alpha\sqrt{\pi}} \tag{7–4}$$

Equation (7–4) shows that Gaussian derivatives of higher order are characterized by higher peak frequencies. Differentiation is thus a way to move energy to higher frequency bands. Figure 7–3 shows the behavior of peak frequencies for the first 15 derivatives of the Gaussian pulse as a function of the shape factor α, based on the computation of Eq. (7–4).

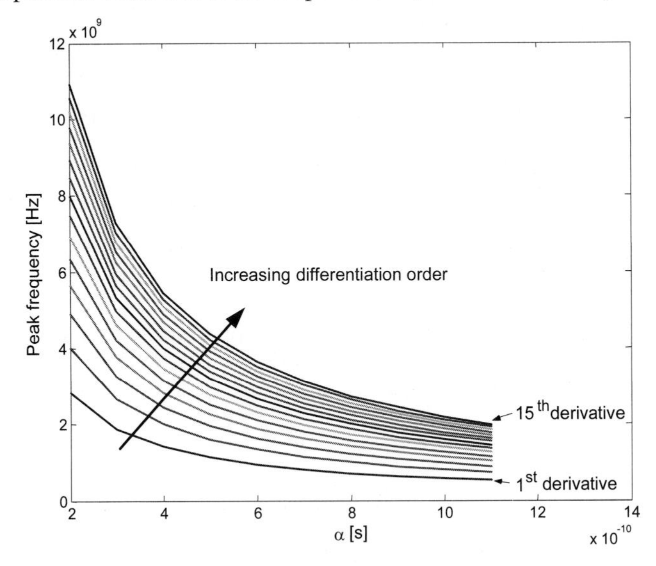

Figure 7–3 Variation of peak frequency with α for the first 15 derivatives of the Gaussian pulse of Eq. (7–1), according to Eq. (7–4) (by permission from *Journal of Communications and Networks*, ISSN 1229-2370, © Korean Institute of Communication Sciences).

CHECKPOINT 7–2

In this checkpoint, we will evaluate the effect of differentiation on pulse waveform and ESD. As already stated, the Gaussian pulse can be differentiated infinite times. In the following, we will limit the analysis to the first 15 derivatives.

To analyze the effect of differentiation in the time domain, we introduce two MATLAB functions: Function 7.2, which plots the waveforms of the Gaussian pulse and its 15 derivatives, and Function 7.3, which generates the analytical expressions of all the waveforms to be plotted.

Function 7.2 (see Appendix 7.A) first normalizes in amplitude the waveforms of the Gaussian pulse and its first 15 derivatives and then plots them together. The function receives in input the shape factor α through the parameter `alpha` and does not return any output. The command line to invoke Function 7.2 is:

```
cp0702_Gaussian_derivatives(alpha);
```

Function 7.3 (see Appendix 7.A) generates the first 15 derivatives of a Gaussian pulse. The function receives the following as input: a vector `t` defining the time axis, the order `k` of the derivative of the Gaussian pulse in the range [1,15], and the value of the shape factor `alpha`. Function 7.3 returns vector `deriv`, representing the derivative of order `k` of the Gaussian pulse calculated over the time axis `t`. The command line to invoke Function 7.3 is:

```
[deriv] = cp0702_analytical_waveforms(t, k, alpha);
```

Figure 7–4 shows the output of Function 7.2 for $\alpha = 0.714$ ns.

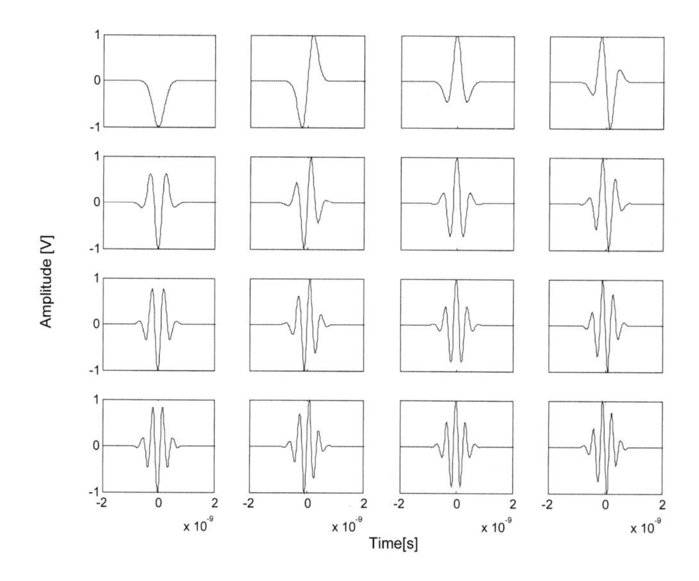

Figure 7–4 Waveforms of the Gaussian pulse and its first 15 derivatives.

The spectral analysis of the effect of pulse differentiation is performed by introducing new MATLAB functions.

Function 7.4 (see Appendix 7.A) evaluates and plots the ESD of the first 15 derivatives of the Gaussian pulse. The function receives the shape factor `alpha` as input and does not return any output.

Function 7.4 is invoked as follows:

```
cp0702_Gaussian_derivatives_ESD(alpha);
```

The output of the function is presented in Figure 7–5 for $\alpha = 0.714$ ns.

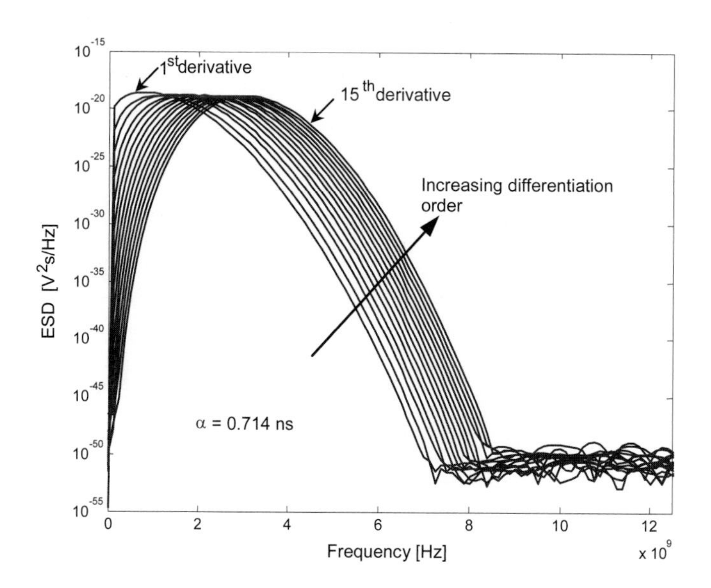

Figure 7–5 ESD of the first 15 derivatives of the Gaussian pulse (by permission from *Journal of Communications and Networks*, ISSN 1229-2370, © Korean Institute of Communication Sciences).

As predicted by Eq. (7–4), simulations show that higher derivatives are characterized by a shift of ESD to higher frequencies.

Function 7.5 (see Appendix 7.A) numerically determines the peak frequency for each derivative as a function of α. The function receives the following as input: the minimum α value `alphamin`, the increment step `alphastep`, and the number of α values `N_alphavalues` to be analyzed. The function does not return any output.

The command line to invoke Function 7.5 is:

```
cp0702_Gaussian_derivatives_peak_frequency...
   (alphamin, alphastep, N_alphavalues);
```

The peak frequency for the first 15 derivatives as determined by Function 7.5 is presented in Figure 7–6 for α ranging between 0.2 ns and 1.1 ns, that is, for the same values used for the analytical results in Figure 7–3. Note that Figures 7–3 and 7–6, which represent theoretical vs. simulated peak frequencies, are extremely close.

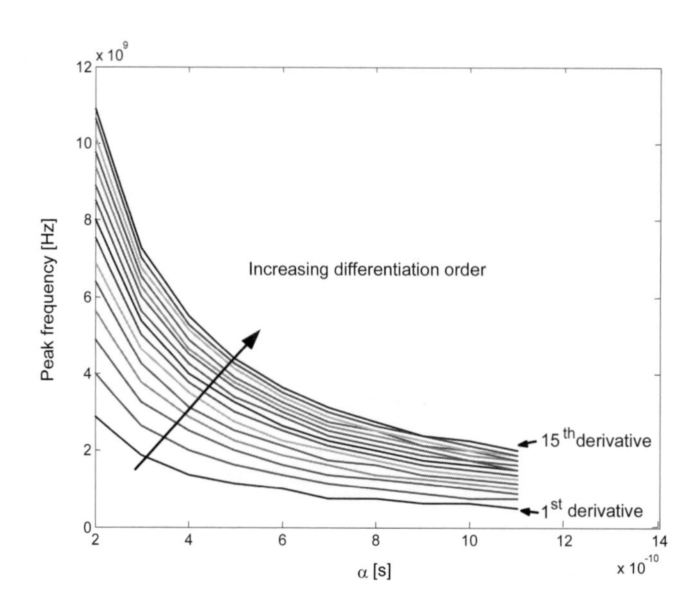

Figure 7–6 Variation of peak frequency with α for the first 15 derivatives of the Gaussian pulse.

In Chapter 1, "Ultra Wide Band Radio Definition," the definition of -10 dB bandwidth for UWB signals was introduced. The relation between differentiation order and -10 dB bandwidth is analyzed by means of Functions 7.6 and 7.7.

Function 7.6 evaluates and plots the -10 bandwidth of the first 15 derivatives of the Gaussian pulse. The function receives the following as input: the minimum α value `alphamin`, the increment step `alphastep`, and the number of α values `N_alphavalues` to be analyzed. The function does not return any output. Function 7.6 is called as follows:

```
cp0702_Gaussian_derivatives_10dB_bandwidth(alphamin,...
   alphastep, N_alphavalues);
```

Function 7.7 evaluates the bandwidth of a generic input signal according to a given threshold. The function receives vector `signal` as input representing the waveform to be analyzed, the value of the sampling period `dt`, and the value in dB of the threshold to be considered `threshold`. Function 7.7 returns the single-sided ESD `ss_ESD`, the bandwidth `BW`, and the values of the high and low frequencies `f_high` and `f_low`, respectively. Function 7.7 is called as follows:

```
[Ess,f_high,f_low,BW] = ...
   cp0702_bandwidth(signal,dt,threshold);
```

Figure 7–7 shows the variation of the -10 dB bandwidth for α ranging between 0.2 ns and 1.1 ns for the set of 15 derivatives of the Gaussian pulse.

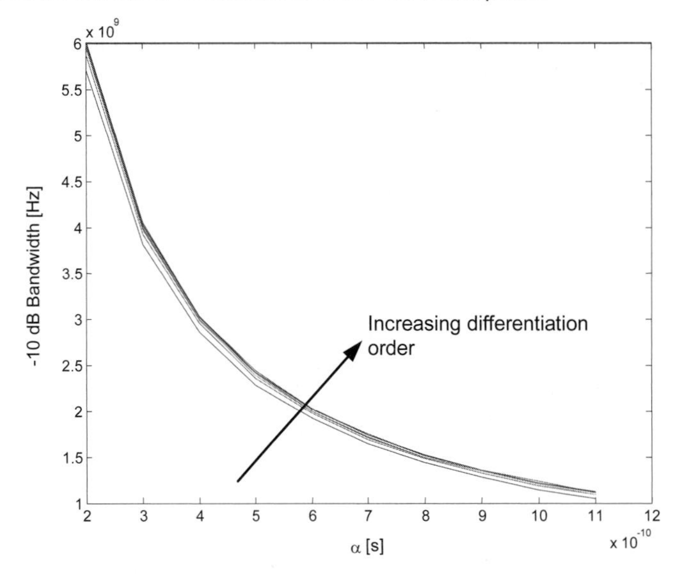

Figure 7–7 Variation of -10 dB bandwidth with α for the first 15 derivatives of the Gaussian pulse.

The effect of differentiation on the signal ESD is then to move the energy at higher frequencies, while slightly increasing the signal bandwidth.

CHECKPOINT 7–2

7.3 Meeting the Emission Masks

It was shown in the previous section that both differentiation and pulse width variation affect the ESD of the Gaussian pulse and can be used to shape the transmitted waveform. In most cases, however, the flexibility in shaping the spectrum guaranteed by a single waveform may not be sufficient to fulfill specific requirements.

Meeting emission masks set by regulation authorities is a typical task demanded of the pulse shaper. In particular, the release of the FCC emission masks for UWB devices, introduced in Chapter 6 and reported in Table 6–1 and Figure 6–1 for the indoor case, stimulated research in designing pulse shaping techniques capable of closely approximating the masks, that is, with the maximum possible transmit power under FCC limits.

In this section, we will analyze the possibility of tuning the ESD of a generated pulse by combining a few single reference pulse waveforms — for example, the Gaussian pulse and its derivatives — to adjust the ESD to the masks. Note that by doing so, we implicitly assume that the generated waveform can be assimilated to one pulse instead of several. This is valid in first approximations as shown in Chapters 3, "The PSD pf TH-UWB Signals," and 4, "The PSD of DS-UWB Signals." Moreover, the validity of the approach is reinforced by results presented in Checkpoint 3–1, Figure 3–36, which shows that despite modulation and coding the masks are always met provided that this holds true for the single pulse.

(Parr et al., 2003) proposed an algorithm for generating a family of pulses capable of meeting the FCC masks and suitable for multi-user scenarios, thanks to the short time duration and orthogonality properties between different pulses. The performance of PPM-TH systems adopting this pulse family was analyzed in (Bin et al., 2003).

In a recent solution based on Gaussian waveforms, (Sheng et al., 2003) proposed an algorithm to select the best pulse differentiation order and shape factor value for fitting the mask in the bandwidth 3.1–10.6 GHz. A limitation of this approach is in the difficulty of meeting the mask outside the above bandwidth by using a single derivative.

In this section, we will therefore investigate the possibility of obtaining the optimal waveform as a combination of different derivative functions of the Gaussian pulse, to approximate the mask at all frequencies, including the 0–0.96 GHz band.

A possible approach is to use linear combinations of N Gaussian derivatives, each being characterized by a given α value (different derivatives may have different α values), which can be thought of as independent Base Functions (BFs) in a space of N dimensions. The choice of the coefficients of the linear combination must be made depending on a design objective such as meeting a given PSD.

A procedure for selecting the coefficients can be described as follows:

1. Choose a set of BFs.
2. Generate in a random way a set of coefficients, named S.
3. Check if the PSD of the linear combination of the functions obtained with coefficients S satisfies the emission limits.
4. If the emission limits in Step 3 are met and this is the first set S verifying the limits, then initialize the procedure by setting SB = S. If the emission limits in Step 3 are met and the procedure was already initialized, then compare S with SB; if S leads to a better waveform than SB according to well-defined distance metrics, set SB = S.
5. Repeat Steps 1–3 until the distance between the mask and PSD of the generated waveform falls below a fixed threshold.

Note that the combination of N derivatives and the possibility of choosing different α values for different derivatives provides a high degree of flexibility in the generation of pulse waveforms. The algorithm may, however, require a high number of iterations before converging to a solution that is capable of guaranteeing the requested distance between the synthetic and reference target.

CHECKPOINT 7–3

In this checkpoint, we will analyze the problem of the approximation of emission masks with combinations of Gaussian derivatives. We will combine the first 15 derivatives and perform the approximation through two approaches: in the first case, all derivatives have the same shape factor α, while in second case, different derivatives adopt different α values. In the following, we will identify a set of α values for the 15 derivative as an "α vector."

To select the coefficients of the combination of the Gaussian derivatives, we introduce a new MATLAB function, Function 7.8. This function implements the random coefficient selection procedure described in Section 7.3 by using three subsidiary MATLAB functions: Functions 7.9, 7.10, and 7.11.

Function 7.8 implements the procedure described in Section 7.3 in a simplified form, that is, by fixing the number of iterations to be performed. It is assumed that different α vectors are available for the selection process. The α vector, which must be considered on each run of the code is specified as input through the parameter `i`. In addition, Function 7.8 receives the value of the pulse repetition period `Ts` as input, and the maximum allowed number of iterations `attempts`. The command to invoke Function 7.8 is:

```
cp0703_random_pulse_combination(i,Ts,attempts);
```

Function 7.9 provides the α values required for generating the derivatives within Function 7.8. The function receives the index `i` as input, indicating the set of α values to be considered. This set is returned in the output vector `alphavector`. Function 7.9 is executed as follows:

```
alphavector = cp0703_get_alpha_value(i);
```

Function 7.10 generates the emission mask. It receives as input the number of points `N` to be used in the frequency domain to represent signals in the interval [-fs/2, fs/2], and returns a vector of `N`/2 elements representing the emission mask in the frequency range [0, fs/2]. The reference emission mask is generated with the following command:

```
[emissionmask] = cp0703_emission_mask(N,fs);
```

Function 7.11 selects the coefficients for the combination of the derivatives. It receives the following as input: the number of iterations to be performed `attempts`, the set of base functions `basefunction`, the values of the sampling period `dt`, the number of samples in the time domain `smp`, the pulse repetition period `Ts`, the frequency smoothing factor `freqsmoothfactor`, the target mask `emissionmask`, and two integers `lowerbasefunction` and `higherbasefunction`, representing the range of derivatives to be used in the combination. Function 7.11 returns vector `c`, containing the best set of coefficients, and the flag `result`, which indicates if at least one valid set of coefficients was discovered by the function.

Function 7.11 is invoked with the following command:

```
[c,result] = cp0703_random_coefficients(attempts,...
   basefunction, dt, smp, Ts, freqsmoothfactor,...
   emissionmask, lowerbasefunction, higherbasefunction);
```

For example, let us consider the first 15 derivatives of the Gaussian pulse. Figure 7–8 shows the PSD of a waveform obtained by linear combination of the above BFs plotted against the FCC emission mask, according to the algorithm proposed in this section, for the case of the first set of α values stored in Function 7.9 (invoked with input parameter i = 1), which corresponds to α values all equal to 0.714 ns, `Ts = 1e-7` and `attempts = 100`.

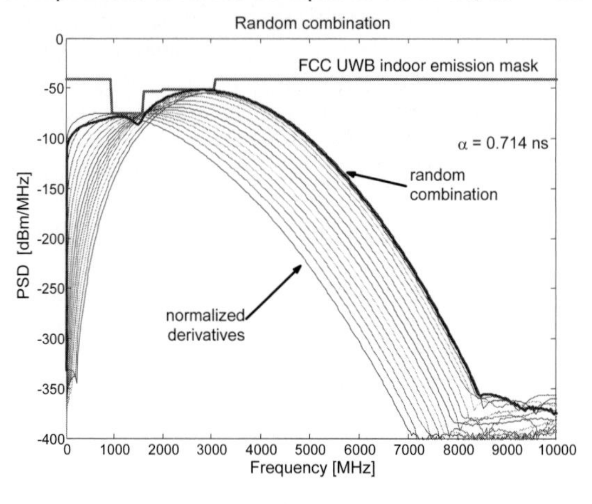

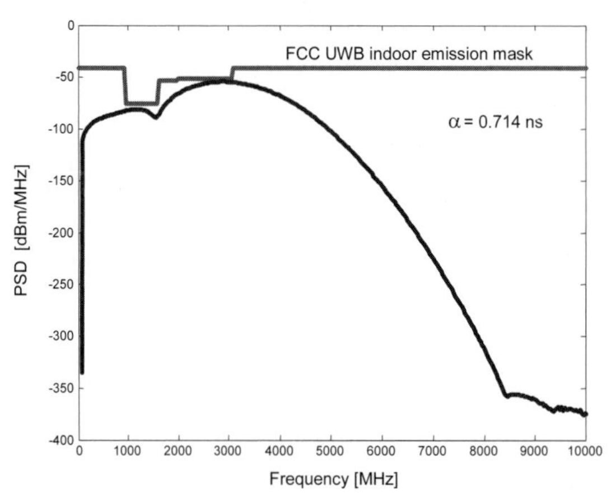

Figure 7–8 PSD of the base functions (upper plot) and of the combined waveform (lower plot) with α = 0.714 ns for all derivatives (by permission from *Journal of Communications and Networks*, ISSN 1229-2370, © Korean Institute of Communication Sciences).

Figure 7–8 shows that the combination of several BFs leads to a good approximation of the emission mask with the adopted α, in particular in the band 0.96 GHz–3.6 GHz. Outside this band, power is less efficiently used.

The adoption of a different α vector can improve performance. The second set of α values included in Function 7.9 is characterized by a high value of α (1.5 ns) for the first derivative and smaller values (0.314 ns) for the other derivatives. The user can select the second set by invoking Function 7.8 with the input parameter `i` set to 2. This alternative set leads to the PSD shown in Figure 7–9. Note the relatively narrow shape of the first derivative on Figure 7–9 (curve with the first maximum at the very left of the plot) due to the high α value.

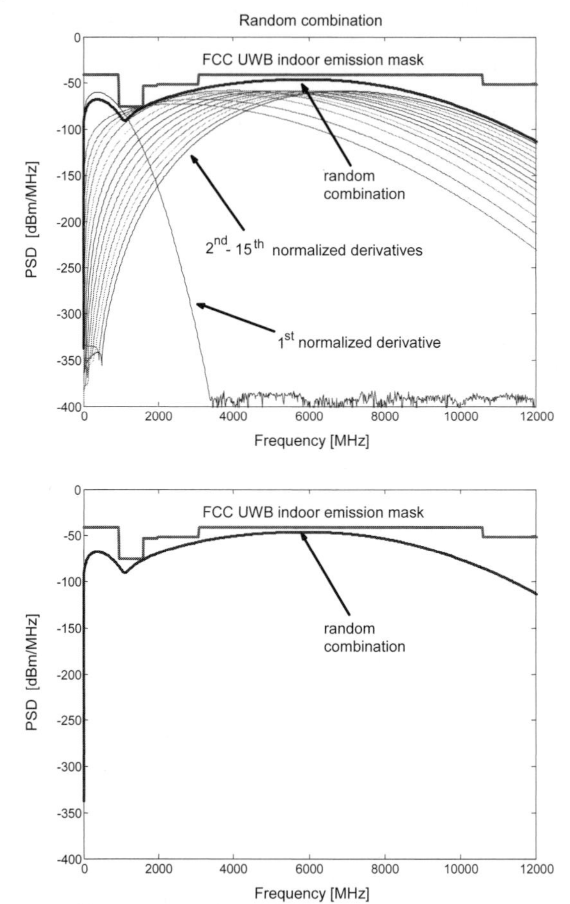

Figure 7–9 PSD of the base functions (upper plot) and of the combined waveform (lower plot) with α = 1.5 ns for the first derivative and α = 0.314 ns for higher derivatives (by permission from *Journal of Communications and Networks*, ISSN 1229-2370, © Korean Institute of Communication Sciences).

The second α set provides a better approximation of the mask than the first, thanks to the larger bandwidth of the higher derivatives. Note that the selection of a relatively high α for the first derivative improves efficiency of power utilization in the low frequency band. An upper bound for α is, however, given by waveform duration, and determined by the chip time T_c (see Chapter 2, "The UWB Radio Signal").

CHECKPOINT 7–3

Random selection is only one of the possible strategies for the set of coefficients in the linear combination. A more systematic way of selecting such coefficients is to apply standard procedures for error minimization such as the Least Square Error (LSE), in which the following error function must be minimized:

$$e_s(t) = \int_{-\infty}^{+\infty} |e(t)|^2 \, dt = \int_{-\infty}^{+\infty} \left| f(t) - \sum_{k=1}^{N} a_k f_k(t) \right|^2 dt \tag{7–5}$$

In Eq. (7–5), $f(t)$ is the target function.

Note that since requirements are specified in terms of meeting a PSD, the error of Eq. (7–5) rewrites as follows:

$$e = \int_{-\infty}^{+\infty} \left| P_M(f) - F(f) \right|^2 df \tag{7–6}$$

where $P_M(f)$ represents the emission mask and $F(f)$ the PSD of the linear combination.

Equivalently, one can consider the corresponding autocorrelation functions $R_M(t)$ and $R_F(t)$ and minimize an error defined as follows:

$$e = \int_{-\infty}^{+\infty} \left| R_M(t) - R_F(t) \right|^2 dt = \int_{-\infty}^{+\infty} \left| R_M(t) - \left[\sum_{k=1}^{N} a_k^2 \int_{-\infty}^{+\infty} f_k(\xi) f_k^*(\xi + t) \, d\xi \right] \right|^2 dt \tag{7–7}$$

Note that both Eqs. (7–6) and (7–7) take into account the PSD of the transmitted signal as defined in Chapters 3 and 4. In Chapters 3 and 4, we also concluded that roughly the shape of the PSD and in particular the PSD envelope are mainly determined by the Fourier transform of the impulse response of the pulse shaper $P(f)$. Using $P(f)$ in place of the exact PSD simplifies the optimization problem if one computes the voltage $m(t)$, which represents the mask bound in the time domain. The target emission voltage mask can be obtained by dividing the power emission mask, normalized by $1/T_s$, by the free space impedance, taking the square root, and applying the Fourier anti-transform. In this case, the error is then defined as follows:

$$e = \int_{-\infty}^{+\infty} \left| m(t) - \sum_{k=1}^{N} a_k f_k(t) \right|^2 dt \tag{7–8}$$

As a general remark, note that the LSE method cannot guarantee by itself compliance with the emission mask. The optimization procedure is based on an average quadratic distance and does not impose bounds on a frequency-by-frequency basis. To guarantee compliance with the mask for each frequency, error minimization must be performed by imposing a bound as set in Eq. (6–4).

CHECKPOINT 7–4

In this checkpoint, we will apply the voltage mask as defined in Eq. (7–8).

We introduce two MATLAB functions: Function 7.12, which implements the LSE approximation of the FCC emission mask according to the voltage mask approach, and Function 7.13, which generates the voltage mask.

Function 7.12 takes the following as input: the integer `i`, indicating the selected α vector (see Checkpoint 7–3), and the pulse repetition period `Ts`. The function returns vector `coefficient` containing the best set of coefficients resulting from the LSE selection algorithm, and matrix `derivative` containing the set of derivatives used for the selection. Function 7.12 is executed with the command:

```
[coefficient, derivative] = ...
    cp0704_LSE_pulse_combination(i,Ts);
```

Function 7.13 generates the voltage mask. It receives the following as input: the lower limit `Tmin` and the upper limit `Tmax` of the time interval to be considered, and the number of samples `smp` of the voltage mask. Function 7.13 returns vector `timeemissionmask`, representing the voltage mask, and it is executed as follows:

```
timeemissionmask = cp0704_time_mask(Tmin,Tmax,smp);
```

Figure 7–10 shows the result of applying Functions 7.12 and 7.13 in the specific case of `i = 1` (corresponding to $\alpha = 0.714$ ns for all derivatives) and `Ts = 1e-7`.

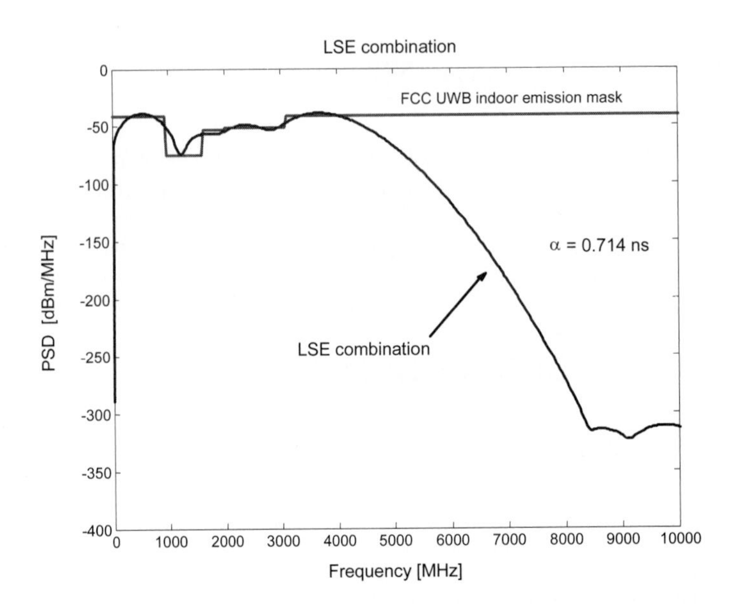

Figure 7–10 Envelope of the PSD of the linear combination of Gaussian waveforms vs. FCC indoor emission mask (by permission from *Journal of Communications and Networks*, ISSN 1229-2370, © Korean Institute of Communication Sciences).

Note the occasional violation of the mask on Figure 7–10.

CHECKPOINT 7–4

Further Reading

Pulse shaping for UWB systems has received increasing interest since the release of the FCC masks in February 2002 (FCC, 2002). The early work presented in (Ghavhami et al., 2002) based on Hermite polynomials was further refined in (Michael et al., 2002), (de Abreu and Kohno, 2003), and (de Abreu et al., 2003), where the distortive effect of antennas was taken into account for pulse waveform design.

A pulse shaping design method based on FIR filters has been recently proposed in (Zeng et al., 2003).

REFERENCES

Bin, L., E. Gunawan, and L.C. Look, "On the BER Performance of TH-PPM UWB Using Parr's Monocycle in the AWGN Channel," *IEEE Conference on Ultra Wideband Systems and Technologies* (November 2003), 403–407.

Conroy, J.T., J.L. LoCicero, and D.R. Ucci, "Communication techniques using monopulse waveforms," *IEEE Military Communications Conference Proceedings*, Volume: 2 (November 1999), 1181–1185.

de Abreu, G.T.F., and R. Kohno, "Design of Jitter-Robust Orthogonal Pulses for UWB Systems," *IEEE Global Telecommunications Conference* (December 2003), 739–743.

de Abreu, G.T.F., G.J. Mitchell, and R. Kohno, "On the design of Orthogonal Pulse-Shape Modulation for UWB Systems Using Hermite Pulses," *Special Issue on Ultra-Wideband Communications of the Journal on Communications and Networks*, Volume: 5, Issue: 4 (December 2003), 328–343.

Di Benedetto, M.-G., and B.R. Vojcic, "Ultra Wide Band (UWB) Wireless Communications: A Tutorial," *Journal of Communication and Networks, Special Issue on Ultra-Wideband Communications*, Volume: 5, Issue: 4 (December 2003), 290–302.

Federal Communications Commission, "Revision of Part 15 of the Commission's rules Regarding Ultra-Wideband Transmission Systems: First report and order," *Technical Report FCC 02-48* (adopted February 14, 2002; released April 22, 2002).

Ghavami, M., L.B. Michael, S. Haruyama, and R. Kohno, "A Novel UWB Pulse Shape Modulation System," *Wireless Personal Communications*, Volume: 23, Issue: 1 (October 2002), 105–120.

Hämäläinen, M., V. Hovinen, J. Iinatti, and M. Latva-aho, "In-band Interference Power Caused by Different Kinds of UWB Signals at UMTS/WCDMA Frequency Bands," *Proceedings of the IEEE Radio and Wireless Conference* (August 2001), 97–100.

Harmuth, H.F., *Radiation of Nonsinusoidal Electromagnetic Waves*, New York: Academic Press, Inc. (1990).

Immoreev, I.I., and N. Sinyavin, "Features of ultra-wideband signals' radiation," *IEEE Conference on Ultra Wideband Systems and Technologies* (May 2002), 345–349.

Kardo-Sysoev, A.F., "Generation and Radiation of UWB Signals," *International Workshop on Ultra Wideband Systems* (June 2003).

Michael, L.B., M. Ghavami, and R. Kohno, "Multiple Pulse Generator for Ultra-Wideband Communication using Hermite Polynomial Based Orthogonal Pulses," *IEEE Ultra Wideband Systems and Technologies* (May 2002), 47–51.

Parr, B., B. Cho, K. Wallace, and Z. Ding, "A novel ultra-wideband pulse design algorithm," *IEEE Communications Letters*, Volume: 7, Issue: 5 (May 2003), 219–221.

Scheers, B., M. Acheroy, and A. Vander Vost, "Time-domain simulation and characterization of TEM horns using a normalized impulse response," *IEE Proc.-Microw. Antennas Propag.,* Volume: 147, Issue: 6 (December 2000), 463–468.

Sheng, H., P. Orlik, A.M. Haimovich, L.J. Cimini Jr., and J. Zhang, "On the Spectral and Power Requirements for Ultra Wideband Transmission," *IEEE International Conference on Communications*, Volume: 1 (May 2003), 738–742.

Win, M.Z., and R.A. Scholtz, "Ultra-Wide Bandwidth Time-Hopping Spread-Spectrum Impulse Radio for Wireless Multiple-Access Communications," *IEEE Transactions on Communications*, Volume: 48, Issue: 4 (April 2000), 679–691.

Zeng, D., A. Annamalai Jr., A.I. Zaghloul, "Pulse Shaping Filter Design in UWB System," *IEEE Conference on Ultra Wideband Systems and Technologies* (November 2003), 66–70.

APPENDIX 7.A

Function 7.1 Shape Variation of the Gaussian Pulse

Function 7.1 generates Gaussian pulse waveforms according to the α values given as input. For each generated signal, the ESD is determined by using the code already introduced in Function 1.2.

```
%
% FUNCTION 7.1 : "cp0701_shape_factor_variation"
%
% Effect of shape factor variation on pulse width and
% ESD of the Gaussian pulse
%
% The pulse amplitude is set to 'A'
% 'smp' samples of the Gaussian pulse are considered in
% the time interval 'Tmax - Tmin'
%
% The function receives as input:
% 1) the minimum value of the shape factor 'alphamin'
% 2) the increase step 'alphastep'
% 3) the number of values to be investigated
%  'N_alphavalues'
%
% The function plots for each value of alpha the waveform
% and corresponding ESD
%
% Programmed by Luca De Nardis

function cp0701_shape_factor_variation(alphamin,...
   alphastep, N_alphavalues)

% ---------------------------------------------
% Step Zero - Input parameters and initialize
% ---------------------------------------------

A = 1;                      % pulse amplitude [V]
smp = 1024;                 % number of samples
Tmin = -4e-9;               % lower time interval limit
Tmax = 4e-9;                % upper time interval limit
```

```
alpha = alphamin;             % initialization of the shape
                              % factor
t=linspace(Tmin,Tmax,smp);  % initialization of the time
                              % axis

for i=1:N_alphavalues

% -----------------------------------------------
% Step One - Pulse waveform in the time domain
% -----------------------------------------------

    % pulse waveform definition
    pulse=-A*exp(-2*pi*(t/alpha).^2);

% -----------------------------------------------
% Step Two - Analysis in the frequency domain
% -----------------------------------------------

    dt = (Tmax-Tmin) / smp;      % sampling period
    fs = 1/dt                    % sampling frequency
    N = smp;                     % number of samples (i.e.,
                                 % size of the FFT)
    df = 1 / (N * dt);           % fundamental frequency

    X=fft(pulse);                % double-sided MATLAB
                                 % amplitude spectrum
    X=X/N;                       % conversion from MATLAB
                                 % spectrum to Fourier
                                 % spectrum
    E = fftshift(abs(X).^2/(df^2)); % double-sided ESD
    Ess = 2*E((N/2+1):N);           % single-sided ESD

% ------------------------------
% Step Three - Graphical output
% ------------------------------

% Time domain representation

    figure(1);
    PT=plot(t,pulse);
    set(PT,'LineWidth',[2]);
    AX=gca;
    set(AX,'FontSize',12);
    T=title('Time domain');
```

```
    set(T,'FontSize',14);
    X=xlabel('Time [s]');
    set(X,'FontSize',14);
    Y=ylabel('Amplitude [V]');
    set(Y,'FontSize',14);
    alphabehaviour = {'Increasing \alpha'};
    text(0.75e-9, -0.5, alphabehaviour,...
       'BackgroundColor', [1 1 1]);
    axis([-2e-9 2e-9 -1.2 1.2]);
    hold on

% frequency domain representation

    figure(2);
    positivefrequency=linspace(0,(fs/2),N/2);
    PF=semilogy(positivefrequency,Ess);
    set(PF,'LineWidth',[2]);
    AX=gca;
    set(AX,'FontSize',12);
    T=title('Frequency domain');
    set(T,'FontSize',14);
    X=xlabel('Frequency [Hz]');
    set(X,'FontSize',14);
    Y=ylabel('ESD  [(V^2)*sec/Hz]');
    set(Y,'FontSize',14);
    axis([0 20e9 1e-60 1e-10]);
    text(7.5e9, 1e-25, alphabehaviour,...
       'BackgroundColor', [1 1 1]);
    hold on

    alpha = alpha + alphastep;% increase of alpha value for
                              % the next step
end
```

Function 7.2 Differentiation of the Gaussian Pulse

Function 7.2 normalizes in amplitude and plots a Gaussian pulse and its 15 derivatives. The generation of the waveforms to be plotted is performed by invoking Function 7.3.

```
%
% FUNCTION 7.2 : "cp0702_Gaussian_derivatives"
%
% Analysis of waveforms of the Gaussian pulse and its first
% 15 derivatives
%
% The pulse amplitude is set to 'A'
% 'smp' samples of the Gaussian pulse are considered in
% the time interval 'Tmax - Tmin'
%
% The function receives as input the value of the shape
% factor 'alpha'
%
% The function plots in a 4 x 4 grid the waveform of the
% Gaussian pulse and of its first 15 derivatives for the
% 'alpha' received as input
%
% Programmed by Luca De Nardis

function cp0702_Gaussian_derivatives(alpha)

% ------------------------------------------------
% Step Zero - Input parameters and initialization
% ------------------------------------------------

A = 1;                              % pulse amplitude [V]
smp = 1024;                         % number of samples
Tmin = -4e-9;                       % lower time limit
Tmax = 4e-9;                        % upper time limit

t=linspace(Tmin,Tmax,smp);          % initialization of the
                                    % time axis
pulse=-A*exp(-2*pi*(t/alpha).^2);   % pulse waveform
                                    % definition

F=figure(1);
set(F,'Position',[100 190 850 450]);
subplot(4,4,1);
```

```
PT=plot(t,pulse);
axis([-2e-9 2e-9 -1 1]);
set(gca,'XTick',0);
set(gca,'XTickLabel',{});

for i=1:15
    % determination of the i-th derivative
    derivative(i,:) = ...
       cp0702_analytical_waveforms(t,i,alpha);

    % amplitude normalization of the i-th derivative
    derivative(i,:) = derivative(i,:) / ...
       max(abs(derivative(i,:)));

% ---------------------------------------------------
% Step One - Graphical output
% ---------------------------------------------------

    subplot(4,4,i+1);
    PT=plot(t,derivative(i,:));
    axis([-2e-9 2e-9 -1 1]);
    if(i < 12)
        set(gca,'XTick',0);
        set(gca,'XTickLabel',{});
    end
    if(mod(i,4) ~= 0)
        set(gca,'YTickLabel',{});
    end
end
h = axes('Position',[0 0 1 1],'Visible','off');
set(gcf,'CurrentAxes',h);
text(.5,0.02,'Time[s]','FontSize',12)
text(0.05,0.4,'Amplitude [V]','FontSize',12,...
   'Rotation', 90);
```

Function 7.3 Derivatives of the Gaussian Pulse

Function 7.3 contains the analytical expressions of the first 15 derivatives of the Gaussian pulse. These expressions were obtained from the original waveform in Eq. (7–1) by using the symbolic toolbox of MATLAB.

```
%
% FUNCTION 7.3 : "cp0702_analytical_waveforms"
%
% Definition of the analytical expression for the first 15
% derivatives of the Gaussian pulse
%
% The function receives as input:
% 1) the time axis vector 't'
% 2) the order of the derivative 'k'
% 3) the value of the shape factor 'alpha'
%
% The function returns the vector representing the
% derivative of order 'k' of the Gaussian pulse calculated
% over the time axis 't'
%
% Programmed by Luca De Nardis

function [deriv] = cp0702_analytical_waveforms(t,k,alpha)
switch(k)
    case 1
        deriv = 4*pi*t/alpha^2.*exp(-2*pi*t.^2/alpha^2);
    case 2
        deriv = -4*pi*exp(-2*pi*(t.^2)/alpha^2).*...
            (-alpha^2+4*pi*(t.^2))/alpha^4;
    case 3
        deriv = 16*pi^2*t.*exp(-2*pi*(t.^2)/alpha^2).*...
            (-3*alpha^2+4*pi*(t.^2))/alpha^6;
    case 4
        deriv = -16*pi^2*exp(-2*pi*(t.^2)/alpha^2).*...
            (3*alpha^4-24*pi*(t.^2)*alpha^2+16*pi^2*...
            (t.^4))/alpha^8;
    case 5
        deriv = 64*pi^3*t.*exp(-2*pi*(t.^2)/alpha^2).*...
           (15*alpha^4-40*pi*(t.^2)*alpha^2+16*pi^2*...
           (t.^4))/alpha^10;
    case 6
        deriv = -64*pi^3*exp(-2*pi*(t.^2)/alpha^2).*...
```

```
        (-15*alpha^6+180*pi*(t.^2)*alpha^4-240*...
        pi^2*(t.^4)*alpha^2+64*pi^3*(t.^6))/alpha^12;
  case 7
      deriv = 256*pi^4*t.*exp(-2*pi*(t.^2)/alpha^2).*...
         (-105*alpha^6+420*pi*(t.^2)*alpha^4-336*pi^2*...
         (t.^4)*alpha^2+64*pi^3*(t.^6))/alpha^14;
  case 8
      deriv = -256*pi^4*exp(-2*pi*(t.^2)/alpha^2).*...
        (105*alpha^8-1680*pi*(t.^2)*alpha^6+3360*pi^2*...
        (t.^4)*alpha^4-1792*pi^3*(t.^6)*alpha^2+...
        256*pi^4*(t.^8))/alpha^16;
  case 9
      deriv = 1024*pi^5*t.*exp(-2*pi*(t.^2)/alpha^2).*...
        (945*alpha^8-5040*pi*(t.^2)*alpha^6+6048*pi^2*...
        (t.^4)*alpha^4-2304*pi^3*(t.^6)*alpha^2+256*...
        pi^4*(t.^8))/alpha^18;
  case 10
      deriv = -1024*pi^5*exp(-2*pi*(t.^2)/alpha^2).*...
         (-945*alpha^10+18900*pi*(t.^2)*alpha^8-50400*...
         pi^2*(t.^4)*alpha^6+40320*pi^3*(t.^6)*....
         alpha^4-11520*pi^4*(t.^8)*alpha^2+1024*pi^5*...
         (t.^10))/alpha^20;
  case 11
      deriv = 4096*pi^6*t.*exp(-2*pi*(t.^2)/alpha^2).*...
         (-10395*alpha^10+69300*pi*(t.^2)*alpha^8-...
         110880*pi^2*(t.^4)*alpha^6+63360*pi^3*(t.^6)*...
         alpha^4-14080*pi^4*(t.^8)*alpha^2+1024*pi^5*...
        (t.^10))/alpha^22;
  case 12
      deriv = -4096*pi^6*exp(-2*pi*(t.^2)/alpha^2).*...
        (10395*alpha^12-249480*pi*(t.^2)*alpha^10+...
        831600*pi^2*(t.^4)*alpha^8-887040*pi^3*(t.^6)*...
        alpha^6+380160*pi^4*(t.^8)*alpha^4-67584*pi^5*...
       (t.^10)*alpha^2+4096*pi^6*(t.^12))/alpha^24;
  case 13
      deriv = 16384*pi^7*t.*exp(-2*pi*(t.^2)/alpha^2)...
        .*(135135*alpha^12-1081080*pi*(t.^2)*alpha^10+...
        2162160*pi^2*(t.^4)*alpha^8-1647360*pi^3*...
        (t.^6)*alpha^6+549120*pi^4*(t.^8)*alpha^4-...
        79872*pi^5*(t.^10)*alpha^2+4096*pi^6*...
        (t.^12))/alpha^26;
  case 14
      deriv = -16384*pi^7*exp(-2*pi*(t.^2)/alpha^2).*...
         (-135135*alpha^14+3783780*pi*(t.^2)*alpha^12-...
         15135120*pi^2*(t.^4)*alpha^10+20180160*pi^3*...
```

```
            (t.^6)*alpha^8-11531520*pi^4*(t.^8)*alpha^6+...
            3075072*pi^5*(t.^10)*alpha^4-372736*pi^6*...
            (t.^12)*alpha^2+16384*pi^7*(t.^14))/alpha^28;
    case 15
        deriv = 65536*pi^8*t.*exp(-2*pi*(t.^2)/alpha...
            ^2).*(-2027025*alpha^14+18918900*pi*(t.^2)*...
            alpha^12-45405360*pi^2*(t.^4)*alpha^10+...
            43243200*pi^3*(t.^6)*alpha^8-19219200*pi^4*...
            (t.^8)*alpha^6+4193280*pi^5*(t.^10)*alpha^4-...
            430080*pi^6*(t.^12)*alpha^2+16384*pi^7*...
            (t.^14))/alpha^30;
end
```

Function 7.4 Spectral analysis of Gaussian Pulses

Function 7.4 evaluates and plots the ESD of the first 15 derivatives of the Gaussian pulse. The procedure for evaluating the ESD for each derivative is the same as in Function 7.1.

```
%
% FUNCTION 7.4 : "cp0702_Gaussian_derivatives_ESD"
%
% Analysis of ESDs of the first 15
% derivatives of the Gaussian pulse
%
% 'smp' samples of the Gaussian pulse are considered in
% the time interval 'Tmax - Tmin'
%
% The function receives as input the value of the shape
% factor 'alpha'
%
% The function computes and plots the ESDs
% of the first 15 derivatives of the Gaussian
% pulse for the 'alpha' value received in input
%
% Programmed by Luca De Nardis

function cp0702_Gaussian_derivatives_ESD(alpha)

% ----------------------------------------------
% Step Zero - Input parameters and initialization
% ----------------------------------------------

smp = 1024;                % number of samples
Tmin = -4e-9;              % lower time limit
Tmax = 4e-9;               % upper time limit

N = smp;                   % number of samples (i.e., size of
                           % the FFT)
dt = (Tmax-Tmin) / N;      % sampling period
fs = 1/dt;                 % sampling frequency
df = 1 / (N * dt);         % fundamental frequency

t=linspace(Tmin,Tmax,smp); % initialization of the time
                           % axis

F=figure(1);
```

```
set(F,'Position',[100 190 650 450]);

for i=1:15

% ----------------------------------------------------------
% Step One - Amplitude-normalized pulse waveforms in the
%            time domain
% ----------------------------------------------------------

    % determination of the i-th derivative
    derivative(i,:)=cp0702_analytical_waveforms(t,i,alpha);
    % amplitude normalization of the i-th derivative
    derivative(i,:)=derivative(i,:) / ...
       max(abs(derivative(i,:)));

% ---------------------------------------------------------
% Step Two - Analysis in the frequency domain and data
%            plotting
% ---------------------------------------------------------

    % double-sided MATLAB amplitude spectrum
    X=fft(derivative(i,:),N);
    % conversion from MATLAB spectrum to Fourier spectrum
    X=X/N;
    % double-sided ESD
    E = fftshift(abs(X).^2/(df^2));
    % single-sided ESD
    Ess = 2 * E((N/2+1):N);
    frequency=linspace((-fs/2),(fs/2),N);

    % positive frequency axis
    positivefrequency=linspace(0,(fs/2),N/2);
    PF=semilogy(positivefrequency,Ess);
    set(PF,'LineWidth',[2]);
    hold on
end

% ------------------------------------------
% Step Three - Graphical output formatting
% ------------------------------------------

axis([0 1.25e10 1e-55 1e-15]);
AX=gca;
set(AX,'FontSize',12);
GT=title('Frequency domain');
```

```
set(GT,'FontSize',14);
X=xlabel('Frequency [Hz]');
set(X,'FontSize',14);
Y=ylabel('ESD  [(V^2)*sec/Hz]');
set(Y,'FontSize',14);
alphavalue = {'\alpha = 0.714 ns'};
derivebehaviour = {'Increasing differentiation order'};
text(7e9, 1e-28, derivebehaviour,'BackgroundColor',...
   [1 1 1]);
text(0.5e9, 3e-17, '1^{st} derivative',...
   'BackgroundColor', [1 1 1]);
text(4e9, 3e-17, '15^{th} derivative',...
   'BackgroundColor', [1 1 1]);
text(2e9,10e-48,alphavalue,'BackgroundColor', [1 1 1]);
```

Function 7.5 Peak frequency of Gaussian Pulses

Function 7.5 determines the peak frequency of the first 15 derivatives of the Gaussian pulse. The evaluation of the ESD for each derivative is performed as in Function 7.4.

```
%
% FUNCTION 7.5 : "cp0702_Gaussian_derivatives_peak
% frequency"
%
% Analysis of peak frequency of the first 15 derivatives of
% the Gaussian pulse as a function of the shape factor
%
% 'smp' samples of the Gaussian pulse are considered in
% the time interval 'Tmax - Tmin'
%
% The function receives as input:
% 1) the minimum value of the shape factor 'alphamin'
% 2) the increase step 'alphastep'
% 3) the number of values to be investigated
%    'N_alphavalues'
%
% The function computes the ESDs of
% the first 15 derivatives of the Gaussian pulse for the
% 'alpha' value received as input, and then evaluates and
% plots the peak frequency for each derivative
%
% Programmed by Luca De Nardis

function cp0702_Gaussian_derivatives_peak_frequency...
   (alphamin, alphastep, N_alphavalues)

% ----------------------------
% Step Zero - Input parameters
% ----------------------------

smp = 1024;                 % number of samples
alpha = alphamin;           % Gaussian pulse form factor
Tmin = -4e-9;               % lower time limit
Tmax = 4e-9;                % upper time limit

dt = (Tmax-Tmin) / smp; % sampling period
fs = 1/dt;                  % sampling frequency
N = smp;                    % number of samples (i.e., size of
```

```
                              %  the FFT)
df = 1 / (N * dt);            % fundamental frequency

x=linspace(Tmin,Tmax,smp);
F=figure(1);

for j=1:N_alphavalues

    factor(j)=alpha;
    for i=1:15

% ---------------------------------------------
% Step One - Amplitude-normalized pulse waveform
%            in the time domain
% ---------------------------------------------

    derivative(i,:) = ...
       cp0702_analytical_waveforms(x,i,alpha);
    derivative(i,:) = derivative(i,:) / ...
       max(abs(derivative(i,:)));

% ------------------------------------------------------
% Step Two - Analysis in the frequency domain and peak
%            frequency evaluation
% ------------------------------------------------------

        % double-sided MATLAB amplitude spectrum
        X=fft(derivative(i,:),N);
        % conversion from MATLAB spectrum to Fourier
        % spectrum
        X=X/N;
        % double-sided ESD
        E = fftshift(abs(X).^2/(df^2));
        % single-sided ESD
        Ess = 2 * E((N/2+1):N);

        positivefrequency=linspace(0,(fs/2),N/2);

        % evaluation of the peak frequency (frequency at
        % which the ESD assumes the maximum value)
        [peak,peakelementindex]=max(Ess);

   peakfrequency(i,j) = ...
      positivefrequency(peakelementindex);
   end
```

```
    % increase of alpha value for the next step
    alpha = alpha + alphastep;
end

% ----------------------------------------
% Step Three - Graphical output formatting
% ----------------------------------------

PT=plot(factor,peakfrequency');
set(PT,'LineWidth',[2]);
AX=gca;
set(AX,'FontSize',12);
X=xlabel('\alpha [s]');
set(X,'FontSize',14);
Y=ylabel('Peak frequency [Hz]');
set(Y,'FontSize',14);
axis([2e-10 14e-10 0 12e9]);
derivebehaviour = {'Increasing differentiation order'};
text(5e-10, 6e9, derivebehaviour,'BackgroundColor',...
   [1 1 1]);
text(1.15e-9, 0.5e9, '1^{st} derivative',...
   'BackgroundColor', [1 1 1]);
text(1.15e-9, 2.4e9,'15^{th} derivative','BackgroundColor',
[1 1 1]);
```

Function 7.6 Bandwidth of Gaussian Pulses

Function 7.6 evaluates and plots the -10 bandwidth of the first 15 derivatives of the Gaussian pulse. The evaluation of the -10 bandwidth is obtained by invoking Function 7.7.

```
%
% FUNCTION 7.6 : "cp0702_Gaussian_derivatives_
%  10dB_bandwidth"
%
% Analysis of -10 dB of the first 15 derivatives of the
% Gaussian pulse as a function of the shape factor
%
% 'smp' samples of the Gaussian pulse are considered in
% the time interval 'Tmax - Tmin'
%
% The function receives as input:
% 1) the minimum value of the shape factor 'alphamin'
% 2) the increase step 'alphastep'
% 3) the number of values to be investigated
%    'N_alphavalues'
%
% The function computes the ESDs of
% the first 15 derivatives of the Gaussian pulse for the
% 'alpha' value received as input, and then evaluates and
% plots the -10 dB bandwidth for each derivative
%
% Programmed by Luca De Nardis

function cp0702_Gaussian_derivatives_10dB_bandwidth(...
   alphamin, alphastep, N_alphavalues)

% ----------------------------------------------
% Step Zero - Input parameters and initialization
% ----------------------------------------------

smp = 4096;                % number of samples
alpha = alphamin;          % Gaussian pulse form factor
Tmin = -4e-9;              % lower time limit
Tmax = 4e-9;               % upper time limit
threshold = -10;           % threshold (in dB) used to compute
                           % the bandwidth

t=linspace(Tmin,Tmax,smp);% initialization of the time axis
```

```
dt = (Tmax - Tmin) / smp; % sampling period

for j=1:N_alphavalues

    factor(j)=alpha;

    for i=1:15

% ------------------------------------------
% Step One - Pulse waveform in the time domain
% ------------------------------------------
        derivative(i,:) =...
           cp0702_analytical_waveforms(t,i,alpha);
        derivative(i,:) = derivative(i,:) / ...
           max(abs(derivative(i,:)));

% ------------------------------------------
% Step Two - Analysis in the frequency domain and
%            evaluation of -10 dB bandwidth
% ------------------------------------------

         [Ess,f_high,f_low,BW] = ...
            cp0702_bandwidth(derivative(i,:),dt,threshold);
        minus10dbBand(i,j)=BW;

    end
    %increase of alpha value for the next step
    alpha = alpha + alphastep;
end

% ----------------------------
% Step Three - Graphical output
% ----------------------------

F=figure(1);
plot(factor,minus10dbBand');
axis([2e-10 12e-10 1e9 6e9]);
AX=gca;
set(AX,'FontSize',12);
X=xlabel('\alpha [s]');
set(X,'FontSize',14);
Y=ylabel('-10 dB Bandwidth [Hz]');
set(Y,'FontSize',14);
grid on
derivebehaviour = {'Increasing differentiation order'};
```

```
text(7e-10, 3e9, derivebehaviour,'BackgroundColor',...
   [1 1 1]);
```

Function 7.7 Bandwidth Evaluation

The approach used for bandwidth evaluation is very close to that of Function 1.2 with two minor modifications: the graphical output is removed, and a higher number of points is adopted for the FFT to achieve a smoother plot in the frequency domain.

```
%
% FUNCTION 7.7 : "cp0702_bandwidth"
%
% Evaluates the bandwidth of the input 'signal' with
% sampling rate 'dt'
% Bandwidth is evaluated according to the given 'threshold'
% (in dB)
% 'BW' is the bandwidth
% 'f_high' is the higher limit
% 'f_low' is the lower limit
%
% Programmed by Guerino Giancola / Luca De Nardis
%

function [Ess,f_high,f_low,BW] = ...
   cp0702_bandwidth(signal,dt,threshold)

% ----------------------------------------------------------
% Step One - Evaluation of the single-sided ESD
% ----------------------------------------------------------

% sampling frequency
fs = 1 / dt;
% frequency smoothing factor
frequencysmoothingfactor = 8;
% number of samples (i.e., size of the FFT)
N = frequencysmoothingfactor * length(signal);
% fundamental frequency
df = 1 / (N * dt);

% double-sided MATLAB amplitude spectrum
X = fft(signal, N);
% conversion from MATLAB spectrum to Fourier spectrum
X = X/N;
% double-sided ESD
E = abs(X).^2/(df^2);
% single-sided ESD
```

```
Ess = 2.*E(1:floor(N/2));

% --------------------------------------------------
% Step Two - Evaluation of the frequency bandwidth
% --------------------------------------------------

% Epeak is the peak value of the ESD
[Epeak,index] = max(Ess);
% peak frequency
f_peak = index * df;

% Eth is the value of the ESD corresponding to the given
% threshold
Eth = Epeak*10^(threshold/10);

% iterative algorithm for evaluating high and low
% frequencies

imax = index;
E0h = Ess(index);

while (E0h>Eth)&(imax<=(N/2))
    imax = imax + 1;
    E0h = Ess(imax);
end

f_high = (imax-1) * df;              % high frequency

imin = index;
E0l = Ess(index);

while (E0l>Eth)&(imin>1)&(index>1)
    imin = imin - 1;
    E0l = Ess(imin);
end

f_low = (min(index,imin)-1) * df;    % low frequency

% end of iterative algorithm

% signal frequency bandwidth
BW = f_high - f_low;
```

Function 7.8 Random Selection Algorithm

Function 7.8 performs three basic tasks. First, it generates the Gaussian derivatives to be used in the procedure. Next, it generates the target emission mask. Finally, it determines the coefficients for the combination of the Gaussian derivatives and plots the PSD of the resulting signal. Function 7.8 uses three subsidiary functions for implementing the random selection algorithm: Functions 7.9, 7.10, and 7.11. The input parameter i is passed to Function 7.9 for selecting one of the two possible sets of α values that can be considered. The reference emission mask is generated by invoking Function 7.10. Finally, Function 7.11 performs the selection of coefficients.

```
%
% FUNCTION 7.8 : "cp0703_random_pulse_combination"
%
% This function implements the random selection algorithm
% described in Section 7.2 for the determination of a
% combination of the first 15 Gaussian derivatives fitting
% the FCC emission mask
%
% 'smp' samples of the Gaussian pulse are considered in
% the time interval 'Tmax - Tmin'
%
% The function receives as input:
% 1) the index 'i' indicating which setting must be adopted
%    for the shape factors of the derivatives
% 2) the pulse repetition period Ts
% 3) the number of attempts in the random selection of the
%    coefficients 'attempts'
%
% The function returns:
% 1) the best coefficient set 'coefficient'
% 2) the coefficients for the set formed  by each single
%    derivative 'singlederivativeset'
% 3) the set of derivatives 'derivative'
% 4) a flag on the validity of the returned vectors
%    'validresult'
% 5) the fundamental frequency df

% The function singles out the best coefficient set within
% the sets found during the 'attempts' iterations and the
% best coefficient for the solutions based on each single
% derivative
% The function then plots the target mask, the solutions
```

```
% based on each single derivative, and the solution based
% on the set of the 15 derivatives of the Gaussian pulse
%
% Programmed by Luca De Nardis

function [coefficients, singlederivativecoeff,...
   derivative, validresult,df] = ...
   cp0703_random_pulse_combination(i,Ts,attempts)

% ---------------------------------------------
% Step Zero - Input parameters and initialization
% ---------------------------------------------

% lower time limit
Tmin=-4e-9;
% upper time limit
Tmax=4e-9;
% number of samples
smp = 1024;

% sampling period
dt = (Tmax-Tmin) / smp;
% sampling frequency
fs = 1/dt;
frequencysmoothingfactor = 8;
% number of samples (i.e., size of the FFT)
N = frequencysmoothingfactor * smp;
% fundamental frequency
df = 1 / (N * dt);

% initialization of the positive frequency axis
positivefrequency=linspace(0,(fs/2),N/2);
% initialization of the time axis
t=linspace(Tmin,Tmax,smp);

% loading the alpha vector depending on the input 'i'
alpha=cp0703_get_alpha_value(i);
% loading the emission mask on N points
emissionmask = cp0703_generate_mask(N, fs);

for i=1:15

% ---------------------------------------------
% Step One - Pulse waveforms in the time domain
% ---------------------------------------------
```

```
   % determination of the i-th derivative
   derivative(i,:) = cp0702_analytical_waveforms(t, i,...
      alpha(i));
   % amplitude normalization of the i-th derivative
   derivative(i,:) = derivative(i,:) /
max(abs(derivative(i,:)));
end

for i=1:15

   % determination of coefficients for each single
   % derivative considered as a stand alone set
   [i_th_derivative_coeff,validresult] = ...
      cp0703_random_coefficients(attempts, derivative,...
      dt, smp, Ts, frequencysmoothingfactor,...
      emissionmask, i, i);
   % application of coefficient to the i-th derivative
   normalizedderivative(i,:) = i_th_derivative_coeff(i)...
      * derivative(i,:);
   % recording coefficient for the function output
   singlederivativecoeff(i)= i_th_derivative_coeff(i);
   if(validresult)

% --------------------------------------------------------
% Step Two - Evaluation of PSDs of normalized waveforms
% --------------------------------------------------------

        % double-sided MATLAB amplitude spectrum
        X=fft(normalizedderivative(i,:),N);
        % conversion from MATLAB spectrum to Fourier
        % spectrum
        X=X/N;
        % double-sided ESD
        E = fftshift(abs(X).^2/(df^2));
        % single-sided ESD
        Ess = 2.*E((N/2+1):N);
        % PSD of the i-th normalized derivative in dBm/MHz
        singlederivativePSD(i,:) = 10 * log10 ((1/Ts)...
           * Ess / 377) + 90;
   end
% end of section dedicated to mask fitting through single
% derivatives
end
```

```
% ---------------------------------------------------------
% Step Three - Evaluation of mask fitting combination of
% pulse waveforms
% ---------------------------------------------------------

[coefficients,validresult] = ...
   cp0703_random_coefficients(attempts, derivative,...
   dt, smp, Ts, frequencysmoothingfactor, ...
   emissionmask,1,15);
if(validresult)
    % double-sided MATLAB amplitude spectrum
    X=fft(coefficients*derivative,N);
    % conversion from MATLAB spectrum to Fourier spectrum
    X=X/N;
    % double-sided ESD
    E = fftshift(abs(X).^2/(df^2));
    % single-sided ESD
    Ess = 2.*E((N/2+1):N);
    % PSD of the combination in dBm/MHz
    PSD = 10 * log10 ((1/Ts) * Ess / 377) + 90;

% ---------------------------
% Step Four - Graphical output
% ---------------------------

    figure(1);
    plot(positivefrequency/1e6,...
       emissionmask,'r','Linewidth',[1]);
    hold on;
    plot(positivefrequency/1e6, singlederivativePSD);
    PF = plot(positivefrequency/1e6, PSD);
    set(PF,'LineWidth',[2]);
    AX=gca;
    set(AX,'FontSize',12);
    T=title('Random combination');
    set(T,'FontSize',14);
    X=xlabel('Frequency [MHz]');
    set(X,'FontSize',14);
    Y=ylabel('PSD  [dBm/MHz]');
    set(Y,'FontSize',14);
    axis([0 12e3 -400 0]);
    alphavalue = '\alpha = 0.714 ns';
    text(8e3, -100, alphavalue,'BackgroundColor', [1 1 1]);
    text(2e3, -300,...
       'normalized derivatives','BackgroundColor',[1 1 1]);
```

```
    text(7e3, -150, 'random combination',...
        'BackgroundColor', [1 1 1]);
    text(5e3, -25, 'FCC UWB indoor emission mask',...
        'BackgroundColor', [1 1 1]);
end
```

Function 7.9 Selection of Alpha Values

Two sets of α values are provided within Function 7.9. The reader can easily introduce new sets by editing the function.

```
%
% FUNCTION 7.9 : "cp0703_get_alpha_value"
%
% This function is used to select the set of alpha values
% for each of the first 15 derivatives of the Gaussian
% pulse
%
% The function receives as input the index 'i' indicating
% which setting must be adopted for the shape factors of
% the derivatives
%
% The function returns a vector of 15 elements
% corresponding to the value of the shape factor for each
% of the derivatives
%
% Programmed by Luca De Nardis

function alphavector = cp0703_get_alpha_value(i)
switch (i)
    case 1
        %vector characterized by a constant value of alpha
for all derivatives
        alphavector = [0.714e-9 0.714e-9 0.714e-9 0.714e-9
0.714e-9 0.714e-9 0.714e-9 0.714e-9 0.714e-9 0.714e-9
0.714e-9 0.714e-9 0.714e-9 0.714e-9 0.714e-9];

    case 2
        %vector characterized by a high value of alpha for
the first derivative and a small value for derivatives 2-15
        alphavector = [1.5e-9 0.314e-9 0.314e-9 0.314e-9
0.314e-9 0.314e-9 0.314e-9 0.314e-9 0.314e-9 0.314e-9
0.314e-9 0.314e-9 0.314e-9 0.314e-9 0.314e-9];
end
```

Function 7.10 Generation of the Emission Mask

```
%
% FUNCTION 7.10 : "cp0703_generate_mask"
%
% This function generates a discrete vector representing
% the FCC indoor
% emission mask for UWB devices
%
% The function receives as input:
% 1) the number of points in the frequency domain 'N'
% 2) the sampling frequency 'fs'
%
% The function returns a vector of N/2 points
% 'emissionmask', representing the emission mask in the
% frequency range [0 , fs/2]
%
% Programmed by Luca De Nardis

function [emissionmask] = cp0703_generate_mask(N,fs)

df = fs / N;                    % fundamental frequency
emissionmask = zeros(N/2);
n1 = round(0.96e9/df);
n2 = round(1.61e9/df);
n3 = round(1.99e9/df);
n4 = round(3.1e9/df);
n5 = round(10.6e9/df);

a(1:n1)=-41.3;
b(1:(n2-n1))=-75.3;
c(1:(n3-n2))=-53.3;
d(1:n4-n3)=-51.3;
e(1:n5-n4)=-41.3;
f(1:(N/2-n5))=-51.3;
emissionmask=[a b c d e f];
positivefrequency=linspace(0, fs/2, N/2);
semilogx(positivefrequency,emissionmask);
axis([8e8 20e9 -78 -40]);
```

Function 7.11 Random Coefficient Selection

The output of this function is used in Function 7.8. The possibility of using only one base function for the selection process is obtained by setting `lowerbasefunction = higherbasefunction` for each possible value in the range [1–15].

```
%
% FUNCTION 7.11 : "cp0703_random_coefficients"
%
% This function selects coefficients for a set of BFs to
% fit a given emission mask
%
% The function receives as input:
% 1) the number of attempts in the random selection of the
%    coefficients 'attempts'
% 2) the set of BFs 'basefunction'
% 3) the sampling period 'dt'
% 4) the number of samples in the time domain 'smp'
% 5) the pulse repetition period 'Ts'
% 6) the frequency smoothing factor 'freqsmoothfactor'
% 7) the target emission mask
% 8) and 9) the range of BFs to be used in the
%    mask fitting, given by the values 'lowerbasefunction'
%    and 'higherbasefunction'
%
% The function returns:
% 1) the best coefficient set 'c'
% 2) a flag on the validity of the returned set 'result'

% The function singles out the best coefficient set for
% the BF set given as input within the sets
% found during the 'attempts' iterations by comparing the
% PSD of the resulting waveform for each iteration with the
% target emission mask
% After 'attempts' iterations, the function returns
% the best set, defined as the set leading
% to the waveform with maximum power within all sets
% fitting the mask
%
% Programmed by Luca De Nardis

function [c,result] = cp0703_random_coefficients...
   (attempts, basefunction, dt, smp, Ts,...
```

```
   freqsmoothfactor, emissionmask, lowerbasefunction,...
   higherbasefunction)

% -----------------------------------------------
% Step Zero - Input parameters and initialization
% -----------------------------------------------

% sampling frequency
fs = 1 / dt;
% number of samples (i.e., size of the FFT)
N = freqsmoothfactor * smp;
% fundamental frequency
df = 1 / (N * dt);

% initialization of the positive frequency axis
positivefrequency=linspace(0,(fs/2),N/2);
% initialization of the coefficient set vector
a=zeros(1,15);
% initialization of the random number generator
rand('state',sum(100*clock));

% ----------------------------------------------------------
% Step One - Evaluation of the best combination through
%           random search
% ----------------------------------------------------------

for numattempts=1:attempts

  % initialization of the power vector component for the
  % actual attempt
  P(numattempts)=0;
  % initialization of the coefficient set vector for the
  % actual attempt
  C(numattempts,1:15)=nan;
  for i=lowerbasefunction:higherbasefunction
        count = 0;
        while (count < 100)
            count=count+1;
            if rand < (0.5)
                a(i) = rand;
            else
                a(i) = -rand;
            end
            % generation of the waveform associated with
          % the actual coefficient set
```

```
            combo=a * basefunction;
            % double-sided MATLAB amplitude spectrum
            X=fft(combo,N);
            % conversion from MATLAB spectrum to Fourier
            % spectrum
            X=X/N;
            % double-sided ESD of the waveform
            E = fftshift(abs(X).^2/(df^2));
            % single-sided ESD of the waveform
            Ess = 2.*E((N/2+1):N);
            % PSD of the combination in dBm/MHz
            PSD = 10 * log10 ((1/Ts) * Ess / 377) + 90;

            % comparison between the PSD and the mask
            if all(PSD < emissionmask)
                % recording the power associated with the
                % actual set
                found=1;
                % recording the actual set
                P(numattempts)=sum(1/Ts .* Ess.*df / 377);
                C(numattempts,1:15)=a;
                count=100;
            end
        end
    end
end

result = found;  % setting the flag for function output
if(found==1)
    [m,h]=max(P);% selection of the set leading to the
                 % waveform at highest power
    c=C(h,1:15); % recording the set for function output
end
```

Function 7.12 LSE Selection Algorithm

To implement the LSE algorithm, Function 7.12 uses the voltage mask provided by Function 7.13. Function 7.12 determines the coefficients corresponding to the LSE optimal approximation of the voltage mask by using the `lsqlin` MATLAB function. Finally, Function 7.12 plots the envelope of the PSD of the resulting combination versus the power mask.

```
%
% FUNCTION 7.12 : "cp0704_LSE_pulse_combination"
%
% This function implements the LSE
% selection algorithm described in Section 7.2 for the
% determination of a combination of the first 15 Gaussian
% derivatives fitting the FCC indoor emission mask
%
% 'smp' samples of the Gaussian pulse are considered in
% the time interval 'Tmax - Tmin'
%
% The function receives as input:
% 1) the index 'i' indicating which setting must be adopted
% for the shape factors of the derivatives
% 2) the pulse repetition period Ts
%
% The function returns:
% 1) the best coefficient set 'coefficient'
% 2) the set of derivatives 'derivative'

% The function singles out the best coefficient set in the
% sense of the LSE minimization between the combination of
% the first 15 derivatives of the Gaussian pulse and the
% energy signal in the time domain corresponding to the
% FCC emission mask.
%
% The function then plots the target mask and PSD of
% the waveform obtained through the LSE minimization
%
% Programmed by Luca De Nardis

function [coefficient, derivative] = ...
   cp0704_LSE_pulse_combination(i,Ts)

% --------------------------------------------------
```

```
% Step Zero - Input parameters and initialization
% ------------------------------------------------

Tmin = -4e-9;
% lower time interval limit
Tmax = 4e-9;
% upper time interval limit
smp = 1024;
% number of samples
frequencysmoothingfactor = 4;
% frequency smoothing factor

dt = (Tmax - Tmin) / smp;
% sampling period
fs = 1/dt;
% sampling frequency
N = frequencysmoothingfactor * smp;
% number of samples (i.e., size of the FFT)
df = 1 / (N * dt);
% fundamental frequency

positivefrequency=linspace(0,(fs/2),N/2);
% initialization of the positive frequency axis
t=linspace(Tmin,Tmax,smp);
% initialization of the time axis
alpha=cp0703_get_alpha_value(i);
% loading the alpha vector depending on the input 'i'


for i=1:15

% ----------------------------------------------
% Step One - Pulse waveforms in the time domain
% ----------------------------------------------

    derivative(i,:) = ...
       cp0702_analytical_waveforms(t,i,alpha(i));
    derivative(i,:) = (derivative(i,:)...
       / max(abs(derivative(i,:))));

end

% ----------------------------------------------
% Step Two - Determination of the LSE solution to the
% approximation problem
```

```
% ----------------------------------------------

timeemissionmask = cp0704_time_mask(Tmin,Tmax,smp);
% determination of the signal generating the mask in the
% frequency domain
% application of the LSE method
coefficient = sqrt(Ts * 377) * ...
   lsqlin(derivative',timeemissionmask');
X = fft(coefficient'*derivative,N);
% double-sided MATLAB amplitude spectrum
X = X / N;
% conversion from MATLAB spectrum to fourier spectrum
E = fftshift(abs(X).^2/(df^2));
% double-sided ESD
Ess = 2 * E((N/2+1):N);
% single-sided ESD
PSD = 10 * log10 ((1/Ts) * Ess / 377) + 90;
% PSD of the combination in dBm/MHz

% ----------------------------
% Step Three - Graphical output
% ----------------------------

emissionmask = cp0703_generate_mask(N, fs);
% loading the emission mask on N/2 points

figure(1);
plot(positivefrequency/1e6,...
   emissionmask,'r','Linewidth',[2]);
hold on;
PF = plot(positivefrequency/1e6, PSD);
set(PF,'LineWidth',[2]);
AX=gca;
set(AX,'FontSize',12);
set(AX,'FontSize',12);
T=title('LSE combination');
set(T,'FontSize',14);
X=xlabel('Frequency [MHz]');
set(X,'FontSize',14);
Y=ylabel('PSD  [dBm/MHz]');
set(Y,'FontSize',14);
axis([0 10e3 -400 0]);
alphavalue = '\alpha = 0.714 ns';
text(8e3, -100, alphavalue,'BackgroundColor', [1 1 1]);
text(3.5e3, -250, 'LSE combination', 'BackgroundColor',...
```

```
    [1 1 1]);
text(5e3, -25, 'FCC UWB indoor emission mask',...
    'BackgroundColor', [1 1 1]);
```

Function 7.13 Generation of the Voltage Mask

```
%
% FUNCTION 7.13 : "cp0704_time_mask"
%
% This function defines the signal in the time domain
% corresponding to the FCC indoor emission mask for UWB
% devices in the frequency domain
%
% The function receives as input:
% 1) and 2) the time interval to be considered through the
% lower limit  'Tmin' and the upper limit 'Tmax'
% 3) the number of samples in the time domain 'smp'
%
% The function returns a vector 'timemissionmask' of 'smp'
% points representing the Fourier anti-transform of the
% square root of the FCC mask
%
% The constant values used in the function have been
% analytically derived by representing the mask as the sum
% of translated rects in the frequency domain, based on the
% following formula:

% A*rectT(t) <--> A * T * sin(pi*f*T)/(pi*f*T)
%
% and on the duality and translation properties of the
% Fourier transform
%
% Programmed by Luca De Nardis

function[timeemissionmask]=cp0704_time_mask(Tmin,Tmax,smp)


% -------------------------------------------------
% Step Zero - Input parameters and initialization
% -------------------------------------------------

% sampling period
dt = (Tmax-Tmin)/smp;
% sampling frequency
fs = 1/dt;
% initialization of the time axis
```

```
x=linspace(Tmin,Tmax,smp);

% periods of the sin(x)/x functions in the time domain

t(1)=1/(0.96e9);
t(2)=1/(0.65e9);
t(3)=1/(0.38e9);
t(4)=1/(1.11e9);
t(5)=1/(7.5e9);


% adjusting the last value depending on fs
t(6)=1/(fs/2-10.6);


% amplitudes of the rects functions in the frequency domain

a(1)=261.408;
a(2)=3.53;
a(3)=25.99;
a(4)=95.571;
a(5)=2042.25;

% adjusting the last value depending on fs
a(6)=8.61e-8/t(6);

% evaluation of the frequency shifts

for i=1:6

    f0(i)=0;
    if i>1
        for k=1:(i-1)
            f0(i)=f0(i)+1/t(k)
        end
        f0(i)=f0(i)+1/(2*t(i));
    else
        f0(i)=1/(2*t(i));
    end

    s(i,:)=a(i)*(sin(pi*x/t(i))./...
    (pi*x/t(i))).*exp(j*2*pi*f0(i)*x);

% performing the Fourier AT
end
```

```
comb=ones(1,6);
timeemissionmask=comb * s;
```

CHAPTER 8

The UWB Channel and Receiver

Properties of IR-UWB signals and related spectral characteristics were investigated in previous chapters. The analyses also included the two recent proposals in the United States within the IEEE 802.15 Task Group 3a, specifically DS-UWB (Roberts, 2003) and the MB (Batra et al., 2003) approach, which adopts continuous transmission based on OFDM rather than IR.

This chapter analyzes the signal at the receiver, that is, after propagation over the radio channel, as shown in Figure 8–1.

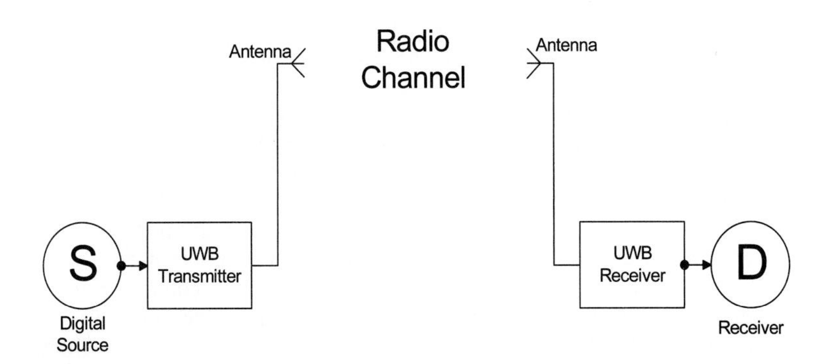

Figure 8–1 System model for a single-user UWB communication.

For continuous transmission techniques, and in particular OFDM, the theory forming the basis for channel modeling and optimal receiver design is well-consolidated and extensively documented in the literature (Proakis, 1995; Hanzo et al., 2003). The purpose of this chapter is to investigate how the above models translate in the case of communication systems using IR-UWB. IR-UWB incorporates peculiar features that need to be taken into account for a comprehensive analysis of system design.

We first hypothesize perfect synchronization between transmitter and receiver, and proceed in Section 8.1 to give a detailed analysis of receiver structures for different IR-UWB modulation formats, and in particular PPM-TH, PAM-TH, and DS-UWB. The IR-UWB transmitted signal described in Chapter 2, "The UWB Radio Signal," is ideally composed of a sequence of pulses that do not overlap in time. Each pulse is confined within a specific time interval, and the pulse itself has finite duration. While ISI among pulses belonging to the same transmission is ideally absent in the transmitted signal, it might not be so after the signal has traveled through a real channel. Pulses might in fact be delayed by different amounts, and replicas of pulses due to multiple paths might cause ISI. Moreover, in the case of the presence of several users transmitting over the same channel, pulses originating in other transmission links may collide with pulses belonging to a reference transmission, giving rise to an interference noise called Multi-User Interference (MUI).

At the receiver, the reference signal, indicated as a useful signal, is corrupted by mainly two additive noise components. The first is thermal noise generated in the receiver antenna and the receiver circuitry. The second is MUI due to the presence of multiple users in the system. The problem of receiver design thus states as follows: finding a good, when possible optimal, way for extracting the useful signal from the received signal. Solving the general problem is a complicated task leading to complex receiver structures and requiring good modeling for the noise components. While it is well-known that thermal noise is well-represented by a white Gaussian random process, MUI noise characteristics depend on the number of users in the system, that is, the number and features of the interferers, as will be addressed in further detail in Chapter 9, "Multi-User Wireless Communications." The problem of optimal receiver design is greatly simplified under the hypothesis of the absence of ISI and MUI.

We will examine in Section 8.1 receiver structures for both PAM and PPM isolated pulses, and then extend the analysis to the case of multi-pulse. As we will see, the IR-UWB optimum receiver schemes can be derived in a quite straightforward way from the traditional Statistical Theory of Reception (Middleton, 1960).

The main focus of this chapter is Section 8.2, which deals with channel modeling and multi-path fading. System performance is in fact significantly degraded by the distortion of pulses due to propagation over a real channel. Electro-magnetic waves traveling in indoor environments, for example, are most often reflected, diffracted, and scattered by structures and obstacles so that the received signal ends up being the superimposition of several contributions. After a review of traditional approaches, which are typically adopted in continuous transmission multi-path propagation models, the section contains a description and simulation of the UWB channel model recently proposed by the IEEE 802.15.SG3a Study Group (IEEE 802.15.SG3a, 2003). The section concludes with an analysis of advanced receiver design for multi-path propagation. While for continuous transmission multi-path causes rapid fluctuations in the received signal envelope and thus severe degradation in performance (Hashemi, 1993a), in the IR-UWB case, multiple paths reflect in a sequence of delayed and attenuated replicas of the transmitted pulse, which can eventually be successfully separated at the receiver. By combining multiple pulses, the energy used in the decision process is increased, that is, the presence of multi-path improves system performance. The analysis of UWB receiver structures, and in particular the RAKE receiver for multi-path environments, is presented in Section 8.2.3 and simulated in Checkpoint 8–2.

We finally address the problem of synchronization between transmitter and receiver. General issues related to synchronization for digital transmission systems can be found, for example, in (Mengali and D'Andrea, 1997) and (Bregni, 2002). The receiver must, in general, extract discrete time information from the received continuous waveform to reconstruct at best the sequence of symbols emitted by the source. Precise information about frequency and phase of the emitted symbols, after distortion by the channel, must be either available or recovered at the receiver. Due to the very short duration of the transmitted pulses, receiver performance is in fact very sensitive to synchronization errors. As shown by (Lovelace and Townsend, 2002), system throughput and multiple access capabilities may degrade significantly even for relatively small time jitters and modest tracking errors. Systems based on TH show, in general, little robustness against acquisition errors (Tian and Giannakis, 2003a). Synchronization issues for the UWB case will be addressed in Section 8.3.

8.1 MULTI-PATH-FREE AWGN CHANNEL

The effect of the propagation of IR signals over the AWGN channel was discussed in Chapter 6, "Performance Analysis of the UWB Radio Link," and is further analyzed in this section. The useful signal at the receiver $r_u(t)$ is corrupted by additive noise $n(t)$, typically thermal noise, which is assumed to be a realization of a stochastic Gaussian process with bilateral PSD $N_0/2$. The received signal is expressed by:

$$r(t) = r_u(t) + n(t) \tag{8–1}$$

$r_u(t)$ is an attenuated and delayed version of the transmitted signal $s(t)$, that is:

$$r_u(t) = \alpha s(t - \tau) \tag{8–2}$$

Both channel gain α and channel delay τ in Eq. (8–2) depend on distance of propagation D between transmitter and receiver. For α, one can assume:

$$\alpha = \frac{c_0}{\sqrt{D^\gamma}} \tag{8–3}$$

where γ is the exponent of the power attenuation law, that is, the path loss as defined in Chapter 6, and c_0 is a constant that can be tuned to obtain a reference gain α_0 at a reference distance $D_0 = 1$ m. The γ value is equal to 2 for propagation over the free space; it is higher than 2 for typically Non-Line of Sight (NLOS) propagation. Note that for multi-path-affected channels (see Section 8.2) with Line of Sight (LOS) over short distances, γ might be lower than 2, as shown by (Ghassemzadeh and Tarokh, 2003). With reference to c_0, it is important to note that for having A_{dB}=10log$_{10}$(E_{TX}/E_{RX}) at $D_0 = 1$ m, one must impose:

$$c_0 = 10^{-A_{dB}/20} \tag{8–4}$$

Regarding delay τ, one can write:

$$\tau = D / c \tag{8–5}$$

where c is the speed of light in vacuum ($c \approx 3 \cdot 10^8$ m/s). We assume perfect synchronization between transmitter and receiver, that is, the value of τ is known at the receiver side. This hypothesis will be maintained throughout this section. Synchronization issues will be discussed in Section 8.3.

As is well-known (Proakis, 1995; Lee and Messerschmitt, 1994), the optimum receiver for the AWGN channel is composed of two systems: the correlator and the detector. The role of the correlator is to convert the received signal of Eq. (8–1) into a set of decision variables $\{\mathbf{Z}\}$. The role of the detector is to decide which signal waveform was transmitted based on the observation of $\{\mathbf{Z}\}$.

Assume that the transmitter sends information in a digital form using M different waveform $s_m(t)$, with $m = 0, \ldots, M\text{-}1$. Within an interval of duration T, called the symbol interval, one of the $s_m(t)$ waveforms is transmitted. The $s_m(t)$ waveforms belong to an ensemble that can be generated by N orthonormal basis functions $\{\psi_k(t)\}$, with $k = 0, \ldots, N\text{-}1$. One can thus write:

$$\begin{aligned} & s_m(t) = \sum_{k=0}^{N-1} s_{mk}\psi_k(t) \qquad \text{with } t \in [0, T] \\ & \text{where } s_{mk} = \int_0^T s_m(t)\psi_k(t)\,dt \quad \text{with } k = 0, \ldots, N-1 \end{aligned} \tag{8–6}$$

Based on Eq. (8–6), the energy E_m of $s_m(t)$ over time T is:

$$E_m = \int_0^T (s_m(t))^2\,dt = \int_0^T \left(\sum_{k=0}^{N-1} s_{mk}\psi_k(t)\right)^2 dt = \sum_{k=0}^{N-1} s_{mk}^2 \quad \forall m = 0, \ldots, M-1 \tag{8–7}$$

By introducing Eq. (8–6) into Eqs. (8–1) and (8–2), the received signal in $[0, T]$ corresponding to a generic transmitted $s_m(t)$ is rewritten as:

$$r(t) = \sum_{k=0}^{N-1} \alpha s_{mk}\psi_k(t-\tau) + n(t) \tag{8–8}$$

The correlator receiver for $r(t)$ of Eq. (8–8) is composed of a bank of N cross-correlators, which output N decision variables Z_k given by:

$$\begin{aligned} & Z_k = \int_\tau^{T+\tau} r(t)\psi_k(t-\tau)\,dt = \alpha s_{mk} + n_k, \quad k = 0, \ldots, N-1 \\ & \text{with } n_k = \int_\tau^{T+\tau} n(t)\psi_k(t-\tau)\,dt \end{aligned} \tag{8–9}$$

Figure 8–2 shows the reference scheme for the correlator described by Eq. (8–9).

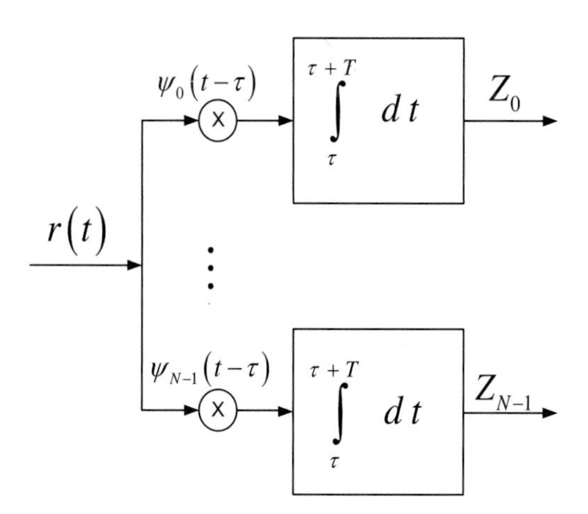

Figure 8–2 Signal correlator.

The detector estimates which waveform was transmitted based on the set $\{\mathbf{Z}\} = \{Z_0, ..., Z_{N-1}\}$. It can be shown that the noise components n_k of Eq. (8–9) are uncorrelated Gaussian variables with zero mean and an equal variance σ_n^2 (see, for example, Proakis, 1995) of:

$$\sigma_n^2 = N_0/2 \tag{8–10}$$

The optimum detector for the signal of Eq. (8–8) applies the Maximum Likelihood (ML) criterion; it selects, among the M possible transmitted waveforms, the one which maximizes the conditional probability $p(\mathbf{Z}|\, s_m(t))$. This is equivalent to saying that the selected $s_m(t)$ is characterized by a set of coefficients $s_m=\{s_{m0}, \dots, s_{m(N-1)}\}$, which is closest to the set $\{\mathbf{Z}\} = \{Z_0, ..., Z_{N-1}\}$. By application of the ML criterion, the following function is minimized:

$$\begin{aligned} e\left(r(t), s_m\right) &= \sum_{k=0}^{N-1} \left(Z_k - s_{mk}\right)^2 = \sum_{k=0}^{N-1} Z_k^2 + \sum_{k=0}^{N-1} s_{mk}^2 - 2\sum_{k=0}^{N-1} s_{mk} Z_k = \\ &= \sum_{k=0}^{N-1} Z_k^2 - 2\left(\sum_{k=0}^{N-1} s_{mk} Z_k - \frac{1}{2} \sum_{k=0}^{N-1} s_{mk}^2 \right) \end{aligned} \tag{8–11}$$

Equation (8–11) indicates that the waveform that maximizes $p(\mathbf{Z}|\, s_m(t))$ is the one that maximizes the so-called correlation metric C:

$$C(r(t),s_m)=\sum_{k=0}^{N-1} s_{mk}Z_k - \frac{1}{2}\sum_{k=0}^{N-1} s_{mk}^2 = \sum_{k=0}^{N-1} s_{mk}Z_k - \frac{E_m}{2} = \\ = \int_{\tau}^{T+\tau} r(t)s_m(t-\tau)dt - \frac{E_m}{2} \tag{8–12}$$

The optimum detector therefore selects the $s_m(t)$, which maximizes $C(r(t),s_m)$. Equation (8–12) also suggests that the optimum receiver operates as follows: 1) the received waveform $r(t)$ is cross-correlated with the M possible transmitted waveforms $s_m(t)$; 2) $E_m/2$ is subtracted from each correlation output; and 3) the maximum over the M resulting values is selected. A reference scheme for this receiver is shown in Figure 8–3.

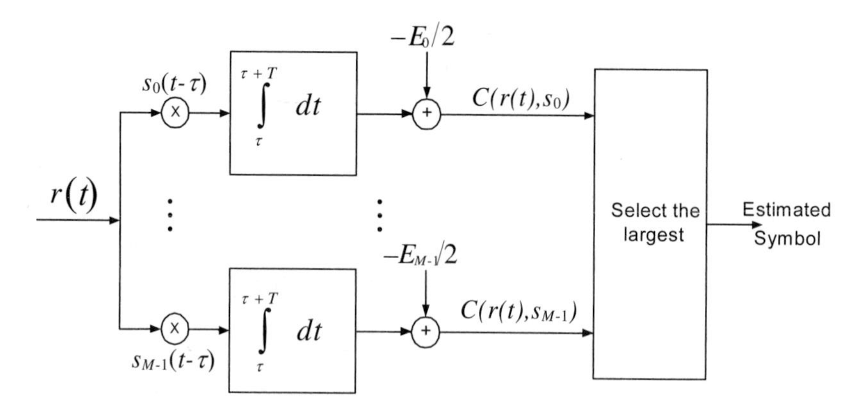

Figure 8-3 Optimum receiver for AWGN channels.

Let us now apply the above analysis to the case of IR-UWB signals. In particular, we will first examine receiver structures for orthogonal and non-orthogonal PPM and PAM-isolated pulses, and then extend the analysis to the multi-pulse case.

8.1.1 The Isolated Pulse Receiver for Binary Orthogonal PPM

In binary orthogonal PPM, $M = 2$ and the two possible transmitted signals are:

$$s_m(t)=\begin{cases}\sqrt{E_{TX}}\,p_0(t) & for\ b=0\\ \sqrt{E_{TX}}\,p_1(t)=\sqrt{E_{TX}}\,p_0(t-\varepsilon) & for\ b=1\end{cases} \tag{8–13}$$

where $p_0(t)$ is the energy-normalized waveform of the basic pulse, E_{TX} is the transmitted energy per pulse, and ε is the time shift introduced by PPM. If ε is larger than pulse duration T_M, the set of orthonormal functions can be formed by $p_0(t)$ and $p_1(t)$, that is:

$$s_m(t) = s_{m0}p_0(t) + s_{m1}p_1(t) \quad m = 0,1$$

$$\text{where} \begin{cases} s_{00} = \sqrt{E_{TX}} \\ s_{01} = 0 \\ s_{10} = 0 \\ s_{11} = \sqrt{E_{TX}} \end{cases} \tag{8–14}$$

The optimum receiver scheme for the above signal format is formed by a bank of two correlators, as shown in Figure 8–4.

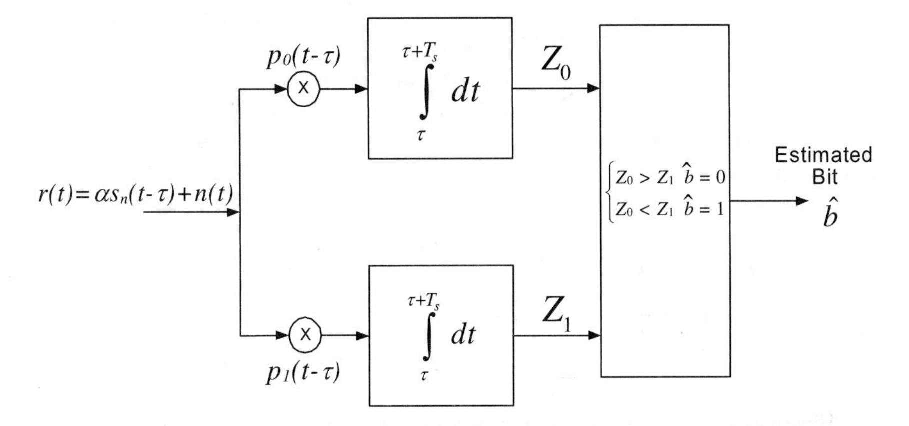

Figure 8–4 Optimum receiver for orthogonal 2PPM.

The decision variables at the output of the correlators are expressed as follows:

$$\begin{aligned} Z_0 &= \alpha s_{m0} + n_0 \\ Z_1 &= \alpha s_{m1} + n_1 \end{aligned} \tag{8–15}$$

where n_0 and n_1 are two independent and equally distributed Gaussian random variables with zero mean and variance $N_0/2$. In the presence of TH coding, the correlator scheme of Figure 8–4 may be modified into Figure 8–5, where the c_j term is the generic j-th coefficient of the TH code assigned to the user under examination.

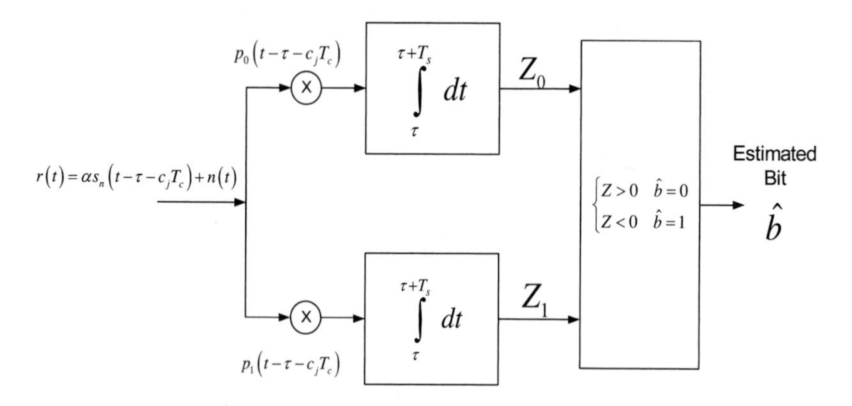

Figure 8-5 Optimum receiver for orthogonal 2PPM-TH.

Note that the scheme of Figure 8–5 can be simplified into an equivalent scheme that uses only one correlator, as shown in Figure 8–6.

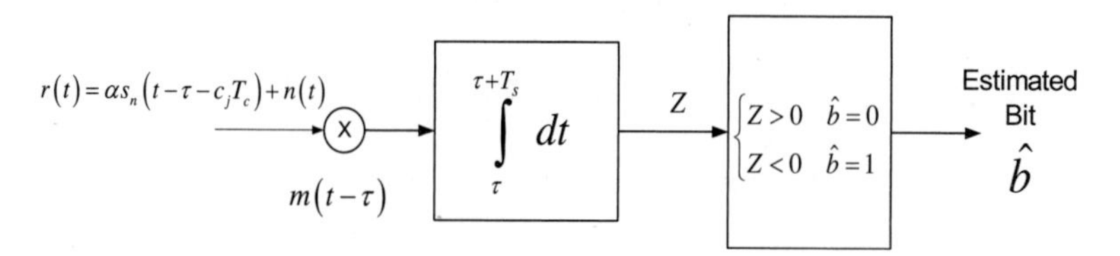

Figure 8–6 2PPM-TH optimum receiver scheme based on a single correlator.

In the one-correlator scheme of Figure 8–6, the incoming signal is multiplied by $m(t)$, indicated as the correlation mask, where $m(t) = p_0(t-\tau-c_jT_c) - p_0(t-\tau-c_jT_c-\varepsilon)$.

At the output of the correlator, one has:

$$Z = \alpha s_m + n_0 - n_1$$
$$\text{where } s_m = \begin{cases} s_0 = +\sqrt{E_{TX}} & for\ b = 0 \\ s_1 = -\sqrt{E_{TX}} & for\ b = 1 \end{cases} \tag{8–16}$$

For independent and equally probable transmitted bits, the average probability of error $\Pr_b$ for an optimum receiver is given by:

$$\begin{aligned} \Pr_b &= \frac{1}{2}\text{Prob}(Z > 0|b=1) + \frac{1}{2}\text{Prob}(Z < 0|b=0) = \text{Prob}(Z < 0|b=0) = \\ &= \text{Prob}\left(\alpha\sqrt{E_{TX}} + n_0 - n_1 < 0\right) = \text{Prob}\left(\sqrt{E_{RX}} + n_0 - n_1 < 0\right) \end{aligned} \tag{8–17}$$

where $E_{RX} = \alpha^2 E_{TX}$ is the received energy per pulse. Since n_0 and n_1 are independent and equally distributed Gaussian random variables, Eq. (8–17) can be rewritten as follows:

$$\Pr_b = \text{Prob}\left(x > \sqrt{E_{RX}}\right) \tag{8–18}$$

where $x = n_1 - n_0$ is a Gaussian random variable with zero mean and variance N_0. $\Pr_b$ can thus be expressed as follows:

$$\Pr_b = \frac{1}{2} erfc\left(\sqrt{\frac{E_{RX}}{2N_0}}\right)$$
$$\text{where} \quad erfc(y) = \frac{2}{\sqrt{\pi}} \int_y^{+\infty} e^{-\xi^2} \, d\xi \tag{8–19}$$

which, as already indicated in Chapter 6 (see Eq. 6–30) represents the average error probability on the bit for binary orthogonal PPM signals. Figure 8–7 shows $\Pr_b$ as a function of E_{RX}/N_0.

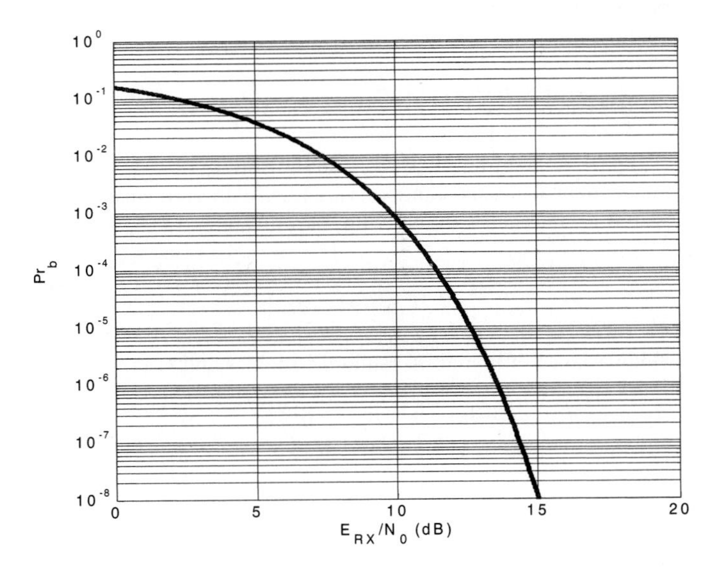

Figure 8–7 Average probability of error $\Pr_b$ for binary orthogonal PPM.

8.1.2 The Isolated Pulse Receiver for Non-Orthogonal Binary PPM

The receiver structure of Figure 8–6 is also valid in the non-orthogonal binary PPM case, in which the PPM shift ε is smaller than pulse duration T_M. The transmitted waveform is still expressed by Eq. (8–14), but the output of the signal correlator in this case is:

$$Z = \alpha s_m + n_0 - n_1$$
$$\text{where } s_m = \begin{cases} s_0 = +\sqrt{E_{TX}}\left(1 - R_0(\varepsilon)\right) & \text{for } b = 0 \\ s_1 = -\sqrt{E_{TX}}\left(1 - R_0(\varepsilon)\right) & \text{for } b = 1 \end{cases} \quad (8\text{–}20)$$

where $R_0(t)$ is the autocorrelation function of the pulse waveform $p_0(t)$. The average bit error probability in this case is:

$$\Pr_b = \text{Prob}\left(Z < 0 \middle| b = 0\right) = \text{Prob}\left(\sqrt{E_{RX}}\left(1 - R_0(\varepsilon)\right) + n_0 - n_1 < 0\right) \quad (8\text{–}21)$$

which leads to the following expression:

$$\Pr_b = \text{Prob}\left(n_1 - n_0 > \sqrt{E_{RX}}\left(1 - R_0(\varepsilon)\right)\right) = \text{Prob}\left(x > \sqrt{E_{RX}}\left(1 - R_0(\varepsilon)\right)\right) \quad (8\text{–}22)$$

The variable x in Eq. (8–22) can be interpreted as the sum of two equally distributed but non-statistically independent Gaussian random variables with zero mean and variance $N_0/2$. The variable x is thus a Gaussian variable with zero mean, and the variance is given by:

$$\sigma_x^2 = 2\sigma_{n_0}^2 + 2Cov\left(n_1, -n_0\right) \quad (8\text{–}23)$$

where $Cov(n_1,-n_0)$ is the covariance of n_1 and $-n_0$ random variables and is expressed by:

$$Cov\left(n_1, -n_0\right) = \left\langle -n_1 n_0 \right\rangle - \left\langle n_1 \right\rangle \left\langle -n_0 \right\rangle = -\left\langle n_1 n_0 \right\rangle \quad (8\text{–}24)$$

which yields:

$$\begin{aligned} Cov(n_1,-n_0) &= -\int_{\tau}^{T_s+\tau}\int_{\tau}^{T_s+\tau} \langle n(x)n(y)\rangle\, p_0(x-\tau)\, p_1(y-\tau)\,dx\,dy \\ &= -\frac{N_0}{2}\int_{\tau}^{T_s+\tau}\int_{\tau}^{T_s+\tau} \delta(x-y)\, p_0(x-\tau)\, p_1(y-\tau)\,dx\,dy \\ &= -\frac{N_0}{2}\int_{\tau}^{T_s+\tau} p_0(x-\tau)\, p_1(x-\tau)\,dx \\ &= -\frac{N_0}{2}\int_{\tau}^{T_s+\tau} p_0(x-\tau)\, p_0(x-\tau-\varepsilon)\,dx \\ &= -\frac{N_0}{2}R_0(\varepsilon) \end{aligned} \tag{8–25}$$

leading to:

$$\sigma_x^2 = 2\frac{N_0}{2} - 2\frac{N_0}{2}R_0(\varepsilon) = N_0\left(1-R_0(\varepsilon)\right) \tag{8–26}$$

The average probability of error on the bit is thus expressed by:

$$\Pr_b = \frac{1}{2}erfc\left(\sqrt{\frac{E_{RX}\left(1-R_0(\varepsilon)\right)^2}{2N_0\left(1-R_0(\varepsilon)\right)}}\right) = \frac{1}{2}erfc\left(\sqrt{\frac{E_{RX}\left(1-R_0(\varepsilon)\right)}{2N_0}}\right) \tag{8–27}$$

Equation (8–27) simplifies into Eq. (8–19) when $R_0(\varepsilon) = 0$, that is, when the PPM modulator uses orthogonal signaling. If $R_0(\varepsilon) > 0$, the receiver experiences a loss in performance, that is, an increased signal energy is required to achieve the same error probability as orthogonal signaling. If $R_0(\varepsilon) < 0$, performance is improved showing the role of the PPM shift ε in the design of a PPM modulator. The optimum PPM shift ε_{opt} is the ε value, which satisfies the following condition:

$$R_0(\varepsilon_{opt}) \le R_0(\varepsilon) \quad \forall\varepsilon \tag{8–28}$$

8.1.3 The Isolated Pulse Receiver for Orthogonal M-ary PPM

The case of M-ary PPM can be considered an extension of binary PPM. The ε value is assumed to be larger than pulse duration T_M. The structure of the optimum receiver is shown in Figure 8–8, and the decision variables at the output of the signal correlator are expressed as follows:

$$
\begin{aligned}
Z_0 &= \alpha\, s_{m0} + n_0 \\
&\vdots \\
Z_{M-1} &= \alpha s_{m(M-1)} + n_{M-1}
\end{aligned}
\tag{8–29}
$$

$$
\text{where } s_{mk} = \sqrt{E_{TX}} \int_0^{T_s} p_0(t - m\varepsilon)\, p_0(t - k\varepsilon)\, dt
$$

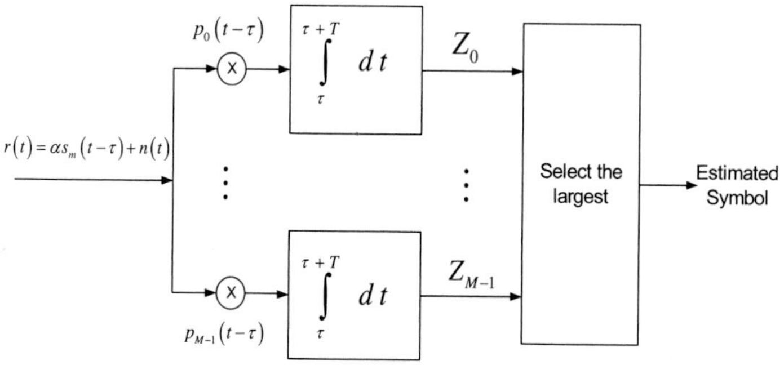

$r(t) = \alpha s_m(t-\tau) + n(t)$
$p_0(t-\tau)$
$\int_\tau^{\tau+T} dt$
Z_0
$p_{M-1}(t-\tau)$
$\int_\tau^{\tau+T} dt$
Z_{M-1}
Select the largest
Estimated Symbol

Figure 8-8 Optimum receiver for orthogonal M-PPM with TH.

If the M possible generated waveforms have equal probability $1/M$, the average error probability on the symbol $\Pr_e$ is the probability of misdetecting one of the symbols. Assume, for example, that $s_0(t)$ was transmitted. An error occurs at the receiver if at least one of the M-1 outputs Z_k, with $k \neq 0$, is larger than Z_0. The probability for this event to occur, that is, the average error probability on the symbol $\Pr_e$, is complementary to the probability of correct decision $\Pr_c$, which can be evaluated by averaging over all possible Z_0 values the joint probability that all noise components $n_1, \ldots, n_{(M-1)}$ are smaller than the useful output, or:

$$
\begin{aligned}
\Pr_e &= 1 - \Pr_c = 1 - \int_{-\infty}^{+\infty} \text{Prob}\left(n_1 < Z_0, \ldots, n_{M-1} < Z_0 \mid Z_0\right) p(Z_0)\, dZ_0 = \\
&= 1 - \int_{-\infty}^{+\infty} \text{Prob}\left(n_x < Z_0 \mid Z_0\right)^{M-1} p(Z_0)\, dZ_0
\end{aligned}
\tag{8–30}
$$

where n_x is a Gaussian random variable with zero mean and variance $N_0/2$. Given the hypothesis of Gaussian noise, it is easily shown that:

$$
\Pr_e = \frac{1}{\sqrt{2\pi}} \int_{-\infty}^{+\infty} \left(1 - \left(\frac{1}{\sqrt{2\pi}} \int_{-\infty}^{y} e^{\frac{-x^2}{2}} dx \right)^{M-1} \right) e^{\left[-\frac{1}{2}\left(y - \sqrt{\frac{2E_{RX}}{N_0}} \right)^2 \right]} dy \tag{8–31}
$$

Equation (8–31) provides the average error probability on the symbol $\Pr_e$ for orthogonal M-ary PPM. This probability can be evaluated using numerical methods.

8.1.4 The Isolated Pulse Receiver for Binary Antipodal PAM

In the case of binary antipodal PAM, the two possible signals generated by the transmitter are:

$$s_m(t) = \begin{cases} \sqrt{E_{TX}}\, p_0(t) & \text{if} \quad b = 1 \\ \sqrt{E_{TX}}\, p_1(t) = -\sqrt{E_{TX}}\, p_0(t) & \text{if} \quad b = 0 \end{cases} \tag{8–32}$$

where $p_0(t)$ is the energy-normalized waveform of the basic pulse and E_{TX} is the transmitted energy per pulse. Since $p_1(t) = -p_0(t)$, these signals are called antipodal. Both $s_1(t)$ and $s_2(t)$ are proportional to the same basic signal $p_0(t)$, that is:

$$s_m(t) = s_m p_0(t) \quad m = 0,1$$
$$\text{with} \begin{cases} s_0 = -\sqrt{E_{TX}} \\ s_1 = +\sqrt{E_{TX}} \end{cases} \tag{8–33}$$

The optimum receiver for the above signal format is thus composed of one correlator, as shown in Figure 8–9.

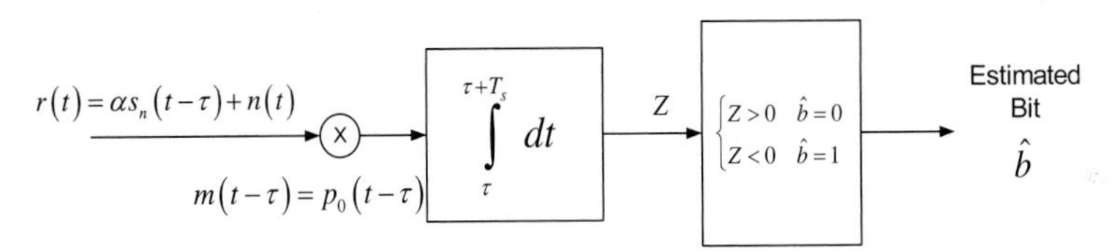

Figure 8–9 Optimum receiver for antipodal 2PAM.

The output of the correlator is given by:

$$Z = \alpha s_m + n \tag{8–34}$$

where n is a Gaussian random variable with zero mean and variance $N_0/2$.

When using a DS code rather than a TH code, the scheme in Figure 8–9 modifies into Figure 8–10, in which the c_j term indicates the generic j-th binary antipodal coefficient of the DS code assigned to the user under examination and $m(t) = c_j p_0(t-\tau)$ is the correlation mask of the correlator.

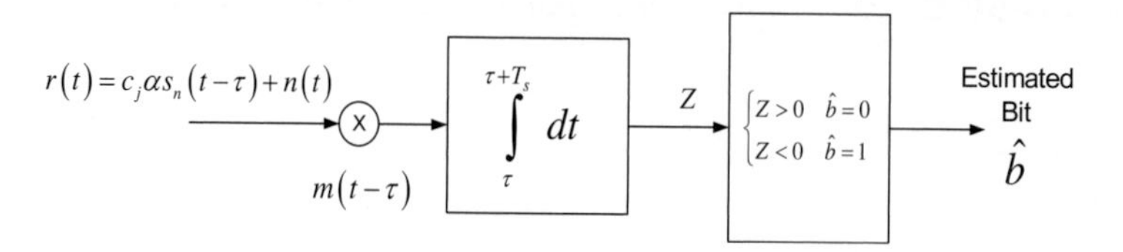

Figure 8–10 Optimum receiver for antipodal 2PAM-DS.

The output of the correlator of Figure 8–10 is given by:

$$Z = \alpha s_m + n$$
$$\text{where} \quad s_m = \begin{cases} s_0 = +\sqrt{E_{TX}} & for\, b=1 \\ s_1 = -\sqrt{E_{TX}} & for\, b=0 \end{cases} \tag{8–35}$$

where n is a Gaussian random variable with zero mean and variance $N_0/2$. For independent and equally probable transmitted bits, the average probability of error on the bit $\Pr_b$ can be expressed as follows:

$$\Pr_b = \frac{1}{2}\text{Prob}\left(Z<0\middle|b=1\right)+\frac{1}{2}\text{Prob}\left(Z>0\middle|b=0\right)=\text{Prob}\left(Z>0\middle|b=0\right)=$$
$$=\text{Prob}\left(-\alpha\sqrt{E_{TX}}+n>0\right)=\text{Prob}\left(n>\alpha\sqrt{E_{TX}}\right)=\text{Prob}\left(n>\sqrt{E_{RX}}\right) \tag{8–36}$$

where $E_{RX} = \alpha^2 E_{TX}$ is the received energy per pulse. $\Pr_b$ is thus expressed as follows:

$$\Pr_b = \frac{1}{2}erfc\left(\sqrt{\frac{E_{RX}}{N_0}}\right) \tag{8–37}$$

which represents the average error probability for binary antipodal signals. Figure 8-11 shows this probability as a function of the ratio E_{RX}/N_0.

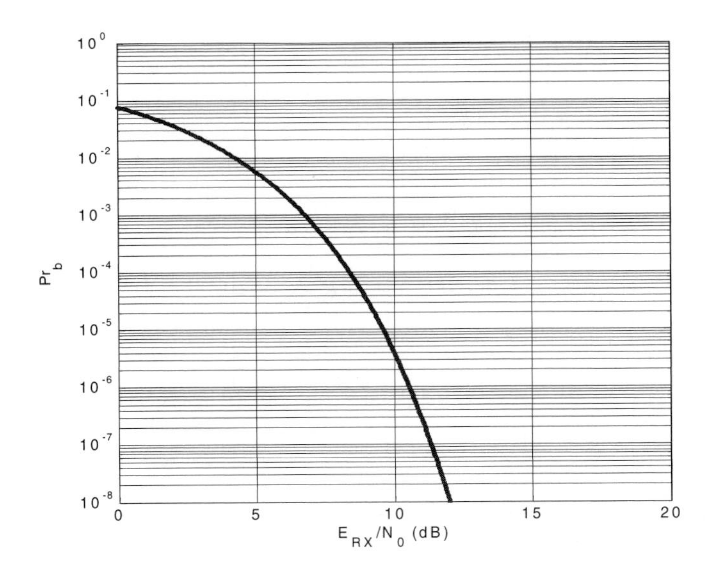

Figure 8–11 Average probability of error for binary antipodal signaling.

As anticipated in Chapter 6, Eq. (8–37) compared to Eq. (8–19) indicates that binary antipodal PAM requires half the energy of binary orthogonal PPM for achieving the same Pr_b.

8.1.5 The Isolated Pulse Receiver for M-ary PAM

In the M-ary PAM case, the received signal can be expressed as follows:

$$\begin{aligned} & r(t) = s_m(t-\tau) + n(t) \\ & \text{where} \quad s_m(t) = A_m \sqrt{E_{RX}}\, p_0(t) \end{aligned} \tag{8–38}$$

$E_{RX} = \alpha^2 E_{TX}$ is the received energy within a basic pulse. The term A_m is given by:

$$A_m = \frac{2m - M + 1}{2}, \quad m = 0, \ldots, M-1 \tag{8–39}$$

The optimum receiver has in this case the structure of Figure 8–12, where E_m is defined as follows:

$$E_m = (A_m)^2 E_{RX} = \left(\frac{2m - M + 1}{2}\right)^2 E_{RX} \qquad (8\text{–}40)$$
$$\forall m = 0, \ldots, M-1$$

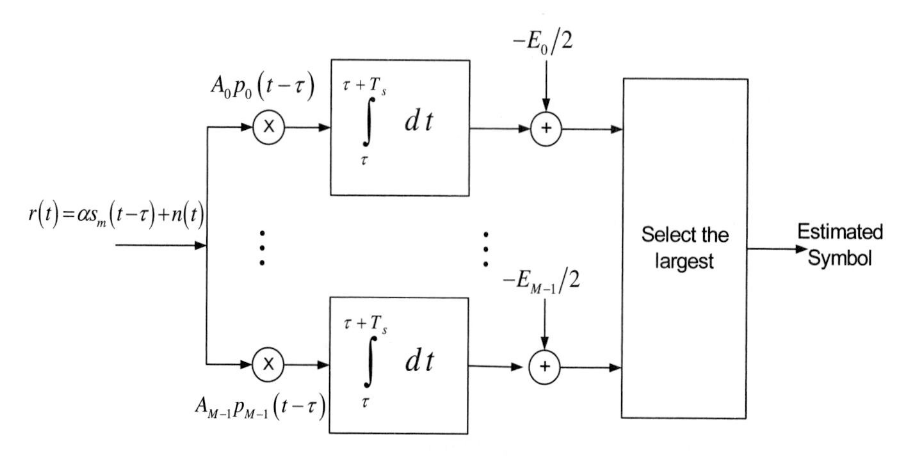

Figure 8–12 Optimum receiver for M-ary PAM.

A receiver scheme that is equivalent to Figure 8–12 is shown in Figure 8–13. This scheme is characterized by the presence of only one correlator, followed by a detector that compares Z with a set of M-1 thresholds w_m. The thresholds are defined as follows:

$$w_m = \sqrt{E_{RX}} \frac{A_m + A_{m+1}}{2}, \quad m = 0, \ldots, M-2 \qquad (8\text{–}41)$$

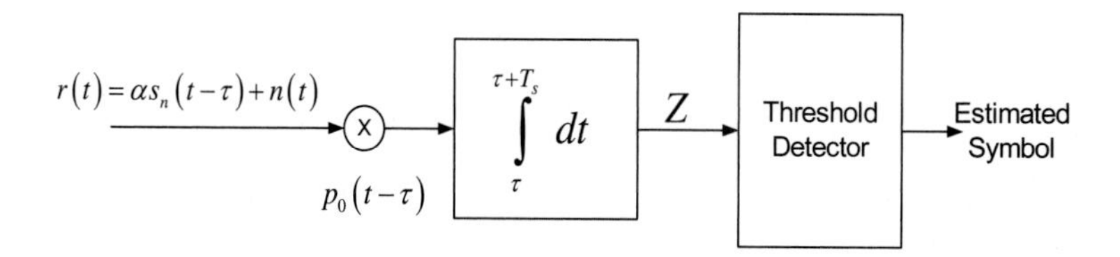

Figure 8–13 Alternative scheme for the optimum receiver for M-ary PAM.

The decision rule can be expressed as follows:

$$\begin{cases} Z < w_0 & \Rightarrow m = 0 \\ w_0 \le Z < w_1 & \Rightarrow m = 1 \\ \vdots & \vdots \\ w_{M-2} \le Z < w_{M-1} & \Rightarrow m = M-2 \\ w_{M-1} \le Z & \Rightarrow m = M-1 \end{cases} \tag{8–42}$$

Note that in both schemes, multiple correlators and single correlator, the knowledge of E_{RX}, that is, of channel gain α, is required. On the contrary, knowing α is not required in PPM since PPM conveys information by modifying pulse position rather than pulse amplitude.

If all A_m are equally probable, the average probability of error on the symbol $\Pr_e$ for M-ary PAM can be expressed as follows:

$$\Pr_e = (M-2)\text{Prob}(|n| > \eta)\frac{1}{M} + \text{Prob}(n > \eta)\frac{1}{M} + \text{Prob}(n < -\eta)\frac{1}{M} \tag{8–43}$$

where n is the noise variable at the output of the correlator and η is a constant term that corresponds to half the distance between two adjacent threshold values, that is:

$$\eta = \frac{1}{2}(w_1 - w_0) = \frac{1}{2}\sqrt{E_{RX}}\left(\frac{A_1 + A_2}{2} - \frac{A_0 + A_1}{2}\right) = \frac{1}{4}\sqrt{E_{RX}}\,(A_2 - A_0) = \frac{\sqrt{E_{RX}}}{2} \tag{8–44}$$

Equation (8–43) indicates that for the (M-2) internal symbols, that is, those symbols that do not correspond to the extreme values of the range of input amplitudes, an error occurs when the absolute value of n exceeds the η value of Eq. (8–44). For the lower (higher) extreme symbol, that is, the lower (higher) limit of the range of input amplitudes, an error occurs when the n value is positive (negative) and greater than η. Due to the symmetry of the probability distribution of n around its average value (which is zero), Eq. (8–43) can be rewritten as follows:

$$\begin{aligned} \Pr_e &= (M-2)\text{Prob}(|n| > \eta)\frac{1}{M} + \frac{1}{2}\text{Prob}(|n| > \eta)\frac{1}{M} + \frac{1}{2}\text{Prob}(|n| > \eta)\frac{1}{M} = \\ &= \left(\frac{(M-2)}{M} + \frac{1}{M}\right)\text{Prob}(|n| > \eta) = \left(1 - \frac{1}{M}\right)\text{Prob}(|n| > \eta) \end{aligned} \tag{8–45}$$

leading to the following expression of the average symbol error probability $\Pr_e$:

$$\Pr_e = \left(1 - \frac{1}{M}\right) erfc\left(\sqrt{\frac{E_{RX}}{2N_0}}\right) \tag{8–46}$$

Note that E_{RX} in Eq. (8–46) has a different meaning from E_{RX} of Eq. (8–37). In the case of M-ary PAM, in fact, E_{RX} is the received energy in a pulse and does not coincide with the average received energy per pulse $\overline{E}_{RX}$, which is expressed by:

$$\overline{E}_{RX} = \frac{1}{M} E_{RX} \sum_{m=0}^{M-1} \left(\frac{2m - M + 1}{2} \right)^2 = \frac{E_{RX}}{4M} \sum_{m=0}^{M-1} \left(2m - M + 1\right)^2 \qquad (8\text{–}47)$$

8.1.6 Receiver Schemes for Multi-Pulse Signals

Results presented in Sections 8.1.1–8.1.5 refer to transmissions characterized by one pulse per symbol. In all cases, receiver performance improves with E_{RX}/N_0. In other words, when only thermal noise corrupts the received signal, it is always possible to find a value for the transmitted energy per pulse that guarantees a target value of the symbol error rate $\Pr_e$. To improve performance, the transmitter can, however, also introduce redundancy by increasing the number of pulses per symbol.

In the presence of multiple pulses per symbol, two possible strategies can be adopted at the receiver: soft decision detection and hard decision detection.

In soft decision detection, the signal formed by N_s pulses is considered by the receiver as a single multi-pulse signal $s_{mp}(t)$. The received signal is cross-correlated with a correlation mask, which is matched with the train of pulses representing the entire symbol. Consider, for example, the case of binary orthogonal PPM with TH coding. In the presence of N_s pulses per bit, two possible signals are generated by the transmitter in the reference bit period $[0,T_b]$:

$$s_{mp}(t) = \begin{cases} \sqrt{E_{TX}} \sum_{j=0}^{N_s-1} p_0\left(t - jT_s - c_jT_c\right) & for\ b = 0 \\ \sqrt{E_{TX}} \sum_{j=0}^{N_s-1} p_0\left(t - jT_s - c_jT_c - \varepsilon\right) & for\ b = 1 \end{cases} \qquad (8\text{–}48)$$

where $p_0(t)$ is the energy-normalized pulse with time duration T_M, E_{TX} is the transmitted energy per pulse, T_s is the average pulse repetition period, T_c is the chip period, c_j is the j-th coefficient of the TH code assigned to the user, and $\varepsilon > T_M$ is the time shift introduced by the PPM modulator. The multi-pulse signal of Eq. (8–48) can be represented as the linear combination of two multi-pulse orthonormal functions $p_{mp0}(t)$ and $p_{mp1}(t)$, and one has:

$$s_{mp}(t) = s_{m0}p_{mp0}(t) + s_{m1}p_{mp1}(t) \quad m = 0,1$$

$$\text{where} \begin{cases} p_{mp0}(t) = \dfrac{1}{\sqrt{N_s}} \displaystyle\sum_{j=0}^{N_s-1} p_0(t - jT_s - c_jT_c) \\ p_{mp1}(t) = p_{mp0}(t - \varepsilon) \end{cases} \tag{8–49}$$

$$\text{and} \begin{cases} s_{00} = \sqrt{N_s E_{TX}} = \sqrt{E_b} \\ s_{01} = 0 \\ s_{10} = 0 \\ s_{11} = \sqrt{N_s E_{TX}} = \sqrt{E_b} \end{cases}$$

where $E_b = N_sE_{TX}$ is the transmitted energy per bit. Following the same approach of Section 8.1.1, the optimum receiver for the signal of Eq. (8–49) is as shown in Figure 8–5, where the correlation mask $m(t)$ is a multi-pulse signal defined as follows:

$$\begin{aligned} m(t) &= p_{mp0}(t) - p_{mp1}(t) \\ &= \frac{1}{\sqrt{N_s}} \sum_{j=0}^{N_s-1} \left(p_0(t - jT_s - c_jT_c) - p_0(t - jT_s - c_jT_c - \varepsilon) \right) \end{aligned} \tag{8–50}$$

To evaluate the average bit error probability $\Pr_b$, we can repeat step-by-step the same analysis of Section 8.1.1, which leads to:

$$\Pr_b = \frac{1}{2} erfc\left(\sqrt{\frac{E_{bRX}}{2N_0}} \right) = \frac{1}{2} erfc\left(\sqrt{\frac{N_s E_{RX}}{2N_0}} \right) \tag{8–51}$$

where E_{bRX} represents the received energy per bit and E_{RX} is the received energy per pulse. Equation (8–51) shows that by increasing the number of pulses per bit, the received energy is increased by a factor N_s, with a consequent reduction in the bit error probability. Note that this is obtained without increasing the average transmitted power, $P_{av} = E_{RX}/T_s$, but has the cost of dividing the bit rate by N_s.

The above result was derived in the case of binary orthogonal PPM, but it immediately extends to all modulation formats discussed in Sections 8.1.1–8.1.5.

In hard decision detection, the receiver implements N_s independent decisions over the N_s pulses that represent one bit. The final decision is obtained by applying a simple majority criterion. Given the number of pulses falling over a threshold and comparing this number with the number of pulses falling below the same threshold, the estimated bit corresponds to the highest of these two numbers. An error occurs if more than half of the pulses are misinterpreted, and the symbol error probability is given by:

$$\Pr_e = \sum_{j=\left\lceil \frac{N_s}{2} \right\rceil}^{N_s} \binom{N_S}{j} \Pr_{e0}{}^{j} \left(1 - \Pr_{e0}\right)^{N_s - j} \tag{8–52}$$

where $\Pr_{e0}$ is the probability of error on the single decision, that is, the probability of error for an isolated pulse receiver as derived in Sections 8.1.1–8.1.5 for the different modulation formats.

Performance comparison between soft and hard detection is a typical problem of digital communications. A detailed analysis can be found in (Proakis, 1995) for different coding schemes (remember that the presence of N_s pulses per bit derives from the application of a code repetition coder before modulation). It can be shown, in general, that for a variety of code families, soft decision outperforms hard decision when considering propagation over an AWGN channel. This result will be shown on simulation data in Checkpoint 8–1.

The above performance loss in hard decision versus soft decision refers to the Gaussian noise case. In the presence of non-Gaussian noise, as with interference noise, for example, an opposite trend might be observed. With reference to the IR-UWB case, in particular (Weeks et al., 1999), showed that hard decision can perform better than soft decision when several UWB interfering signals are present at the receiver. Performance of hard decision in this case is mainly affected by the total number of interferers, that is the symbol error rate decreases due to the presence of collisions when deciding for a single pulse, while performance of the soft decision scheme is mainly affected by the average interfering power that is collected at the receiver during an entire symbol period. In the presence of great variations in the interfering signal powers, as occurs for example in the case of a few close interferers and several distant interferers, hard decision can perform better than soft decision.

CHECKPOINT 8–1

In this checkpoint, we will simulate the behavior of the receivers described in Section 8.1. New MATLAB functions are required for simulating both the propagation of the transmitted signal over the AWGN channel and detection at the receiver. Performance is evaluated in terms of measured average error probability at the output of the detector.

Regarding the transmitter, PPM and PAM signals can be generated by using the MATLAB functions introduced in Chapter 2, in particular Functions 2.6 and 2.9 which generate PPM-TH and PAM-DS signals, respectively.

To simulate propagation of these signals over the AWGN channel, three MATLAB functions are defined: Function 8.1 for emulating the attenuation of the signal energy with distance and Functions 8.2 and 8.3 for introducing thermal noise at the receiver input. We assume perfect synchronization between transmitter and receiver and $\tau = 0$, that is, no additional MATLAB functions are required for simulating the effect of channel delay.

Function 8.1 (see Appendix 8.A) attenuates the input signal `tx` according to the rule of Eq. (8–3), that is, the channel gain is evaluated as a function of the distance `d` between transmitter and receiver, and two constant terms `gamma` and `c0` representing the power decay with distance and the reference attenuation at 1 meter. Function 8.1 returns vector `rx`, representing the attenuated signal and the value of the channel gain `attn`.

The command line for executing Function 8.1 is:

```
[rx,attn] = cp0801_pathloss(tx,c0,d,gamma);
```

Function 8.2 (see Appendix 8.A) generates a white Gaussian noise signal and adds it to the input signal given via vector `input`. The noise level is determined according to a specified E_b/N_0 ratio, where E_b is the energy per bit of the useful signal at the receiver. Function 8.2 can manage multiple E_b/N_0 values, which must be given as input to the function in vector `ebno`. Values in `ebno` are in logarithmic units. In the presence of multiple E_b/N_0 values, Function 8.2 generates multiple outputs. For a given E_b/N_0 value, Function 8.2 measures the E_b value from the input signal and determines the corresponding N_0 value. Measuring the E_b value, in particular, requires knowledge of the number of bits that are conveyed by the input signal. This value is given as input by `numbits`. Function 8.2 returns the array `output`, which contains all signals resulting from the introduction of Gaussian noise over the input signal. Each row of the array output corresponds to one E_b/N_0 value. Function 8.2 also returns the array `noise`, which contains different realizations of the Gaussian noise, one realization per row.

The command line for executing Function 8.2 is:

```
[output,noise] = cp0801_Gnoise1(input,ebno,numbits);
```

The above command stores the array `output` in memory, which contains all signals generated by Function 8.2 for the different E_b/N_0 values contained in the input vector `ebno`.

We can use Functions 8.1 and 8.2 for evaluating the effect of propagation of 2PPM-TH signals over AWGN channels. First, the transmitted signal is generated by executing Function 2.6 (see Checkpoint 2–1) with the following parameters: `Pow=-30`; `fc=50e9`; `numbits=2`; `Ts=3e-9`; `Ns=1`; `Tc=1e-9`; `Nh=3`; `Np=2`; `Tm=0.5e-9`; `tau=0.25e-9`; `dPPM=0.5e-9`; `G=1`. The command line for generating the 2PPM-TH signal is:

```
[bits,THcode,stx0,ref]=cp0201_transmitter_2PPM_TH;
```

The above command stores vector `bits` in memory, which contains the binary sequence generated by the source and vector `stx0` representing the transmitted signal, and produces the graphical output of Figure 8–14.

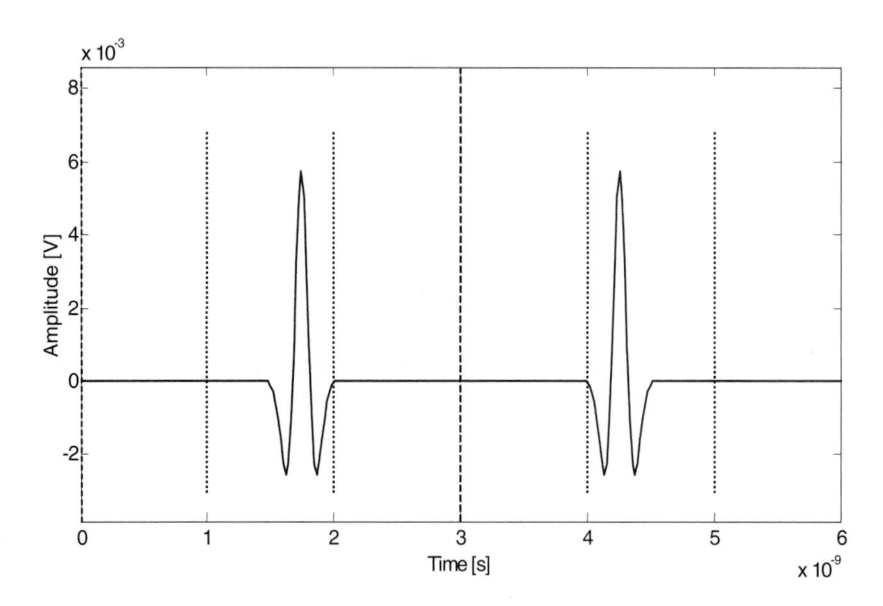

Figure 8–14 Transmitted 2PPM-TH signal `stx0`.

Figure 8–14 shows that the transmitted signal `stx0` consists of two pulses, located in the central slot of the corresponding frame. The peak value of each pulse is on the order of $6{\cdot}10^{-3}$ V.

Function 8.1 can now be used for simulating the attenuation suffered by signal `stx0` during propagation. In particular, we assume the receiver to be 10 meters away from the transmitter in a free space scenario with 30 dB of power loss at a reference distance of 1 meter. The following quantities are thus defined:

```
d = 10;
gamma = 2;
c0 = 10^(-30/20);
```

and then the following command is executed:

```
[srx0,attn] = cp0801_pathloss(stx0,c0,d,gamma);
```

The above command stores vector `srx0` in memory, which represents the attenuated version of signal `stx0`. Figure 8–15 shows the envelope of signal `srx0`.

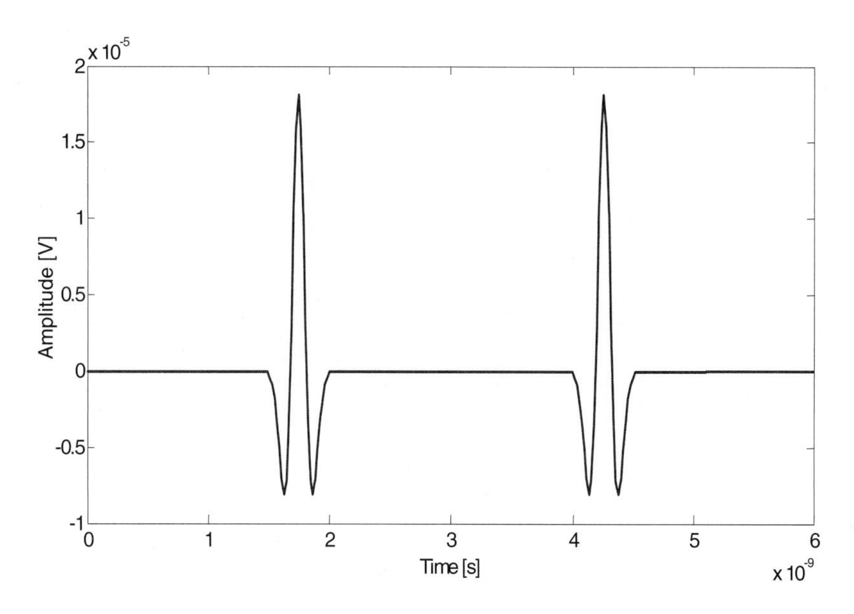

Figure 8–15 Signal `srx0`, that is, signal `stx0` after 10 meters propagation over the free space.

The peak value of the two pulses in Figure 8–15 are on the order of $2 \cdot 10^{-5}$ V. The channel gain α is thus about 0.0033.

Thermal noise can be introduced in signal `srx0` by using Function 8.2. Four different E_b/N_0 values are considered: 0 dB, 10 dB, 20 dB, and 30 dB. The following quantities are thus defined:

```
numbits = length(bits);
ebno = [0 10 20 30];
```

and the following commands are executed:

```
[rx0,noise] = cp0801_Gnoise1(srx0,ebno,numbits);
rx1 = rx0(1,:);
rx2 = rx0(2,:);
rx3 = rx0(3,:);
rx4 = rx0(4,:);
```

The above commands store four different signals in memory, `rx1`, `rx2`, `rx3` and `rx4`, which correspond to the introduction of increasing levels of thermal noise at the receiver. The graphical representation of these signals is presented in Figures 8–16, 8–17, 8–18, and 8–19.

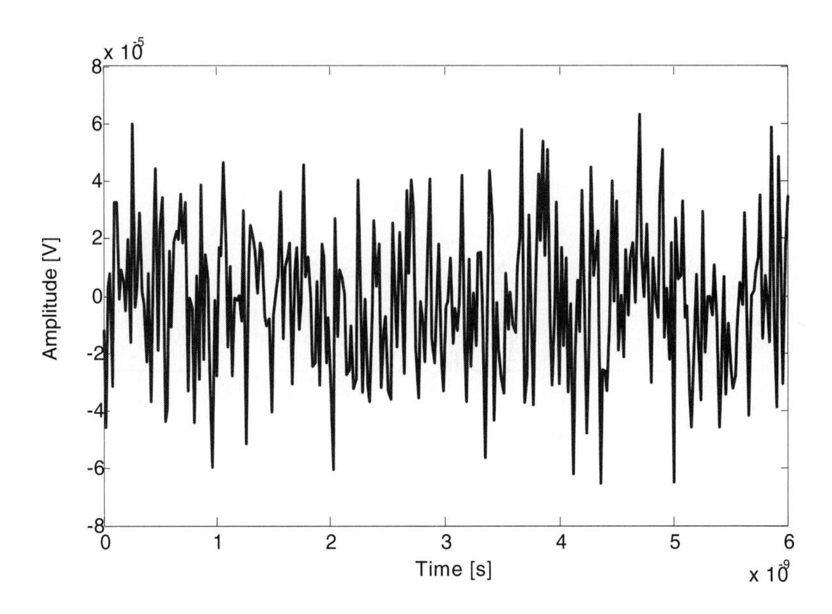

Figure 8–16 Signal rx1 ($E_b/N_0 = 0$ dB).

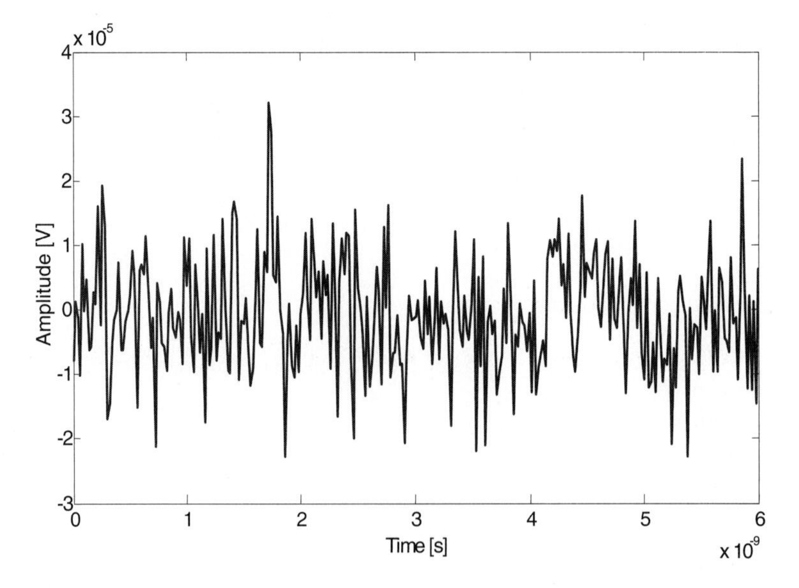

Figure 8–17 Signal rx2 ($E_b/N_0 = 10$ dB).

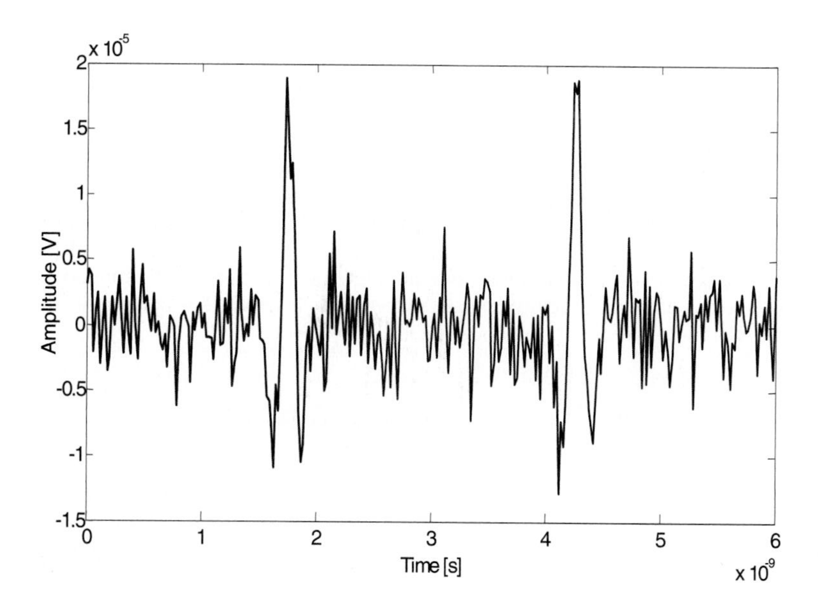

Figure 8–18 Signal rx3 (E_b/N_0 = 20 dB).

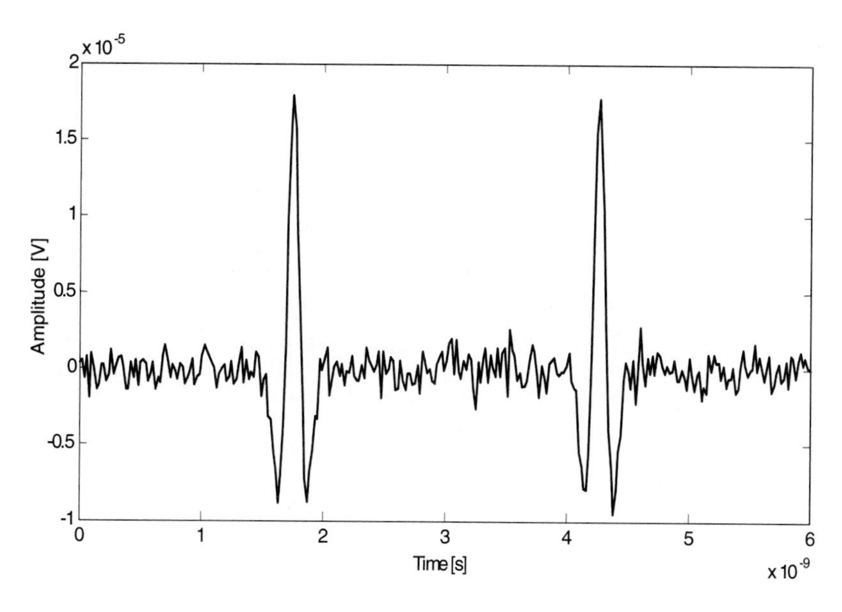

Figure 8–19 Signal rx4 (E_b/N_0 = 30 dB).

Figures 8–16 to 8–19 show the effect over the useful signal waveform of the presence of thermal noise at the receiver. As expected, distortion is more evident for low E_b/N_0 values.

The above simulation, which has focused on binary PPM, can be obtained in a similar way for PAM signals.

Function 8.3 (see Appendix 8.A) is a second MATLAB function that can be used for generating Gaussian noise. Function 8.3 is similar to Function 8.2, except for the fact that the noise level is evaluated according to a target E_x/N_0 value rather than a target E_b/N_0 value, where E_x is the received energy within a single pulse. The E_x value is determined within the function by dividing the total measured energy by the number of transmitted pulses. The number of pulses, which is eventually higher than the number of transmitted bits (in the presence of code repetition coding), must be given as input via the parameter `numpulses`. The command line for executing Function 8.3 is:

```
[output,noise] = cp0801_Gnoise2(input,exno,numpulses);
```

Once again, the array `output` contains signals corresponding to different input values of E_x/N_0, one signal per row.

We can now introduce the MATLAB functions used for simulating receiver activity.

According to the analysis of Section 8.1, the optimum receiver for the AWGN channel consists of a signal correlator followed by a ML detector. As shown in Figure 8–3, the correlator cross-correlates the signal at the receiver input with multiple waveforms to obtain the correlation metrics. The detector then applies a decision rule by observing the correlation metrics. We simulate the above process in the specific cases of binary orthogonal PPM-TH, and of binary antipodal PAM-DS. Functions 8.4 and 8.5 are introduced for the PPM-TH case, while Functions 8.6 and 8.7 serve for the PAM-DS case.

Function 8.4 (see Appendix 8.A) evaluates the correlation mask m(t) of the correlator of Figure 8–6. To determine the correlation mask, the receiver must know the exact position of all pulses within each frame of the transmitted signal. When a PPM-TH signal is generated via Function 2.6 (see Appendix 2.A), this information is available in the output vector `ref`. Vector `ref`, in fact, is a copy of the transmitted signal `Stx`, except for the absence of PPM modulation. Function 8.4 receives as input vector `ref` together with the information on sampling frequency `fc`, number of transmitted pulses `numpulses`, and value of the PPM shift `dPPM` used for generating the transmitted signal. Function 8.4 returns vector `mask`, which represents the waveform of the correlation mask and is executed as follows:

```
[mask] = cp0801_PPMcorrmask(ref,fc,numpulses,dPPM);
```

Note that vector `mask` represents the correlation mask for the whole transmitted signal. To operate the decision over a single bit period, the code that implements the receiver divides the above vector in different parts, each part corresponding to the correlation mask for a single bit period.

Function 8.5 (see Appendix 8.A) simulates the activity of both correlator and detector in the receiver scheme of Figure 8–6, and evaluates receiver performance in terms of average probability of error on the bit Pr_b as measured after decision. Multiple signals corresponding to the same binary stream can be analyzed all together, that is, the Pr_b for different values of noise level can be evaluated with a single run of the code.

The following inputs are required: array `R` representing the different waveforms to be considered (one waveform per row), vector `mask` representing the correlation mask (the same for all signals), the value of the sampling frequency `fc`, the original binary stream generated by the transmitter `bits`, the number of pulses per bit `Ns`, and the average pulse repetition period in seconds `Ts`. Function 8.5 returns two outputs: the array `RXbits` containing the different binary streams at the receiver output (one stream on each row) and vector `BER` containing the measured Pr_b values. The command line for executing Function 8.5 is:

```
[RXbits,BER] = cp0801_PPMreceiver(R, mask, fc,...
    bits, numbit, Ns, Ts);
```

Function 8.6 (see Appendix 8.A) evaluates the correlation mask m(t) in the case of PAM-DS signals (see Figure 8–10). The algorithm implemented in Function 8.6 is similar to Function 8.4. It requires the following inputs: the reference vector `ref`, which is provided by Function 2.9 when generating the PAM-DS signal; the value of the sampling frequency `fc`; and the number of transmitted pulses `numpulses`. Function 8.6 returns vector `mask`, which represents the waveform of the correlation mask and is executed as follows:

```
[mask] = cp0801_PAMcorrmask(ref,fc,numpulses);
```

Once again, note that vector `mask` represents the correlation mask for the whole transmitted signal. To operate the decision over a single bit period, the receiver divides the above vector into different parts, each part corresponding to the correlation mask for a single bit period.

Function 8.7 (see Appendix 8.A) simulates the activity of both correlator and detector for the scheme in Figure 8–10, and evaluates receiver performance in terms of measured Pr_b at the receiver output. Function 8.7 is similar to Function 8.5, except for the decision rule. In the PAM case, in fact, the detector decides for a 0 bit in the presence of negative values at the correlation output, and for a 1 bit for positive values at the correlation output. The command line for executing Function 8.7 is:

```
[RXbits,BER] = ...
    cp0801_PAMreceiver(R,mask,fc,bits,numbit,Ns,Ts);
```

Functions 8.5 to 8.7 can be used to evaluate by simulation the performance of the receiver structures described in Section 8.1. We start by considering the case of binary orthogonal PPM-TH. Two UWB signals with different N_s values are generated: signal `s_ppm1` with `Ns=1` and signal `s_ppm2` with `Ns=3`. The remaining parameters of Function 2.6 are the same for both signals: `Pow=-30`; `fc=50e9`; `numbits=10000`; `Ts=3e-9`; `Tc=1e-9`; `Nh=3`; `Np=30000`; `Tm=0.5e-9`; `tau=0.25e-9`; `dPPM=0.5e-9`; `G=0`.

The output parameters for the two signals are obtained as follows:

```
[bits1,THcode,s_ppm1,ref1]=cp0201_transmitter_2PPM_TH;
[bits2,THcode,s_ppm2,ref2]=cp0201_transmitter_2PPM_TH;
```

Note that in the case of signal `s_ppm2`, the transmitter generates three pulses for each bit. The number of transmitted bits is thus the same as signal `s_ppm1`, but the number of transmitted pulses is multiplied by three.

Given `s_ppm1` and `s_ppm2`, we can determine the corresponding correlator masks, `mppm1` for signal `s_ppm1` and `mppm2` for `s_ppm2`:

```
fc = 50e9;
dPPM = 0.5e-9;
Ts = 3e-9;
numbit = 10000;
numpulses1 = 10000;
[mppm1] = cp0801_PPMcorrmask(ref1,fc,numpulses1,dPPM);
numpulses2 = 30000;
[mppm2] = cp0801_PPMcorrmask(ref2,fc,numpulses2,dPPM);
```

We can now simulate the effect of propagation over the AWGN channel. Our aim is to evaluate performance for the different signal formats as a function of the signal-to-noise ratio at the receiver. In particular, we are interested in evaluating Pr_b vs. E_x/N_0, where E_x is the energy received within a single pulse. Note that both Functions 8.2 and 8.3 introduce thermal noise by measuring the energy of the received signal, that is, either E_b or E_x, and by then generating a noise signal with the corresponding N_0 value. As a consequence, in this particular analysis, channel gain does not affect receiver performance and we can introduce thermal noise directly in the transmitted signal. To evaluate Pr_b when E_x/N_0 is in the range between 0 dB and 8 dB, we define the following vector `exno`:

```
exno = [ 0 2 4 6 8 ];
```

Then, we generate noise at the receiver by executing Function 8.3:

```
[rx_ppm1,noise] = cp0801_Gnoise2(s_ppm1, exno,...
   numpulses1);
[rx_ppm2,noise] = cp0801_Gnoise2(s_ppm2, exno,...
   numpulses2);
```

Both arrays `rx_ppm1` and `rx_ppm2` are composed of five rows. Each row is associated with one E_x/N_0 value.

The final step of the simulation requires the introduction of Function 8.5. In the case of signal `s_ppm1`, the parameter `HDSD` does not affect performance since hard decision is equivalent to soft decision given that one pulse is transmitted per bit. The code for evaluating Pr_b in the case of signal `s_ppm1` is:

```
[RXbits1, BER1] = cp0801_PPMreceiver(rx_ppm1,...
   mppm1, fc, bits1, numbit, 1, Ts);
```

The above command stores the following in memory: the five estimated binary streams `RXbits1` and vector `BER` of the corresponding Pr_b values. Figure 8–20 illustrates the result of the simulation by representing Pr_b for different values of E_x/N_0. As expected, Pr_b decreases with E_x/N_0 in dB. Moreover, since the PPM shift of signal `s_ppm1` is equal to the time duration of the pulse, the result in Figure 8–20 corresponds to the case of binary transmission with orthogonal signaling. By comparing Figure 8–20 with Figure 8–7, we verify the agreement of the simulation results with the values derived analytically.

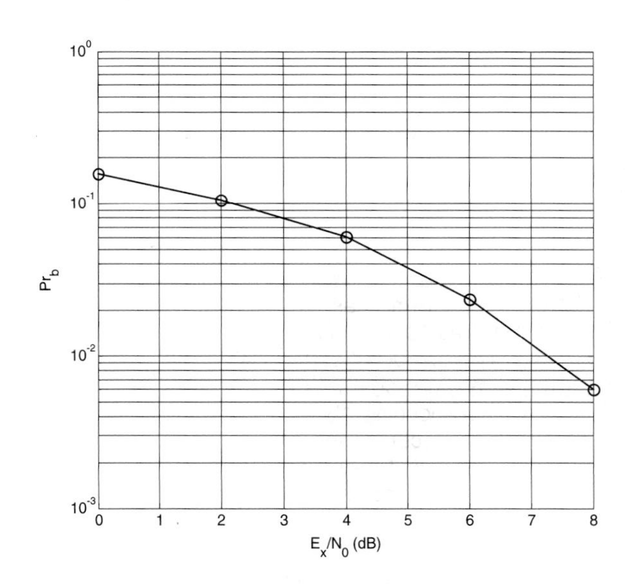

Figure 8–20 Pr_b vs. E_x/N_0 for signal s_ppm1.

In the case of signal s_ppm2, we first execute Function 8.5 with HDSD = 1 and then with HDSD = 2. In both cases, the command line is the following:

```
[RXbits2,BER2] = cp0801_PPMreceiver(rx_ppm2, mppm2,...
    fc, bits2, numbit, 3, Ts);
```

Figure 8–21 compares Pr_b for signal s_ppm1 with Pr_b for signal s_ppm2. Regarding s_ppm2, both hard decision and soft decision cases are represented. As expected, the probability of error on the bit decreases with increasing redundancy represented by the N_s parameter of the code repetition coder. Moreover, the result in Figure 8–21 confirms that in the presence of Gaussian noise, soft decision performs better than hard decision. Note that for $E_x/N_0 > 6$ dB and $N_s = 3$, the measured Pr_b values for hard and soft decision are not represented since the simulation for the given number of simulated bits returns a 0 value.

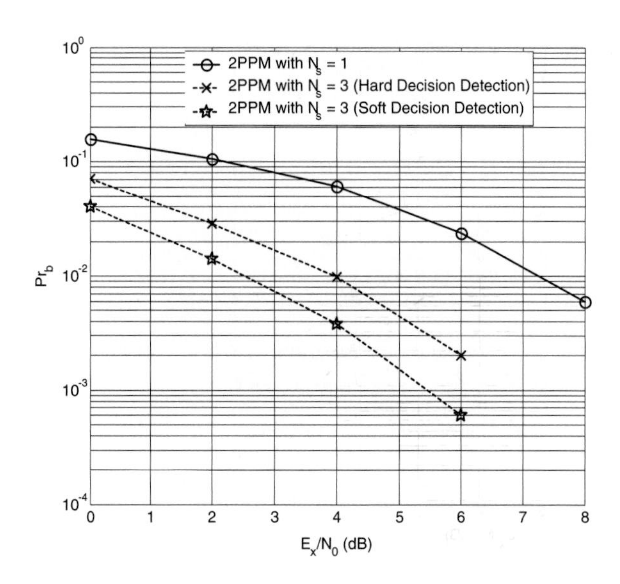

Figure 8–21 Pr_b vs. E_x/N_0 for a 2PPM-TH-UWB signal with one pulse per bit (solid line), three pulses per bit and hard decision detection (crosses), and three pulses per bit and soft decision detection (stars).

The same analysis can be repeated for the PAM-DS case, with the only difference that Functions 8.6 and 8.7 are executed in place of Functions 8.4 and 8.5. Figure 8–22 compares the average Pr_b of 2PAM with 2PPM in the presence of only one pulse per bit, or $N_s = 1$. The result in Figure 8–22 confirms that for the same transmitted signal energy, the use of antipodal signals results in improved performance when using PAM. Simulation results also confirm a gain of 3 dB in transmitted power.

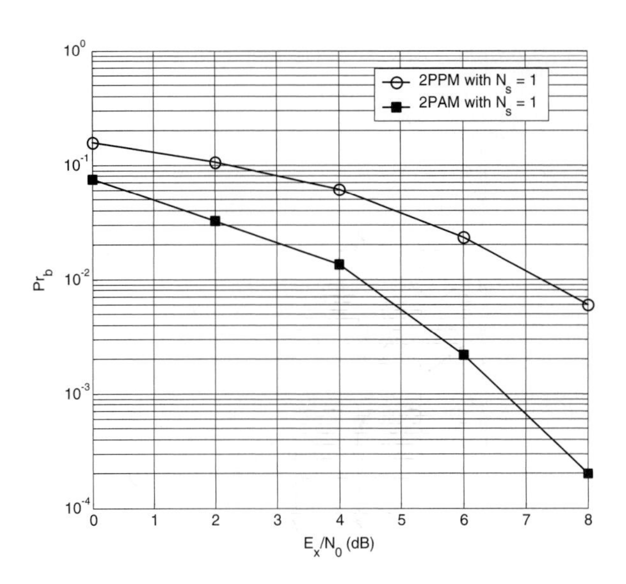

Figure 8–22 Pr_b vs. E_x/N_0 for a 2PPM-TH-UWB signal (white circles), and for a 2PAM-DS-UWB signal (black squares). In both cases, one pulse is transmitted for each bit.

Finally, Figure 8–23 compares the Pr_b versus E_x/N_0 curve for a 2PAM signal with different N_s values and hard versus soft decoding. Once again we observe that Pr_b drastically decreases when increasing redundancy of the code repetition coder, and that soft decision outperforms hard decision.

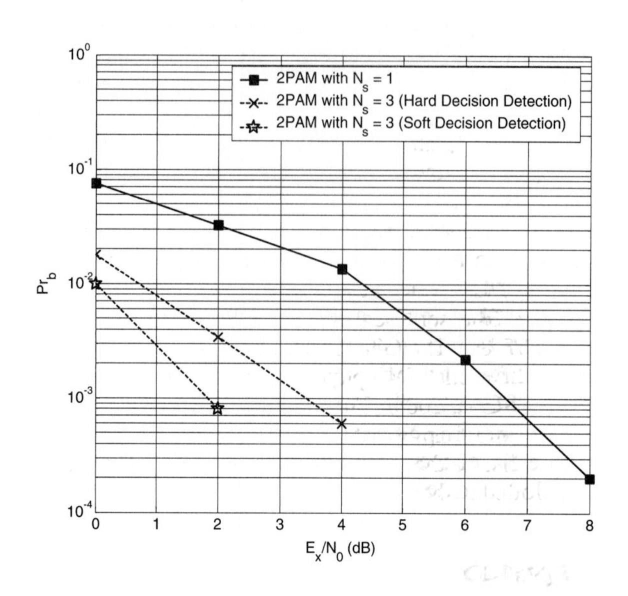

Figure 8–23 Pr_b vs. E_x/N_0 for a 2PAM-DS-UWB signal with one pulse per bit ($N_s = 1$, solid line), three pulses per bit ($N_s = 3$) with hard decision detection (crosses), and three pulses per bit ($N_s = 3$) with soft decision detection (stars).

CHECKPOINT 8–1

8.2 PROPAGATION OVER A MULTI-PATH-AFFECTED UWB RADIO CHANNEL

Propagation over the AWGN channel was analyzed in Section 8.1. As previously discussed, the AWGN channel is characterized by two parameters, channel gain and channel delay, and by AWGN, typically thermal noise, at the receiver. In this case, the structure of the optimum receiver is relatively simple. Basically, the receiver has the task of selecting the one waveform out of M that best matches the received signal.

The presence of multiple paths between transmitter and receiver introduces complexity in both channel model and receiver structure. The channel exhibits time-varying properties that must be taken into account in the channel model. Due to distortion, the received signal often has little resemblance with the transmitted waveform. This is particularly true for indoor transmissions, where propagation is perturbed by a number of interfering objects. The presence of multi-path imposes severe limitations on receiver performance, which can be mitigated if a detailed characterization of the multi-path channel is available at the receiver. The general theory for characterizing multi-path channels is presented in Sections 8.2.1 and 8.2.2.

The first studies of the indoor multi-path channel date back to 1959 (Rice, 1959), and since then, several models have been made available in the literature for indoor narrowband transmissions (Hashemi, 1993a; Saleh and Valenzuela, 1987; Rappaport et al., 1991). With reference to the UWB case, several propagation measurement campaigns for the indoor environment have been carried out, and have given rise to different channel models (Win and Scholtz, 2002; Cassioli et al., 2002b; Cramer et al., 2002a; Turin et al., 2002; Qiu, 2002; Ghassemzadeh and Tarokh, 2003; Ghassemzadeh et al., 2003). In 2003, the channel modeling sub-committee of the IEEE 802.15.SG3a issued an UWB-tailored channel model to be used in the evaluation of the different UWB physical layer proposals submitted in the 802.15.3 IEEE group (IEEE 802.15.SG3a, 2003). This channel model, here called the IEEE channel model, will be described and implemented by simulation in Section 8.2.3 and Checkpoint 8–2. A comprehensive list of the main results published in the past few years regarding the UWB channel is included at the end of Section 8.2.3.

8.2.1 The Impulse Response

Propagation over an indoor channel has the main effect of introducing multiple delayed and attenuated replicas of a transmitted pulse, corresponding to the different propagation paths between transmitter and receiver. The received signal $r(t)$ can be expressed as follows:

$$r(t) = \sum_{n=1}^{N(t)} a_n(t)\, p\left(t - \tau_n(t)\right) + n(t) \tag{8–53}$$

where $a_n(t)$ and $\tau_n(t)$ are the channel gain and delay measured at time t for the n-th path, $N(t)$ is the number of paths observed at time t, and $n(t)$ is the additive noise at the receiver. Equation (8–53) shows that the channel is completely characterized by $N(t)$, $a_n(t)$, and $\tau_n(t)$. The impulse response of the channel can be derived from Eq. (8–53) and can be written as follows:

$$\begin{aligned} r(t) &= s(t) * h(t) + n(t) \\ \text{where} \quad h(t) &= \sum_{n=1}^{N(t)} a_n(t)\, \delta\left(t - \tau_n(t)\right) \end{aligned} \tag{8–54}$$

where $\delta(t)$ is the Dirac function.

Note that the parameters characterizing the channel impulse response of Eq. (8–54) are time-varying, to take into account possible modifications in the propagation environment that are typically provoked by the motion of either the transmitter or the receiver. We can generally assume, however, that this rate of variation is slow compared to the pulse rate. In other words, we assume the channel to be stationary within an observation time T, which is larger than the average pulse repetition period. Under this assumption, Eq. (8–53) can be rewritten as follows:

$$r(t)=\sum_{n=1}^{N} a_n p(t-\tau_n)+n(t) \tag{8–55}$$

and the channel impulse response can be given by:

$$h(t)=\sum_{n=1}^{N} a_n \delta(t-\tau_n) \tag{8–56}$$

The multi-path channel model of Eq. (8–56) was first suggested by Turin in 1956 (Turin, 1956), and is known as the *Turin model.* The Turin model assumes that all parameters characterizing the channel are random variables with specific distributions. Statistical information is required regarding the distribution of channel gains a_n, pulse arrival times τ_n, and number of different paths N, which can be resolved at the receiver.

As suggested in (Cramer et al., 2002; Qiu, 2002), the model of Eq. (8–55) has a major drawback when applied to IR. This model does not account for pulse shape modifications due to reflections or penetration through materials. The shape of a pulse should be dependent on the propagation path, each path being characterized by its own impulse response. Accordingly, the received signal should be written as follows:

$$r(t)=\sum_{n=1}^{N} a_n p_n(t-\tau_n)+n(t) \tag{8–57}$$

where a specific pulse waveform $p_n(t)$ is associated with each path n. As suggested in (Cramer et al., 2002), the multi-path model of Eq. (8–57) can also be used to take into account direction-dependent distortions. Different paths correspond to different angles of arrival at the receiving antenna and to different angles of transmission at the transmitting antenna. In the presence of non-uniform radiation patterns, these paths may therefore be characterized by different distortions. The model of Eq. (8–57) is of great interest to future characterizations of pulse distortion. Due to its complexity, we will restrict the analysis of this section to the simpler model of Eq. (8–56).

In the next three subsections, we describe in further detail three parameters that can be derived from Eq. (8–56) and that characterize multi-path: the total multi-path gain, the root mean square delay spread, and the power delay profile.

8.2.1.1 The Total Multi-Path Gain

A first parameter that is usually derived from the impulse response in Eq. (8–56) is the total multi-path gain G, which measures the total amount of energy collected over the N received pulses when a pulse with unitary energy is transmitted. The G parameter can be determined as follows:

$$G = \sum_{n=1}^{N} |a_n|^2 \tag{8–58}$$

Given the G value, one can write:

$$h(t) = \sqrt{G} \sum_{n=1}^{N} \alpha_n \delta(t - \tau_n) \tag{8–59}$$

where $\alpha_1, \ldots, \alpha_N$ are the energy-normalized channel gain parameters verifying:

$$\sum_{n=1}^{N} |\alpha_n|^2 = 1 \tag{8–60}$$

Note that $G \leq 1$ and is related to the attenuation suffered by the transmitted pulses during propagation. In multi-path environments, G decreases with distance according to the following law:

$$G = \frac{G_0}{D^{\gamma}} \tag{8–61}$$

where G_0 is the reference value for power gain evaluated at $D = 1$ m and γ is the exponent of the power or energy attenuation law. The G_0 value can be evaluated as follows:

$$G_0 = 10^{-A_0/10} \tag{8–62}$$

where A_0 (in dB) represents the path loss at a reference distance $D_0 = 1$ m, that is, $A_0 = 10\log_{10}(E_{TX}/E_{RX0})$. E_{RX0} is the energy of a single pulse at D_0. Values for both A_0 and γ are suggested in (Ghassemzadeh and Tarokh, 2003) for different propagation environments: $A_0 = 47$ dB and $\gamma = 1.7$ for a LOS environment, and $A_0 = 51$ dB and $\gamma = 3.5$ for a NLOS environment. Due to the rapid fluctuations of the a_n values with small movements of the receivers, both A_0 and γ are evaluated by averaging several observations in the neighborhood of the location of interest.

8.2.1.2 The Root Mean Square Delay Spread

A second parameter that is usually derived from the impulse response in Eq. (8–56) is the root mean square (rms) delay spread τ_{rms}, that is:

$$\tau_{rms} = \sqrt{\frac{\sum_{n=1}^{N} \tau_n^2 |a_n|^2}{G} - \left(\frac{\sum_{n=1}^{N} \tau_n |a_n|^2}{G} \right)^2} \tag{8–63}$$

Equation (8–63) measures the effective duration of the channel impulse response. It is a fundamental parameter for evaluating the presence of ISI at the receiver. If the time interval separating two pulses is smaller than τ_{rms}, ISI is present. The reception of one pulse is in fact affected by late arrivals of replicas of previous pulses. In the presence of dense multi-path, as in indoor propagation, the channel impulse response shows significant energy contributions for high τ_n. τ_{rms} is thus high and the time separation between subsequent pulses must be increased accordingly to control ISI. For an environment with small multi-path, as in outdoor environments, the energy of the channel impulse response is concentrated on the first paths, that is, those paths corresponding to the smallest τ_n values. In this case, τ_{rms} is low and the pulse repetition period can be reduced.

8.2.1.3 The Power Delay Profile (PDP)

The Power Delay Profile (PDP) of an impulse response given by Eq. (8–56) is a graphical representation that shows time of arrival of the different contributions versus received power. Time of arrival of a generic path is usually indicated relative to the LOS contribution, which has a time of arrival fixed at 0. An example of a PDP will be shown in Checkpoint 8–2.

8.2.2 The Discrete Time Impulse Response

As suggested by (Hashemi, 1993a), a convenient model for characterizing channels affected by multi-path is to introduce a discrete impulse response model. In this model, the time axis is divided into small time intervals, called *bins*, which are assumed to contain either one multi-path component, or no multi-path component. Having multiple paths in one single bin is not allowed. The bin can thus be interpreted as the largest time interval over which the receiver is not capable of distinguishing two separate paths, or in other words, the resolution of the devices that are used for channel estimation.

The introduction of a discrete time channel model simplifies both analysis and simulation of system performance for multi-path-affected propagations, as will be shown in Checkpoint 8–2. Following a discrete time channel model, the impulse response in Eq. (8–56) can be rewritten as follows:

$$h(t) = \sum_{n=1}^{N_{\max}} a_n \delta(t - n\Delta\tau) \quad (8\text{–}64)$$

where N_{max} is the maximum number of bins considered within a single observation interval, and $\Delta\tau$ is the time duration of the bin.

8.2.3 The UWB Channel Model Proposed by the IEEE 802.15.3a

In July 2003, the Channel-Modeling sub-committee of study group IEEE 802.15.SG3a published the final report regarding the UWB indoor multi-path channel model (IEEE 802.15.SG3a, 2003). This model should be used for evaluating the performance of different physical layers as submitted to the IEEE 802.15.3 task group. Different contributions were considered for developing the proposed model: a) the statistical path loss model for indoor UWB signals presented by (Ghassemzadeh et al., 2002); b) channel measurements and modeling described by (Pendergrass and Beeler, 2002); c) the Intel proposal (Foerster and Li, 2002); d) the results of the measurement campaign performed at Oulu University by (Hovinen et al., 2002a); e) the radio channel model proposed by (Kunisch and Pamp, 2002b) after a channel sounding campaign in an office environment; f) the statistical path loss model proposed by (Ghassemzadeh and Tarokh, 2002) after the analysis of over 300,000 frequency responses at 712 locations and 23 homes; g) the channel model submitted by Mitsubishi after a measurement campaign in an office building (Molisch et al, 2002); h) the analysis on the UWB propagation channel performed by (Cramer et al., 2002b) by applying the CLEAN algorithm; i) and the studies by (Siwiak, 2002a and 2002b).

The IEEE Channel-Modeling sub-committee finally converged on a model based on the cluster approach proposed by Turin and others in 1972 (Turin et al., 1972), and further formalized by Saleh and Valenzuela in 1987 (Saleh and Valenzuela, 1987) in a seminal work on statistical modeling for indoor multi-path propagation. Although derived on the basis of a measurement campaign using low-power radar-like pulses, the Saleh and Valenzuela channel model (S-V model) is not UWB-specific.

The S-V model is based on the observation that usually multi-path contributions generated by the same pulse arrive at the receiver grouped into clusters. The time of arrival of clusters is modeled as a Poisson arrival process with rate Λ:

$$p\left(T_n \middle| T_{n-1}\right) = \Lambda e^{-\Lambda\left(T_n - T_{n-1}\right)} \tag{8–65}$$

where T_n and T_{n-1} are the times of arrival of the n-th and the $(n-1)$-th clusters, respectively. We set $T_1 = 0$ for the first cluster.

Within each cluster, subsequent multi-path contributions also arrive according to a Poisson process with rate λ:

$$p\left(\tau_{nk} \middle| \tau_{(n-1)k}\right) = \lambda e^{-\lambda\left(\tau_{nk} - \tau_{(n-1)k}\right)} \tag{8–66}$$

where τ_{nk} and $\tau_{(n-1)k}$ are the time of arrival of the n-th and the $(n-1)$-th contributions within cluster k. The time of arrival of the first contribution within each cluster, that is, τ_{n1} for $n = 1, \ldots, N$, is set to zero.

In the S-V model, the gain of the n-th ray of the k-th cluster is a complex random variable a_n with modulus β_{nk} and phase θ_{nk}. The β_{nk} values are assumed to be statistically independent and Rayleigh distributed positive random variables, while the θ_{nk} values are assumed to be statistically independent uniform random variables over $[0, 2\pi)$, or:

$$p(\beta_{nk}) = \frac{2\beta_{nk}}{\langle|\beta_{nk}|^2\rangle} e^{-\frac{\beta_{nk}^2}{\langle|\beta_{nk}|^2\rangle}}$$

$$p(\theta_{nk}) = \frac{1}{2\pi} \quad \text{with } 0 \le \theta_{nk} < 2\pi \tag{8–67}$$

where $\langle x \rangle$ is the expected value of x and where:

$$\langle|\beta_{nk}|^2\rangle = \langle|\beta_{00}|^2\rangle e^{-\frac{T_n}{\Gamma}} e^{-\frac{\tau_{nk}}{\gamma}} \tag{8–68}$$

The term β_{00} in Eq. (8–68) represents the average energy of the first path of the first cluster, while Γ and γ are the power decay coefficients for clusters and multi-path, respectively. According to Eq. (8–68), the average PDP is characterized by an exponential decay of the amplitude of the clusters, and a different exponential decay for the amplitude of the received pulses within each cluster (see Figure 8–24).

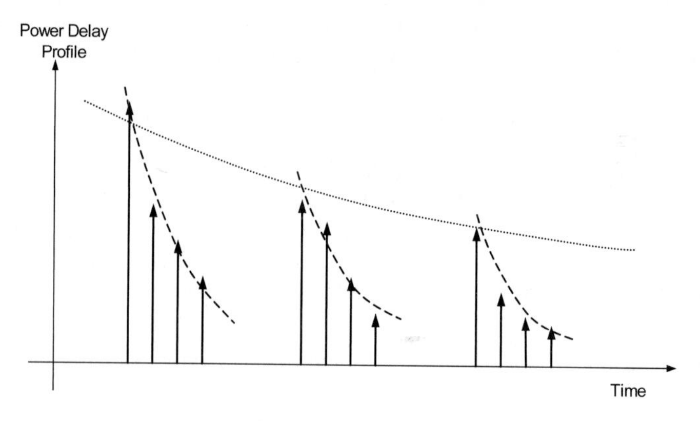

Figure 8–24 Typical PDP for S-V channel model.

To better fit the data resulting from UWB measurement campaigns, the IEEE group proposed a few modifications to the S-V model. In particular, a log-normal distribution was suggested for characterizing the multi-path gain amplitudes, and an additional log-normal variable was introduced for representing the fluctuations of the total multi-path gain. Finally, the channel coefficients were assumed to be real variables rather than complex variables, that is the phase term θ_{nk} will assume $\pm\pi$ with equal probability for representing pulse inversion due to reflection from dielectric surfaces.

The channel impulse response of the IEEE model can be expressed as follows:

$$h(t) = X\sum_{n=1}^{N}\sum_{k=1}^{K(n)} \alpha_{nk}\delta(t - T_n - \tau_{nk}) \quad (8\text{–}69)$$

where X is a log-normal random variable representing the amplitude gain of the channel, N is the number of observed clusters, $K(n)$ is the number of multi-path contributions received within the n-th cluster, α_{nk} is the coefficient of the k-th multi-path contribution of the n-th cluster, T_n is the time of arrival of the n-th cluster, and τ_{nk} is the delay of the k-th multi-path contribution within the n-th cluster.

The channel coefficient α_{nk} can be defined as follows:

$$\alpha_{nk} = p_{nk}\beta_{nk} \quad (8\text{–}70)$$

where p_{nk} is a discrete random variable assuming values ±1 with equal probability and β_{nk} is the log-normal distributed channel coefficient of multi-path contribution k belonging to cluster n. The β_{nk} term can thus be expressed as follows:

$$\beta_{nk} = 10^{\frac{x_{nk}}{20}} \quad (8\text{–}71)$$

where x_{nk} is assumed to be a Gaussian random variable with mean μ_{nk} and standard deviation σ_{nk}. Variable x_{nk}, in particular, can be further decomposed as follows:

$$x_{nk} = \mu_{nk} + \xi_n + \zeta_{nk} \quad (8\text{–}72)$$

where ξ_n and ζ_{nk} are two Gaussian random variables that represent the fluctuations of the channel coefficient on each cluster and on each contribution, respectively. We indicate the variance of ξ_n and ζ_{nk} by σ_ξ^2 and σ_ζ^2. The μ_{nk} value is determined to reproduce the exponential power decay for the amplitude of the clusters and for the amplitude of the multi-path contribution within each cluster. One can thus write:

$$\left\langle |\beta_{nk}|^2 \right\rangle = \left\langle \left| 10^{\frac{\mu_{nk}+\xi_n+\zeta_{nk}}{20}} \right|^2 \right\rangle \triangleq \left\langle |\beta_{00}|^2 \right\rangle e^{-\frac{T_n}{\Gamma}} e^{-\frac{\tau_k}{\gamma}}$$

$$\Rightarrow \mu_{nk} = \frac{10\log_e\left(\left\langle |\beta_{00}|^2 \right\rangle\right) - 10\frac{T_n}{\Gamma} - 10\frac{\tau_k}{\gamma}}{\log_e 10} + \quad (8\text{–}73)$$

$$-\frac{\left(\sigma_\xi^2 + \sigma_\zeta^2\right)\log_e 10}{20}$$

The total energy contained in the β_{nk} terms must be normalized to unity for each realization, that is:

$$\sum_{n=1}^{N}\sum_{k=1}^{K(n)}|\beta_{nk}|^2 = 1 \qquad (8\text{–}74)$$

According to the S-V model, the arrival time variables T_n and τ_{nk} are assumed to be modeled by two Poisson processes with average rates Λ and γ, respectively (see Eqs. (8–65) and (8–66)).

The amplitude gain X is assumed to be a log-normal random variable:

$$X = 10^{\frac{g}{20}} \qquad (8\text{–}75)$$

where g is a Gaussian random variable with mean g_0 and variance σ_g^2. The g_0 value depends on the average total multi-path gain G, which is measured at the location under examination, that is:

$$g_0 = \frac{10\log_e G}{\log_e 10} - \frac{\sigma_g^2 \log_e 10}{20} \qquad (8\text{–}76)$$

The G value can be determined as indicated in Eq. (8–61) for a given average attenuation exponent γ.

According to the above definitions, the channel model represented by the impulse response of Eq. (8–69) is fully characterized when the following parameters are defined:

- The cluster average arrival rate Λ
- The pulse average arrival rate λ
- The power decay factor Γ for clusters
- The power decay factor γ for pulses within a cluster
- The standard deviation σ_ξ of the fluctuations of the channel coefficients for clusters
- The standard deviation σ_ζ of the fluctuations of the channel coefficients for pulses within each cluster
- The standard deviation σ_g of the channel amplitude gain

The IEEE suggested an initial set of values for the above parameters. These values were tuned to fit some of the measurement data submitted to the IEEE. The list of parameters for different environmental scenarios as defined by the IEEE is provided in Table 8–1.

Table 8–1 Parameter Settings for the IEEE UWB Channel Model

Scenario	Λ (1/ns)	λ (1/ns)	Γ	γ	σ_{ξ} (dB)	σ_{ζ} (dB)	σ_g (dB)
Case A LOS (0–4 m)	0.0233	2.5	7.1	4.3	3.3941	3.3941	3
Case B NLOS (0–4 m)	0.4	0.5	5.5	6.7	3.3941	3.3941	3
Case C NLOS (4–10 m)	0.0667	2.1	14	7.9	3.3941	3.3941	3
Case D Extreme NLOS Multi-path Channel	0.0667	2.1	24	12	3.3941	3.3941	3

Note that the channel impulse response of Eq. (8–69) defines a continuous time multi-path channel model. It is possible to evaluate an equivalent discrete time channel model by simply adding all contributions received within the same bin.

CHECKPOINT 8–2

In this checkpoint, we will simulate the channel impulse response resulting from the statistical model proposed by the IEEE 802.15.SG3a. The MATLAB function to be used for generating the channel impulse response is Function 8.8.

Function 8.8 (see Appendix 8.A) generates the channel impulse response for a multi-path UWB channel according to the IEEE 802.15.SG3a model presented in Section 8.2.3. This function receives as input the value `TMG` of the average total multi-path gain at the distance under examination and the value of the sampling frequency `fc`. The `TMG` value can be determined by applying Eq. (8–61) for a given pair (G_0,γ). Within the function, the user can select the value of the observation time `OT` in seconds; the time resolution `ts` of the equivalent discrete time impulse response; the values for the statistical parameters characterizing the model (see Table 8–1), that is, the cluster arrival rate `LAMBDA`; the ray arrival rate `lambda`; the cluster decay factor `GAMMA`; the ray decay factor `gamma`; the standard deviation `sigma1` of the cluster fading; the standard deviation `sigma2` of the amplitude of the multi-path contributions within each cluster; and the standard deviation `sigmax` of the log-normal total multi-path gain. In addition, the user can also select the parameter `rdt`, which is used within the simulation for limiting the generation of the multi-path contributions within each cluster. It is assumed in the simulation model that the received pulses can be neglected when the term $e^{-\tau_{nk}/\gamma}$ in Eq. (8–68) becomes smaller than a given quantity `rdt` (note that the `rdt` value is not present in Eq. (8–68), which refers to the mathematical model of the channel impulse response with an infinite number of multi-path contributions). The presence of the `rdt` term is suggested in (Saleh and Valenzuela, 1987). `PT` is an additional input parameter that can be used for limiting the number of multi-path components of the channel impulse response. The presence of such a parameter is suggested in the definition of the model as described in (IEEE 802.15.SG3a, 2003). The final input parameter is `G`, which is set equal to 1 when graphical output is required. Function 8.8 returns the following outputs: vector `h0` representing the continuous time channel impulse response, vector `hf` representing the equivalent discrete time channel impulse response, the observation time `OT` in seconds, the time resolution `ts` (bin duration) in seconds, and the channel gain `X`.

The command line for executing Function 8.8 is:

```
[h0,hf,OT,ts,X] = cp0802_IEEEuwb(fc,TMG);
```

The above command line stores the output variables in memory, and if the input variable `G` is set to 1, it generates two plots representing the continuous time and the discrete time versions of the channel impulse response.

A second MATLAB function, Function 8.9, is introduced for evaluating the root mean square delay spread of a given channel impulse response h(t).

Function 8.9 (see Appendix 8.A) evaluates the rms delay spread by implementing Eq. (8–63). The function receives as input the input channel response `h` sampled at frequency `fc`, and returns the rms delay spread `rmsds`. The command line for executing Function 8.9 is:

```
[rmsds] = cp0802_rmsds(h,fc);
```

A third MATLAB function, Function 8.10, is introduced for evaluating the PDP of a given channel impulse response h(t).

Function 8.10 (see Appendix 8.A) receives as input the input channel response `h` sampled at frequency `fc`, and returns the PDP `PDP`. The command line for executing Function 8.10 is:

```
[PDP] = cp0802_PDP(h,fc);
```

In the following, we use Functions 8.8, 8.9, and 8.10 for analyzing the four different scenarios suggested by the IEEE and summarized in Table 8-1.

A first simulation assumes the position of the receiver to be two meters away from the transmitter. The receiver and transmitter are in LOS. For such a scenario (denoted as "Case A"), we assume a reference attenuation A_0 = 47 dB and an attenuation exponent γ = 1.7 (Ghassemzadeh and Tarokh, 2003). The value of `TMG` is determined by computing the square of the average channel gain `ag`, which can be evaluated via Function 8.1:

```
tx=1;
c0=10^(-47/20);
d=2;
gamma=1.7;
[rx,ag] = cp0801_pathloss(tx,c0,d,gamma)
>> rx = 0.0025
>> ag = 0.0025
TMG = ag^2;
```

A realization of the channel impulse response for Case A is obtained by executing Function 8.8 with the following parameters: `OT = 300e-9; ts = 2e-9; LAMBDA = 0.0233*1e9; lambda = 2.5e9; GAMMA = 7.1e-9; gamma = 4.3e-9; sigma1 = 10^(3.3941/10) ; sigma2 = 10^(3.3941/10); sigmax = 10^(3/10) ; rdt = 0.001; PT = 50; G = 1`. The command line for evaluating the channel impulse response is:

```
fc=50e9;
[h0,hf,OT,ts,X] = cp0802_IEEEuwb(fc,TMG);
```

The above command lines store the output variables of Function 8.8 in memory and produce the plots in Figures 8–25 and 8–26, corresponding to the continuous time and the discrete time channel impulse responses. Figure 8–25 shows that the transmission of a pulse generates multiple contributions at the receiver. As expected, the first multi-path component conveys the highest contribution of energy, even if we can recognize two other energy concentrations after 25 ns and 50 ns from reception of the first multi-path contribution. The presence of different clusters is not evident, and probably only one cluster characterizes the impulse response under examination. The same conclusion is suggested by the discrete time impulse response in Figure 8–26.

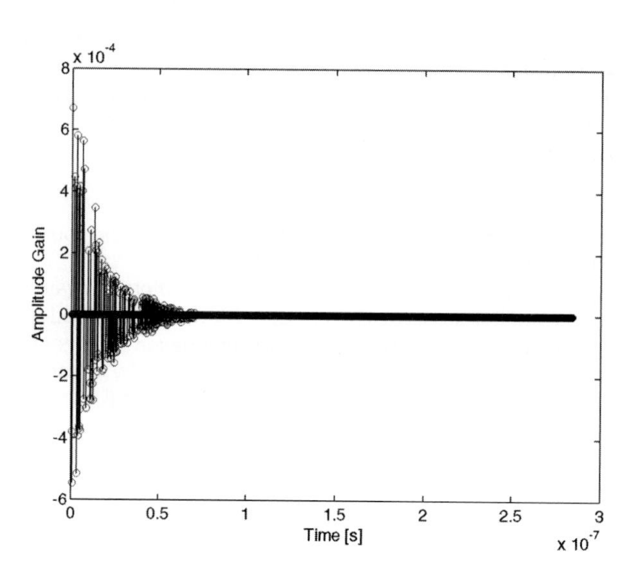

Figure 8–25 Channel impulse response for Case A — LOS.

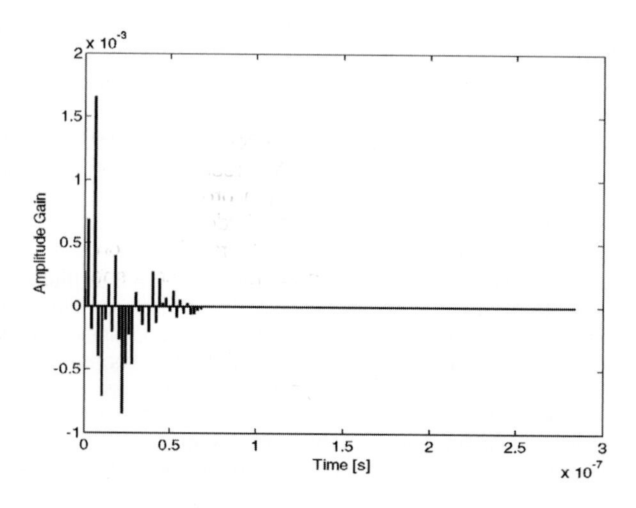

Figure 8–26 Discrete time channel impulse response for Case A — LOS.

We can apply Functions 8.9 and 8.10 for evaluating the rms delay spread and the PDP of the channel impulse response in Figure 8–25:

```
[rmsds] = cp0802_rmsds(h0,fc)
>> rmsds =  8.8711e-009
[PDP] = cp0802_PDP(h0,fc);
```

The graphical output provided by Function 8.10 is shown in Figure 8–27, where we can see the exponential decay of the received power.

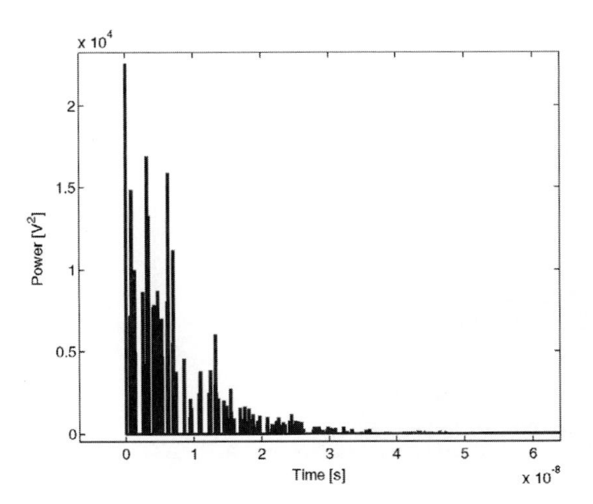

Figure 8–27 PDP of the channel impulse response in Figure 8–25 (Case A — LOS).

Note that the results deriving from the execution of Function 8.8 can appreciably vary from one simulation to another. A statistical analysis of the results obtained from several simulations is thus necessary to verify the robustness of the proposed algorithm. We can verify, for example, the statistical characteristics of the amplitude gain `X` by measuring the distribution of the `X` values that result from the execution of several simulations. We generate 1,000 different channel impulse responses for Case A and store the corresponding `X` values in memory with the following command:

```
for j = 1:1000
[h0,hf,OT,ts,X] = cp0802_IEEEuwb(fc,TMG);
G(j)=X;
end
```

The distribution of the `X` values can be easily visualized with the MATLAB function `hist(A,N)`, which produces a histogram bar plot of the elements of vector `A` with `M` equally spaced bins. The command line for representing the histogram of `X` is:

```
hist(G,30);
```

The above command produces the plot of Figure 8–28, in which we recognize the shape of the log-normal probability density function. This distribution is centered around an average value equal to 0.0025, which is the value of the average amplitude gain when the receiver is two meters away from the transmitter.

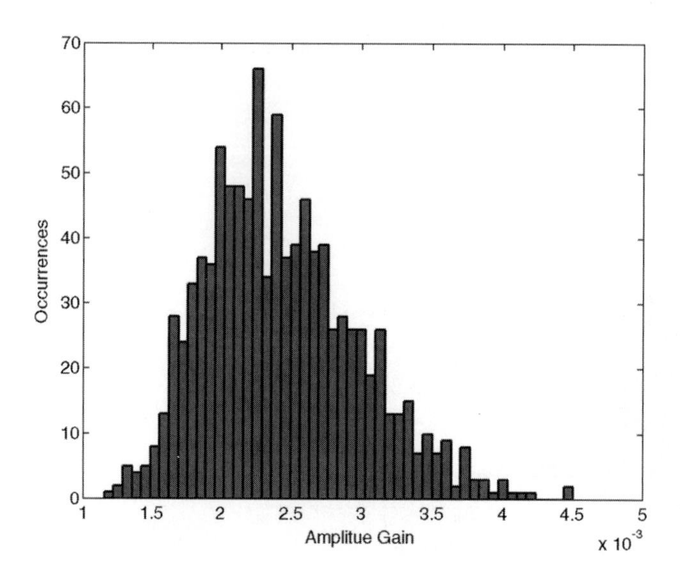

Figure 8–28 Histogram of occurrences of the amplitude gain for Case A — LOS.

The reader can perform a similar analysis to evaluate the average values of all the other parameters characterizing the model, such as the rms delay spread.

Case B corresponds to a scenario where the receiver is located at the same distance from the transmitter as Case A, but for NLOS between transmitter and receiver. We assume a reference attenuation A_0 = 51 dB, and a path loss exponent γ= 3.5 (Ghassemzadeh and Tarokh, 2003). The value of `TMG` is determined as follows:

```
tx=1;
c0=10^(-51/20);
d=2;
gamma=3.5;
[rx,ag] = cp0801_pathloss(tx,c0,d,gamma)
>> rx = 8.3791e-004
>> ag = 8.3791e-004
TMG = ag^2;
```

The parameters that must be considered within Function 8.8 are: `OT = 300e-9; ts = 2e-9; LAMBDA = 0.4*1e9; lambda = 0.5e9; GAMMA = 5.5e-9; gamma = 6.7e-9;`

`sigma1 = 10^(3.3941/10); sigma2 = 10^(3.3941/10); sigmax = 10^(3/10); rdt = 0.001; PT = 50; G = 1.` The continuous time channel impulse response, the discrete time channel impulse response, and the PDP for Case B are shown in Figures 8–29, 8–30, and 8–31, respectively.

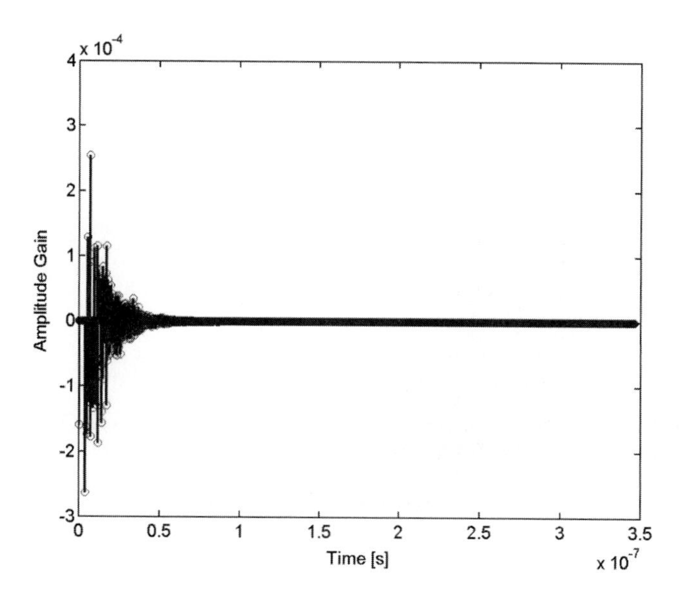

Figure 8–29 Channel impulse response for Case B — NLOS; distance between TX and RX is 2 m.

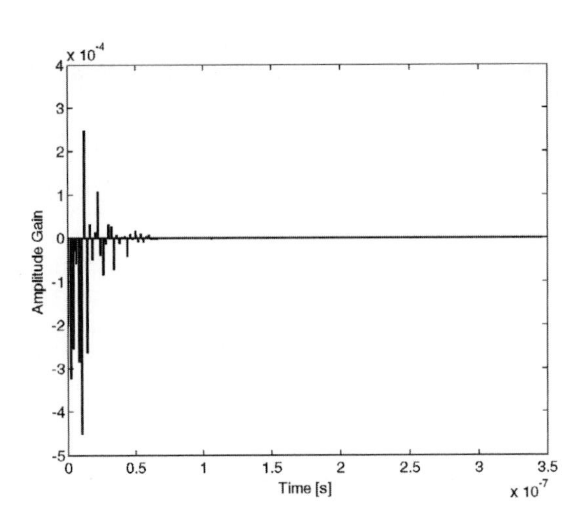

Figure 8–30 Discrete time channel impulse response for Case B — NLOS; distance between TX and RX is 2 m.

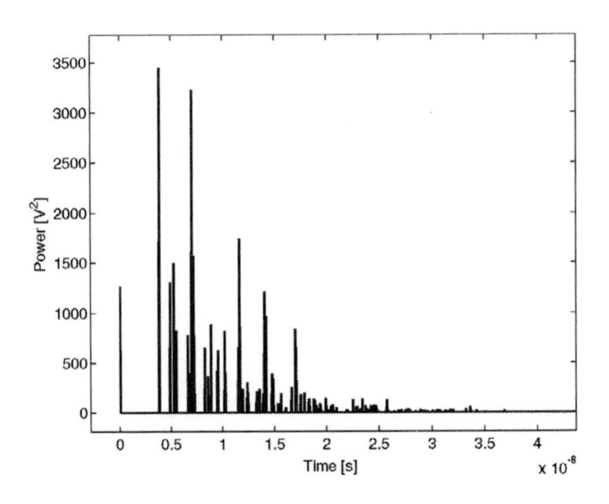

Figure 8–31 PDP for the channel impulse response of Figure 8–29 (Case B — NLOS; distance between TX and RX is 2 m).

Figures 8–29, 8–30, and 8–31 show that the multi-path impulse response of Case B is composed of the superimposition of several overlapping clusters. This is particularly evident in the PDP of Figure 8–31, where we recognize the presence of several strong peaks surrounded by multiple smaller peaks. In addition, we observe that the strongest peak is not the first component reaching the receiver, but arrives about 4 ns after the first contribution. This is a typical result of NLOS scenarios, that is, where obstacles are located between the transmitter and the receiver. In this case, the strongest peaks arrive at the receiver after reflections or diffractions, while the first peak reaches the antenna after penetrating the obstacles. The attenuation due to penetration is in general higher than the attenuation due to reflection or diffraction.

We do not repeat here the statistical analysis of the `X` term, given that the part of the code in Function 8.8 that emulates the fluctuations of the channel gain is independent of the particular distribution of the paths.

Case C analyzes a scenario where the receiver is moved 8 meters away from the transmitter and is NLOS. For such a case, both the reference attenuation and attenuation exponent are as in Case B. The `TMG` value is determined as follows:

```
tx=1;
c0=10^(-51/20);
d=8;
gamma=3.5;
[rx,ag] = cp0801_pathloss(tx,c0,d,gamma)
>> rx = 7.4062e-005
>> ag = 7.4062e-005
TMG = ag^2;
```

The parameters to be introduced within Function 8.8 are: `OT = 300e-9; ts = 2e-9; LAMBDA = 0.0667*1e9; lambda = 2.1e9; GAMMA = 14e-9; gamma = 7.9e-9; sigma1 = 10^(3.3941/10); sigma2 = 10^(3.3941/10); sigmax = 10^(3/10); rdt = 0.001; PT = 50; G = 1`. The continuous time channel impulse response, the discrete time channel impulse response, and the PDP for Case C are shown in Figures 8–32, 8–33, and 8–34, respectively.

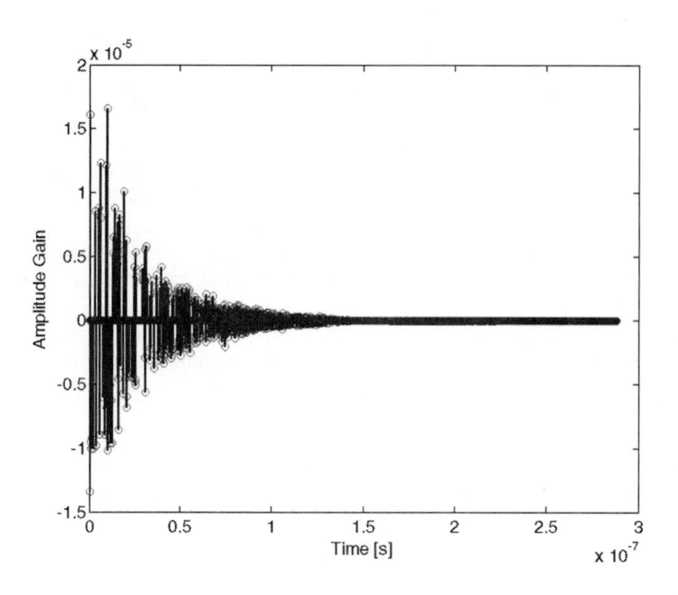

Figure 8–32 Channel impulse response for Case C — NLOS; distance between TX and RX is 8 m.

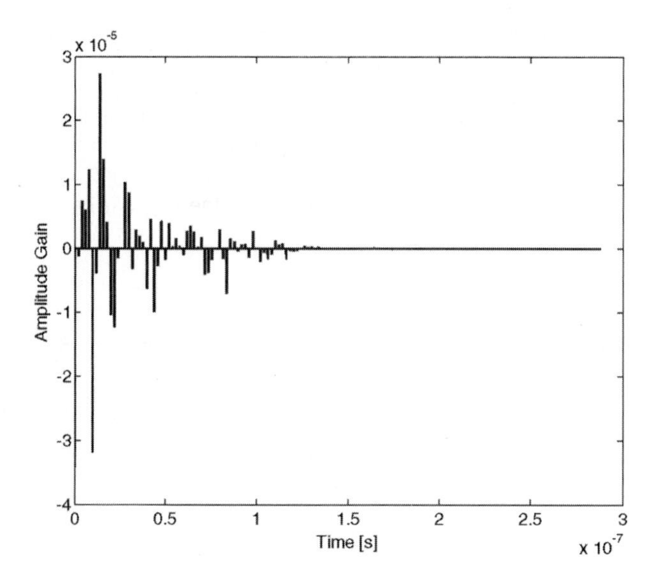

Figure 8–33 Discrete time channel impulse response for Case C — NLOS; distance between TX and RX is 8 m.

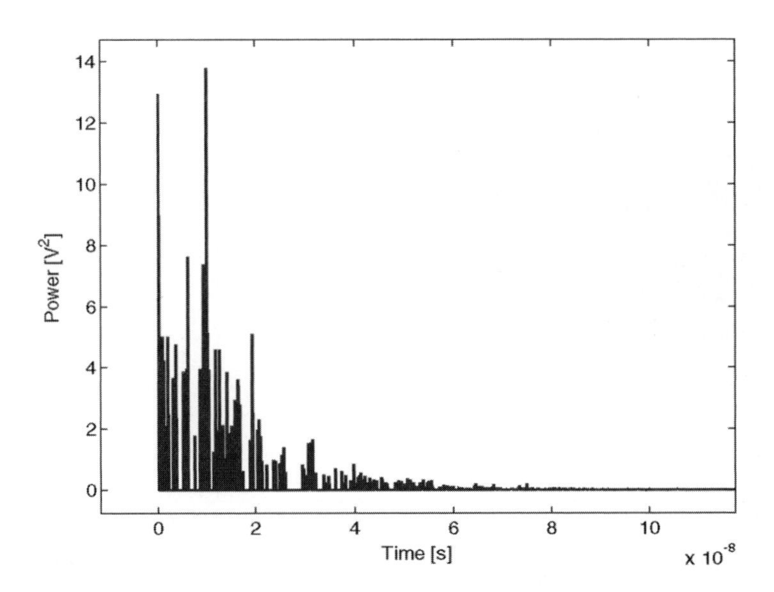

Figure 8–34 PDP of the channel impulse response of Fig. 8–32 (Case C — NLOS; distance between TX and RX is 8 m).

Figures 8–32, 8–33, and 8–34 show that Case C is characterized by a higher temporal dispersion of the transmitted energy than in previous cases. Note the presence of pulses in a time interval that is much longer than in Cases A and B (40–50 ns in Cases A and B; 60–80 ns in Case C). Once again, the strongest peak is not the first component reaching the receiver antenna.

A final simulation takes into account the "extreme NLOS" scenario suggested by the IEEE (Case D). Both the distance between the receiver and the transmitter and `TMG` value are as in Case C. The following parameters are considered within Function 8.8: `OT = 300e-9; ts = 2e-9; LAMBDA = 0.0667*1e9; lambda = 2.1e9; GAMMA = 24e-9; gamma = 12e-9; sigma1 = 10^(3.3941/10); sigma2 = 10^(3.3941/10); sigmax = 10^(3/10); rdt = 0.001; PT = 50; G = 1.`

Results of simulating Case D are shown in Figures 8–35, 8–36, and 8–37.

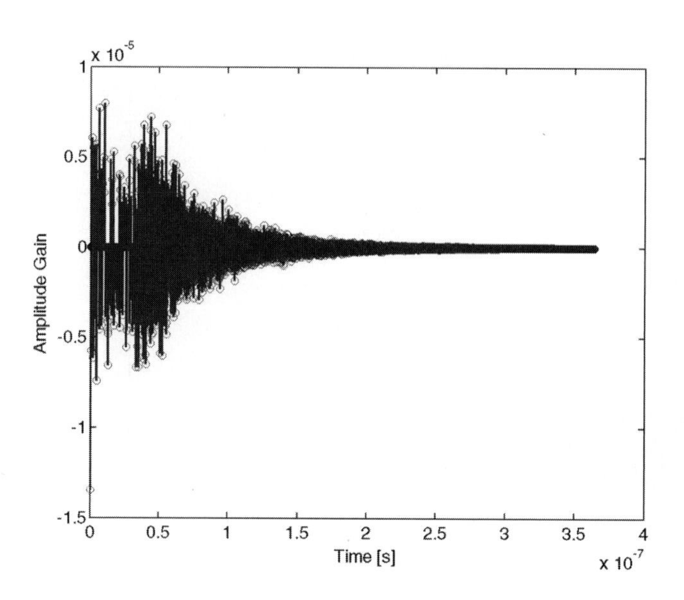

Figure 8–35 Channel impulse response for Case D — extreme NLOS.

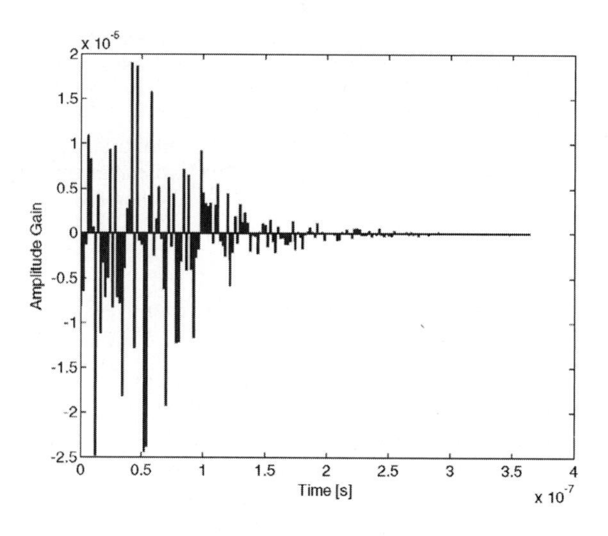

Figure 8–36 Discrete time channel impulse response for Case D — extreme NLOS.

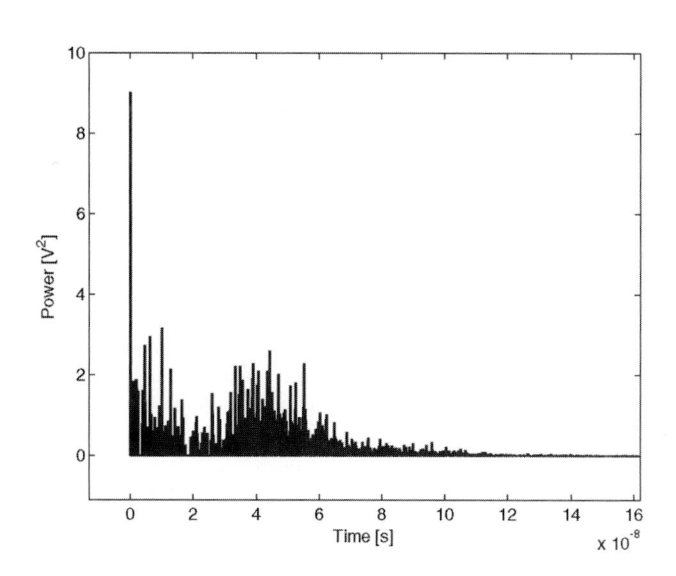

Figure 8–37 PDP of the channel impulse response of Fig. 8–35 (Case D — extreme NLOS).

Figures 8–35, 8–36, and 8–37 show that in the case of extreme NLOS, the effect of time dispersion of the transmitted energy is more evident than in previously analyzed cases. We observe in particular a considerable concentration of received energy even after 50 ns from reception of the first contribution. Figure 8–37 shows that we may control ISI only by allowing pulses to be spaced by at least 100 ns. Simulations also showed that after propagation, transmitted energy spreads in time over a large temporal window. The receiver should try to collect all of energy to improve the correct detection probability. The problem of designing a receiver that is capable of capturing and utilizing different received replicas of the same transmitted pulse will be analyzed in Section 8.2.4.

CHECKPOINT 8–2

Other models for the UWB channel have been presented in recent years. Two simple models for evaluating receiver performance in the presence of multi-path were presented by (Lee et al., 2000) and (Siwiak and Petroff, 2001). The first introduces a two-path model for characterizing fading at the receiver; the second applies the free space law for determining Pr_b as a function of the time delay difference between the multi-path components, the path gains, the value of the PPM shift, and the characteristics of the pulse waveform. On the basis of a measurement campaign carried out at the University of Oulu, (Hovinen et al., 2002) propose a statistical characterization of the exponential decay of the signal power with distance. The UWB indoor channel was also characterized through the POCA-NAZU fading distributions in (Zhang et al., 2002), while (Zhung et al., 2002) suggested a combination of a two-state Markov model and the Δ-K model for estimating the arrival of the multi-path components and the Gamma distribution for estimating the power of the signal arriving in each bin. In (Prettie et al., 2002), the authors analyze the spatial correlation of the received signal when an UWB receiver is moving inside a residential environment. This analysis is based on propagation measurements and shows that spatial correlation is present within a range of 0.05–0.15 m for both LOS and NLOS scenarios. The effect of the antennas on the design of an UWB communication link was analyzed by (Licul et al., 2003) and (Dabin et al., 2003). A deterministic UWB multi-path channel model was developed by (Yao et al., 2003). The model applies the Time Domain Uniform Theory of Diffraction, which is based on the assumption that the UWB signal propagates along discrete ray paths from transmitter to receiver. Such an approach leads to the possibility of applying ray-tracing methods for the characterization of signal attenuations and distortions. A similar approach was also presented in (Uguen et al., 2002). Interesting works about modeling UWB diffraction in the presence of buried objects were presented by (Jaureguy and Boredies, 1996) and (Cherniakov and Donskoi, 1999). Here, the authors proposed the UWB technique for the detection and reconnaissance of abandoned mines. In the work by (Suzuki and Kobayashi, 2003), UWB propagation in desktop environments was considered, and channel performance was analyzed in the case of human arms blocking the paths or human palms covering the antennas.

Interesting results on the characterization of the UWB channel have been also produced within the European Union research project U.C.A.N. (Ultra-wideband Concepts for Ad-hoc Networks) and can be found in (Alvarez et al., 2003). Here, the authors present a model that is based on the cluster approach proposed by (Saleh and Valenzuela, 1987), with the introduction of a dual slope power decay for each cluster. In addition, new parameters are introduced for characterizing the channel impulse response both in the frequency and time domains. This model was derived after a measurement campaign of the UWB channel performed in the frequency domain with a vector network analyzer. A description of the above methodology for the characterization of the UWB channel is also present in (Keignart and Daniele, 2002; Denis and Keignart, 2003).

In the context of the analyses of the UWB channel supported by the European Union, we also cite the measurement campaigns performed within the "Whyless.com" project and documented in (Kunisch and Pamp, 2002a and 2002b). On the basis of these measurements, Kunisch and Pamp recently proposed a hybrid statistical/deterministic multi-path channel model that is capable of producing space-variant impulse responses (Kunisch and Pamp, 2003). The same measurements were also used by (Romme and Kull, 2003) to analyze the existing tradeoff between robustness and throughput in UWB indoor communications. Regarding UWB outdoor propagation, we can cite the statistical model proposed by (Win et al., 1997).

8.2.4 Temporal Diversity and the RAKE Receiver

As described in Section 8.2.1, the multi-path-affected received signal $r(t)$ consists of the superimposition of several attenuated, delayed, and eventually distorted replicas of a transmitted waveform $s_m(t)$. When propagation fluctuations within an observation time $T >> T_b$ and path-dependent distortions can be neglected, $r(t)$ can be expressed as follows:

$$r(t) = \sum_j a_j s_m(t - \tau_j) + n(t) \tag{8–77}$$

where $n(t)$ is the AWGN at the receiver input.

Equation (8–77) can be rewritten for IR transmissions on the basis of the statistical channel model discussed in Section 8.2.3:

$$r(t) = X\sqrt{E_{TX}} \sum_j \sum_{n=1}^{N} \sum_{k=1}^{K(n)} \alpha_{nk} a_j p_0(t - jT_s - \varphi_j - \tau_{nk}) + n(t) \tag{8–78}$$

where:

- X is the log-normal distributed amplitude gain of the channel.
- E_{TX} is the transmitted energy per pulse.
- N is the number of clusters observed at destination.
- $K(n)$ is the number of multi-path contributions associated with the n-th cluster.
- α_{nk} is the channel coefficient of the k-th path within the n-th cluster.
- a_j is the amplitude of the j-th transmitted pulse ($a_j = 1$ in the case of PPM).
- T_s is the average pulse repetition period.
- φ_j is the time dithering associated to the j-th pulse ($\varphi_j = 0$ in the case of DS-PAM).
- τ_{nk} is the delay of the k-th path within the n-th cluster.

The energy contained in the channel coefficients α_{nk} is normalized to unity for each realization of the channel impulse response, that is:

$$\sum_{n=1}^{N} \sum_{k=1}^{K(n)} |\alpha_{nk}|^2 = 1 \tag{8–79}$$

and Eq. (8–78) can be rewritten as follows:

$$r(t) = \sqrt{E_{RX}} \sum_j \sum_{n=1}^{N} \sum_{k=1}^{K(n)} \alpha_{nk} a_j p_0(t - jT_s - \varphi_j - \tau_{nk}) + n(t) \tag{8–80}$$

where $E_{RX} = X^2 E_{TX}$ is the total received energy for one transmitted pulse. Different from the AWGN channel, E_{RX} is spread in time over the different multi-path contributions and can be used by the detector if the receiver is capable of capturing all replicas of the same pulse. Realistically, the receiver can only analyze a finite subset of N_R contributions, and the effective energy E_{eff}, which is used in the decision process, is smaller than E_{RX}, that is:

$$E_{eff} = E_{RX} \sum_{j=1}^{N_R} \left|\alpha_j\right|^2 \leq E_{RX} \tag{8–81}$$

According to Eq. (8–80), different replicas of the same transmitted pulse overlap at the receiver only when the corresponding inter-arrival time is smaller than pulse duration T_M. In this case, signals associated with different paths are not independent, that is, the amplitude of the pulse observed at time t is affected by the presence of multi-path contributions arriving immediately before or after time t. Given the characteristics of the propagation channel, the number of independent paths at the receiver depends on T_M: the smaller T_M, the higher the number of independent contributions at the receiver input. For IR-UWB systems, the T_M value is on the order of nanoseconds or fractions of nanoseconds, leading to the hypothesis that all multi-path contributions are non-overlapping, so that the received waveform consists of several independent components (Win and Scholtz, 1998a). IR-UWB systems can thus in principle take advantage of multi-path propagation by combining a large number of different and independent replicas of the same transmitted pulse. In this case, we say that the receiver exploits "temporal diversity" of the multi-path channel to improve performance of the decision process.

Different strategies for exploiting diversity can be adopted by the receiver: Selection Diversity (SD), Equal Gain Combining (EGC), and Maximal Ratio Combining (MRC). With the SD method, the receiver selects the multi-path contribution exhibiting the best signal quality and operates the decision on the transmitted symbol based on the observation of this contribution only. Choosing the "best" path guarantees an increase in receiver performance with respect to the simple selection of the first path, deriving from having selected the path with highest instantaneous SNR. A different method for increasing SNR consists of combining multi-path contributions rather than selecting the "best" path. With the EGC method in particular, the different contributions are first aligned in time and then added without any particular weighting. In MRC, the different contributions are weighted before the combination and the weights are determined to maximize the SNR before the decision process. In the presence of Gaussian noise at the receiver, the SNR is maximized by applying to each multi-path contribution a weighting factor that is proportional to the amplitude of the corresponding received signal. In other words, the MRC method adjusts the received contributions before combining them. The adjustment is performed by amplifying the strongest components and by attenuating the weak ones. In a single-user communication system without ISI, the method that achieves the best performance is the MRC, which ensures the largest SNR at the combiner output.

In all the above cases, the receiver takes advantage of multi-path under the hypothesis that different (independent) replicas of the same transmitted pulse can be analyzed separately and eventually combined before decision. The optimum receiver for the AWGN channel described in Section 8.1 is therefore not appropriate here since its structure foresees the presence of a correlator that is matched to one single waveform, whether single-pulse or multi-pulse. The optimum correlator for the present case must include additional correlators associated with different replicas of a same transmitted waveform. Such a scheme was invented by (Price and Green, 1958), and is called the RAKE receiver (see Figure 8–38).

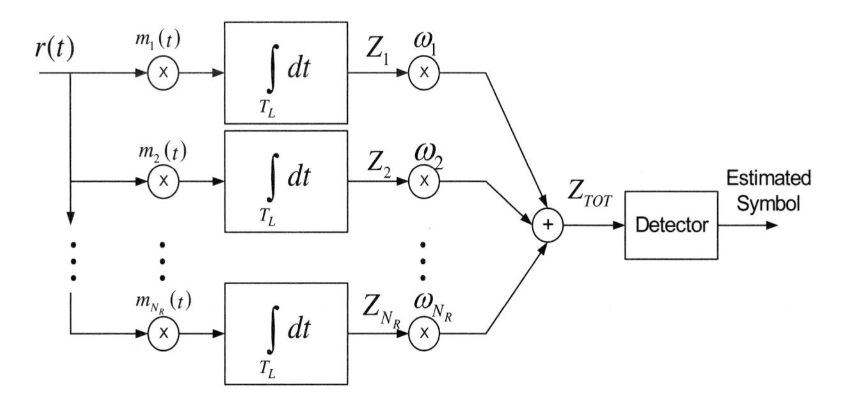

Figure 8–38 RAKE receiver with N_R parallel correlators.

In Figure 8–38, $T_L > T_s$ represents the time duration of the channel impulse response, and Z_{TOT} is the decision variable at the output of the RAKE combiner that enters the detector. Figure 8–38 shows the structure of the RAKE receiver, which consists of a parallel bank of N_R correlators, followed by a combiner that determines the variable to be used for the decision on the transmitted symbol. Each correlator is locked on one of the different replicas of the transmitted symbol, that is, the correlator mask $m_j(t)$ on the j-th branch of the RAKE is aligned in time with the j-th delayed replica of the transmitted symbol, or:

$$m_j(t) = m(t - \tau_j) \tag{8–82}$$

where $m(t)$ is the correlator mask introduced in Section 8.1 for the AWGN case, and τ_j is the propagation delay that characterizes the j-th path. The output of the bank of correlators feeds the combiner. Depending on the diversity method implemented at the receiver, a different set of weighting factors $\{\omega_1,..., \omega_{Nr}\}$ is used to combine the outputs of the correlators. In the SD case, the weighting factors are equal to zero, except for the factor on the branch corresponding to the signal with highest amplitude, which is equal to one. In the case of EGC, all factors are equal to 1, that is, the combiner simply adds the outputs of the correlators without applying any weighting. Finally, in the MRC case, the output of each branch is multiplied by a weighting factor, which is proportional to the signal amplitude on that branch.

An alternative but equivalent implementation of the RAKE receiver is shown in Figure 8–39, where the N_R correlators are preceded by time shift elements. The function of these elements is to align all multi-path contributions in time. The advantage of the solution of Figure 8–39 with respect to the scheme of Figure 8–38 is in the possibility of adopting the same correlator mask $m(t)$ on all branches of the RAKE.

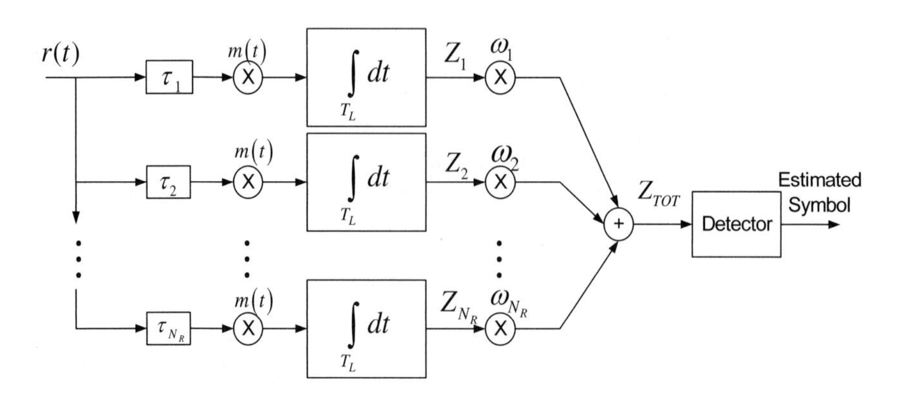

Figure 8–39 RAKE receiver with N_R parallel correlators and time delay units.

According to the schemes of Figures 8–38 and 8–39, the RAKE receiver must know the time distribution for all multi-path contributions composing the received waveform. This task is performed by supplying the RAKE with the capability of scanning the channel impulse response, tracking, and adjusting the delay of a certain number of multi-path components. Time delay synchronization for the different multi-path contributions is based in general on correlation measurements that are performed on the received waveform. In addition, if the SD or MRC methods are adopted within the combiner, the knowledge of the amplitudes of the multi-path components is also required for adjusting the weighting factors. This task is performed in general by using pilot symbols for channel estimation.

The RAKE scheme of Figure 8–39 can be greatly simplified when the channel is modeled with a discrete time impulse response, as discussed in Section 8.2.2. In this case, the different contributions at the receiver are spaced in time by a multiple of the bin duration Δt, and a single correlator structure is possible for the RAKE (Figure 8–40). In Figure 8–40. the correlator integrates the product between $m(t)$ and the received waveform $r(t)$. The output of the correlator is sampled with period $\Delta\tau$ before passing through a delay unit and a combiner, which implements one of the previously described diversity methods: SD, EGC. or MRC.

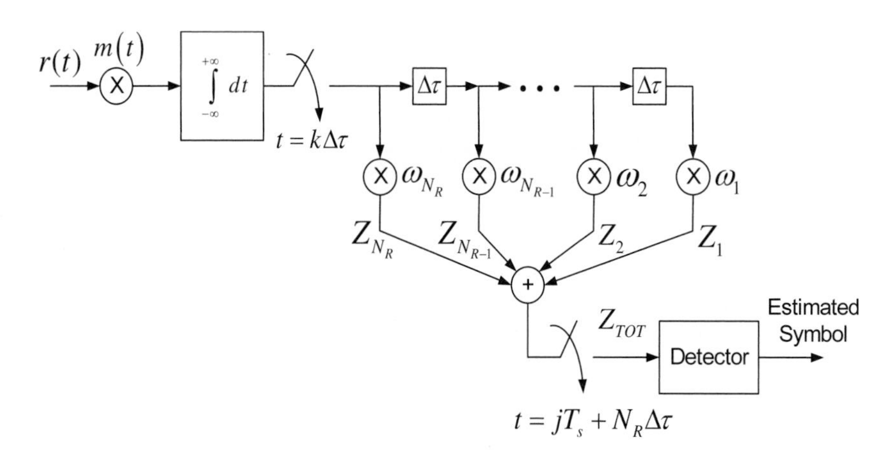

Figure 8–40 RAKE receiver for discrete-time channel models.

Performance of the RAKE receiver for propagations over a multi-path channel can be evaluated by first assuming a specific model for the channel impulse response, and by then evaluating the probability of error on the symbol $\Pr_e$ as a function of the E_{RX}/N_0 ratio for the different diversity methods. This analysis is performed in general under the hypothesis of perfect knowledge of the coefficients of the channel impulse response, or perfect channel estimation. A detailed review of reference results concerning performance of the RAKE is presented in the "Further Reading" section. In Checkpoint 8–3, we will evaluate a RAKE receiver implementing MRC when considering the transmission of IR-UWB signals over the multi-path channel presented in Section 8.2.3.

The adoption of a RAKE considerably increases the complexity of the receiver. This complexity increases with the number of multi-path components analyzed and combined before decision, and can be reduced by decreasing the number of components processed by the receiver. According to Eq. (8–81), however, a reduction of the number of paths leads to a decrease of energy collected by the receiver. A quasi-analytical investigation of the existing tradeoff between receiver complexity and percentage of captured energy in a RAKE receiver for IR-UWB systems is presented in (Win and Scholtz, 1998b). The results of this analysis show that a RAKE receiver operating in a typical modern office building requires about 50 different RAKE branches to capture about 60% of the total energy of the received waveform. In (Cassioli et al., 2002a), different strategies for reducing the complexity of the RAKE are presented and analyzed in terms of $\Pr_e$ degradation. The first strategy, called Selective RAKE (SRake), consists of selecting the L_B best components among the L_{TOT} available at the receiver input. The number of branches of the RAKE is reduced, but the receiver still must keep track of all the multi-path components to perform the selection. A second and simpler solution, called Partial RAKE (PRake), combines the first arriving L_P paths without operating any selection among all available multi-path components. As expected, SRake outperforms PRake since it achieves higher SNR at the output of the combiner. The gap in performance, however, decreases when the best paths are located at the beginning of the channel impulse response as it happens, in general, when considering LOS scenarios (see Checkpoint 8–2). An example of a comparison between SRake and PRake will be presented in Checkpoint 8–3.

CHECKPOINT 8–3

In this checkpoint, we will present an algorithm that simulates the activity of a RAKE receiver in the case of a multi-path propagation channel. This simulated receiver implements the MRC method. Different solutions for reducing the complexity of the receiver are also investigated.

The basic idea of the proposed algorithm consists of simplifying the scheme of the RAKE receiver in Figure 8–38, and can be obtained at the price of an increased complexity of the correlation mask m(t). Consider the basic scheme of Figure 8–38. The variable Z_{TOT} at the output of the combiner can be expressed as follows:

$$\begin{aligned} Z_{TOT} &= \sum_{j=1}^{N_R} \omega_j \int_{T_L} r(t) m_j(t) dt \\ &= \sum_{j=1}^{N_R} \omega_j \int_{T_L} r(t) m(t-\tau_j) dt \end{aligned} \qquad (8\text{–}83)$$

where T_L is the observation interval, N_R is the number of branches of the RAKE receiver, ω_j is the weighting coefficient of the *j*-th component, m(t) is the correlation mask for the transmitted symbol, and τ_j is the delay of the multi-path component, which is processed on the *j*-th branch.

Equation (8–83) can also be expressed as follows:

$$\begin{aligned} Z_{TOT} &= \int_{T_L} r(t) m_R(t) dt \\ \text{where} \quad m_R(t) &= \sum_{j=1}^{N_R} \omega_j m(t-\tau_j) \end{aligned} \qquad (8\text{–}84)$$

which leads to a receiver structure as shown in Figure 8–41. Such a receiver consists of one single correlator and does not require the introduction of a combiner. Note that the scheme of Figure 8–41 has the same complexity as Figure 8–38 since accurate algorithms for channel estimation and tracking are still necessary for evaluating the waveform $m_R(t)$. The main advantage of the proposed approach consists of using the same functions as the AWGN case, that is, Function 8.5 for PPM and Function 8.7 for PAM.

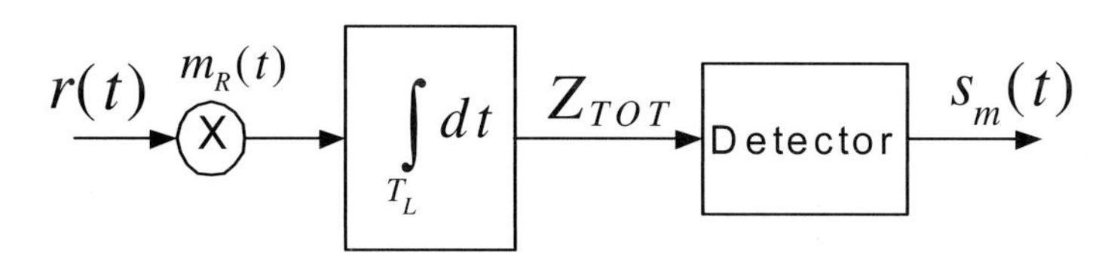

Figure 8–41 Equivalent RAKE receiver structure.

The first MATLAB function introduced for implementing the RAKE algorithm is Function 8.11, which operates channel estimation and generates the weighting coefficients to be used in the definition of the mask $m_R(t)$.

Function 8.11 (see Appendix 8.A) operates channel estimation in the case of a discrete time channel model. The function receives as input vector `hf` containing the coefficients of the channel impulse response. According to Section 8.2.2, these coefficients represent the amplitude gain of the multi-path contributions in the different bins in which the time axis is subdivided. The time duration in seconds of the bin interval is given as input via the parameter `ts`. The value of the sampling frequency `fc` is also required at the input of the function to operate the conversion between the continuous time and the discrete time domains. In addition to channel estimation, Function 8.11 evaluates the weighting factors to be used. The MRC strategy where the weighting factors coincide with the coefficients of the impulse response is adopted. The possibility of reducing the number of branches of the RAKE is also considered. Both PRAKE and SRAKE are simulated. The function in fact receives as input the number of components to be considered in both cases, that is, the number `L` of multi-path components to be processed in the case of PRAKE and the number `S` of multi-path components to be used in the case of SRAKE. Remember that PRAKE processes the first `L` multi-path components of the signal that arrive at the receiver input, while SRAKE operates the decision on the basis of the `S` multi-path components with the highest energy. Function 8.11 returns the following outputs: vector `G`, vector `T`, number of multi-path components `NF`, and vectors `Arake`, `Srake`, and `Prake`.

The command line for executing Function 8.11 is:

```
[G,T,NF,Arake,Srake,Prake] = cp0803_rakeselector(hf,...
   fc, ts, L, S);
```

We can test Function 8.11 by the following example. We first generate a channel impulse response by executing Function 8.8. The case of extreme NLOS is considered (see Table 8.1). We then execute Function 8.11 with `L=5` and `S=5`. Results of simulation are shown in Figures 8–42 to 8–45. Figure 8–42 represents the continuous time channel impulse response, and Figure 8–43 represents its equivalent discrete time version. Since the ideal RAKE takes into account all the non-zero components of the channel impulse response, the results in Figure 8–43 coincide with vector `Arake`, which is generated by Function 8.11. In the case under examination, vector `Arake` contains 185 non-zero elements, which leads to an unacceptable level of complexity for the receiver. Figure 8-44 represents channel estimation for a SRAKE with five branches. As expected, the components of vector `Srake` correspond to the five components of the discrete time impulse response having the highest absolute value. Finally, Figure 8–45 shows channel estimation for a PRAKE with five branches. When comparing Figures 8–44 and 8–45, we recognize that the gain in complexity due to the absence of the selection process for PRAKE is paid at the price of a reduction in the amount of energy collected by the receiver. In particular, we can verify that in the case under examination, only one out of five components arriving first at the receiver belongs to the set of the best five multi-path contributions.

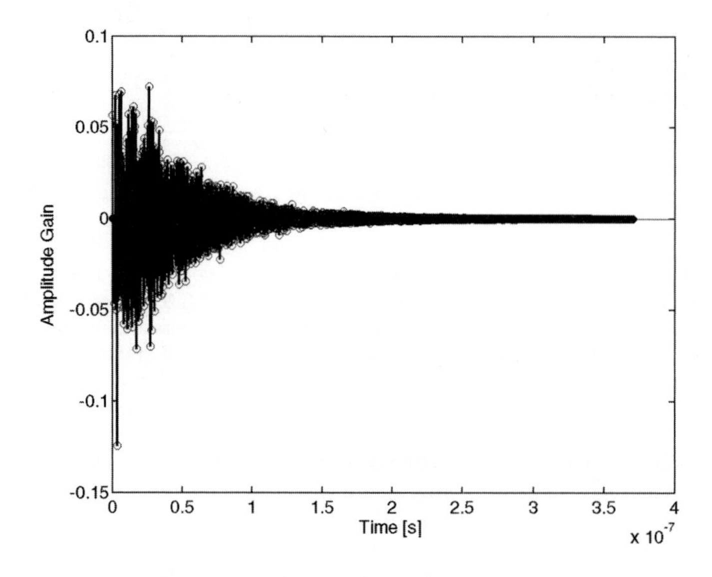

Figure 8–42 Channel impulse response (extreme NLOS scenario).

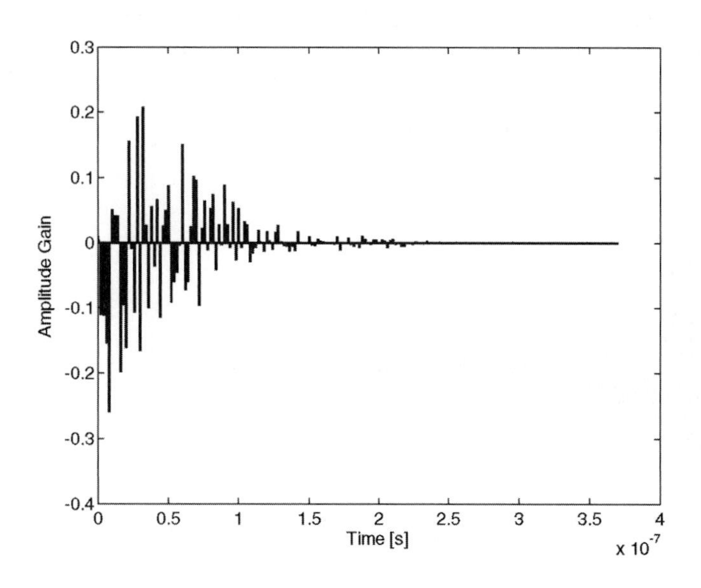

Figure 8–43 Equivalent discrete time channel impulse response (extreme NLOS scenario; see Figure 8–42).

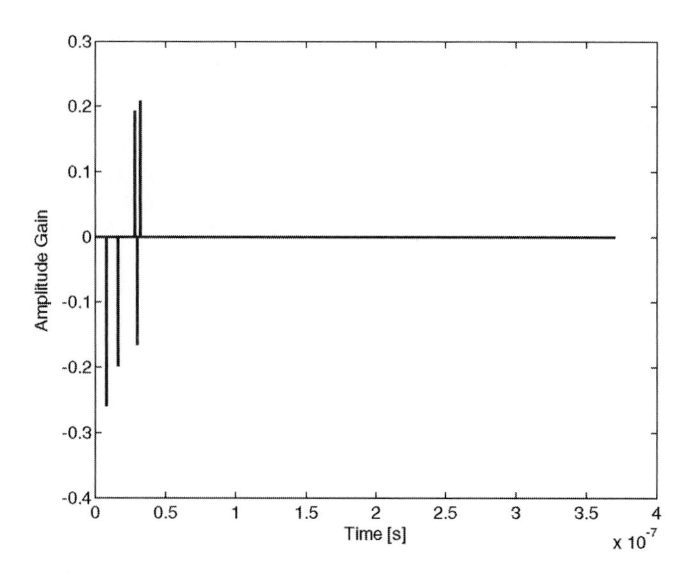

Figure 8–44 Channel estimation with SRAKE.

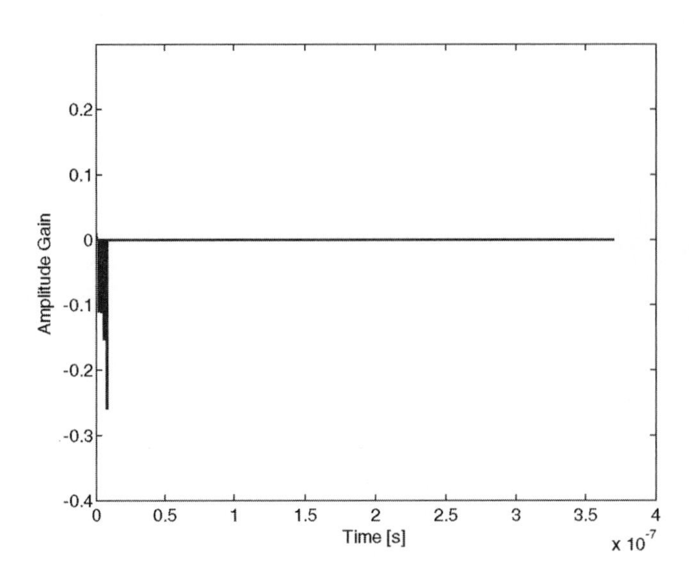

Figure 8–45 Channel estimation with PRAKE.

Given the coefficients of the channel impulse response, we will introduce two MATLAB functions that evaluate the correlator mask $m_R(t)$ in the case of PPM and PAM transmission.

Function 8.12 (see Appendix 8.A) evaluates the correlation mask $m_R(t)$ in the case of PPM-TH-UWB signal. The RAKE receiver is assumed to apply the MRC method. The same function can be used for simulating the activity of an ideal RAKE or the activity of a PRAKE or SRAKE, which processes only a subset of these components. To evaluate the correlation mask, Function 8.12 requires knowledge of the TH code assigned to the user under examination. This information is available in vector `ref`, which is produced by Function 2.6 during the generation of the signal. This vector is thus given as input to Function 8.12, together with the value of the sampling frequency `fc`, the number of pulses that must be analyzed `numpulses`, the value of the PPM shift in seconds `dPPM`, and the vector `rake`, which contains the information on both position and amplitude of the multi-path components to be considered. In the case of an ideal RAKE, vector `rake` coincides with vector `Arake` generated by Function 8.11, while in the case of SRAKE and PRAKE, vector `rake` coincides with `Srake` and `Prake`, respectively. Function 8.12 returns vector `mask`, which represents the correlator mask $m_R(t)$ to be used in the RAKE receiver. The command line for executing Function 8.12 is:

```
[mask] = cp0803_PPMcorrmask_R(ref,fc,numpulses,...
   dPPM,rake);
```

Function 8.13 (see Appendix 8.A) evaluates the correlation mask $m_R(t)$ in the case of PAM-DS-UWB signals. Function 8.13 receives the same inputs of Function 8.12, except for the absence of the PPM shift, and returns vector mask, which represents the correlator mask $m_R(t)$ to be used in the RAKE receiver. Function 8.13 is executed as follows:

```
[mask] = cp0803_PAMcorrmask_R(ref,fc,numpulses,rake);
```

Given Functions 8.11, 8.12, and 8.13, we are ready to perform a global simulation for analyzing the performance of a RAKE receiver. We consider the case of binary PPM only; the proposed procedure can be, however, easily duplicated for the PAM case.

The first step is the generation of the transmitted signal. This task can be performed by using Function 2.6 (see Appendix 2.A). The following input parameters are considered: Pow=-30; fc=50e9; numbits=2000; Ts=60·1e-9; Ns=1; Tc=1e-9; Nh=5; Np=2000; Tm=0.5e-9; tau=0.2e-9; dPPM=0.5e-9; G = 0. The resulting signal conveys 2,000 bits transmitted through 2,000 pulses, that is, code repetition coding is not applied. The average pulse repetition period is 60 ns, which should guarantee the absence of ISI in the presence of LOS (see Checkpoint 8–2). Orthogonal PPM is considered. The UWB signal under examination is generated by executing the following command:

```
[bits,THcode,Stx,ref]=cp0201_transmitter_2PPM_TH;
```

The above command stores the following in memory: the stream of bits generated by the source bits, the TH code assigned to the user THcode, the waveform Stx representing the UWB signal at the transmitter output, and the reference signal ref, which will be used for evaluating the correlation mask at the receiver.

In the following step of the simulation, we evaluate the effect of propagation over the multi-path channel. A simple scenario where the receiver is in LOS with the transmitter is considered. The receiver is located two meters away from the transmitter, as specified in "Case A" and previously considered in Checkpoint 8–2, for which we assumed a reference path loss A_0 = 47 dB and an exponent γ = 1.7. The total multi-path gain TMG of the channel is determined as follows:

```
tx=1;
c0=10^(-47/20);
d=2;
gamma=1.7;
[rx,ag] = cp0801_pathloss(tx,c0,d,gamma);
TMG = ag^2;
```

The channel impulse response is generated by executing Function 8.8 with the following parameters: OT = 200e-9; ts = 1e-9; LAMBDA = 0.0233*1e9; lambda = 2.5e9; GAMMA = 7.1e-9; gamma = 4.3e-9; sigma1 = 10^(3.3941/10); sigma2 = 10^(3.3941/10); sigmax = 10^(3/10); rdt=0.001; PT=50; G = 1. Note that the bin duration is set to 1 ns, which is equal to the chip time. The command line for evaluating the channel impulse response is:

```
fc=50e9;
[h0,hf,OT,ts,X] = cp0802_IEEEuwb(fc,TMG);
```

The continuous time and discrete time versions of the channel impulse response generated with the above command are shown in Figures 8–46 and 8–47, respectively.

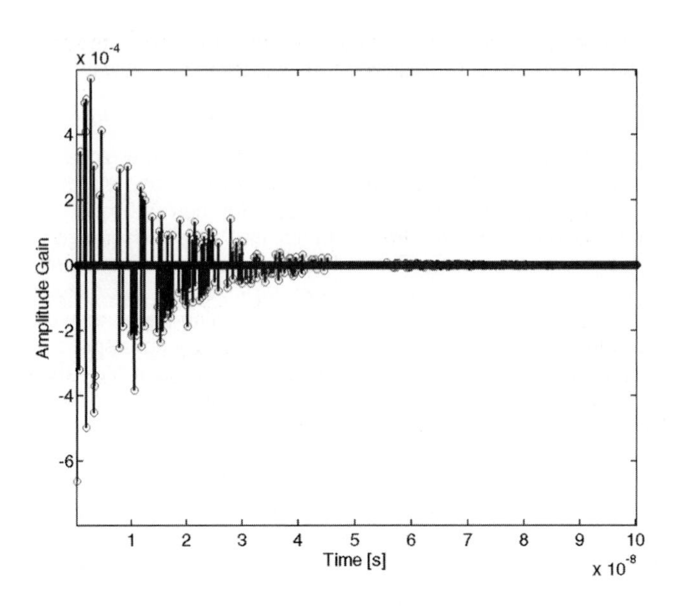

Figure 8–46 Continuous time channel impulse response — Case A, LOS (same channel parameters as Figure 8–25).

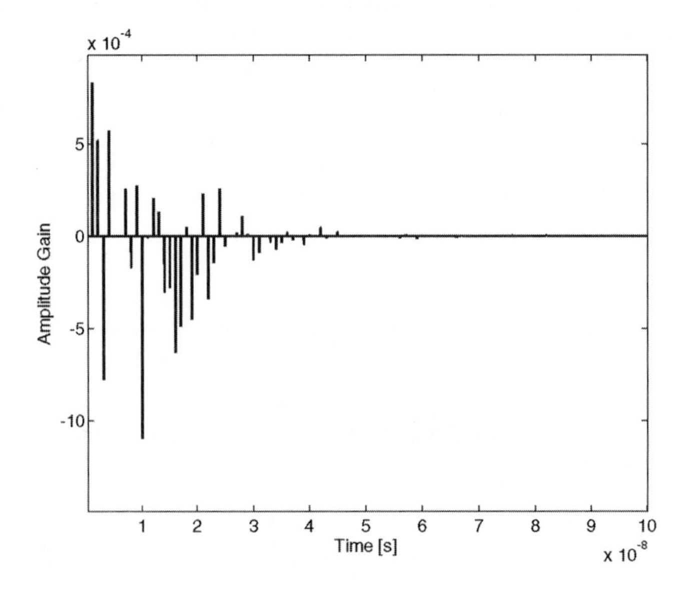

Figure 8–47 Discrete time channel impulse response — Case A, LOS (see Figure 8–46).

The next step is the generation of the Gaussian noise at the receiver. This task is performed by first attenuating the transmitted signal and then by executing Function 8.3 (see Appendix 8.A):

```
SRX0 = Stx.*ag;
numpulses = 2000;
exno = [0 3 6 9];
[output,noise] = cp0801_Gnoise2(SRX0,exno,numpulses);
```

The above command lines generate four different noise signals, characterized by decreasing values of noise power.

The waveform at the receiver input can now be evaluated by first convolving the transmitted signal with the discrete time channel impulse response, and by then adding Gaussian noise:

```
SRX = conv(Stx,hf);
SRX = SRX(1:length(Stx));
RX(1,:)=SRX+noise(1,:);
RX(2,:)=SRX+noise(2,:);
RX(3,:)=SRX+noise(3,:);
RX(4,:)=SRX+noise(4,:);
```

The above commands generate matrix `RX`, which represents the signal at the receiving antenna after propagation over the multi-path channel and the addition of Gaussian noise. This matrix is composed of four rows, each corresponding to a different value of the E_{RX}/N_0 ratio.

We can now proceed by simulating the activity of the RAKE receiver. Five different cases are considered: an ideal RAKE that processes all the multi-path components (`rec_A`), a SRAKE with `S=5` (`rec_B`), a SRAKE with `S=2` (`rec_C`), a PRAKE with `L=5` (`rec_D`), and a PRAKE with `L=2` (`rec_E`).

The first step for simulating the receiver is channel estimation:

```
L=5;
S=5;
[G,T,NF,rec_A,rec_B,rec_D] = ...
   cp0803_rakeselector(hf,fc,ts,L,S);
L=2;
S=2;
[G,T,NF,rec_A,rec_C,rec_E] = ...
   cp0803_rakeselector(hf,fc,ts,L,S);
```

For each receiver, we can construct the corresponding correlator mask:

```
dPPM = 0.5e-9;
[mask_A] = ...
   cp0803_PPMcorrmask_R(ref,fc,numpulses,dPPM,rec_A);
[mask_B] = ...
```

```
   cp0803_PPMcorrmask_R(ref,fc,numpulses,dPPM,rec_B);
[mask_C] = ...
   cp0803_PPMcorrmask_R(ref,fc,numpulses,dPPM,rec_C);
[mask_D] = ...
   cp0803_PPMcorrmask_R(ref,fc,numpulses,dPPM,rec_D);
[mask_E] = ...
   cp0803_PPMcorrmask_R(ref,fc,numpulses,dPPM,rec_E);
```

Finally, signal correlation and the decision process are simulated with Function 8.5:

```
numbit = 2000;
Ns = 1;
Ts = 60e-9;
[RXbits,ABER] = cp0801_PPMreceiver(RX,...
    mask_A,fc,bits,numbit,Ns,Ts);
[RXbits,BBER]=cp0801_PPMreceiver(RX,...
    mask_B,fc,bits,numbit,Ns,Ts);
[RXbits,CBER]=cp0801_PPMreceiver(RX,...
    mask_C,fc,bits,numbit,Ns,Ts);
[RXbits,DBER]=cp0801_PPMreceiver(RX,...
    mask_D,fc,bits,numbit,Ns,Ts);
[RXbits,EBER]=cp0801_PPMreceiver(RX,...
    mask_E,fc,bits,numbit,Ns,Ts);
```

The result of the simulation can be visualized by typing:

```
semilogy(exno,ABER,exno,BBER,exno,CBER,exno,...
   DBER,exno,EBER);
```

Figure 8–48 shows the plot generated by running the above command, completed with axes labels, legend, and change of style for the three lines.

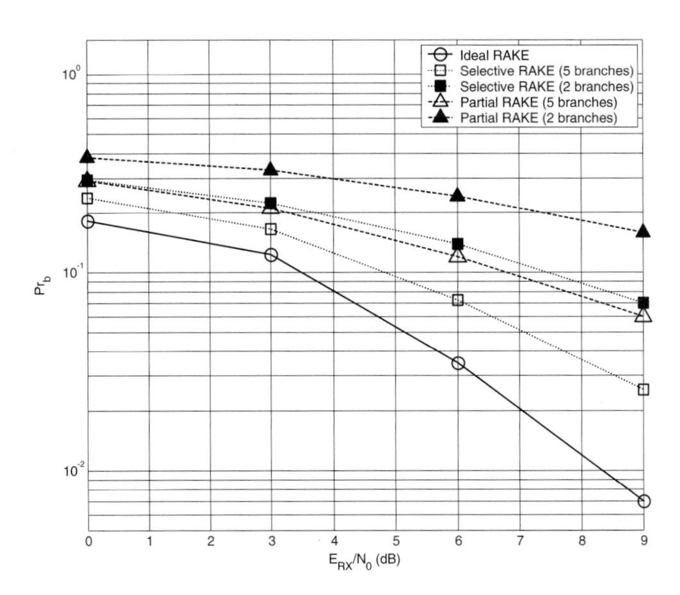

Figure 8–48 Pr_b vs. E_{RX}/N_0 for: a) an ideal RAKE that processes all the multi-path contributions (solid line and circles); b) a SRAKE that processes the best five multi-path contributions (dotted line and white squares); c) a SRAKE that processes the best two multi-path contributions (dotted line and black squares); d) a PRAKE that processes the first five multi-path contributions (dashed line and white triangles); e) a Partial RAKE that processes the first two multi-path contributions (dashed line and black triangles).

Figure 8–48 compares the performance of the five analyzed RAKE receivers. As expected, the best results are obtained with the ideal RAKE, which processes all the multi-path contributions being resolved at the receiver. In the case under examination `NF=104`, more than 100 components of the channel impulse response must be first estimated and then processed by the receiver. The dotted lighter line in Figure 8–48 refers to the case of SRAKE, which reduces the complexity of the receiver by decreasing the number of multi-path components taken into account. In particular, white squares in Figure 8–48 correspond to a SRAKE that processes the best five components, while black squares correspond to a simpler solution where only the two best components are processed. The loss in performance of the SRAKE can be quantified in about 3 dB for the solution with five branches, and in about 6 dB for the solution with only two branches. Remember that the reduction in complexity of the SRAKE refers only to a reduced number of branches within the receiver. Because of the selection process, the complexity of both channel estimation and channel tracking is the same as in the ideal RAKE. The simplest solution is in fact the PRAKE, which eliminates the selection process and accounts for only the first multi-path components that arrive at the receiver. In Figure 8–48, dashed bolder lines and white triangles refer to the case of a PRAKE that processes the first five components arriving at the receiver, while dashed bolder lines and black triangles refer to the case of a PRAKE that processes only the first two components arriving at the receiver. The last solution is the simplest among the five considered in the simulation. The loss in performance of the PRAKE with respect to the ideal RAKE is emphasized compared to the SRAKE. In particular, we observe a 6 dB loss in the case of a PRAKE with five branches, and a 9 dB loss in the case of a PRAKE with two branches. Moreover, it is interesting to observe that performance of the PRAKE with five branches is similar to that of a SRAKE with two branches, i.e. the energy collected with the first five paths is nearly the same as the energy associated with the best two paths.

As a final comment, one should keep in mind that results depend on the particular realization of the channel impulse response. To derive the average performance of the RAKE, one should average Pr_b values over several realizations of the channel.

CHECKPOINT 8–3

8.3 Synchronization Issues in IR-UWB Communications

8.3.1 Signal Acquisition

Different algorithms for UWB signal acquisition have recently been proposed. In (Junk, 2003; Di Benedetto et al, 2003), a synchronization scheme that accounts for the way information is typically structured into packets by the MAC module is introduced. The proposed algorithm is based on the presence of a fixed synchronization trailer at the beginning of each packet, and assumes that synchronization can be maintained for the entire duration of a packet. For each single packet, synchronization must be re-established. This scheme accounts for the presence of clock drifts between the two stations involved in the synchronization process. Within a packet, the procedure keeps track of synchronization, while in between two packets, when tracking is not active, synchronization might be lost after a time lag that depends on time drifts of local oscillators. Each data packet generated by the UWB transmitter is composed of three parts: a trailer for synchronization, a header that contains control information plus signaling symbols for channel estimation and error detection, and a payload for the user data (see Figure 8–49).

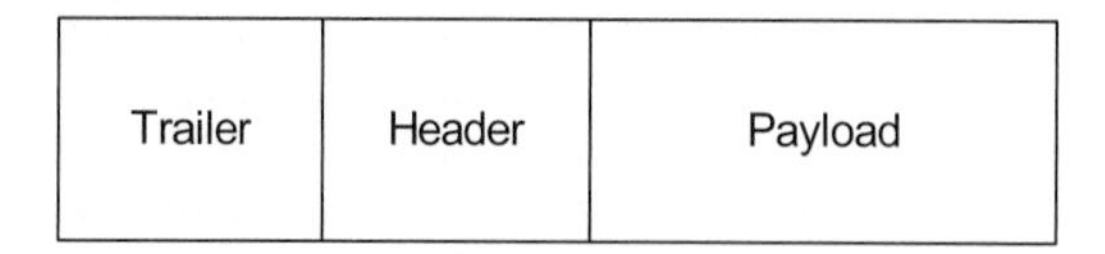

Figure 8–49 Data packet structure (by permission from *Mobile Networks and Applications Journal*, Kluwer Academic Publishers).

The synchronization trailer consists of a sequence of M pulses and is known *a priori* by all network participants. The trailer allows a receiver to detect incoming packets. Correlation filters that are matched to the synchronization trailer are in fact implemented in all receiving stations.

Synchronization performance is directly related to the M value. A rough appraisal of the requested M value can be obtained by simulation as a function of the SNR at the receiver. As indicated in (Di Benedetto et al., 2003), estimated M values fall in the range of 40–100 pulses for a wide range of applications.

The algorithm proposed by (Yang and Giannakis, 2003) introduces additional cross-correlations of the received waveform to increase performance in the presence of multi-path. The proposed scheme is based on the transmission of training sequences and exploits multi-path diversity of UWB channels.

The problem of synchronization acquisition in the presence of multi-path is also addressed in (Tian and Giannakis, 2003a). Here, the authors propose a timing acquisition algorithm based on a maximum likelihood approach.

A relatively different approach that reduces complexity is presented in (Reggiani and Maggio, 2003). The acquisition procedure correlates the received signal with an expected pulse sequence for any possible position within the observation window. The proposed approach reduces the complexity of the searching algorithm by adopting a different quantization level of the time axis for each run of the algorithm. In other words, the algorithm starts with a coarse acquisition and then iteratively refines the result with increasing precision.

The algorithm for timing acquisition proposed in (Carbonelli et. al, 2003) is divided into two steps. First, a periodic sequence of training pulses is transmitted at a high pulse repetition frequency to perform channel estimation. Then, data is transmitted and the receiver exploits the previous channel estimate to better recover symbol timing.

The scheme proposed in (Ma, et al. 2002) consists of a non-consecutive search algorithm based on the presence of a parallel bank of matched filters. For each symbol period, the largest output of the bank of filters is compared with a detection threshold. If the signal is over the threshold, a longer observation interval and a higher detection threshold is used to decrease the probability of false alarms.

The work in (Hong and Scaglione, 2003) extends the issue of pulse detection to the network level and presents an adaptive and distributed time synchronization method that emulates the mechanism of pulse coupling in biological systems.

8.3.2 Tracking

When the receiver has achieved synchronization with the reference transmitter, algorithms for tracking the correct timing of the pulses are required to compensate for timing errors due to clock jitters and drifts. (Lovelace and Townsend, 2002) analyzed the problem of tracking in terms of performance loss due to the presence of time jitters in both receiver and transmitter clocks. Performance loss is evaluated in different perturbation environments and as a function of the amount of jitter. When MUI noise is predominant, the analysis shows that even in the case of modest jitter values, for example 30 ps of total jitter, the reduction of the maximum number of users that must be introduced for a target $Pr_b = 10^{-3}$ is on the order of 30%. When thermal noise dominates MUI, it is shown that timing errors can be compensated by an increase in the transmitted power. As expected, the amount of additional power required for a given jitter depends on the cumulative system throughput, that is, the higher the throughput, the higher the increase of power required for compensating for the same clock jitter. In addition, the power loss due to timing jitter also depends on the modulation scheme that is implemented. Orthogonal PPM, in particular, is more robust than non-orthogonal PPM systems, and consequently requires a smaller increase of power for compensating for the same timing jitter. (Tian and Giannakis, 2003b) highlight that the promising capability of UWB receivers to exploit multi-path diversity can be severely compromised by timing errors.

An analytical method for investigating UWB receiver sensitivity to clock jitters is presented in (Pelissier et al. 2003). Here, jitter is introduced within the Delay Locked Loop (DLL) and Phase Locked Loop (PLL) of the receiver structure, and the effect of jitter is evaluated in terms of the undesired variance observed when sampling or correlating an input waveform.

In (Chui and Scholtz, 2003), the relationship between pulse waveform and timing jitter, when considering correlating detectors, is analyzed. Results show that in the special case of n–th order derivative Gaussian pulses, small n values should be chosen for the purpose of acquiring synchronization, while higher n values should be considered for reducing the variance of timing jitter due to additive noise in the input.

CHECKPOINT 8–4

In this checkpoint, we will present and simulate a simple solution for solving the problems of signal acquisition and temporal synchronization in the case of IR-UWB transmission. This solution is based on the transmission of a pilot sequence of pulses that is known *a priori* at the receiver. A correlator filter that is matched to this sequence is thus introduced at the receiver side. The presence of the pilot sequence can be estimated by observing the signal at the output of the correlator. In addition, the peak of the correlator output allows the receiver to achieve time alignment with the transmitter.

Two new MATLAB functions are required for the simulation. The first, Function 8.14, operates a circular shift of the elements of an input vector. The scope of this function is twofold. It is introduced for simulating the delay of propagation and for implementing the correlation at the receiver. The second function, Function 8.15, simulates the activity of the correlator.

Function 8.14 (see Appendix 8.A) applies a circular shift of `t` seconds to the signal represented by the input vector `in`. Function 8.14 receives as input the original signal `in`, the value of the sampling frequency `fc`, and the value in seconds of the time shift to be applied `t`. The function returns the shifted signal `out`. The command line for executing Function 8.14 is:

```
[out] = cp0804_signalshift(in,fc,t);
```

Function 8.15 (see Appendix 8.A) implements the correlation between vector `signal`, representing the signal at the receiver input, and a reference vector `template`, representing the pilot sequence of pulses. Both the vectors representing the received signal and the reference waveform must be given as input to the function, together with the value of the sampling frequency `fc`. Function 8.15 returns signal `C`, representing the correlator output. The function is executed with the following command:

```
[C] = cp0804_corrsyn(signal,template,fc);
```

Given Functions 8.14 and 8.15, we can simulate the generation of the pilot sequence, the propagation of this sequence over the radio channel, and the algorithm for signal acquisition and temporal synchronization.

The generation of the pilot sequence can be performed via Function 2.6 (see Appendix 2.A), which generates a train of pulses with binary PPM and TH coding. The following parameters are considered within the simulation: `Pow=-30; fc=1e11; numbits=1; Ts=10e-9; Ns=10; Tc=1e-9; Nh=10; Np=10; Tm=0.5e-9; tau=0.2e-9; dPPM=0.5e-9; G=0`. The pilot signal is generated as follows:

```
[bits,THcode,Stx,ref]=cp0201_transmitter_2PPM_TH;
n = length(Stx);
xn = 3*n;
pilot = zeros(1,xn);
pilot(1:n) = Stx;
```

The above command lines store the vector `pilot` in memory, which represents a sequence of 10 pulses with random positions on the time axis. Note that `pilot` contains non-zero elements only in the first part of the vector. This way, we are able to introduce the temporal shift due to propagation without the risk of truncation errors for the useful waveform. The pilot signal generated above is shown in Figure 8–50.

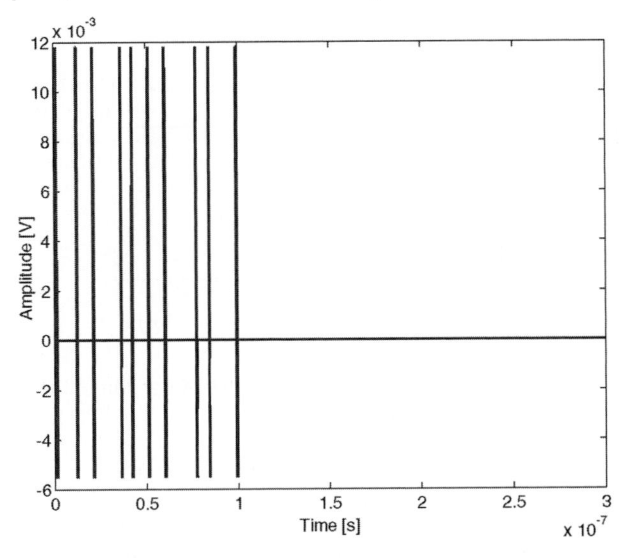

Figure 8–50 Pilot sequence of pulses for signal acquisition and receiver synchronization (signal pilot).

The delay of propagation is applied to the generated signal `pilot` by introducing Function 8.14:

```
fc = 1e11;
dt = 1/fc;
Tb = 100e-9;
delay = dt*floor((Tb*rand)/dt);
[rx0] = cp0804_signalshift(pilot,fc,delay);
```

The above commands store signal `rx0` in memory, which represents the time-shifted copy of the original signal `pilot`, and the value of the time delay `delay`, which is applied by Function 8.14. Note that `delay` is assumed to be a random variable uniformly distributed within the time duration of the original sequence. In the current example, this variable has a value of `delay=5.0280e-008`, that is, the time shift introduced by propagation is equal to 50.28 ns. To simulate the process of signal acquisition and temporal synchronization, we assume that the value of `delay` is unknown at the receiver. The time-shifted signal `rx0` is shown in Figure 8–51.

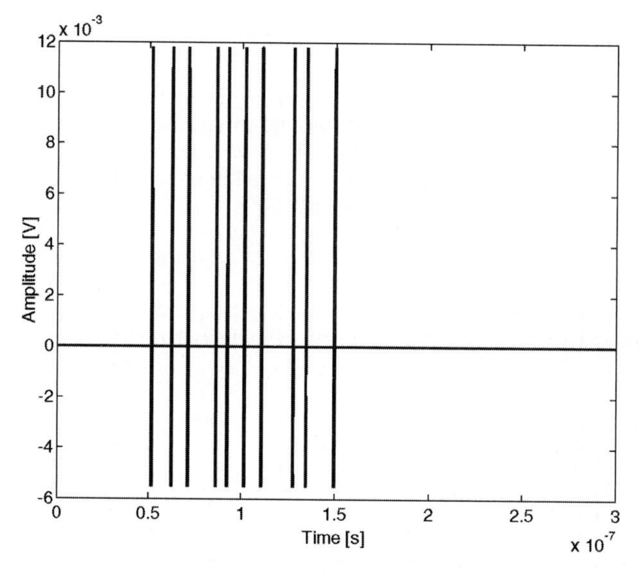

Figure 8–51 Signal rx0, corresponding to the time-delayed version of the signal pilot of Fig. 8–50.

As a first example, we can simulate the propagation of signal `pilot` over an AWGN channel. The signal at the receiver input can be evaluated through Function 8.3, which generates white Gaussian noise according to some target values of the E_{RX}/N_0 ratio, where E_{RX} is the received energy per pulse. Different cases will be analyzed corresponding to decreasing values of E_{RX}/N_0: a quasi-ideal scenario with $E_{RX}/N_0 = 50$ dB, and two realistic scenarios with E_{RX}/N_0 = 0 dB and E_{RX}/N_0 = -10 dB. The received signals for the three cases under examination can be generated as follows:

```
numpulses = 10;
exno = [50 0 -10];
[rxn,noise] = cp0801_Gnoise2(rx0,exno,numpulses);
```

The above command lines store matrices `rxn` and `noise` in memory. Matrix `rxn`, in particular, contains the different noisy signals at the input of the correlator, one signal per row. A graphical representation of these signals is shown in Figures 8–52, 8–53, and 8–54. Note that in the case of Figure 8–52, which corresponds to E_{RX}/N_0 = 50 dB, the pilot sequence can be easily recognized within the received waveform. In this case, we expect that the procedure for signal acquisition and temporal synchronization will manage to detect the presence of the sequence and recover synchronization. Figures 8–53 and 8–54, corresponding to E_{RX}/N_0 = 0 dB and E_{RX}/N_0 = -10 dB, are different. Here, the useful signal is completely blurred by noise, and the possibility for the receiver to detect and estimate the correct position of the pilot sequence is less evident than in the previous case.

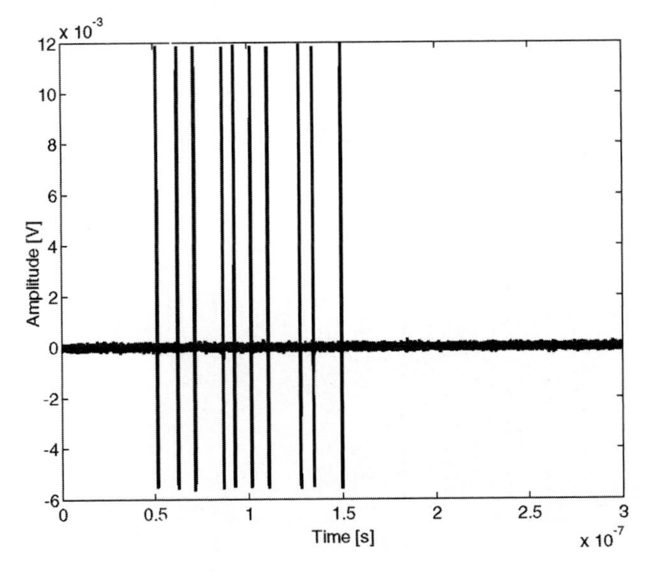

Figure 8–52 Received sequence at the input of the correlator in the case E_{RX}/N_0 = 50 dB.

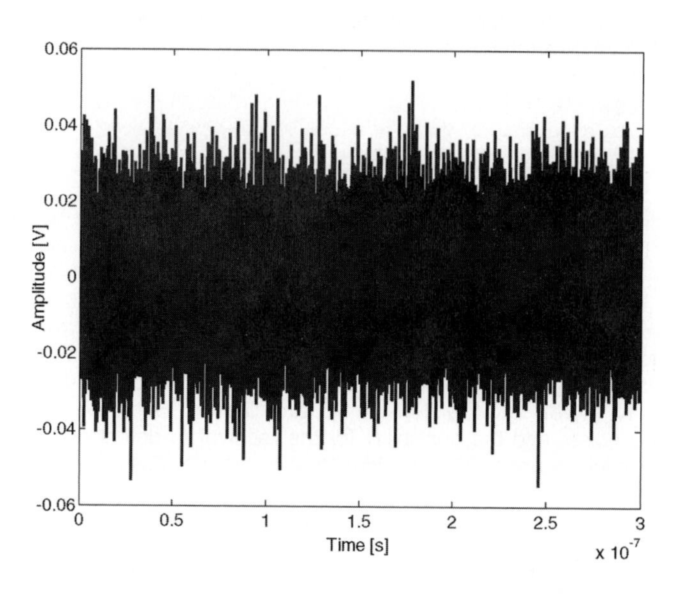

Figure 8–53 Received sequence at the input of the correlator in the case $E_{RX}/N_0 = 0$ dB.

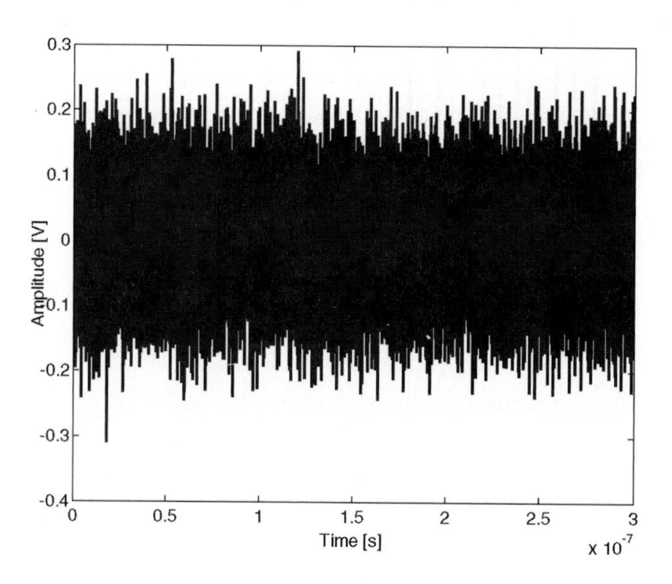

Figure 8–54 Received sequence at the input of the correlator in the case $E_{RX}/N_0 = -10$ dB.

As anticipated at the beginning of this checkpoint, the algorithm for signal acquisition and temporal synchronization cross-correlates the received signal with a copy of the original sequence of pulses, denoted in the following as the template signal. The correlation integral must be evaluated by considering all possible positions in the time of the template signal. This way, the output of the correlator varies in amplitude depending on the similarity of the received signal with the time-shifted version of the template signal. The higher the correlator output, the greater the similarity between the received signal and the time-shifted template signal. If the correlator output overcomes a given threshold, the receiver forms the hypothesis that the pilot signal is present at the antenna (signal acquisition). In this case, the receiver evaluates the peak of the correlator output and computes the corresponding delay of the template signal. Such a delay is an estimate of the delay introduced by propagation. In an ideal scenario where the signal is not affected by noise and distortions, the above estimation leads to correct synchronization between transmitter and receiver (temporal synchronization). In realistic scenarios, however, both signal acquisition and temporal synchronization can be affected by errors. In the following, we will focus on the effect of noise and distortions on the evaluation of the propagation delay.

The activity of the correlation filter for the three cases shown in Figures 8–52, 8–53, and 8–54 can be simulated as follows:

```
[C1] = cp0804_corrsyn(rxn(1,:),pilot,fc);
[C2] = cp0804_corrsyn(rxn(2,:),pilot,fc);
[C3] = cp0804_corrsyn(rxn(3,:),pilot,fc);
```

The above commands store the correlator outputs `C1`, `C2` and `C3` in memory. These correspond to the three cases E_{RX}/N_0 = 50 dB, E_{RX}/N_0 = 0 dB, and E_{RX}/N_0 = -10 dBm, respectively.

Figure 8–55 represents the correlator output `C1`. Here, we observe a strong peak located at about 50 ns, that is, in the starting position of signal `rx0` in Figure 8–51. The exact estimation of the peak of the correlator output can be derived using the MATLAB function `[Y,I] = MAX(X)`, which returns quantity `Y` representing the largest element in `X`, and integer `I` corresponding to the index of `Y` in vector `X`.

```
[peak,index]=max(C1);
estimated_delay_1 = index*dt
>> estimated_delay_1 = 5.0280e-008
```

As expected, the position in the time of the peak of signal `C1` provides the exact estimation of the delay of the pilot sequence after propagation. Moreover, no ambiguity is present in the estimation process since the peak of the correlator output clearly overcomes the smaller peaks of signal `C1`.

Figure 8–55 also highlights the importance of a correct choice for the threshold value. If it is too high — for instance, it is higher than 10 mV in the case shown in Figure 8–55 — the receiver is not capable of detecting the presence of the pilot sequence. If the threshold is too low — for instance, it is lower than 3 mV in the case shown in Figure 8–55 — the process of signal acquisition is triggered in the presence of all the small peaks of the correlator output. This problem, however, can be solved by selecting appropriate sequences for the pilot signal, that is, sequences leading to an autocorrelation function with a strong peak in the origin and quasi-zero values outside the origin.

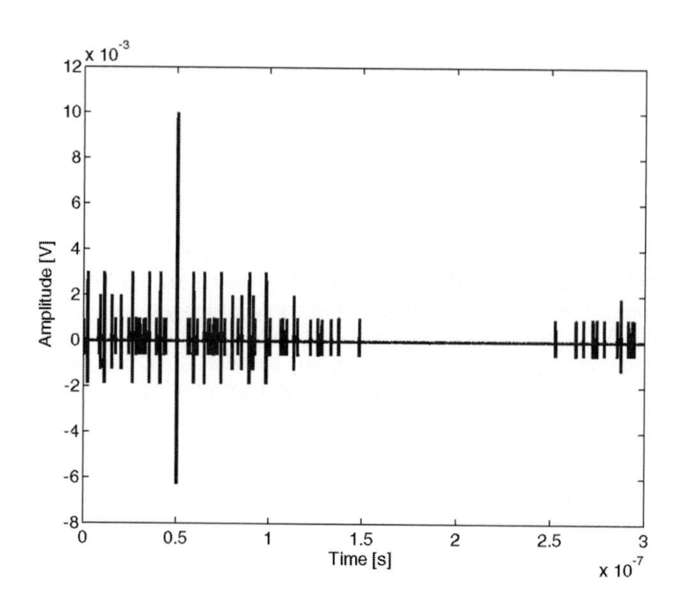

Figure 8–55 Correlator output C1, corresponding to $E_{RX}/N_0 = 50$ dB

The correlator output C2, corresponding to a more realistic case $E_{RX}/N_0 = 0$ dB, is shown in Figure 8–56. In this case, the signal at the output of the correlator is affected by the presence of noise, but we can still recognize the presence of a strong peak at about 50 ns. We derive that the correlation filter at the receiver is capable of catching the presence of the pilot sequence even in the case where the noise signal has an energy contribution comparable to that of the single pulse. The exact estimation of the peak of the correlator is evaluated as follows:

```
[peak,index]=max(C2);
estimated_delay_2 = index*dt
>> estimated_delay_2 = 5.0280e-008
```

Once again, the analysis of the signal at the output of the correlator gives the possibility to evaluate the exact position in time of the received pilot signal, that is, to achieve temporal synchronization between transmitter and receiver.

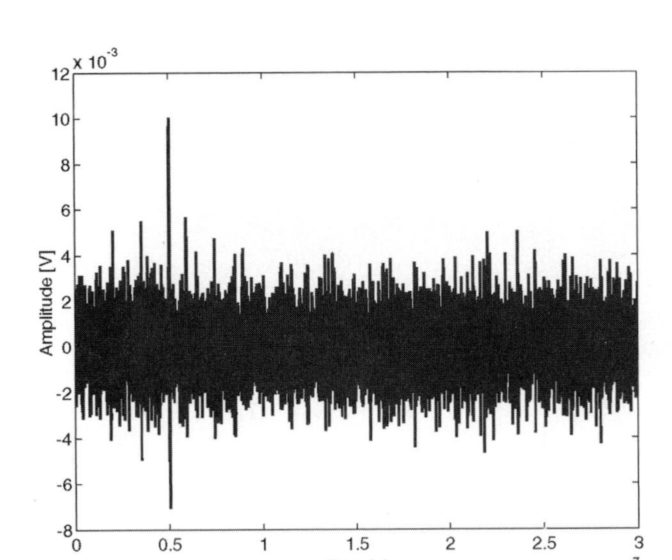

Figure 8–56 Correlator output C2, corresponding to $E_{RX}/N_0 = 0$ dB.

The correlator output C3, corresponding to the scenario with $E_{RX}/N_0 = -10$ dB, is shown in Figure 8–57. In this case, the peak at 50 ns is still recognizable, but it is dangerously close to the floor level of the noise signal. In particular, we observe the presence of several other peaks that have an amplitude comparable with that of the useful peak. In such a condition, signal detection can be affected by several false alarms due to the presence of spurious peaks that trigger the synchronization process. To avoid false alarms, one could increase the detection threshold, but this introduces an increased risk of missing the detection of the pilot sequence. In such a condition, a possible solution is to increase the length of the pilot signal. Note that increasing the length of the pilot sequence has the effect of increasing both energy and time, which must be used for sending the pilot sequence. The tradeoff between energy and performance of the detection process depends on the requirements of the application under consideration. Regarding the performance of the estimation process for temporal synchronization, it is also evident that the presence of a high level of noise has the additional effect of introducing errors in the evaluation of the delay for the pilot sequence at the receiver, even when the sequence is correctly detected. In the case of signal C3, in fact, one has:

```
[peak,index]=max(C3);
estimated_delay_3 = index*dt
>> estimated_delay_3 = 5.0100e-008
```

The estimated time of arrival of the pilot sequence in the case of signal C3 is affected by an error of 0.18 ns. Such an error can lead to a loss in performance for the decision process when data symbols are transmitted.

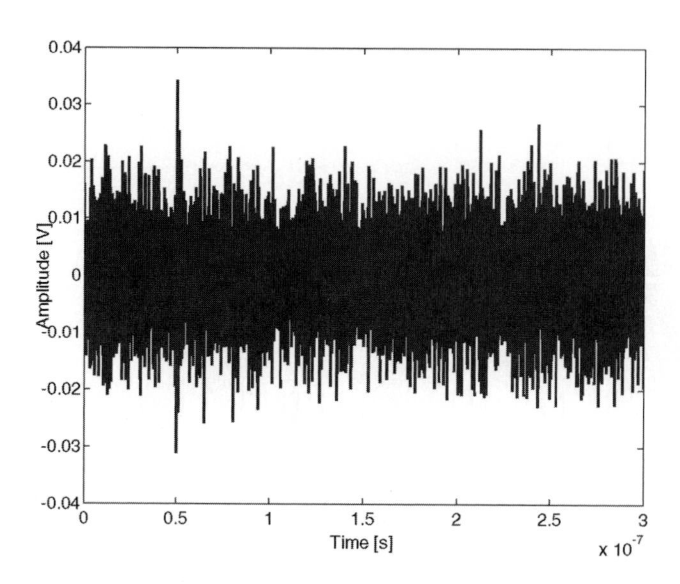

Figure 8–57 Correlator output C3, corresponding to E_{RX}/N_0 = -10 dB.

As a final example, we can analyze the effect on synchronization performance of the propagation of the pilot sequence over multi-path channels. In particular, we will consider the case where the spreading in time operated by the channel impulse response produces ISI at the receiver, or when the temporal separation of the pulses within the pilot sequence is smaller than the delay spread introduced by multi-path. Two different cases are considered, corresponding to different scenarios of propagation.

In the first case, we assume the receiver to be in LOS with the transmitter. We can thus consider the first scenario of the IEEE channel model described in Section 8.2.3, Case A in Checkpoint 8–2. The channel impulse response for such a scenario is obtained by executing Function 8.8 with the following parameters: `OT = 200e-9; ts = 1e-9; LAMBDA = 0.0233*1e9; lambda = 2.5e9; GAMMA = 7.1e-9; gamma = 4.3e-9; sigma1 = 10^(3.3941/10); sigma2 = 10^(3.3941/10) ; sigmax = 10^(3/10); rdt = 0.001; PT = 50; G = 1`. The command line for generating the channel impulse response is:

```
[h01,hf1,OT,ts,X] = cp0802_IEEEuwb(fc,1);
```

which produces the plots in Figures 8–58 and 8–59.

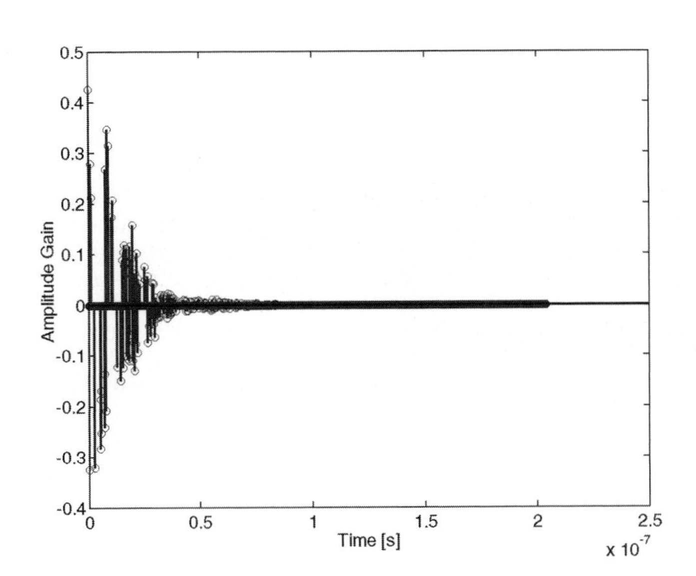

Figure 8–58 Channel impulse response for LOS propagation — Case A.

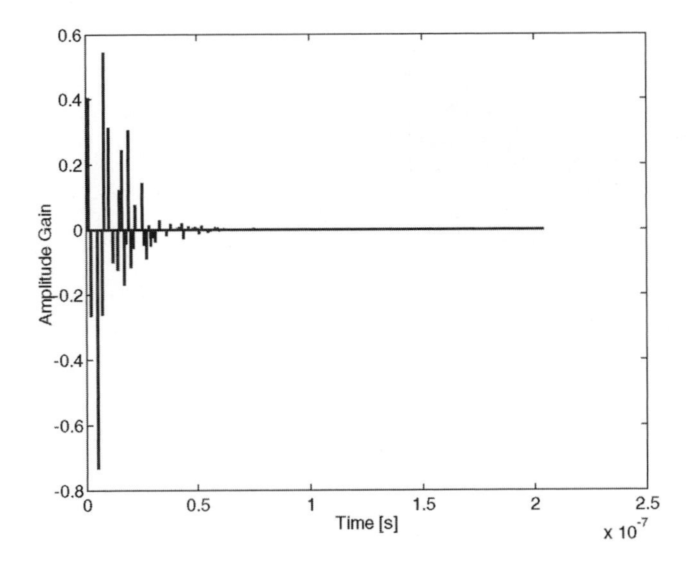

Figure 8–59 Discrete time channel impulse response for LOS propagation — Case A.

The signal received at the input of the correlator can be evaluated as follows:

```
rxc1  = conv(rx0,hf1);
rxc1  = rxc1(1:length(rx0));
rxc1a = rxc1 + noise(1,:);
rxc1b = rxc1 + noise(2,:);
rxc1c = rxc1 + noise(3,:);
```

The above commands store the three signals rxc1a, rxc1b, and rxc1c in memory. These correspond to the propagation of the pilot sequence over the multi-path channel of Figure 8–59, for different values of the E_{RX}/N_0 ratio at the receiver. Signal rxc1a corresponds to the quasi-ideal scenario with $E_{RX}/N_0 = 50$ dB, while signals rxc1b and rxc1c correspond to the cases $E_{RX}/N_0 = 0$ dB and $E_{RX}/N_0 = -10$ dB. Note that E_{RX} is the total received energy per pulse, or the energy that can be collected by capturing all replicas of the same pulse that arrive at the receiver after multi-path propagation.

Signals rxc1a, rxc1b, and rxc1c at the correlator input are affected by both Gaussian noise and distortions due to pulse overlapping. An example of the distortions introduced on the signal by multi-path is represented in Figure 8–60, which shows the waveform rxc1a at the receiver input when $E_{RX}/N_0 = 50$ dB.

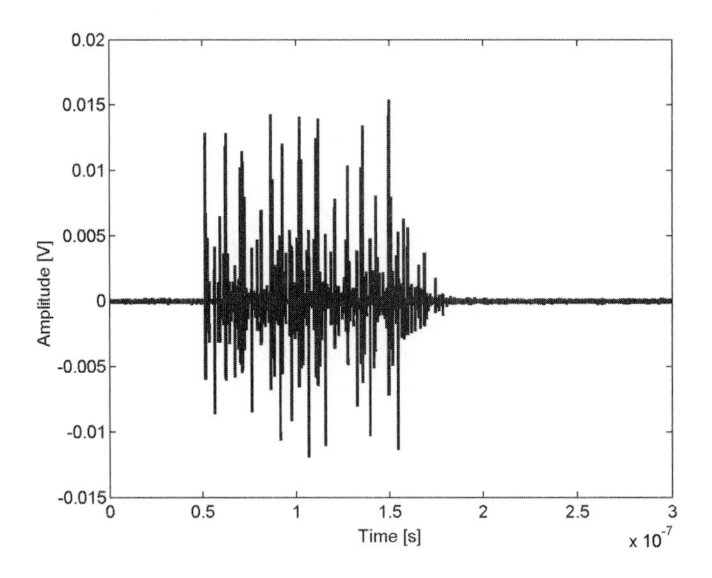

Figure 8–60 Received signal rxc1a at the input of the correlator ($E_{RX}/N_0 = 50$ dB).

Given waveforms rxc1a, rxc1b, and rxc1c, the activity of the correlation filter can be simulated as in the case of AWGN:

```
[Cm1] = cp0804_corrsyn(rxc1a,pilot,fc);
[Cm2] = cp0804_corrsyn(rxc1b,pilot,fc);
[Cm3] = cp0804_corrsyn(rxc1c,pilot,fc);
```

```
[peak,index]=max(Cm1);
estimated_delay_m1 = index*dt
>> estimated_delay_m1 = 5.0280e-008
[peak,index]=max(Cm2);
estimated_delay_m2 = index*dt
>> estimated_delay_m2 = 5.0280e-008
[peak,index]=max(Cm3);
estimated_delay_m3 = index*dt
>> estimated_delay_m3 = 5.0100e-008
```

The result of this simulation is the same as the AWGN case, that is, we obtain a perfect synchronization with E_{RX}/N_0 = 50 dB and E_{RX}/N_0 = 0 dB, while a synchronization error of 0.18 ns occurs in the case of E_{RX}/N_0 = -10 dB. We conclude that in the case under examination, the presence of multi-path does not introduce an additional loss in performance due to the synchronization process. The presence of multi-path, however, renders the signal at the output of the correlator more sensitive to noise, as one can verify by further decreasing the value of E_{RX}/N_0.

Different results are observed when considering a NLOS scenario, that is, when the delay spread of the multi-path channel is increased. We can generate an impulse response for a NLOS scenario by executing Function 8.8 with the following parameters: `OT = 200e-9; ts = 1e-9; LAMBDA = 0.0667*1e9; lambda = 2.1e9; GAMMA = 14e-9; gamma = 7.9e-9; sigma1 = 10^(3.3941/10); sigma2 = 10^(3.3941/10); sigmax = 10^(3/10); rdt = 0.001; PT = 50; G = 1`. This scenario coincides with the Case C scenario introduced in Checkpoint 8–2. The command line for generating the channel impulse response is:

```
[h02,hf2,OT,ts,X] = cp0802_IEEEuwb(fc,1);
```

which generates in Figures 8–61 and 8–62.

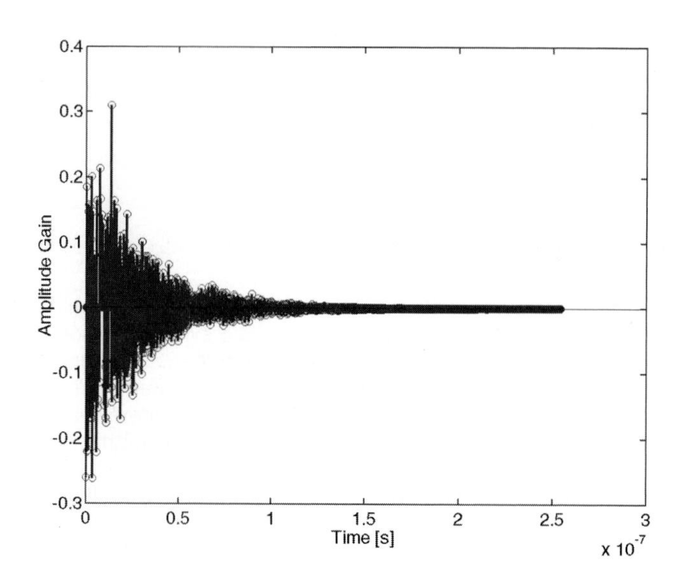

Figure 8–61 Channel impulse response for NLOS propagation — Case C.

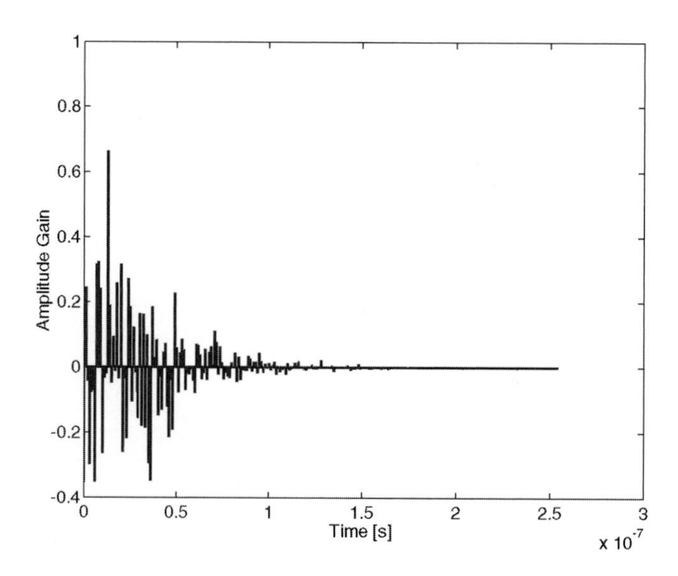

Figure 8–62 Discrete time channel impulse response for NLOS propagation — Case C.

The signal received at the input of the correlator can be evaluated as follows:

```
rxc2  = conv(rx0,hf2);
rxc2  = rxc2(1:length(rx0));
rxc2a = rxc2 + noise(1,:);
rxc2b = rxc2 + noise(2,:);
rxc2c = rxc2 + noise(3,:);
```

As in the previous example, signals `rxc2a`, `rxc2b`, and `rxc2c` correspond to the propagation of the pilot sequence over the multi-path channel in Figure 8–62 for different values of the E_{RX}/N_0 ratio at the receiver. Signal `rxc2a` corresponds to the quasi-ideal scenario $E_{RX}/N_0 = 50$ dB, while signals `rxc2b` and `rxc2c` correspond to the cases $E_{RX}/N_0 = 0$ dB and $E_{RX}/N_0 = -10$ dB. Due to the longer delay spread introduced by the propagation in the NLOS scenario, the distortion introduced by multi-path on the pilot sequence is higher than in the previous case. This increased effect can be observed in Figure 8–63, which shows the waveform `rxc2a` at the receiver input in presence of $E_{RX}/N_0 = 50$ dB. When comparing the plot in Figure 8–63 with the one in Figure 8–60, we recognize that the spreading effect due to multi-path propagation is more evident in the NLOS case than in the LOS case.

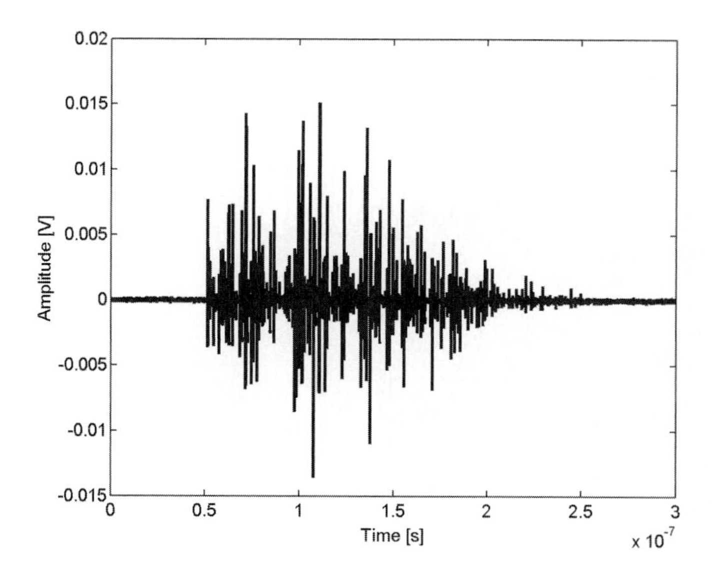

Figure 8–63 Received signal rxc2a at the input of the correlator ($E_{RX}/N_0 = 50$ dB).

Performance of the correlation filter in the case of NLOS propagation is derived as follows:

```
[Cn1] = cp0804_corrsyn(rxc2a,pilot,fc);
[Cn2] = cp0804_corrsyn(rxc2b,pilot,fc);
```

```
[Cn3] = cp0804_corrsyn(rxc2c,pilot,fc);
[peak,index]=max(Cn1);
estimated_delay_n1 = index*dt
>> estimated_delay_n1 = 5.0280e-008
[peak,index]=max(Cn2);
estimated_delay_n2 = index*dt
>> estimated_delay_n2 = 5.9200e-008
[peak,index]=max(Cn3);
estimated_delay_n3 = index*dt
>> estimated_delay_n3 = 5.9280e-008
```

According to the above results, the distortions suffered by the received sequence in the NLOS case generate a synchronization error of about 9 ns for both cases $E_{RX}/N_0 = 0$ dB and $E_{RX}/N_0 = -10$ dB. The reason for this gap between the propagation delay and the delay estimated at the receiver can be found in the shape of the channel impulse response in Figure 8–62. In this figure, we can verify the presence of a dominant path located 9 ns after the first path arriving at the receiver. The effect of this path is shown in Figure 8–64, which represents the output of the correlator in the case of $E_{RX}/N_0 = 0$ dB. In this figure, the useful peak at 50.28 ns is exceeded in amplitude by a strongest peak at 59.2 ns.

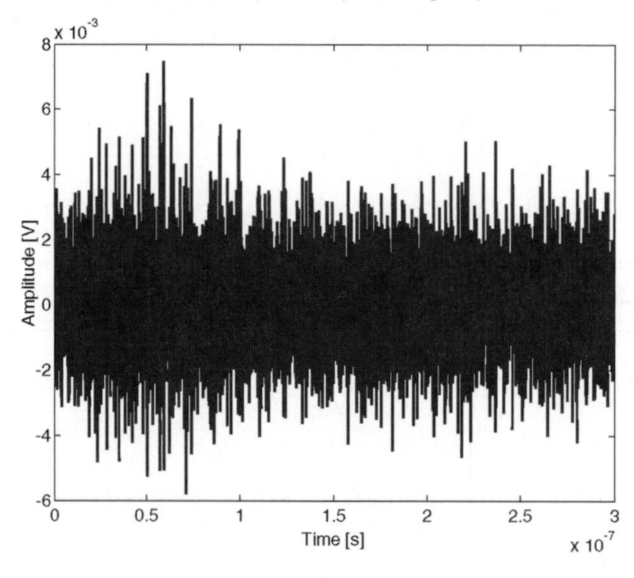

Figure 8–64 Correlator output Cn2, corresponding to $E_{RX}/N_0 = 0$ dB.

CHECKPOINT 8–4

FURTHER READING

As a direct consequence of the introduction and large success of commercial mobile radio applications, numerous studies were presented in the 1990s regarding the characterization of the mobile radio channel. Reference studies on the temporal variations of the mobile radio channel were published by Pahlavan et al. in (Howard and Pahlavan, 1990; Ganesh and Pahlavan, 1991; Ali, Parker, and Pahlavan, 1994). Here, statistical parameters for evaluating the multi-path profiles in different propagation environments are derived from propagation measurements in the 900MHz band, that is, in the portion of the frequency spectrum that was reserved for the European GSM system. Regarding outdoor propagation, a simulation model for the evaluation of the channel impulse response was presented by (Hashemi, 1979). For such a scenario, a statistical model for characterizing the PDP was presented by (Hendrikson et al., 1999), and a characterization of the rms delay spread was performed by (Rappaport et al., 1990).

The case of propagation in factories and open plan offices is addressed by (Rappaport et al., 1991) on the basis of wideband channel measurements in the 1.3 GHz band. A reference review of narrowband and wideband radio propagation into and within buildings was performed by (Molkdar, 1991). With reference to indoor propagation, we can also cite the theoretical and experimental results carried out by Hashemi reported in (Hashemi, 1993b; Hashemi et al., 1994; Hashemi and Tholl, 1994). In these papers, the statistical characterization of the radio channel is derived on the basis of extensive measurements carried out in office environments. Detailed analyses of the indoor radio channel can also be found in (Sexton and Pahlavan, 1989; Howard and Pahlavan, 1992; Cartwright and Sowerby, 1995).

A different approach for the characterization of the mobile radio channel is based on the application of the ray-tracing/ray-launching concepts. A reference work on the advantages offered by ray-tracing when developing site-specific channel predictions can be found in (Seidel et al., 1994). Interesting results regarding indoor radio propagation derived through the application of a deterministic ray-tracing model were published by (Yang et al., 1992), while an application of ray-launching for estimating the propagation of wideband radio signals was presented in (Lawton and McGeehan, 1994). In the work by (Feuerstein and Rappaport, 1993), the results obtained through ray-tracing were also used for developing a theoretical model of propagation for indoor communications.

The first analyses on the robustness of IR-UWB signals in dense multi-path environments can be found in (Win and Scholtz, 1998a). Here, the authors present the results of a measurement campaign performed in a typical modern office building, which indicate that the fading margin required to guarantee reliable communications is considerably smaller than in the case of narrowband transmissions. On the basis of these measurements, an extensive report on the characterization of the UWB channel was presented in (Win and Scholtz, 2002). (Ramirez-Mireles, 2002) proposes a signal design optimization for PPM-TH in a dense multi-path channel. Regarding the robustness of UWB to multi-path fading, we can also cite the recent works by (Muqaibel et al., 2003) and (Buehrer et al., 2003). For UWB channel estimation, the use of a Sensor-CLEAN algorithm is presented in (Scholtz et al., 1998) and (Cramer et al., 2002). This algorithm analyzes UWB signals by processing the data collected with an array of sensors in an indoor environment. This approach showed to guarantee the possibility of identifying co-incident pulses both in the case of equal amplitudes and in the presence of an amplitude difference of 40 dB. A different channel sounding scheme, which operates in the frequency domain, was recently proposed in (Haneda and Takada, 2003).

Receiver design is another topic that garnered significant research efforts for UWB communication systems. The first U.S. patent on a receiver structure for sub-nanosecond electromagnetic pulses is dated 1972 by (Robbins, 1972). Two other interesting patents are dated 1994 (McEwan, 1994) and 1996 (McEwan, 1996). Regarding the theoretical analysis of the UWB receiver, some interesting papers regarding the performance of the RAKE scheme in the presence of realistic propagation environments have been recently published. A reference study on the performance of the RAKE with M-ary time shift-modulated signals can be found in (Ramirez-Mireles et al., 1997). In the work presented by (Lottici et al., 2002), the authors analyze the effect of channel estimation errors on the performance of the RAKE receiver. Two channel estimation schemes are presented, which are demonstrated to control performance degradation in the presence of several asynchronous users. The problem of channel estimation for multi-path-affected UWB channel is also addressed in (Baccarelli and Biagi, 2004). The work by (Durisi and Romano, 2002) analyzes the performance of a RAKE receiver for PPM-UWB systems when a RAKE receiver with a variable number of fingers is used. The analysis is performed by assuming the amplitude gain of each multi-path component to be Nakagami-distributed, and confirms that a high number of branches is required for obtaining performance comparable to the ideal RAKE. In particular, at least 30 multi-path components must be processed within the RAKE to reduce the difference with the performance of the ideal case to 0.5 dB. Performance of the RAKE for a high data rate UWB system implementing both EGC and MRC was recently analyzed by (Rajeswaran et al., 2003).

With reference to multi-user detection for UWB systems, we can indicate the results published by (Li and Rusch, 2002), (Muqaibel et al., 2002), and (Liu et al., 2003). Several other contributions were presented on the application of the UWB concept to MIMO (Multiple Input Multiple Output) or SIMO (Single Input Multiple Output) systems. In particular, we can mention the results derived by (Biagi and Baccarelli, 2003), (Funk and Lee, 1996), (Huang et Letaief, 2002), (Bi and Hui, 2002), (Pajusco and Pagani, 2003), (Ezaki and Ohtsuki, 2003), (Weisenhorn and Hirt, 2003), (Kaiser, 2003), and (Tan et al., 2003). The Transmitted Reference (TR) approach (Rushforth, 1964; Hingorani and Hancock, 1965) was applied to UWB systems in (Choi and Stark, 2002; Franz and Mitra, 2003; Chao and Scholtz, 2003; Zhang and Goeckel, 2003). Finally, advanced architectures for the design of UWB receivers can be found in (Narayanan and Dawood, 2000), (Namgoong, 2003), (Feng and Namgoong, 2003), (Lee et al., 2003), (Hoyos et al., 2003), (Sheng et al., 2003), and (de Rivaz et al., 2003).

With reference to the general problem of signal acquisition and code synchronization, we mention the works of (Polydoros and Glisic, 1994) and (Corazza and Polydoros, 1998). Regarding the UWB case, the problem of signal acquisition is taken into account under a different perspective by addressing it in the MAC protocol by (Kolenchery et al. 1998). Timing jitter effect on the PSD of a TH-UWB signal is analyzed in (Win, 2002).

REFERENCES

Ali, M.H., A.S. Parker, and K. Pahlavan, "Frequency domain model for standard simulation of wideband radio propagation for personal communications," *Electronics Letters*, Volume: 30, Issue: 25 (December 1994), 2103–2104.

Alvarez, A., G. Valera, M. Lobeira, R. Torres, and J.L. Garcia, "Ultra Wideband Channel Model for Indoor Environments," *Journal of Communications and Networks*, Volume: 5, Issue: 4 (December 2003), 309–318.

Batra, A., J. Balakrishnan, A. Dabak, R. Gharpurey, J. Lin, P. Fontaine, J.-M. Ho, S. Lee, M. Frechette, S. March, M. Yamaguchi, and Texas Instruments et al., *Multi-band OFDM Physical Layer Proposal for IEEE 802.15 Task Group 3a*, Available at *www.multibandofdm.org/papers/15-03-0268-01-003a-Multi-band-CFP-Document.pdf* (September 2003).

Bi, C., and J.Y. Hui, "Multiple Access Capacity for Ultra-Wide Band Radio with Multi-Antenna Receivers," *IEEE Conference on Ultra Wideband Systems and Technologies* (May 2002), 151–155.

Baccarelli, E., and M. Biagi, "A Novel self-Pilot based Transmit-Receiving Architecture for Multipath-Impaired UWB systems," in press in *IEEE Transactions on Communications* (June 2004).

Biagi, M., and E. Baccarelli, "A Simple Multiple-Antenna UWB Transceiver Scheme for 4th generation WLAN," *IEEE Vehicular Technology Conference Proceedings* (October 2003), 1903–1907.

Bregni, S., *Synchronization of Digital Telecommunications Networks*, Chichester, UK: J. Wiley and Sons (2002).

Buehrer, R.M., W.A. Davis, A. Safaai-Jazi, and D. Sweeney, "Characterization of the Ultra-wideband Channel," *IEEE Conference on Ultra Wideband Systems and Technologies* (November 2003), 26–31.

Carbonelli, C., U. Mengali, and U. Mitra, "Synchronization and channel estimation for UWB signals," *IEEE Global Telecommunications Conference* (December 2003), 764–768.

Cartwright, P.M., and K.W. Sowerby, "Uplink and downlink analysis of a three-dimensional wireless indoor communication system," *Electronics Letters*, Volume: 31, Issue: 7 (March 1995), 538–539.

Cassioli, D., M.Z. Win, F. Vatalaro, and A.F. Molisch, "Performance of Low-Complexity Rake Reception in a Realistic UWB Channel," *IEEE International Conference on Communications*, Volume: 2 (April 2002a), 763–767.

Cassioli, D., M.Z. Win, and A.F. Molisch, "The Ultra-Wide Bandwidth Indoor Channel: From Statistical Model to Simulations," *IEEE Journal on Selected Areas in Communications*, Volume: 20 , Issue: 6 (August 2002b), 1247–1257.

Chao, Y.L., and R.A. Scholtz, "Optimal and Suboptimal Receivers for Ultra-wideband Transmitted Reference Systems," *IEEE Global Telecommunications Conference* (December 2003), 759–763.

Cherniakov, M., and L. Donskoi, "Frequency Band Selection of Radars for Buried Object Detection," *IEEE Transaction on Geoscience and Remote Sensing*, Volume: 37, Issue: 2 (March 1999), 838–845.

Choi, J.D., and W.E. Stark, "Performance of Ultra-Wideband Communications With Suboptimal Receivers in Multipath Channels," *IEEE Journal on Selected Areas in Communications*, Volume: 20, Issue: 9 (December 2002), 1754–1766.

Chui, C.-C., and R.A. Scholtz, "Optimizing Tracking Loops for UWB Monocycles," *IEEE Global Telecommunications Conference* (December 2003), 425–430.

Corazza, G.E., and A. Polydoros, "Code Acquisition in CDMA Cellular Mobile Networks. Part I: Theory," *Proceedings of the IEEE 5th International Symposium on Spread Spectrum Techniques and Applications*, Volume: 2 (September 1998), 454–458.

Cramer, R.J-M., R.A. Scholtz, and M.Z. Win, "Evaluation of an Ultra-Wide-Band Propagation Channel," *IEEE Transactions on Antennas and Propagation*, Volume: 50, Issue: 5 (May 2002a), 561–570.

Cramer, R.J-M., R.A. Scholtz, and M.Z. Win, *Evaluation of an Indoor Ultra-Wideband Propagation Channel,* Available at *http://grouper.ieee.org/groups/802/15/ pub/2002/Jul02/02286r0P802-15_SG3a-Evaluation-of-an-Indoor-Ultra-Wideband-Propagation-Channel.doc* (June 2002b).

Dabin, J.A., N. Ni, A.M. Haimovich, E. Niver, and H. Grebel, "The Effects of Antenna Directivity on Path Loss and Multipath Propagation in UWB Indoor Wireless Channels," *IEEE Conference on Ultra Wideband Systems and Technologies* (November 2003), 305–309.

de Rivaz, S., B. Denis, J. Keignart, M. Pezzin, N. Daniele, and D. Morche, "Performances Analysis of a UWB Receiver using Complex Processing," *IEEE Conference on Ultra Wideband Systems and Technologies* (November 2003), 229–233.

Denis, B., and J. Keignart, "Post-Processing Framework for Enhanced UWB Channel Modeling from Band-limited Measurements," *IEEE Conference on Ultra Wideband Systems and Technologies* (November 2003), 260–264.

Di Benedetto, M.-G., L. De Nardis, M. Junk, and G. Giancola, "$(UWB)^2$: Uncoordinated, Wireless, Baseborn medium access for UWB communication networks," in press in *Mobile Networks and Applications special issue on WLAN Optimization at the MAC and Networks Levels*, (2004).

Durisi, G., and G. Romano, "Simulation Analysis and Performance Evaluation of an UWB system in indoor multipath channel," *IEEE Conference on Ultra Wideband Systems and Technologies* (May 2002), 255–258.

Ezaki, T., and T. Ohtsuki, "Diversity Gain in Ultra Wideband Impulse Radio (UWB-IR)," *IEEE Conference on Ultra Wideband Systems and Technologies* (November 2003), 56–60.

Feng, L., and W. Namgoong, "Joint Estimation and Detection of UWB Signals with Timing Offset Error and Unknown Channel," *IEEE Conference on Ultra Wideband Systems and Technologies* (November 2003), 152–156.

Feuerstein, M.J., and T.S. Rappaport, *Wireless Personal Communications*, Boston, MA: Kluwer Academic Publishers (1993), 225–249.

Foerster, J., and Q. Li, *UWB Channel Modeling Contribution from Intel*, Available at *http://grouper.ieee.org/groups/802/15/pub/2002/Jul02/02279r0P802-15_SG3a-Channel-Model-Cont-Intel.doc* (June 2002).

Franz, S., and U. Mitra, "On Optimal Data Detection for UWB Transmitted Reference Systems," *IEEE Global Telecommunications Conference* (December 2003), 744–748.

Funk, E.E., and C.H. Lee, "Free-Space Power Combining and Beam Steering of Ultra-Wideband Radiation Using an Array of Laser-Triggered Antennas," *IEEE Transactions on Microwave Theory and Techniques*, Volume: 44, Issue: 11 (November 1996), 2039–2044.

Ganesh, R., and K. Pahlavan, "Statistical modeling and computer simulation of indoor radio channel," *IEE Proceeding-I*, Volume: 138, Issue: 3 (June 1991), 153–161.

Ghassemzadeh, S.S., L.J. Greenstein, A. Kavcic, T. Sveinsson, and V. Tarokh, "An Empirical Indoor Path Loss Model for Ultra-Wideband Channels", *Journal of Communications and Networks*, Volume: 5, Issue: 4 (December 2003), 303–308.

Ghassemzadeh, S.S., L.J. Greenstein, and V. Tarokh, *The Ultra-Wideband Indoor Multipath Model*, Available at *http://grouper.ieee.org/groups/802/15/pub/2002/Jul02/02282r1P802-15_SG3a-802-15-UWB-Multipath-Model.doc* (July 2002).

Ghassemzadeh, S.S., and V. Tarokh, *The Ultra-wideband Indoor Path Loss model*, Available at *http://grouper.ieee.org/groups/802/15/pub/2002/Jul02/02277r1P802-15_SG3a-802.15-UWB-Propagation-Path%20Loss-Model.doc* (July 2002).

Ghassemzadeh, S.S., and V. Tarokh, "UWB Path Loss Characterization In Residential Environments," *IEEE Radio Frequency Integrated Circuits Symposium* (June 2003), 501–504.

Haneda, K., and J. Takada, "An Application of SAGE Algorithm for UWB Propagation Channel Estimation," *IEEE Conference on Ultra Wideband Systems and Technologies* (November 2003), 483–487.

Hanzo, L., M. Münster, B.J. Choi, and T. Keller, *OFDM and MC-CDMA for Broadband Multi-User Communications, WLANs and Broadcasting*, Chichester, West Sussex, England: John Wiley and Sons, Inc. (2003).

Hashemi, H., "Simulation of the Urban Radio Propagation Channel," *IEEE Transactions on Vehicular Technology*, Volume: 28, Issue: 3 (August 1979), 213–225.

Hashemi, H., "The Indoor Radio Propagation Channel," *Proceedings of the IEEE*, Volume: 81, Issue: 7 (July 1993a), 943–968.

Hashemi, H., "Impulse Response Modeling of Indoor Radio Propagation Channels," *IEEE Journal on Selected Areas in Communications*, Volume: 11, Issue: 7 (September 1993b), 967–978.

Hashemi, H., and D. Tholl, "Statistical Modeling and Simulation of the RMS Delay Spread of Indoor Radio Propagation Channels," *IEEE Transactions on Vehicular Technology*, Volume: 43, Issue: 1 (February 1994), 110–119.

Hashemi, H., M. McGuire, T. Vlasschaert, and T. Tholl, "Measurements and Modeling of Temporal Variations of the Indoor Radio Propagation Channel," *IEEE Transactions on Vehicular Technology*, Volume: 43, Issue: 3 (August 1994), 733–737.

Hendrickson, C., G. Gerace, C. Yerkes, and J. Forgy, "Wideband Wireless Peer to Peer Propagation Measurements," *Conference Record of the Thirty-Third Asilomar Conference on Signals, Systems, and Computers*, Volume: 1 (October 1999), 183–189.

Hingorani, G.D., and J.C. Hancock, "A Transmitted Reference System for Communication in Random of Unknown Channels," *IEEE Transactions on Communication Technology*, Volume: 13, Issue: 3 (September 1965), 293–301.

Hong, Y.W., and A. Scaglione, "Time synchronization and reach-back communications with pulse-coupled oscillators for UWB wireless ad hoc networks," *IEEE Conference on Ultra Wideband Systems and Technologies* (November 2003), 190–194.

Hovinen, V., M. Hamalainen, R. Tesi, L. Hentila, N. Laine, D. Porcino, and G. Shor, *A proposal for a selection of indoor UWB path loss model*, Available at *http://grouper.ieee.org/groups/802/15/pub/2002/Mar02/02119r0P802-15_SG3a-Response-to-CFA-ULTRAWAVES.ppt* (June 2002a).

Hovinen, V., M. Hamalainen, and T. Pätsi, "Ultra Wideband Indoor Radio Channel Models: Preliminary Results," *IEEE Conference on Ultra Wideband Systems and Technologies* (May 2002b), 75–79.

Howard, S.J., and K. Pahlavan, "Doppler Spread Measurements of Indoor Radio Channel," *Electronics Letters*, Volume: 26, Issue: 2 (January 1990), 107–108.

Howard, S.J., and K. Pahlavan, "Autoregressive Modeling of Wide-Band Indoor Radio Propagation," *IEEE Transactions on Communications*, Volume: 40, Issue: 9 (September 1992), 1540–1552.

Hoyos, S., B.M. Sadler, and G.R. Arce, "Dithering and $\Sigma\Delta$ Modulation in Mono-bit Digital Receivers for Ultra-Wideband Communications," *IEEE Conference on Ultra Wideband Systems and Technologies* (November 2003), 71–75.

Huang, D., and K.B. Letaief, "A Reduced Complexity Coded OFDM System with MIMO Antennas for Broadband Wireless Communications," *IEEE Global Telecommunications Conference,* Volume:1 (November 2002), 661–665.

IEEE 802.15.SG3a, "Channel modeling Sub-committee Report Final," *IEEE P802.15-02/490r1-SG3a* (February 2003).

Jaureguy, M., and P. Borderies, "Modelling and Processing of Ultra Wide Band Scattering of Buried Targets," *IEE Conference Publication No.431, "Detection of abandoned land mines"* (October 1996), 119–123.

Junk, M.D., *Synchronization Issues in Ultra Wideband Communication Networks*, M.S. Thesis, Fachgebiet Nachrichtentechnische Systeme der Universitat Duisburg-Essen Standort Duisburg (2003).

Kaiser, T., "On UWB Beamforming," *Proceedings of Kleinheubacher Tagung* (October 2003).

Keignart, J., and N. Daniele, "Subnanosecond UWB Channel Sounding in Frequency and Temporal Domain," *IEEE Conference on Ultra Wideband Systems and Technologies* (May 2002), 25–30.

Kolenchery, S.S., J.K. Townsend, and J.A. Freebersyer, "A novel impulse radio network for tactical military wireless communications," *IEEE Military Communications Conference,* Volume: 1 (October 1998), 59–65.

Kunisch, J., and J. Pamp, "Measurement results and Modeling Aspects for the UWB Radio Channel," *IEEE Conference on Ultra Wideband Systems and Technologies* (May 2002a), 19–23.

Kunisch, J., and J. Pamp, *Radio Channel Model for Indoor UWB WPAN Environments*, Available at *http://grouper.ieee.org/groups/802/15/pub/2002/Jul02/02281r0P802-15_SG3a-IMST-Response-Call-Contributions-UWB-Channel-Models.pdf* (June 2002b), 290–294.

Kunisch, J., and J. Pamp, "An Ultra-Wideband Space-Variant Multipath Indoor Radio Channel Model," *IEEE Conference on Ultra Wideband Systems and Technologies* (November 2003).

Lawton, M.C., and J.P. McGeehan, "The Application of a Deterministic Ray Launching Algorithm for the Prediction of Radio Channel Characteristics in Small-Cell Environments," *IEEE Transactions on Vehicular Technology*, Volume: 43, Issue: 4 (November 1994), 955–969.

Lee, E.A., and D.G. Messerschmitt, *Digital Communication*, 2nd Edition, Boston, Massachusetts: Kluwer Academic Publishers (1994).

Lee, H., B. Han, Y. Shin, and S. Im, "Multipath Characteristics of Impulse Radio Channels," *IEEE Vehicular Technology Conference Proceedings*, Volume: 3 (May 2000), 2487–2491.

Lee, H.J., D.S. Ha, and H.S. Lee, "A Frequency-Domain Approach for all-digital CMOS Ultra Wideband Receivers," *IEEE Conference on Ultra Wideband Systems and Technologies* (November 2003), 86–90.

Li, Q., and L.A. Rusch, "Multiuser Detection for DS-CDMA UWB in the Home Environment," *IEEE Journal on Selected Areas in Communications*, Volume: 20, Issue: 9 (December 2002), 1701–1711.

Licul, S., W.A. Davis, and W.L. Stutzman, "Ultra-Wideband (UWB) Communication Link Modeling and Characterization," *IEEE Conference on Ultra Wideband Systems and Technologies* (November 2003), 310–314.

Liu, P., Z. Xu, and J. Tang, "Subspace Multiuser Receivers for UWB Communication Systems," *IEEE Conference on Ultra Wideband Systems and Technologies* (November 2003), 116–120.

Lottici, V., A. D'Andrea, and U. Mengali, "Channel Estimation for Ultra-Wideband Communications," *IEEE Journal on Selected Areas in Communications*, Volume: 20, Issue: 9 (December 2002), 1638–1645.

Lovelace, W.M., and J.K. Townsend, "The Effects of Timing Jitter and Tracking on the Performance of Impulse Radio," *IEEE Journal on Selected Areas in Communications,* Volume: 20, Issue: 9 (December 2002), 1646–1651.

Ma, Y., F. Chin, B. Kannan, and S. Pasupathy, "Acquisition Performance of an Ultra Wide-band Communications System Over a Multiple-Access Fading Channel," *IEEE Conference on Ultra Wideband Systems and Technologies* (May 2002), 99–103.

McEwan, T.E., *Ultra-Wideband Receiver*, United States Patent number 5345471 (September 1994).

McEwan, T.E., *Ultra-Wideband Receiver*, United States Patent number 5523760 (June 1996).

Mengali, U., and A. D'Andrea, *Synchronization Techniques for Digital Receivers*, 1st Edition, New York : Plenum Press/Kluwer Academic Publishers (1997).

Middleton, D., *An introduction to Statistical Communication Theory*, New York: McGraw-Hill Book Company (1960).

Molisch, A.F., M.Z. Win, and D. Cassioli, *The Ultra-Wide Bandwidth Indoor Channel: from Statistical Model to Simulations*. Available at *http://grouper.ieee.org/groups/802/15/pub/2002/Jul02/02284r0P802-15_SG3a-The-Ultra-Wide-Bandwidth-Indoor-Channel-from-Statistical-Model-to-Simulations.pdf* (June 2002).

Molkdar, D., "Review on radio propagation into and within buildings," *IEE Proceedings-H*, Volume: 138, Issue: 1 (February 1991), 61–73.

Muqaibel, A., B. Woerner, and S. Riad, "Application of Multi-User Detection Techniques to Impulse Radio Time Hopping Multiple Access Systems," *IEEE Conference on Ultra Wideband Systems and Technologies* (May 2002), 169-173.

Muqaibel, A.H., A. Safaai-Jazi, A.M. Attiya, A. Bayram, and S. Riad, "Measurement and Characterization of Indoor Ultra-Wideband Propagation," *IEEE Conference on Ultra Wideband Systems and Technologies* (November 2003), 295–299.

Namgoong, W., "A Channalized Digital Ultrawideband Receiver," *IEEE Transactions on Wireless Communications*, Volume: 2, Issue: 3 (May 2003), 502–510.

Narayanan, R.M., and M. Dawood, "Doppler Estimation Using a Coherent Ultrawide-Band Random Noise Radar," *IEEE Transactions on Atennas and Propagation*, Volume: 48, Issue: 6 (June 2000), 868–878.

Pajusco, P., and P. Pagani, "Extension of SIMO Wideband Channel Sounder for UWB Propagation Experiment," *IEEE Conference on Ultra Wideband Systems and Technologies* (November 2003), 250–254.

Pelissier, M., B. Denis, and D. Morche, "A methodology to investigate UWB digital receiver sensitivity to clock jitter," *IEEE Conference on Ultra Wideband Systems and Technologies* (November 2003), 126–130.

Pendergrass, M., and W.C. Beeler, *Empirically Based Statistical Ultra-Wideband (UWB) Channel Model*, Available at *http://grouper.ieee.org/groups/802/15/pub/2002/Jul02/02294r1p802-15_SG3a-Empirically_Based_UWB_Channel_Model.ppt* (July 2002).

Polydoros, A., and S. Glisic, "Code synchronization a review of principles and techniques," *IEEE Third International Symposium on Spread Spectrum Techniques and Applications*, Volume: 1 (July 1994), 115-137.

Prettie, C., D. Cheung, L. Rusch, and M. Ho, "Spatial Correlation of UWB Signals in a Home Environment," *IEEE Conference on Ultra Wideband Systems and Technologies* (May 2002), 65–69.

Price, R., and P.E. Green Jr., "A communication technique for multipath channels," *Proc. IRE*, Volume: 46 (March 1958), 555–570.

Proakis, J.G., *Digital Communications*, 3rd Edition, New York: McGraw-Hill International Editions (1995).

Qiu, R.C., "A Study of the Ultra-Wideband Wireless Propagation Channel and Optimum UWB Receiver Design," *IEEE Journal on Selected Areas in Communications,* Volume: 20, Issue: 9 (December 2002), 1628–1637.

Rajeswaran, A., V.S. Somayazulu, and J.R. Foerster, "Rake Performance for a Pulse Based UWB System in a Realistic UWB Indoor Channel," *IEEE Conference on Communications*, Volume: 4 (2003), 2879–2883.

Ramirez-Mireles, F., "Signal Design for Ultra-Wide-Band Communications in Dense Multipath," *IEEE Transactions on Vehicular Technology*, Volume: 51, Issue: 6 (November 2002), 1517–1521.

Ramirez-Mireles, F., M.W. Win, and R.A. Scholtz, "Performance of Ultra-Wideband Time-Shift-Modulated Signals in the Indoor Wireless Impulse Radio Channel," *Conference Record of the Thirty-First Asilomar Conference on Signals, Systems, and Computers*, Volume: 1 (November 1997), 192–196.

Rappaport, T.S., S.Y. Seidel, and R. Singh, "900-MHz Multipath Propagation Measurements for U.S. Digital Cellular Radiotelephone," *IEEE Transactions on Vehicular Technology*, Volume: 39, Issue: 2 (May 1990), 132–139.

Rappaport, T.S., S.Y. Seidel, K. Takamizawa, "Statistical Channel Impulse Response Models for Factory and Open Plan Building Radio Communication System Design," *IEEE Transactions on Communications*, Volume: 39, Issue: 5 (May 1991), 794–807.

Reggiani, L., and G.M. Maggio, "A Reduced-Complexity Acquisition Algorithm for UWB Impulse Radio," *IEEE Conference on Ultra Wideband Systems and Technologies* (November 2003), 131–135.

Rice, L.P., "Radio transmission into buildings at 35 and 150 mc," *Bell Syst. Tech. J.*, Volume: 38, Issue: 1 (January 1959), 197–210.

Robbins, K.W., Short Base-Band Pulse Receiver, United States Patent number 3662316 (May 1972).

Roberts, R., *XtremeSpectrum CFP Document*, Available at *grouper.ieee.org/groups/802/15/pub/2003/Jul03/03154r3P802-15_TG3a-XtremeSpectrum-CFP-Documentation.pdf* (July 2003).

Romme, J., and B. Kull, "On the Relation Between Bandwidth and robustness of indoor UWB Coomunication," *IEEE Conference on Ultra Wideband Systems and Technologies* (November 2003), 255–259.

Rushforth, C.K., "Transmitted-reference Techniques for Random or Unknown Channels," *IEEE Transactions on Information Theory*, Volume: 10, Issue: 1 (January 1964), 39–42.

Saleh, A.A.M., and R.A. Valenzuela, "A Statistical Model for Indoor Multipath Propagation," *IEEE Journal on Selected Areas in Communications*, Volume: 5, Issue: 2 (February 1987), 128–137.

Scholtz, R.A., R.J.-M. Cramer, and M.Z. Win, "Evaluation of the Propagation Characteristics of Ultra-Wideband Communication Channels," *IEEE Transactions on Antennas and Propagation*, Volume: 50 (February 1998), 626–630.

Seidel, S.Y., and T.S. Rappaport, "Site-Specific Propagation Prediction for Wireless In-Building Personal Communication System Design," *IEEE Transactions on Vehicular Technology*, Volume: 43, Issue: 4 (November 1994), 879–891.

Sexton, T.A., and K. Pahlavan, "Channel Modeling and Adaptive Equalization of Indoor Radio Channels," *IEEE Journal on Selected Areas in Communications*, Volume: 7, Issue 1 (January 1989), 114–120.

Sheng, H., A.M. Haimovich, A.F. Molisch, and J. Zhang, "Optimum Combining for Time Hopping Impulse Radio UWB Rake Receivers," *IEEE Conference on Ultra Wideband Systems and Technologies* (November 2003), 224–228.

Siwiak, K., and A. Petroff, "A Path Link Model for Ultra Wide Band Pulse Transmissions," *IEEE Conference on Vehicular Technology*, Volume: 2 (May 2001), 1173–1175.

Siwiak, K., *Propagation notes to P802.15 SG3a from IEEE Tutorial*, Available at *http://grouper.ieee.org/groups/802/15/pub/2002/Jul02/02328r0P802-15_SG3a-Propagation-notes-to-SG3a-from-IEEE-Tutorial.ppt* (July 2002a).

Siwiak, K., *UWB Propagation phenomena*, Available at *http://grouper.ieee.org/groups/802/15/pub/2002/Jul02/02301r3P802-15_SG3a-UWB-Propagation-Phenomena.ppt* (July 2002b).

Suzuki, Y., and T. Kobayashi, "Ultra Wideband Signal Propagation in Desktop Environments," *IEEE Conference on Ultra Wideband Systems and Technologies* (November 2003), 493–497.

Tan, S.S., B. Kannan, and A. Nallanathan, "Performance of UWB Multiple Access Impulse Radio Systems In Multipath Environment with Antenna Array," *IEEE Global Telecommunications Conference* (December 2003), 2182–2186.

Tian, Z., and G.B. Giannakis, "Data-aided ML timing acquisition in ultra-wideband radios," *IEEE Conference on Ultra Wideband Systems and Technologies* (November 2003a), 142–146.

Tian, Z., and G.B. Giannakis, "BER Sensitivity to Mis-Timing in Correlation based UWB Receivers," *IEEE Global Telecommunications Conference* (December 2003b), 441–445.

Turin, G.L., "Communication through Noisy, Random-Multipath Channels," *IRE Convention* (1956), 154–166.

Turin, G.L., F.D. Clapp, T.L. Johnston, S.B. Fine, and D. Lavry, "A Statistical Model of Urban Multipath Propagation," *IEEE Transactions Vehicular Technology*, Volume: 21 (February 1972), 1–9.

Turin, W., R. Jana, S.S. Ghassemzadeh, C.W. Rice, and V. Tarokh, "Autoregressive Modeling of an Indoor UWB Channel," *IEEE Conference on Ultra Wideband Systems and Technologies* (May 2002), 71–74.

Uguen, B., E. Plouhinec, Y. Lostanlen, and G. Chassay, "A Deterministic Ultra Wideband Channel Modeling," *IEEE Conference on Ultra Wideband Systems and Technologies* (May 2002), 1–5.

Weeks, G.D., J.K. Townsend, and J.A. Freebersyser, "Performance of Hard Decision Detection for Impulse Radio," *IEEE Military Communications Conference* (October 1999), 1201–1206.

Weisenhorn, M., and W. Hirt, "Performance of Binary Antipodal Signaling over the Indoor UWB MIMO Channel," *IEEE International Conference on Communications*, Volume: 4 (May 2003), 2872–2878.

Win, M.Z., "Spectral Density of Random UWB Signals," *IEEE Communications Letters*, Volume: 6, Issue: 12 (December 2002), 526–528.

Win, M.Z., F. Ramirez-Mireles, R.A. Scholtz, and M.A. Barnes, "Ultra-Wide Bandwidth (UWB) Signal Propagation for Outdoor Wireless Communications," *IEEE Conference on Vehicular Technology*, Volume: 1 (May 1997), 251–255.

Win, M.Z., and R.A. Scholtz, "On the Robustness of Ultra-Wide Bandwidth Signals in Dense Multipath Environments," *IEEE Communications Letters*, Volume: 2, Issue: 2 (February 1998a), 51–53.

Win, M.Z., and R.A. Scholtz, "On the Energy Capture of Ultrawide Bandwidth Signals in Dense Multipath Environments", *IEEE Communications Letters*, Volume: 2, Issue: 9 (September 1998b), 245–247.

Win, M.Z., and R.A. Scholtz, "Characterization of Ultra-Wide Bandwidth Wireless Indoor Channels: A Communication-Theoretic View," *IEEE Journal on Selected Areas in Communications*, Volume: 20, Issue: 9 (December 2002), 1613–1627.

Yang, G., K. Pahlavan, and T. Holt, "Effects Antenna Sectorisation on Data Rate Limitations of Indoor Radio Modems," *Electronics Letters*, Volume: 28, Issue: 13 (June 1992), 1182–1183.

Yang, L., and G.B. Giannakis, "Low-Complexity Training for Rapid Timing Acquisition in Ultra Wideband Communications," *IEEE Global Telecommunications Conference* (December 2003), 769–773.

Yao, R., G. Gao, Z. Chen, and W. Zhu, "UWB Multipath Channel Model Based on Time-Domain UTD Technique," *IEEE Global Telecommunications Conference* (December 2003), 1205–1210.

Zhang, H., T. Udagawa, T. Arita, and M. Nakagawa, "A Statistical Model for the Small-Scale Multipath Fading Characteristics of Ultra Wideband Indoor Channel," *IEEE Conference on Ultra Wideband Systems and Technologies* (May 2002), 81–85.

Zhang, H., and D.L. Goeckel, "Generalized Transmitted-Reference UWB Systems," *IEEE Conference on Ultra Wideband Systems and Technologies* (November 2003), 147–151.

Zhu, F., Z. Wu, and C.R. Nassar, "Generalized Fading Channel Model with Application to UWB," *IEEE Conference on Ultra Wideband Systems and Technologies* (May 2002), 13–17.

APPENDIX 8.A

Function 8.1 Path Loss

The code of Function 8.1 consists basically of two instructions: First, the channel gain is computed as indicated in Eq. (8–3), then the amplitude of the input signal is scaled by array multiplication.

```
%
% FUNCTION 8.1 : "cp0801_pathloss"
%
% Attenuates the input signal 'tx' according to
% the distance 'd' [m], the decaying factor 'gamma'
% and the constant term 'c0', which represents the
% reference attenuation at 1 meter
%
% The function returns the attenuated signal 'rx'
% and the value of the channel gain 'attn'
%
% Programmed by Guerino Giancola
%

function [rx,attn] = cp0801_pathloss(tx,c0,d,gamma)

% -------------------------------
% Step One - Path loss evaluation
% -------------------------------

attn = (c0/sqrt(d^gamma));
rx = attn .* tx;
```

Function 8.2 Generation of AWGN as a Function of E_b/N_0

The first part of Function 8.2 evaluates N_0 values according to both target E_b/N_0 values contained in vector `ebno`, and the value of E_b, which is computed by observing the input signal. For each N_0 value, the corresponding standard deviation of noise `nstdv` is computed. The second part of the code consists of one main loop for the generation of thermal noise. The noise signals are generated by using the MATLAB function `randn(1,N)`, which produces `N` pseudo-random numbers chosen from a normal distribution with zero mean, variance one, and standard deviation one. At each iteration, the amplitude of noise is scaled by the corresponding value of the standard deviation. Finally, the output signal is generated by adding the noise signal to the input signal.

```
%
% FUNCTION 8.2 : "cp0801_Gnoise1"
%
% Introduces additive white Gaussian noise over signal
%  'input'
% Vector 'ebno' contains the target values of Eb/No (in dB)
% 'numbits' is the number of bits conveyed by the input
%  signal
%
% Multiple output signals are generated, one signal for
% each target value of Eb/No
% The array 'output' contains
% all the signals (input+AWGN), one signal per row
% The array 'noise' contains the different realization of
% the Gaussian noise, one realization per row
%
% Programmed by Guerino Giancola
%

function [output,noise] = ...
   cp0801_Gnoise1(input,ebno,numbits)

% --------------------------------
% Step One - Introduction of AWGN
% --------------------------------

Eb = (1/numbits)*sum(input.^2); % measured energy per bit
EbNo = 10.^(ebno./10);           % Eb/No in linear unit
No = Eb ./ EbNo;                 % unilateral spectral
                                 % density
nstdv = sqrt(No./2);             % standard deviation for
                                 % the noise
```

```
for j = 1 : length(EbNo)

    noise(j,:) = nstdv(j) .* randn(1,length(input));
    output(j,:) = noise(j,:) + input;

end
```

Function 8.3 Generation of AWGN as a Function of E_x/N_0

Function 8.3 is similar to Function 8.2, except for the fact that the noise level is evaluated according to a target E_x/N_0 value rather than a target E_b/N_0 value, where E_x is the received energy within a single pulse.

```
%
% FUNCTION 8.3 : "cp0801_Gnoise2"
%
% Introduces additive white Gaussian noise over signal
% 'input'
% Vector 'exno' contains the target values of Ex/No (in dB)
% 'numpulses' is the number of pulses composing the input
% signal
%
% Multiple output signals are generated, one signal for
% each target value of Ex/No.
% The array 'output' contains
% all the signals (input+AWGN), one signal per row
% The array 'noise' contains the different realization of
% the Gaussian noise, one realization per row
%
% Programmed by Guerino Giancola
%


function [output,noise] = ...
   cp0801_Gnoise2(input,exno,numpulses)


% -------------------------------
% Step One - Introduction of AWGN
% -------------------------------


Ex = (1/numpulses)*sum(input.^2);   % measured energy per
                                    % pulse
ExNo = 10.^(exno./10);              % Ex/No in linear units
No = Ex ./ ExNo;                    % unilateral spectral
                                    % density
nstdv = sqrt(No./2);                % standard deviation for
                                    % the noise


for j = 1 : length(ExNo)
```

```
    noise(j,:) = nstdv(j) .* randn(1,length(input));
    output(j,:) = noise(j,:) + input;

end
```

Function 8.4 Correlation Mask for PPM-TH-UWB Signals

Function 8.4 consists of one single step, which is divided into two blocks of commands. The first part of the function is required for implementing energy normalization on the correlation mask. The second part builds up the mask by first computing a delayed version of vector `ref` (`sref`), and by then performing the subtraction `ref-sref`.

```
%
% FUNCTION 8.4 : "cp0801_PPMcorrmask"
%
% Evaluates the correlation mask ('mask') in the
% case of binary PPM UWB signals
% 'ref' is the reference signal (with no modulation)
% that is produced by the 2PPM+TH transmitter
% 'fc' is the sampling frequency
% 'numpulses' is the number of transmitted pulses
% 'dPPM' is the value of the PPM shift
%
% Programmed by Guerino Giancola
%

function [mask] = cp0801_PPMcorrmask(ref,fc,numpulses,dPPM)

% ---------------------------------------------
% Step One - Evaluation of the correlation mask
% ---------------------------------------------

dt = 1 / fc;

% energy normalization
Epulse = (sum((ref.^2).*dt))/numpulses;
ref = ref./sqrt(Epulse);

% mask construction
PPMsamples = floor (dPPM ./ dt);
sref(1:PPMsamples)=ref(length(ref)- ...
   PPMsamples+1:length(ref));
sref(PPMsamples+1:length(ref)) = ref(1:length(ref)- ...
   PPMsamples);
mask = ref-sref;
```

Function 8.5 Receiver for PPM-TH-UWB Signals

Function 8.5 is composed of three steps. Step Zero defines the receiver setting. The flag HDSD indicates whether hard decision detection (HDSD=1) or soft decision detection (HDSD=2) is selected. Step One implements both correlator and detector, and Step Two generates the statistics. At the beginning of Step One, Function 8.5 evaluates the number N of different signals to be considered. For the generic nth signal, the code starts by computing the array product between the waveform at the receiver input, that is, the one-dimensional vector R(n,:), and the correlation mask. The resulting vector mx must be integrated over different intervals to produce the decision variables. In the case of hard decision detection, the integration period corresponds to the average pulse repetition period T_s, while in the case of soft decision, it coincides with the bit period N_sT_s. In both cases, the above operation is executed iteratively. For each iteration, the code extracts the fragment of vector mx to be analyzed (mxkp for hard decision and mxk for soft decision) and integrates it over the appropriate time period. In the case of hard decision, the result of the integration is the variable zp. To decide on the single bit, the detector collects N_s subsequent zp values and applies the simple majority criterion for estimating the transmitted bit. In the case of soft decision detection, the integration is performed over a bit interval and produces the variable zb. A single decision is then executed for bit estimation. In both cases, positive values at the correlator output are associated with a "0" bit and negative values are associated with a "1" bit. To measure receiver performance, the estimated binary stream at the output of the detector is compared with the original binary stream produced by the transmitter. For each stream of estimated bits, Pr_b is determined by dividing the number of wrong bits WB by the total number of transmitted bits.

```
%
% FUNCTION 8.5 : "cp0801_PPMreceiver"
%
% Simulates the receiver for 2PPM TH UWB signals
% and computes the average BER
% 'R' is an array containing different waveforms
% of the received signal
% 'mask' is the waveform of the correlation mask
% 'fc' is the sampling frequency
% 'bits' is the binary stream generated by the source
% (it is the same stream for all the waveforms in 'R')
% 'Ns' is the number of pulses per bit
% 'Ts' is the average pulse repetition period [s]
%
% The function returns the binary stream after the
% detection process ('RXbits') and the vector 'BER'
% containing the average bit error rates for all
% the signals in the input array 'R'
%
```

```
% Programmed by Guerino Giancola
%

function [RXbits,BER] = ...
   cp0801_PPMreceiver(R,mask,fc,bits,numbit,Ns,Ts)

% -----------------------------
% Step Zero - Receiver settings
% -----------------------------

HDSD = 1;
% HDSD = 1 --> hard decision detection
% HDSD = 2 --> soft decision detection

% -----------------------------------------
% Step One - Implementation of the receiver
% -----------------------------------------

% N is the number of different signals at the receiver
% input L is the number of samples representing each signal
[N,L] = size(R);

RXbits = zeros(N,numbit);

dt = 1 / fc;                          % sampling time
framesamples = floor(Ts ./ dt);       % number of samples per
                                      % frame
bitsamples = framesamples * Ns;       % number of samples per
                                      % bit

for n = 1 : N

    rx = R(n,:);
    mx = rx .* mask;

    if HDSD == 1 % hard decision detection

        for nb = 1 : numbit

            mxk = mx(1+(nb-1)*bitsamples:bitsamples+...
               (nb-1)*bitsamples);
            No0 = 0;
            No1 = 0;

            for np = 1 : Ns
                mxkp = mxk(1+(np-1)*framesamples:...
```

```
                framesamples+(np-1)*framesamples);
            zp = sum(mxkp.*dt);
            if zp > 0
                No0 = No0 + 1;
            else
                No1 = No1 + 1;
            end
        end % for np = 1 : Ns

        if No0 > No1
            % the estimated bit is '0'
            RXbits(n,nb) = 0;
        else
            % the estimated bit is '0'
            RXbits(n,nb) = 1;
        end

    end % for nb = 1 : numbit

end % end of hard decision detection

if HDSD == 2 % soft decision detection

    for nb = 1 : numbit

        mxk = mx(1+(nb-1)*bitsamples:bitsamples+...
            (nb-1)*bitsamples);
        zb = sum(mxk.*dt);

        if zb > 0
            % the estimated bit is '0'
            RXbits(n,nb) = 0;
        else
            % the estimated bit is '1'
            RXbits(n,nb) = 1;
        end

    end % for nb = 1 : numbit

end % end of soft decision detection

end % for n = 1 : N

% ---------------------
% Step Two - Statistics
% ---------------------
```

```
for n = 1 : N
    WB = sum(abs(bits-RXbits(n,:)));
    BER(n) = WB / numbit; % average BER
end
```

Function 8.6 Correlation Mask for PAM-DS-UWB Signals

Function 8.6 consists of one single step, which computes the correlation mask by simply operating energy normalization over the input signal `ref`.

```
%
% FUNCTION 8.6 : "cp0801_PAMcorrmask"
%
% Evaluates the correlation mask ('mask') for
% receiving 2PAM UWB signals
% 'ref' is the reference signal (with no modulation)
% that is produced by the 2PAM+DS transmitter
% 'fc' is the sampling frequency
% 'numpulses' is the number of transmitted pulses
%
% Programmed by Guerino Giancola
%

function [mask] = cp0801_PAMcorrmask(ref,fc,numpulses)

% ----------------------------------------
% Step One - Evaluation of the correlation mask
% ----------------------------------------

dt = 1 / fc;

% energy normalization
Epulse = (sum((ref.^2).*dt))/numpulses;
ref = ref./sqrt(Epulse);

% Mask construction
mask = ref;
```

Function 8.7 Receiver for PAM-DS-UWB Signals

Function 8.7 is similar to Function 8.5, except for the decision rule. In the PAM case, in fact, the detector decides for a 0 bit in the presence of negative values at the correlation output, and for a 1 bit for positive values at the correlation output.

```
%
% FUNCTION 8.7 : "cp0801_PAMreceiver"
%
% Simulates the receiver for 2PAM DS UWB signals
% and computes the average BER
% 'R' is an array containing different waveforms
% of the received signal
% 'mask' is the waveform of the correlation mask
% 'fc' is the sampling frequency
% 'bits' is the binary stream generated by the source
% (it is the same stream for all the waveforms in 'R')
% 'Ns' is the number of pulses per bit
% 'Ts' is the average pulse repetition period [s]
%
% The function returns the binary stream after the
% detection process ('RXbits') and the vector 'BER'
% containing the average bit error rates for all
% the signals in the input array 'R'
%
% Programmed by Guerino Giancola
%

function [RXbits,BER] = ...
   cp0801_PAMreceiver(R,mask,fc,bits,numbit,Ns,Ts)

% -----------------------------
% Step Zero - Receiver settings
% -----------------------------

HDSD = 2;
% HDSD = 1 --> hard decision detection
% HDSD = 2 --> soft decision detection

% ----------------------------------------
% Step One - Implementation of the receiver
% ----------------------------------------

% N is the number of different signals at the receiver
```

```
% input L is the number of samples representing each signal
[N,L] = size(R);

RXbits = zeros(N,numbit);

dt = 1 / fc;                           % sampling time
framesamples = floor(Ts ./ dt);        % number of samples per
                                       % frame
bitsamples = framesamples * Ns;        % number of samples per
                                       % bit

for n = 1 : N

    rx = R(n,:);
    mx = rx .* mask;

    if HDSD == 1 % hard decision detection

        for nb = 1 : numbit

            mxk = mx(1+(nb-1)*bitsamples:bitsamples+...
                (nb-1)*bitsamples);
            No0 = 0;
            No1 = 0;

            for np = 1 : Ns
                mxkp = mxk(1+(np-1)*framesamples:...
                    framesamples+(np-1)*framesamples);
                zp = sum(mxkp.*dt);
                if zp < 0
                    No0 = No0 + 1;
                else
                    No1 = No1 + 1;
                end
            end % for np = 1 : Ns

            if No0 > No1
                % the estimated bit is '0'
                RXbits(n,nb) = 0;
            else
                % the estimated bit is '0'
                RXbits(n,nb) = 1;
            end

        end % for nb = 1 : numbit
```

```
    end % end of hard decision detection

    if HDSD == 2 % soft decision detection

        for nb = 1 : numbit

            mxk = mx(1+(nb-1)*bitsamples:...
               bitsamples+(nb-1)*bitsamples);
            zb = sum(mxk.*dt);

            if zb < 0
                % the estimated bit is '0'
                RXbits(n,nb) = 0;
            else
                % the estimated bit is '1'
                RXbits(n,nb) = 1;
            end

        end % for nb = 1 : numbit

    end % end of soft decision detection

end % for n = 1 : N

% ---------------------
% Step Two - Statistics
% ---------------------

for n = 1 : N
    WB = sum(abs(bits-RXbits(n,:)));
    BER(n) = WB / numbit; % average BER
end
```

Function 8.8 IEEE UWB Channel Impulse Response

Function 8.8 is composed of five steps. Step One contains the basic parameters characterizing the channel impulse response. The user can select the value of the observation time `OT` in seconds, the time resolution `ts` of the equivalent discrete time impulse response, and the values for the statistical parameters characterizing the model (see Table 8–1), that is, the cluster arrival rate `LAMBDA`, the ray arrival rate `lambda`, the cluster decay factor `GAMMA`, the ray decay factor `gamma`, the standard deviation `sigma1` of the cluster fading, the standard deviation `sigma2` of the amplitude of the multi-path contributions within each cluster, and the standard deviation `sigmax` of the log-normal total multi-path gain. In addition, the user can also select the parameter `rdt`, which is used within the simulation for limiting the generation of the multi-path contributions within each cluster. `PT` is an additional input parameter that can be used for limiting the number of multi-path components of the channel impulse response. The final input parameter is `G`, which is set equal to 1 when graphical output is required.

Steps One and Two contain the code for generating the multi-path components. Step One, in particular, generates the time of arrival of each cluster within the observation time. According to Eq. (8–65), the time of arrival of each cluster is determined as an outcome of an exponentially distributed random variable. Step Two generates the multi-path contributions within each cluster determined in Step One. According to Eq. (8–66), the time of arrival of each multi-path contribution within a cluster is determined as the outcome of an exponentially distributed random variable. The amplitude of each ray is evaluated by implementing Eq. (8–71), that is, by determining an outcome of a log-normal distributed random variable having the mean value of Eq. (8–73). After generating the multi-path contributions, the function selects those components above the threshold defined by `PT`. Based on the continuous time impulse response, Step Three derives the equivalent discrete time impulse response. As suggested in (IEEE 802.15.SG3a, 2003), this is obtained by adding all multi-path contributions that fall into the same bin interval `ts`. Finally, both continuous time and discrete time versions of the channel impulse response are normalized in energy. The log-normal shadowing is then applied by scaling vectors `h0` and `hf` with the outcome `X` of a log-normal random variable with a mean value given by Eq. (8–76). Step Four contains the code for the graphical representation of both the continuous time and discrete time channel impulse responses.

```
%
% FUNCTION 8.8 : "cp0802_IEEEuwb"
%
% Generates the channel impulse response for a multi-path
% channel according to the statistical model proposed by
% the IEEE 802.15.SG3a
%
% 'fc' is the sampling frequency
% 'TMG' is the total multi-path gain
%
```

```
% The function returns:
% 1) the channel impulse response 'h0'
% 2) the equivalent discrete time impulse response 'hf'
% 3) the value of the observation time 'OT'
% 4) the value of the resolution time 'ts'
% 5) the value of the total multi-path gain 'X'
%
% Programmed by Guerino Giancola
%

function [h0,hf,OT,ts,X] = cp0802_IEEEuwb(fc,TMG);


% ----------------------------
% Step Zero - Input parameters
% ----------------------------

OT = 300e-9;                  % observation time [s]
ts = 2e-9;                    % time resolution [s]
                              % i.e., the 'bin' duration

LAMBDA = 0.0667*1e9;          % cluster arrival rate (1/s)
lambda = 2.1e9;               % ray arrival rate (1/s)
GAMMA = 24e-9;                % cluster decay factor
gamma = 12e-9;                % ray decay factor
sigma1 = 10^(3.3941/10);      % stdev of the cluster fading
sigma2 = 10^(3.3941/10);      % stdev of the ray fading
sigmax = 10^(3/10);           % stdev of log-normal shadowing


% ray decay threshold
rdt = 0.001;
% rays are neglected when exp(-t/gamma)<rdt

% peak threshold [dB]
PT = 50;
% rays are considered if their amplitude is
% within the -PT range with respect to the peak

G = 1;
% G = 1 graphical output
% G = 0 no graphical output

% ------------------------------------
% Step One - Cluster characterization
% ------------------------------------
```

```
dt = 1 / fc;            % sampling time

T = 1 / LAMBDA;         % average cluster inter-arrival time
                        % [s]
t = 1 / lambda;         % average ray inter-arrival time [s]

i = 1;
CAT(i)=0;               % first cluster arrival time
next = 0;
while next < OT
    i = i + 1;
    next = next + expinv(rand,T);
    if next < OT
        CAT(i)= next;
    end
end % while remaining > 0

% --------------------------------
% Step Two - Path characterization
% --------------------------------

NC = length(CAT);    % number of observed clusters
logvar = (1/20)*((sigma1^2)+(sigma2^2))*log(10);
omega = 1;
pc = 0;                 % path counter

for i = 1 : NC

    pc = pc + 1;
    CT = CAT(i);        % cluster time

    HT(pc) = CT;

    next = 0;
    mx = 10*log(omega)-(10*CT/GAMMA);
    mu = (mx/log(10))-logvar;
    a = 10^((mu+(sigma1*randn)+(sigma2*randn))/20);

    HA(pc) = ((rand>0.5)*2-1).*a;

    ccoeff = sigma1*randn;  % fast fading on the cluster

    while exp(-next/gamma)>rdt

    pc = pc + 1;
```

```
        next = next + expinv(rand,t);
        HT(pc) = CT + next;
        mx = 10*log(omega)-(10*CT/GAMMA)-(10*next/GAMMA);
        mu = (mx/log(10))-logvar;
        a = 10^((mu+ccoeff+(sigma2*randn))/20);
        HA(pc) = ((rand>0.5)*2-1).*a;

        end

    end % for i = 1 : NC

    % Weak peak filtering
    peak = abs(max(HA));
    limit = peak/10^(PT/10);
    HA = HA .* (abs(HA)>(limit.*ones(1,length(HA))));

    for i = 1 : pc
        itk = floor(HT(i)/dt);
        h(itk+1) = HA(i);
    end

    % ------------------------------------------
    % Step Three - Discrete time impulse response
    % ------------------------------------------

    N = floor(ts/dt);
    L = N*ceil(length(h)/N);
    h0 = zeros(1,L);
    hf = h0;
    h0(1:length(h)) = h;
    for i = 1 : (length(h0)/N)
        tmp = 0;
        for j = 1 : N
            tmp = tmp + h0(j+(i-1)*N);
        end
        hf(1+(i-1)*N) = tmp;
    end


    % energy normalization
    E_tot=sum(h.^2);
    h0 = h0 / sqrt(E_tot);
    E_tot=sum(hf.^2);
    hf = hf / sqrt(E_tot);

    % log-normal shadowing
```

```
mux = ((10*log(TMG))/log(10)) - (((sigmax^2)*log(10))/20);

X = 10^((mux+(sigmax*randn))/20);

h0 = X.*h0;
hf = X.*hf;

% -----------------------------
% Step Four - Graphical output
% -----------------------------


if G

    Tmax = dt*length(h0);
    time = (0:dt:Tmax-dt);

    figure(1)
    S1=stem(time,h0);
    AX=gca;
    set(AX,'FontSize',14);
    T=title('Channel Impulse Response');
    set(T,'FontSize',14);
    x=xlabel('Time [s]');
    set(x,'FontSize',14);
    y=ylabel('Amplitude Gain');
    set(y,'FontSize',14);
    figure(2)
    S2=stairs(time,hf);
    AX=gca;
    set(AX,'FontSize',14);
    T=title('Discrete Time Impulse Response');
    set(T,'FontSize',14);
    x=xlabel('Time [s]');
    set(x,'FontSize',14);
    y=ylabel('Amplitude Gain');
    set(y,'FontSize',14);

end
```

Function 8.9 Root Mean Square Delay Spread

Function 8.9 is composed of one single step, which evaluates the rms delay spread `rmsds` of the input channel response `h`, sampled at frequency `fc`.

```
%
% FUNCTION 8.9 : "cp0802_rmsds"
%
% Evaluates the rms dealy spread 'rmsds'
% of a channel impulse response 'h' sampled
% at frequency 'fc'
%
% Programmed by Guerino Giancola
%


function [rmsds] = cp0802_rmsds(h,fc)

% ----------------------------------------------
% Step One - Evaluation of the rms delay spread
% ----------------------------------------------

dt = 1 / fc;          % sampling time

ns = length(h);       % number of samples representing
                      % the channel impulse response


time = (0 : dt : (ns-1)*dt);
den = sum(h.^2);
num1 = sum(time.*(h.^2));
num2 = sum((time.^2).*(h.^2));
rmsds = sqrt((num2/den)-(num1/den)^2);
```

Function 8.10 Power Delay Profile

Function 8.10 is composed of two steps. Step One evaluates the power delay profile `PDP` of the input vector `h`, representing the channel impulse response sampled at `fc`, while Step Two generates the corresponding graphical output.

```
%
% FUNCTION 8.10 : "cp0802_PDP"
%
% Evaluates the power delay profile 'PDP'
% of a channel impulse response 'h' sampled
% at frequency 'fc'
%
% Programmed by Guerino Giancola
%

function [PDP] = cp0802_PDP(h,fc)

% ----------------------------------
% Step One - Evaluation of PDP
% ----------------------------------

dt = 1 / fc;          % sampling time

PDP = (abs(h).^2)./dt;    % PDP

% ------------------------------
% Step Two - Graphical output
% ------------------------------

Tmax = dt*length(h);
time = (0:dt:Tmax-dt);

S1=plot(time,PDP);
AX=gca;
set(AX,'FontSize',14);
T=title('Power Delay Profile');
set(T,'FontSize',14);
x=xlabel('Time [s]');
set(x,'FontSize',14);
y=ylabel('Power [V^2]');
set(y,'FontSize',14);
```

Function 8.11 Channel Estimation

Function 8.11 is composed of two steps. Step One operates channel estimation, while Step Two evaluates the weighting factors of the RAKE combiner. To simplify the evaluation of the weighting factors for the SRAKE, channel estimation is performed by sorting the coefficients of vector `hf` in descending order. This task is executed by using the MATLAB function `[B,C]=sort(A)`. This function receives as input vector `A` and returns two output vectors: vector `B`, which contains the elements of `A` sorted in ascending order, and index vector `C`, which contains the original indexes of the elements of `B`, `B=A(C)`.

Function 8.11 executes the function `sort` on the absolute values of the channel impulse response. The output vectors provided by such an operation are then iteratively analyzed to produce three reference vectors: `G`, `T`, and `I`. Vector `G` contains all the non-zero components of the channel impulse response, sorted in descending order: `G(1)` is the amplitude of the multi-path contribution having the highest energy, and `G(length(G))` is the amplitude of the multi-path contribution having the lowest energy. Vector `T` contains the time of arrival for all the multi-path components collected in `G`: `T(j)` is the time of arrival of the multi-path component having the amplitude gain `G(j)`. Note that the elements of `T` are not sorted in ascending order since the strongest multi-path components are not always the first reaching the receiver (see Checkpoint 8–2). Finally, vector `I` stores the original positions of the elements of `G` in the channel impulse response `hf`: the multi-path contribution having coefficient `G(j)` corresponds to the element `I(j)` of the input vector `hf`.

Step Two of Function 8.11 uses vectors `G`, `T`, and `I` for evaluating three sets of weighting factors. The first set is stored in vector `Arake`, and contains the weighting factors to be used in the ideal RAKE. A second set of coefficients is stored in vector `Srake`, which contains the weighting factors to be used in a SRAKE. Finally, a third set of coefficients is stored in the vector `Prake`, which contains the coefficients to be used in a PRAKE. Note that all vectors, `Arake`, `Srake`, and `Prake`, are created in the form of a discrete time channel impulse response. In other words, these vectors are composed by all zero values, except for those elements that correspond to the presence of a multi-path component. This choice simplifies the creation of the correlator mask (see Function 8.12).

```
%
% FUNCTION 8.11 : "cp0803_rakeselector"
%
% Simulates channel estimation for a discrete time channel
% impulse response 'hf' with time resolution 'ts' in
% seconds
% In addition, the function evaluates the
% weighting factors to be used in a RAKE receiver
% implementing maximal ratio combining
% 'fc' is the value of the sampling frequency
% 'L' is the number of coefficients to be used in the
%  PRake
% 'S' is the number of coefficients to be used in the
```

```
%  SRake
%
% The function returns:
% 1) a vector 'G' containing all the amplitude coefficients
%     of the channel impulse response in descending
%     order
% 2) a vector 'T' containing all the relative delays for
%     the elements in vector 'G', i.e., T(j) represents the
%     relative delay of the multi-path component having
%     amplitude G(j)
% 3) the number 'NF' of non-zero contributions of the
%     channel impulse response
% 4) a vector 'Arake' representing the weighting factors to
%     be used in an ideal RAKE, which processes all the
%     multi-path contributions at the receiver input
% 5) a vector 'Srake' representing the weighting factors to
%     be used in a SRAKE, which processes the best
%     L multi-path contributions at the receiver input
% 6) a vector 'Srake' representing the weighting factors to
%     be used in a SRAKE, which processes the best
%     S multi-path contributions at the receiver input
% 7) a vector 'Prake' representing the weighting factors to
%     be used in a PRAKE, which processes the first L
%     multi-path contributions that arrive at the receiver
%     input
%
% Programmed by Guerino Giancola
%

function [G,T,NF,Arake,Srake,Prake] = ...
   cp0803_rakeselector(hf,fc,ts,L,S)

% ------------------------------
% Step One - Channel estimation
% ------------------------------

dt = 1 / fc;

ahf = abs(hf);
[s_val,s_ind] = sort(ahf);

NF = 0;
i = length(s_ind);
j = 0;
% evaluation of the reference vectors
% for the RAKE combiner
```

```
while (s_val(i)>0)&(i>0)
    NF = NF + 1;
    j = j + 1;
    index = s_ind(i);
    I(j) = index;
    T(j) = (index-1)*dt;
    G(j) = hf(index);
    i = i - 1;
end

% ---------------------------------------------
% Step Two - Evaluation of the weighting terms
% ---------------------------------------------

binsamples = floor(ts/dt);
if S > NF
    S = NF;
end
if L > NF
    L = NF;
end

Arake = zeros(1,NF*binsamples);
Srake = zeros(1,NF*binsamples);
Prake = zeros(1,NF*binsamples);

% SRAKE and all RAKE

for nf = 1 : NF

    x = I(nf);
    y = G(nf);
    Arake(x) = y;
    if nf <= S
        Srake(x) = y;
    end

end % for nf = 1 : NF

% PRAKE

[tv,ti] = sort(T);
TV = tv(1:L);
TI = ti(1:L);
tc = 0;
for nl = 1 : length(TV)
```

```
    index = TI(nl);
    x = I(index);
    y = G(index);
    Prake(x) = y;
    tc = tc + 1;
    L = L - 1;
end
```

Function 8.12 Correlator Mask for PPM-TH-UWB and RAKE Receiver

Function 8.12 is composed of one single step, which first operates energy normalization on the input vector `ref`, and then creates the mask by convolving the energy-normalized vector `mref` with the input vector `rake`. The last commands of Function 8.12 account for the PPM shift.

```
%
% FUNCTION 8.12 : "cp0803_PPMcorrmask_R"
%
% Evaluates the correlation mask ('mask') for a RAKE
% receiver in the case of binary PPM UWB signals
% 'ref' is the reference signal (with no modulation)
%       that is produced by the 2PPM+TH transmitter
% 'fc' is the sampling frequency in Hertz
% 'numpulses' is the number of transmitted pulses
% 'dPPM' is the value of the PPM shift in seconds
% 'rake' represents the estimated discrete-type impulse
%       response
%
% Programmed by Guerino Giancola
%

function [mask] = ...
   cp0803_PPMcorrmask_R(ref,fc,numpulses,dPPM,rake)

% -----------------------------------------------
% Step One - Evaluation of the correlation mask
% -----------------------------------------------

dt = 1 / fc;
LR = length(ref);

% energy normalization
Epulse = (sum((ref.^2).*dt))/numpulses;
nref = ref./sqrt(Epulse);

% RAKE convolution

mref = conv(nref,rake);
mref = mref(1:LR);

% mask construction
```

```
PPMsamples = floor (dPPM ./ dt);
sref(1:PPMsamples)=mref(LR-PPMsamples+1:LR);
sref(PPMsamples+1:LR)=mref(1:LR-PPMsamples);
mask = mref-sref;
```

Function 8.13 Correlator Mask for PAM-DS-UWB and RAKE Receiver

The code of Function 8.13 is the same as Function 8.12 except for the absence of the four last command lines, which take into account the PPM shift.

```
%
% FUNCTION 8.13 : "cp0803_PAMcorrmask_R"
%
% Evaluates the correlation mask ('mask') for a RAKE
% receiver in the case of binary PPM UWB signals
% 'ref' is the reference signal (with no modulation)
%      that is produced by the 2PPM+TH transmitter
% 'fc' is the sampling frequency in Hertz
% 'numpulses' is the number of transmitted pulses
% 'rake' represents the estimated discrete-type impulse
%      response
%
% Programmed by Guerino Giancola
%

function [mask] = ...
   cp0803_PAMcorrmask_R(ref,fc,numpulses,rake)

% -----------------------------------------------
% Step One - Evaluation of the correlation mask
% -----------------------------------------------

dt = 1 / fc;
LR = length(ref);

% energy normalization
Epulse = (sum((ref.^2).*dt))/numpulses;
nref = ref./sqrt(Epulse);

% RAKE convolution

mref = conv(nref,rake);
mref = mref(1:LR);
```

Function 8.14 Circular Vector Shift

Function 8.14 is composed of one single step, which first converts the value of t in the corresponding number of samples ss, and then generates the shifted vector by operating a rotation of ss samples to the elements of vector in.

```
%
% Function 8.14: "cp0804_signalshift"
%
% Applies a circular shift of 't' seconds
% to the input vector 'in'
%
% Programmed by Guerino Giancola
%

function [out] = cp0804_signalshift(in,fc,t)

% ------------------------------------
% Step One - Shifting the input signal
% ------------------------------------

dt = 1 / fc;            % sampling period

% shift samples
ss = mod(floor(t/dt),length(in));
out = in([end-ss+1:end 1:end-ss]);
```

Function 8.15 Cross-Correlation of Two Signals

Function 8.15 is composed of one single step, which iteratively computes the correlation of input waveforms. On each iteration, the position of the template waveform is shifted by one sample using Function 8.14. Then, the corresponding value of the correlator output is evaluated by first multiplying element by element the shifted version of vector `template` with vector `signal`, and then computing the sum of the elements of the resulting vector.

```
%
% Function 8.15: "cp0804_corrsyn"
%
% Implements the correlation of an input vector (signal)
% with a reference vector (template)
% 'fc' represents the sampling frequency
%
% The function returns the output of the correlator (C)
%
% Programmed by Guerino Giancola
%

function [C] = cp0804_corrsyn(signal,template,fc)

% ----------------------
% Step One - Correlation
% ----------------------

dt = 1 / fc;

for s = 1 : length(signal)

    stmp = cp0804_signalshift(template,fc,s*dt);
    % shifted templated waveform

    C(s) = sum(signal.*stmp);
    % value of the correlation

end % for 1 : length(signal)
```

CHAPTER 9

Multi-User UWB Wireless Communications

The design of a multi-user wireless communication system is based on the adoption of a multiple access strategy by means of which users share the Hertzian medium under specified controlled levels of mutual interference. This chapter analyzes multi-user UWB communications. In particular, it focuses on the analysis of the multi-user interference contribution, and extends the results of Chapter 6, "Performance Analysis of the UWB Radio Link," in which only thermal noise was taken into account. Furthermore, it forms the basis for understanding the algorithms that rule the access to the medium, and which are the object of Chapter 11, "UWB Networks: Principled Design of MAC".

9.1 Multiple Access and Multi-User Interference

A communication system provides Multiple Access (MA) when different users are allowed to share the same physical medium for transmitting and receiving different data flows. The separation between users is possible when the transmission resource is shared in a coordinated manner and different users use different so-called "channels". A channel is usually viewed as corresponding to:

1. A time slot in Time Division Multiple Access (TDMA), that is, the resource being partitioned is time.
2. A frequency band in Frequency Division Multiple Access (FDMA), that is, the resource being partitioned is frequency.
3. A code in Code Division Multiple Access (CDMA), that is, the resource being partitioned is a family of codes.

(or a combination of the above). Multiple Access methods have been extensively described for systems based on continuous transmission techniques (see, for example, [Ravi, 1994] for a synthetic and comparative overview). A multi-user communication system based on UWB may adopt any of the above methods. The current trend in the IEEE 802.15.3a is adopting a MA strategy, typically TDMA-based (IEEE, 2003), for the UWB physical layer.

Impulse Radio offers the possibility of partitioning time in a peculiar way because of the intrinsic short and limited duration of the pulses. As derived in Chapter 2, "The UWB Radio Signal," the spectrum of the IR signal is usually shaped by encoding data symbols using TH sequences, which are typically described as pseudorandom PN codes. These same sequences can also serve as users' signatures and ensure access to the medium to multiple users. This manner of partitioning the resource is called Time-Hopping Multiple Access (THMA). Note that in THMA, the resource is formed by combining time and codes. The second IR option, DS-UWB, is a pulsed version of DS-CDMA. As THMA, DS-UWB offers an intrinsic feature for multiple access, since again each user can be assigned a different code (Di Benedetto and Vojcic, 2003).

With regard to the last UWB option, described in Chapter 2, in the non-IR multi-band approach, and in particular OFDM, MA is typically achieved by using TDMA. An alternative to TDMA is Multi-Carrier CDMA (MC-CDMA), by which different users transmit over sub-carriers by using different codes (Hanzo et al., 2003), as mentioned in Chapter 2 (see Eq. (2–20)).

The analysis in this chapter will focus on the IR options (both THMA and CDMA), in an attempt to characterize the physical foundation of a novel system and set the basis for a definition of the MAC module, eventually tailored on IR-specific features.

The selection of a MA strategy is part of a complex process, which consists of defining a MAC module, its ruling algorithms, and related protocols. Issues related to a complete MAC design will be analyzed in detail in Chapter 11.

At first glance, THMA falls in the CDMA category since different users adopt different codes, although in THMA, users are also separated in the time dimension. Each code has the effect of modifying the transmitted signal in such a way that a reference receiver is capable of isolating the useful signal from other users' signals, which are seen by the reference receiver as interfering signals. The possibility of removing these unnecessary contributions mainly depends on the characteristics of the codes used for separating data flows, and on the degree of system-level synchronization. In the ideal condition of perfect system synchronization, ideal channel, and orthogonal codes associated with different data flows, the receiver is not affected by the presence of multiple transmissions. In a realistic scenario, however, where devices do not achieve ideal synchronization and codes lose orthogonality due to different propagation delays on different paths, the receiver might not be capable of completely removing the presence of the undesired signals, and as a consequence, system performance is affected by MUI.

Several methods have been proposed in the recent past for evaluating the effect of MUI on symbol error rate Pr_e for conventional continuous transmission CDMA and OFDM systems implementing single-user reception in an AWGN channel (Lehnert and Pursley, 1987; Crespo, Honig, and Salehi, 1995; Gui and Ng, 1999). Most of these methods are based on the Standard Gaussian Approximation (SGA) hypothesis, where the cumulative effect of all interfering contributions at the receiver is treated as an additive Gaussian noise with uniform PSD over the range of frequencies of interest. Under this hypothesis, the receiver tolerance to MUI can be expressed as a function of the processing gain of the system, that is, the ratio between total transmission bandwidth and user bit rate. The higher the processing gain, the higher the system's robustness against MUI, or the higher the number of devices that can simultaneously access the physical medium for a specific Pr_e. A similar treatise to the continuous transmission case based on the SGA hypothesis was first proposed for the case of IR-UWB systems by (Scholtz, 1993), and further described by (Win and Scholtz, 2000). The analysis of these investigations focuses on asynchronous THMA-UWB, under the hypothesis of perfect synchronization between transmitter and receiver, that is, coherent reception.

In the next section, we will derive system performance and MUI characterization for the IR-UWB options under the SGA hypothesis. The SGA hypothesis is valid only asymptotically and is questionable for both conventional CDMA and IR systems. Regarding THMA-UWB systems, (Durisi and Romano, 2002) showed that for systems implementing perfect power control, the SGA leads to more optimistic predictions of the Pr_e floor than results obtained by simulation. These results were confirmed by (Giancola et al., 2003) in the absence of power control. The limit of SGA for IR systems, with and without perfect power control, will be discussed in the next section, and will be analyzed based on simulations in Checkpoint 9–1.

In Section 9.3, MUI will be described from a different perspective in an attempt to take into account the way the information is structured, in particular by the MAC layer, which typically reorganizes information into packet units. This approach will be further described in Chapter 11.

9.2 Multi-User IR-UWB System Performance Based on the SGA

This section is devoted to deriving multi-user system performance as expressed by the probability of error on the symbol Pr_e in the case of IR transmission, and in particular, for THMA-UWB and DS-UWB. Results presented in Chapter 6 are extended, with the aim of incorporating the presence of several users in the system, and therefore eventually MUI. As discussed in Chapter 6, TH may be used in combination with both PPM and PAM. We derive Pr_e for both modulation options, and finally we analyze the DS-UWB case. All the analyses derived in this section are made under the following hypotheses:

1. All sources produce binary sequences **b** formed by independent and identically distributed random variables with equally probable symbols "0" and "1".
2. All sources use the same pulse repetition frequency $1/T_s$.
3. Codes are independent and equally probable. Each code is randomly generated and corresponds to a PN sequence.

4. For each link between a reference transmitter and a reference receiver, the reference transmitter/receiver pair uses a specific code known at the receiver.
5. The pulse is assumed to have limited duration T_M and a symmetrical shape around its central value.
6. Radio waves propagate over a multi-path-free channel. The multi-path-free impulse response of the channel for the link between user n and a reference receiver is a function of $\alpha^{(n)}$ and $\tau^{(n)}$, which are the path gain and time delay for the n-th user. The time delays are assumed to be independent and identically uniformly distributed random variables over $[0, T_s)$. The impulse response can be written as follows:

$$h^{(n)}(t) = \alpha^{(n)}\delta\left(t - \tau^{(n)}\right) \tag{9–1}$$

7. The channel output is corrupted by thermal noise $n(t)$, characterized by a double-sided spectral density $N_o/2$ (in W/Hz).
8. The receiver implements a single-user coherent correlation structure with soft decision followed by a ML detector, as discussed in Chapter 8, "The UWB Channel and Receiver." The received signal is thus analyzed over T_b by considering all N_s pulses composing each bit.
9. The system is asynchronous, but reference transmitter and receiver of a reference link are supposed to be perfectly synchronized under the coherent detection hypothesis.

9.2.1 Binary Pulse Position Modulation with THMA: 2PPM-THMA

The binary PPM-THMA signal transmitted by user n as expressed by Eq. (2–2) can be rewritten as follows:

$$s_{TX}^{(n)}(t) = \sum_{j=-\infty}^{\infty} \sqrt{E_{TX}^{(n)}}\, p_0\left(t - jT_s - c_j^{(n)}T_c - a_j^{(n)}\varepsilon\right) \tag{9–2}$$

where $p_0(t)$ is the energy-normalized pulse waveform and $E_{TX}^{(n)}$ is the energy transmitted over each single pulse.

As explained in Chapter 2, $c_j^{(n)}T_c$ is the time shift introduced by the TH code; $c_j^{(n)}$ is the j-th coefficient of the TH sequence used by user n; and T_c is the chip duration. Each TH code is a sequence of N_p independent and identically distributed random variables, with probability $1/N_h$ of assuming one of the integer values in the range $[0, N_h\text{-}1]$. Since multiple devices share the same medium, each user is provided with a specific TH code to avoid catastrophic collisions at the receiver.

The $a_j^{(n)}\varepsilon$ term represents the time shift introduced by modulation; ε is the PPM shift; and $a_j^{(n)}$ is the binary value (1 or 0) conveyed by pulse j of user n. The **a** sequence is the output of a $(N_s,1)$ code repetition coder that receives as input the binary source sequence **b**. In the general case, $N_s \geq 1$ pulses carry the information of one bit. The presence of N_s pulses for each bit imposes a constraint on the average pulse repetition period T_s, that is:

$$T_s \leq T_b / N_s \quad \Rightarrow \quad T_s = \gamma_R \left(T_b / N_s\right) \ with \ \gamma_R \leq 1 \tag{9–3}$$

To simplify the analysis, one may assert that the maximum shift introduced by the TH code does not exceed T_s. For a given N_h, this introduces a constraint on the chip duration T_c expressed by:

$$T_c \leq T_s / N_h \quad \Rightarrow \quad T_c = \gamma_C \left(T_s / N_h\right) \ with \ \gamma_C \leq 1 \tag{9–4}$$

Finally, for each pulse to be contained within T_c after modulation, one must have:

$$\varepsilon \leq T_c - T_M \tag{9–5}$$

where T_M is the time duration of $p_0(t)$.

Given the channel model of Eq. (9–1) and the presence of thermal noise at the channel output, the received signal at a reference receiver is given by the sum of all signals originating from the N_u transmitters, and can be written as follows:

$$r(t) = \sum_{n=1}^{N_u} \sum_{j=-\infty}^{\infty} \sqrt{E_{RX}^{(n)}}\, p_0\left(t - jT_s - c_j^{(n)}T_c - a_j^{(n)}\varepsilon - \tau^{(n)}\right) + n(t) \tag{9–6}$$

where $E_{RX}^{(n)} = E_{TX}^{(n)}\left(\alpha^{(n)}\right)^2$

The system schematic model for the asynchronous system under investigation is shown in Figure 9–1.

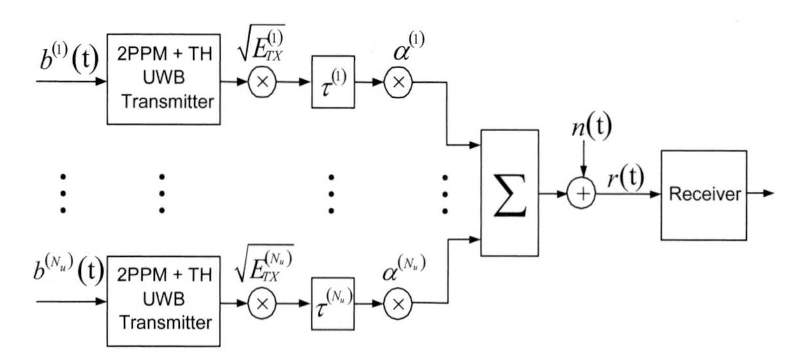

Figure 9–1 System model.

Given the symmetry of the model, we can focus the analysis on one active link. Suppose the reference receiver is listening to the first transmitter (TX1). Under the hypothesis of perfect synchronization between the TX1 and reference receiver, the time delay $\tau^{(1)}$ is known by the receiver, and one can assume $\tau^{(1)} = 0$ given that only relative delays and phases are relevant. The received signal can thus be rewritten as follows:

$$r(t) = r_u(t) + r_{mui}(t) + n(t) \tag{9–7}$$

where $r_u(t)$ and $r_{mui}(t)$ are the useful signal and MUI contributions at the receiver input.

With soft decision at the receiver, the analysis focuses on a bit time interval of duration T_b. Given again the symmetry of the system, the analysis can be focused on the interval $[0, T_b]$. The $r_u(t)$ and $r_{mui}(t)$ contributions can be written as follows:

$$r_u(t) = \sum_{j=0}^{N_s-1} \sqrt{E_{RX}^{(1)}}\, p_0\left(t - jT_s - c_j^{(1)}T_c - a_j^{(1)}\varepsilon\right) \quad \text{for } t \in [0, T_b] \tag{9–8}$$

$$r_{mui}(t) = \sum_{n=2}^{N_u} \sum_{j=-\infty}^{\infty} \sqrt{E_{RX}^{(n)}}\, p_0\left(t - jT_s - c_j^{(n)}T_c - a_j^{(n)}\varepsilon - \tau^{(n)}\right) \quad \text{for } t \in [0, T_b] \tag{9–9}$$

The average symbol error rate coincides with the average bit error rate $\Pr_b$ since modulation is binary and corresponds to the probability of misdetecting a reference bit b transmitted by TX1. The reference bit b is defined as the bit received in the time interval $[0, T_b]$. The soft decision correlation receiver output, as described in Section 8.1, can be expressed as follows:

$$Z = \int_0^{T_b} r(t)\, m(t)\, dt \tag{9–10}$$

where $m(t)$ is the correlation receiver mask and is defined as follows:

$$m(t) = \sum_{j=0}^{N_s-1} v\left(t - jT_s - c_j^{(1)}T_s\right) \quad \text{with } v(t) = p_0(t) - p_0(t - \varepsilon) \tag{9–11}$$

The decision rule based on the ML criterion for both orthogonal and optimum binary PPM implies the comparison of Z against a threshold that is zero in the case of Eqs. (9–10) and (9–11). We know that this detector is optimum if noise is additive and Gaussian, and coincides with the Maximum A Posteriori probability (MAP) if all received signals are equally probable, as in the present case. The ML decision rule can thus be expressed as follows (see also Section 8.1):

$$\textbf{ML receiver decision rule:} \quad if \begin{cases} Z > 0 \Rightarrow \hat{b} = 0 \\ Z < 0 \Rightarrow \hat{b} = 1 \end{cases} \tag{9–12}$$

where $\hat{b}$ indicates the estimated bit. By combining Eqs. (9–7) and (9–10), one can write:

$$Z = Z_u + Z_{mui} + Z_n \tag{9–13}$$

where Z_u, Z_{mui}, and Z_n indicate a useful signal, MUI noise, and thermal noise at the receiver output. Z_{mui} can be removed at the receiver if all codes were orthogonal at the receiver under perfect synchronization of all users in the system. The first term Z_u is deterministic when reference bit b is fixed.

According to the decision rule of Eq. (9–12), and given equally probable source symbols, the probability of bit error $\Pr_b$ (see Section 8.1) is given by:

$$\begin{aligned} \Pr_b &= \frac{1}{2}\Pr\left(\hat{b}=1 \mid b=0\right) + \frac{1}{2}\Pr\left(\hat{b}=0 \mid b=1\right) = \Pr\left(\hat{b}=1 \mid b=0\right) = \\ &= \Pr\left(Z<0 \mid b=0\right) \end{aligned} \tag{9–14}$$

The SGA hypothesis assumes that Z_{mui}, as well as Z_n, is a zero-mean Gaussian random process characterized by variance σ_{mui}^2. σ_n^2 is the variance of thermal noise. Under the SGA hypothesis, the relation between $\Pr_b$ and SNR_{spec} can be extended from Eq. (6–30), which accounted for thermal noise only to also incorporate the current case of MUI, and one can write:

$$\Pr_b = \frac{1}{2} erfc\left(\sqrt{\frac{SNR_{spec}}{2}}\right) \tag{9–15}$$

where SNR_{spec} in Eq. (9–15) accounts for both thermal and interference noise contributions. The useful signal contribution is the bit energy E_b, therefore, one can write:

$$SNR_{spec} = \frac{E_b}{\sigma_n^2 + \sigma_{mui}^2} \tag{9–16}$$

By isolating thermal and MUI contributions, Eq. (9–16) can be rewritten as follows:

$$SNR_{ref} = \left(\left(SNR_n\right)^{-1} + \left(SIR\right)^{-1}\right)^{-1} = \left(\left(\frac{E_b}{\sigma_n^2}\right)^{-1} + \left(\frac{E_b}{\sigma_{mui}^2}\right)^{-1}\right)^{-1} \tag{9–17}$$

where SNR_n and SIR are the signal to thermal noise and signal to MUI ratios, respectively.

The useful signal energy E_b can be derived by computing the energy of the useful component at the output of the receiver for all N_s pulses forming one bit, that is:

$$\begin{aligned} E_b &= (Z_u)^2 \\ &= \left(\sqrt{E_{RX}^{(1)}} \sum_{j=0}^{N_s-1} \int_{jT_s+c_j^{(1)}T_c}^{jT_s+c_j^{(1)}T_c+T_c} p_0\left(t-jT_s-c_j^{(1)}T_c\right) v\left(t-jT_s-c_j^{(1)}T_c\right) dt \right)^2 \\ &= E_{RX}^{(1)} \left(N_s \int_0^{T_c} p_0(t)\left(p_0(t)-p_0(t-\varepsilon)\right) dt \right)^2 \\ &= E_{RX}^{(1)} N_s^2 \left(\int_0^{T_c} p_0(t)\, p_0(t)\, dt - \int_0^{T_c} p_0(t) p_0(t-\varepsilon)\, dt \right)^2 \\ &= E_{RX}^{(1)} N_s^2 \left(1-R_0(\varepsilon)\right)^2 \end{aligned} \tag{9–18}$$

where $R_0(t)$ is the autocorrelation function of the pulse waveform $p_0(t)$. [Remember that given Eq. (9–5), pulse duration T_M is smaller than T_c.]

The variance of thermal noise at the binary PPM receiver output σ_n^2 was derived in Section 8.1.2 and is expressed as follows (see Eq. (8–26)):

$$\sigma_n^2 = N_s N_0 \left(1-R_0(\varepsilon)\right) \tag{9–19}$$

which leads to the following expression for SNR_n:

$$SNR_n = \frac{N_s E_{RX}^{(1)}}{N_0}\left(1-R_0(\varepsilon)\right) = \frac{E_b}{N_0}\left(1-R_0(\varepsilon)\right) \tag{9–20}$$

As already discussed in Section 8.1.2, Eq. (9–20) shows that SNR_n is maximum when $R_0(\varepsilon)$ is minimum, and can therefore be maximized by selecting an optimal ε value. This procedure leads to optimal binary PPM. Note that since $R_0(\varepsilon) \equiv 0$ for $\varepsilon \geq T_M$, then $1\text{-}R_0(\varepsilon) = 1$ for all ε values greater than pulse duration T_M, that is, for orthogonal pulses. The optimum ε value can, however, be smaller than T_M if $R_0(\varepsilon)$ takes on negative values, that is, $1\text{-}R_0(\varepsilon) > 1$. In the orthogonal pulse case, Eq. (9–20) reduces to Eq. (6–30).

Let us now characterize MUI. Since the system is asynchronous, we need to consider as "interfering events" all cases where an "alien" pulse, or a pulse originating by any of the transmitters but TX1, is detected by the receiver. Figure 9–2 shows the effect of the presence of an alien pulse originating from TXn in the receiver, which results in an interference noise amplitude related to the area of the shaded area on the figure. The interference noise provoked by the presence of one alien pulse at the output of the receiver can be written as follows:

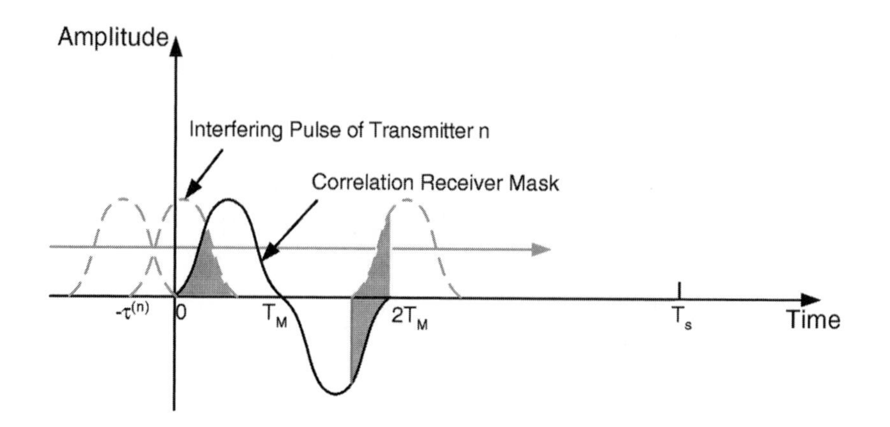

Figure 9–2 Graphical representation of the presence of an interfering pulse at the reference receiver input.

$$mui_p^{(n)}\left(\tau^{(n)}\right)=\sqrt{E_{RX}^{(n)}}\int_0^{2T_M} p_0\left(t-\tau^{(n)}\right)v(t)\,dt \tag{9–21}$$

We can now find the variance of the interference noise of Eq. (9–21). Since according to assumptions delay, $\tau^{(n)}$ is uniformly distributed over $[0, T_s)$, one has:

$$\begin{aligned}\sigma^2_{mui_p^{(n)}} &= \frac{1}{T_s}\int_0^{T_s}\left(\sqrt{E_{RX}^{(n)}}\int_0^{2T_M} p_0\left(t-\tau^{(n)}\right)v(t)\,dt\right)^2 d\tau^{(n)} \\ &= \frac{E_{RX}^{(n)}}{T_s}\int_0^{T_s}\left(\int_0^{2T_M} p_0\left(t-\tau^{(n)}\right)v(t)\,dt\right)^2 d\tau^{(n)}\end{aligned} \tag{9–22}$$

The assumption that all delays and all codes are independent forces pulse collision in one frame to be an independent event of collisions occurring in any other frame regardless of whether the interfering pulses originated from TXn or from any of the other N_u-2 interfering transmitters. Interfering signals therefore sum up in power, that is, in energy and not in equivalent voltage. Also note that an interfering pulse might have originated in a different interval from the observation interval, and nevertheless partly falls into the observation interval due the delayed time of arrival. In particular, this occurs when the interfering pulse originated in the interval previous to the observation interval and τ is such that $T_s - T_M \leq \tau \leq T_s$. The total interfering MUI energy *on the bit*, that is, on N_s pulses at the output of the receiver, can thus be written as follows:

$$\sigma_{mui}^2 = \sum_{n=2}^{N_u} \left(\frac{N_s}{T_s} E_{RX}^{(n)} \int_0^{T_s} \left(\int_0^{2T_M} p_0\left(t - \tau^{(n)}\right) v(t)\, dt \right)^2 d\tau^{(n)} \right) \qquad (9\text{–}23)$$

Since all delays are identically distributed, Eq. (9–23) becomes:

$$\begin{aligned} \sigma_{mui}^2 &= \frac{N_s}{T_s} \left(\int_0^{T_s} \left(\int_0^{2T_M} p_0(t-\tau) v(t)\, dt \right)^2 d\tau \right) \sum_{n=2}^{N_u} E_{RX}^{(n)} \\ &= \frac{N_s}{T_s} \sigma_M^2 \sum_{n=2}^{N_u} E_{RX}^{(n)} \end{aligned} \qquad (9\text{–}24)$$

for τ uniformly distributed over $[0, T_s)$.

The term σ_M^2 of Eq. (9–24) can be written as follows:

$$\begin{aligned} \sigma_M^2 &= \int_0^{T_s} \left(\int_0^{2T_M} p_0(t-\tau) v(t)\, dt \right)^2 d\tau \\ &= \int_0^{T_s} \left(\int_0^{2T_M} p_0(t-\tau) \left(p_0(t) - p_0(t-\varepsilon)\right) dt \right)^2 d\tau \\ &= \int_0^{T_s} \left(\int_0^{T_M} p_0(t-\tau) p_0(t)\, dt - \int_{\varepsilon}^{T_M+\varepsilon} p_0(t-\tau) p_0(t-\varepsilon)\, dt \right)^2 d\tau \\ &= \int_{-T_M}^{2T_M} \left(\int_0^{T_M} p_0(t-\tau) p_0(t)\, dt - \int_{\varepsilon}^{T_M+\varepsilon} p_0(t-\tau) p_0(t-\varepsilon)\, dt \right)^2 d\tau \\ &= \int_{-T_M}^{2T_M} \left(R_0(\tau) - R_0(\tau+\varepsilon)\right)^2 d\tau \end{aligned} \qquad (9\text{–}25)$$

In the case of orthogonal pulses, or $\varepsilon \geq T_M$, Eq. (9–25) can be written as follows:

$$\begin{aligned} \sigma_M^2 &= \int_{-T_M}^{2T_M} \left(\int_0^{T_M} p_0(t-\tau) p_0(t)\, dt - \int_{T_M}^{2T_M} p_0(t-\tau) p_0(t-T_M)\, dt \right)^2 d\tau \\ &= \int_{-T_M}^{2T_M} \left(R_0(\tau) - R_0(\tau+T_M)\right)^2 d\tau \end{aligned} \qquad (9\text{–}26)$$

Note that $R_0(\tau)$ and $R_0(\tau+T_M)$ in Eq. (9–26) overlap for $0 < \tau < T_M$. Equation (9–26) can be conveniently rewritten by separating $R_0(\tau)$ and $R_0(\tau+T_M)$ contributions in the following way:

$$\sigma_M^2 = \int_{-T_M}^{T_M} R_0^2(\tau)\, d\tau + \int_0^{2T_M} R_0^2(\tau+T_M)\, d\tau = 2 \int_{-T_M}^{T_M} R_0^2(\tau)\, d\tau \qquad (9\text{–}27)$$

Note that in the above calculations, we have modified the external integration interval from [0, T_s] to [-T_M, 2T_M] to take into account interference provoked by pulses that fall in the interval of observation when shifted from their original position in the antecedent interval. We incorporated this effect by integrating the autocorrelation function over [-T_M, 2T_M]. Alternatively, we could have written the integral $\int_0^{T_s}\left(R_0^{\,2}(\tau)-R_0^{\,2}(T_s-\tau)\right)d\tau$ in Eq. (9–25) in place of the integral $\int_0^{T_s} R_0^{\,2}(\tau)d\tau$, and a similar expression for the integral of the squared shifted autocorrelation function.

Referring back to Eq. (9–17), we can express the Signal to Interference noise Ratio (*SIR*) as follows:

$$SIR=\frac{E_{RX}^{(1)}N_s^2\left(1-R_0(\varepsilon)\right)^2}{\frac{1}{T_s}N_s\sigma_M^2\sum_{n=2}^{N_u}E_{RX}^{(n)}}=\frac{\left(1-R_0(\varepsilon)\right)^2}{\sigma_M^2}\frac{N_sT_s}{\sum_{n=2}^{N_u}\frac{E_{RX}^{(n)}}{E_{RX}^{(1)}}}=$$
$$=\frac{\left(1-R_0(\varepsilon)\right)^2\gamma_R}{\sigma_M^2}\frac{1}{R_b\sum_{n=2}^{N_u}\frac{E_{RX}^{(n)}}{E_{RX}^{(1)}}} \quad (9\text{–}28)$$

which, in the case of orthogonal pulses, reduces to:

$$SIR=\frac{\gamma_R}{\sigma_M^2}\frac{1}{R_b\sum_{n=2}^{N_u}\frac{E_{RX}^{(n)}}{E_{RX}^{(1)}}}=\frac{\gamma_R}{2\int_{-T_M}^{T_M}R_0^{\,2}(\tau)d\tau}\frac{1}{R_b\sum_{n=2}^{N_u}\frac{E_{RX}^{(n)}}{E_{RX}^{(1)}}} \quad (9\text{–}29)$$

Equations (9–28) and (9–29) indicate that for a given interference scenario — number and position of users — the MUI term can be controlled by controlling the bit rate used by all transmitters. We will come back to this in Chapter 11 where we will describe an admission control rule that makes use of this concept. Under the simplifying hypothesis of perfect power control, for example, that is, all terms $E_{RX}^{(i)}$ are equal for i = 1, …, N_u, one can derive from Eq. (9–28), and similarly from Eq. (9–29), a rule for the evaluation of the maximum allowed bit rate R_b to be used by any of the N_u users for a given *SIR* specification:

$$R_b\left(SIR,N_u\right)=\frac{\left(1-R_0(\varepsilon)\right)^2\gamma_R}{\sigma_M^2}\left(SIR\left(N_u-1\right)\right)^{-1} \quad (9\text{–}30)$$

By combining Eqs. (9–30) and (9–28) the $\Pr_b$ for a 2PPM-THMA system based on the SGA is:

$$\Pr_b = \frac{1}{2} erfc\left(\sqrt{\frac{\left(\left(\frac{E_b^{(1)}}{N_0}\left(1-R_0(\varepsilon)\right)\right)^{-1} + \left(\frac{\left(1-R_0(\varepsilon)\right)^2 \gamma_R}{\sigma_M^2 R_b \sum_{n=2}^{N_u} \frac{E_{RX}^{(n)}}{E_{RX}^{(1)}}}\right)^{-1}\right)^{-1}}{2}}\right) \quad (9\text{–}31)$$

In the PPM orthogonal case, Eq. (9–31) becomes:

$$\Pr_b = \frac{1}{2} erfc\left(\sqrt{\frac{\left(\left(\frac{E_b^{(1)}}{N_0}\right)^{-1} + \left(\frac{\gamma_R}{2R_b \sum_{n=2}^{N_u} \frac{E_{RX}^{(n)}}{E_{RX}^{(1)}} \int_{-T_M}^{T_M} R_0^2(\tau)\, d\tau}\right)^{-1}\right)^{-1}}{2}}\right) \quad (9\text{–}32)$$

Under the hypothesis of perfect power control, Eqs. (9–31) and (9–32) can be rewritten as follows:

$$\Pr_b = \frac{1}{2} erfc\left(\sqrt{\frac{\left(\left(\frac{E_b^{(1)}}{N_0}\left(1-R_0(\varepsilon)\right)\right)^{-1} + \left(\frac{\left(1-R_0(\varepsilon)\right)^2 \gamma_R}{\sigma_M^2 R_b (N_u - 1)}\right)^{-1}\right)^{-1}}{2}}\right) \quad (9\text{–}33)$$

$$\Pr_b = \frac{1}{2} erfc\left(\sqrt{\frac{\left(\left(\frac{E_b^{(1)}}{N_0}\right)^{-1} + \left(\frac{\gamma_R}{2R_b (N_u - 1) \int_{-T_M}^{T_M} R_0^2(\tau)\, d\tau}\right)^{-1}\right)^{-1}}{2}}\right) \quad (9\text{–}34)$$

9.2.2 Binary Antipodal PAM Modulation with THMA: 2PAM-THMA

Binary antipodal PAM-THMA can be analyzed following a similar procedure described in the previous paragraph for binary PPM-THMA.

The signal at the output of the correlation receiver Z is expressed by Eq. (9–10) with a correlation receiver mask *m(t)* defined as follows:

$$m(t)=\sum_{j=0}^{N_s-1} p_0\left(t-jT_s-c_j^{(1)}T_s\right) \tag{9–35}$$

The decision rule at the receiver is based again on the ML criterion and Eq. (9–14), which expresses the probability of error on the bit Pr_b, remains valid. Under the SGA hypothesis, the relation between Pr_b and SNR_{spec} can be derived in the present case from Eq. (6–29), with Eqs. (9–16) and (9–17) remaining valid.

With a correlation mask as of Eq. (9–35), the energy of the useful contribution at the output of the receiver for all N_s pulses forming one bit can be written as follows:

$$\begin{aligned}
E_b &= (Z_u)^2 \\
&= \left(\sqrt{E_{RX}^{(1)}}\sum_{j=0}^{N_s-1}\int_{jT_s+c_j^{(1)}T_c}^{jT_s+c_j^{(1)}T_c+T_c} p_0\left(t-jT_s-c_j^{(1)}T_c\right)p_0\left(t-jT_s-c_j^{(1)}T_c\right)dt\right)^2 \\
&= E_{RX}^{(1)}\left(N_s\int_0^{T_c} p_0(t)\,p_0(t)\,dt\right)^2 \\
&= E_{RX}^{(1)}N_s^{\,2}\left(\int_0^{T_c} p_0(t)\,p_0(t)\,dt\right)^2 \\
&= E_{RX}^{(1)}N_s^{\,2}
\end{aligned} \tag{9–36}$$

The variance of thermal noise at the receiver output σ_n^2 is, in the present case:

$$\sigma_n^2 = N_s\frac{N_0}{2} \tag{9–37}$$

leading to the following expression for SNR_n:

$$SNR_n = \frac{N_sE_{RX}^{(1)}}{\frac{N_0}{2}} = \frac{2E_b}{N_0} \tag{9–38}$$

that is, a specified SNR_n can be obtained with half the energy on the bit compared to orthogonal binary PPM.

Regarding MUI, the same line of reasoning followed for binary PPM can be applied to the binary antipodal PAM case. Supposing that TX1 is the reference transmitter, the interference noise provoked by the presence of one alien pulse generated by TXn at the output of a reference receiver can be written as follows:

$$mui_p^{(n)}\left(\tau^{(n)}\right)=\sqrt{E_{RX}^{(n)}}\int_0^{T_M} p_0\left(t-\tau^{(n)}\right)p_0(t)\,dt \tag{9–39}$$

The variance of the interference noise introduced by the presence of this alien pulse, given that delays are uniformly distributed in $[0, T_s)$, is:

$$\sigma^2_{mui_p^{(n)}} = \frac{1}{T_s}\int_0^{T_s}\left(\sqrt{E_{RX}^{(n)}}\int_0^{T_M} p_0\left(t-\tau^{(n)}\right)p_0(t)\,dt\right)^2 d\tau^{(n)}$$
$$= \frac{E_{RX}^{(n)}}{T_s}\int_0^{T_s}\left(\int_0^{T_M} p_0\left(t-\tau^{(n)}\right)p_0(t)\,dt\right)^2 d\tau^{(n)} \quad (9\text{–}40)$$

Since all delays and codes are supposed to be independent, pulse collision in one frame is assumed to be independent from collisions occurring in all other frames. Recall that this implies that interfering signals sum up in power, that is, in energy rather than in equivalent voltage, and that the total interfering MUI noise energy on the bit, or on N_s pulses at the output of the receiver, can be written as follows:

$$\sigma^2_{mui} = \sum_{n=2}^{N_u}\left(\frac{N_s}{T_s}E_{RX}^{(n)}\int_0^{T_s}\left(\int_0^{T_M} p_0\left(t-\tau^{(n)}\right)p_0(t)\,dt\right)^2 d\tau^{(n)}\right) \quad (9\text{–}41)$$

Since all delays are identical, Eq. (9–41) becomes:

$$\sigma^2_{mui} = \frac{N_s}{T_s}\left(\int_0^{T_s}\left(\int_0^{T_M} p_0(t-\tau)p_0(t)\,dt\right)^2 d\tau\right)\sum_{n=2}^{N_u} E_{RX}^{(n)} = \frac{N_s}{T_s}\sigma_M^2\sum_{n=2}^{N_u} E_{RX}^{(n)} \quad (9\text{–}42)$$

for τ uniformly distributed over $[0, T_s)$.
The σ_n^2 term in Eq. (9–42), in the present 2PAM-THMA case is expressed by:

$$\sigma_M^2 = \int_0^{T_s}\left(\int_0^{T_M} p_0(t-\tau)p_0(t)\,dt\right)^2 d\tau = \int_{-T_M}^{T_M} R^2{}_0(\tau)\,d\tau \quad (9\text{–}43)$$

The *SIR* in the binary antipodal PAM case is therefore expressed as follows:

$$SIR = \frac{N_sT_s}{\sigma_M^2}\frac{E_{RX}^{(1)}}{\sum_{n=2}^{N_u} E_{RX}^{(n)}} = \frac{\gamma_R}{\int_{-T_M}^{T_M} R_0^2(\tau)\,d\tau}\frac{1}{R_b\sum_{n=2}^{N_u}\frac{E_{RX}^{(n)}}{E_{RX}^{(1)}}} \quad (9\text{–}44)$$

which is, as expected, twice the *SIR* for orthogonal binary PPM (+3 dB) for same received energy. In the case of perfect power control, a similar rule of Eq. (9–30) can be derived which is:

$$R_b\left(SIR, N_u\right) = \frac{\gamma_R}{\sigma_M^2}\left(SIR\left(N_u-1\right)\right)^{-1} \quad (9\text{–}45)$$

The $\Pr_b$ for a binary antipodal PAM-THMA system based on the SGA is thus:

$$\Pr_b = \frac{1}{2} erfc\left(\sqrt{\frac{\left(\left(\frac{2E_b^{(1)}}{N_0} \right)^{-1} + \left(\frac{\gamma_R}{R_b \sum_{n=2}^{N_u} \frac{E_{RX}^{(n)}}{E_{RX}^{(1)}} \int_{-T_M}^{T_M} R_0^2(\tau)\, d\tau} \right)^{-1} \right)^{-1}}{2}} \right) \tag{9–46}$$

Under the hypothesis of perfect power control, Eq. (9–46) can be rewritten as follows:

$$\Pr_b = \frac{1}{2} erfc\left(\sqrt{\frac{\left(\left(\frac{2E_b^{(1)}}{N_0} \right)^{-1} + \left(\frac{\gamma_R}{R_b (N_u - 1) \int_{-T_M}^{T_M} R_0^2(\tau)\, d\tau} \right)^{-1} \right)^{-1}}{2}} \right) \tag{9–47}$$

9.2.3 DS-UWB

Direct Sequence UWB (DS-UWB) codes the information bit in a DS-CDMA fashion, by using a binary code. Transmission is based on the emission of N_s pulses for each bit as in the TH scheme. Pulses are amplitude-modulated using binary antipodal PAM, and are as in the TH scheme emitted every T_s seconds, with a rate $R_s = 1/T_s$. Therefore, MA can be based on the association of different transmissions to different codes in agreement with a pure DS-CDMA scheme.

Deriving system performance in the case of DS-UWB follows in a straightforward way the case of binary antipodal PAM-TH, which was examined in Section 9.2.2. The modulation technique, or binary antipodal PAM, as well as the structure of the receiver, is common to the two cases. The expressions of the variance of thermal noise (Eq. (9–37)) and of SNR_n (Eq. (9–38)) remain valid for DS-UWB.

Regarding MUI, the calculation can proceed along very similar lines to those followed in Section 9.2.2. Although the two systems differ in terms of coding, which is binary here, and in terms of multiple access strategy, which here is only based on coding, one should note that MUI noise was characterized in the PAM-TH case in terms of an average behavior of the system, reflected in the average operation over codes and intervals of observation. This is to say that Eqs. (9–38) to (9–46) remain valid for the DS-UWB case.

The derivation of MUI properties in DS-CDMA can also proceed by extending DS-CDMA results, which have been made available in the literature for years, by specifying a pulse shape that is of the UWB type rather than the commonly used DS-CDMA rectangular pulse of duration T_s. Modeling interference for DS-CDMA systems using continuous transmission was first analyzed in great detail by (Pursley, 1977). The specification of the DS-CDMA MUI model to the DS-UWB case can be found in (Vojcic and Pickholtz, 2003).

9.2.4 Limit of Application of the Standard Gaussian Approximation

The SGA hypothesis that formed the basis for the derivation of the probability of error reported in the previous paragraphs is valid only asymptotically and must be questioned for both conventional CDMA and IR systems. Regarding CDMA, results show that the SGA leads to inaccurate predictions when considering scarcely populated systems or with dominating interferers (Sunay and McLane, 1996). The latter condition, in particular, refers to cases in which power control is not implemented, as generally occurs in ad-hoc networks. On this basis, Pr_b derivations for CDMA systems that are not based on the SGA hypothesis were proposed by (Morrow and Lehnert, 1989) and (Sunay and McLane, 1995).

The discussion around the validity of the SGA has recently intensified regarding IR schemes. First, results suggest that the validity of the SGA increases with the number of interfering users (Win and Scholtz, 2000), while the SGA does not drive to adequate Pr_e derivations for low values of user bit rate (Durisi, and Romano, 2002) and pulse repetition frequency (Foerster, 2002). As mentioned above, the analyses presented in (Durisi and Romano, 2002) and (Giancola et al., 2003) show that for TH-UWB systems, with and without perfect power control, the SGA leads to more optimistic predictions of the Pr_e floor than results obtained by simulation. The above studies are based on experimental data obtained by simulation. Recent work published by (Sabattini et al., 2003) and (Hu and Beaulieu, 2003a, 2003b) propose analytical non-Gaussian approaches for the estimation of binary PPM-TH system performance. In particular, (Hu and Beaulieu, 2003a, 2003b) propose the application of the Characteristic Function (CF) technique following a consolidated procedure adopted for traditional DS-CDMA systems (Lehnert and Pursley, 1987). An approximation of the CF is proposed by (Sabattini et al., 2003). A different approach is presented by (Durisi and Benedetto, 2003); here, the probability of error for PPM-TH is estimated by applying the Gaussian quadrature rule (Golub and Welsh, 1969). This method consists of first defining the probability of error conditioned on Z_{mui}, and then in finding the unconditioned probability of error by averaging with respect of Z_{mui}. This averaging, which corresponds to computing an integral, is approximated by means of the Gaussian quadrature rule, which requires prior computation of the moments of the Z_{mui}. These moments are computed following an approach first presented by (Benedetto et al., 1973a, 1973b).

As a general comment, the SGA hypothesis seems, as expected, to cease having validity when the number of pulses in the air is not sufficiently high to fill up the time dimension. A reduced number of pulses traveling over the air interface can be associated with scenarios where a reduced number of transmitters is present over a given geographical area, that is, the density of users is low, transmitters are characterized by a low data rate, the number of pulses per bit is low, there are dominant interferers, or combinations of the above conditions. Examples of such application scenarios are sensor networks, which are typically characterized by low data rates, and sparse topologies.

Characterizing the limit of the SGA hypothesis, and in particular, comparing this limit across the different ways of generating IR signals remains an open question. Asymptotically, we have seen that the three analyzed schemes, 2PPM-THMA,

2PAM-THMA, and DS-UWB, behave in a similar way, except for the expected difference related to how data are modulated, specifically binary antipodal PAM versus binary PPM, that is, the well-known loss of a maximum of 3 dB (exactly 3 dB for orthogonal PPM) in binary PPM. It is not clear, however, whether these systems behave in a similar way when the multi-user signal traveling in the air interface still preserves some impulsive characteristic, and in particular, in which exact conditions the system must be for granting validity to the SGA hypothesis. The cardinality of the code N_h, for example, is a parameter that may play a role in the discussion since, as shown in Eq. (9–4), this parameter regulates the width of the chip time T_c, and therefore indirectly the eventual width of time gaps in between two pulses. Parameter N_h, however, is not defined in DS-UWB, which adopts binary codes. Checkpoint 9–1 will compare performance of the three analyzed schemes, 2PPM-THMA, 2PAM-THMA, and DS-UWB.

CHECKPOINT 9–1

In this checkpoint, we will analyze the performance of an IR-UWB system in the presence of MUI. Both theoretical results and simulation analyses will be presented, to discuss the validity of the standard Gaussian approximation for estimating the probability of error.

In the first part of this checkpoint, we present three MATLAB functions that can be used for evaluating the probability of error Pr_b in the case of TH multiple access for both binary PPM and binary antipodal PAM. In both cases, the Pr_b value is evaluated according to the theoretical results derived in Sections 9.2.1 and 9.2.2 for PPM and PAM, respectively. In particular, the aim of this checkpoint is to compute the expressions in Eq. (9–31) for the Pr_b value in the case of 2PPM-TH systems, and the expression in Eq. (9–46) for the Pr_b value in the case of 2PAM-TH systems. In both Eqs. (9–31) and (9–46), we observe that the evaluation of Pr_b requires the knowledge of several system parameters, and in particular of the quantity σ_M^2, which depends on the characteristics of the pulse waveform $p_0(t)$. We can thus introduce two MATLAB functions for evaluating the σ_M^2 term in the case of 2PPM and 2PAM.

Function 9.1 (see Appendix 9.A), computes the σ_M^2 term for an UWB system implementing binary PPM according to the expression given in Section 9.2.1, Eq. (9–25). Function 9.1 receives three inputs: vector `pulse` representing the pulse waveform; the value in seconds of the PPM shift `PPMshift`; and the value of the sampling frequency `fc`, which is used for representing the pulse waveform in the discrete time domain. Function 9.1 returns the value of the σ_M^2 term `sm2`, and is executed by typing:

```
[sm2] = cp0901_sm2_PPM(pulse,PPMshift,fc);
```

Function 9.2 (see Appendix 9.A), computes the σ_M^2 term for an UWB system implementing binary PAM according to the expression given in Section 9.2.2, Eq. (9–43). Function 9.2 receives two inputs: vector `pulse` representing the pulse waveform; and the value of the sampling frequency `fc`, which is used for representing the pulse waveform in the discrete time domain. Function 9.2 returns the value of the σ_M^2 term (`sm2`), and is executed by typing:

```
[sm2] = cp0901_sm2_PAM(pulse,fc);
```

For both 2PPM and 2PAM, the σ_M^2 term affects the value of the SIR, and consequently, the probability of error. In both cases, the SIR increases in inverse proportion with σ_M^2, that is, the higher the σ_M^2, the smaller the SIR. A comparative analysis of the values of σ_M^2 corresponding to PPM and PAM can be useful to confirm the gain of 3 dB in the SIR, which is achieved by antipodal PAM with respect to orthogonal PPM. This comparison can be made by executing the following command lines:

```
fc=1e11;
Tm=1e-9;
tau=0.25e-9;
[pulse]= cp0201_waveform(fc,Tm,tau);
epsilon = linspace(0,1e-9,100);
for e = 1 : 100
SPPM(e) = cp0901_sm2_PPM(pulse,epsilon(e),fc);
end
SPAM = ones(1,100).*cp0901_sm2_PAM(pulse,fc);
plot(epsilon,SPPM,epsilon,SPAM)
```

The above commands store in memory vectors `SPPM` and `SPAM`, which contain the values of σ_M^2 for different values of the PPM shift in the range [0,1] ns. A pulse having the second derivative Gaussian waveform with a shaping factor of 0.25 ns was also considered. The above command lines also produce the graphical output shown in Figure 9–3 after the introduction of axes labels and a change of style for the different curves. Figure 9–3 shows that the σ_M^2 for binary PPM asymptotically tends to be a value that is twice the σ_M^2 value for antipodal PAM. In the case of orthogonal PPM, or $\varepsilon > 0.5$ ns in the case under examination, the SIR of a system implementing PAM is 3 dB higher than the SIR of a system implementing PPM.

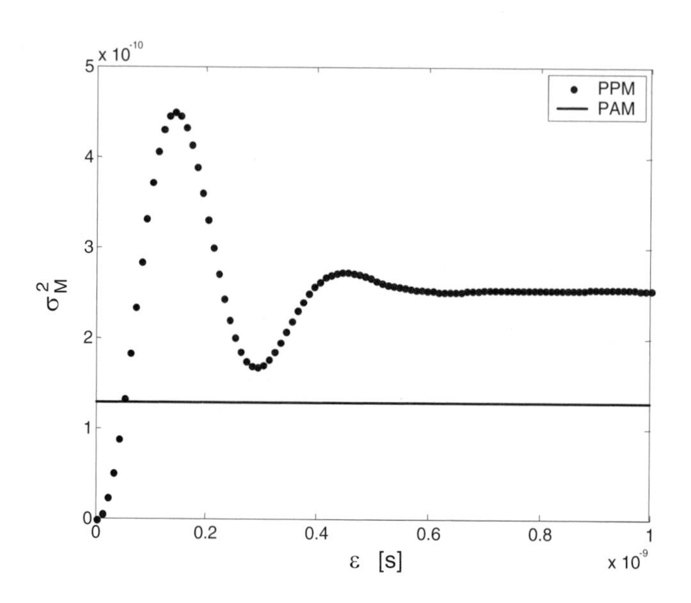

Figure 9–3 Comparison of σ_M^2 vs. ε for binary PPM (dots) and PAM (solid line).

Given the above functions, we can now evaluate the probability of error for both PPM and PAM. We introduce two new MATLAB functions, Function 9.3 and Function 9.4, which compute Pr_b according to Eqs. (9–31) and (9–46) for 2PPM and 2PAM, respectively.

Function 9.3 (see Appendix 9.A) evaluates Pr_b for a multi-user UWB system implementing a TH multiple access with binary PPM. The Pr_b value is determined under the SGA hypothesis. Function 9.3 receives the following inputs: vector `ebno` representing the values in dB of the E_b/N_0 ratio for which the probability of error must be evaluated; the value `erx0` of the received energy per pulse of the useful signal; vector `erxMUI` containing the values of the received energy per pulse for all the interfering users; vector `pulse`, which represents the waveform of the basic pulse; the value in bits/s of the user bit rate `Rb`; the value in seconds of shift introduced by PPM `PPMshift`; the sampling frequency used for the conversion from the continuous time domain to the discrete time domain `fc`; and the constant `gamma_r` representing the fraction of the bit period γ_R, which can be occupied by the transmitted pulses (see Eq. (9–3)). Function 9.3 returns vector `BER` containing the values of the average probability of error vs. E_b/N_0. Function 9.3 is executed by typing the following command:

```
[BER] = cp0901_MUIBER_2PPM(ebno,erx0,erxMUI,...
    pulse,Rb,PPMshift,fc,gamma_r);
```

Function 9.4 (see Appendix 9.A) evaluates Pr_b for a multi-user UWB system implementing a TH multiple access with binary PAM. The Pr_b value is determined under the SGA hypothesis. Function 9.4 receives the same inputs as Function 9.3 (except for the absence of the indication of the PPM shift): vector `ebno`, scalar `erx0`, vector `erxMUI`, vector pulse, scalar Rb, scalar fc, and scalar gamma_r. Function 9.4 returns vector BER containing the values of the average probability of error vs. E_b/N_0. Function 9.4 is executed by typing the following command:

```
>> [BER] = cp0901_MUIBER_2PPM(ebno,erx0,erxMUI,...
      pulse,Rb,PPMshift,fc,gamma_r);
```

We can use Functions 9.3 and 9.4 for comparing the performance of 2PPM and 2PAM in different scenarios corresponding to increased values of the number of interferers: *Case A* with five interferers, *Case B* with 20 interferers, and *Case C* with 50 interferers. The useful transmitter generates a binary stream with R_b = 20 Mbits/s. Perfect power control is assumed. A basic pulse having the second derivative Gaussian waveform with a shaping factor of 0.25 ns is considered. With reference to PPM, orthogonal modulation is considered, ε = 0.5 ns. The command lines for comparing system performance with PPM and PAM in the three cases under examination are:

```
ebno = linspace(0,30,21);
erx0 = 1;
erxMUIa = ones(1,5);
erxMUIb = ones(1,20);
erxMUIc = ones(1,50);
fc = 1e11;
Tm = 1e-9;
tau= 0.25e-9;
[pulse]= cp0201_waveform(fc,Tm,tau);
PPMshift=0.5e-9;
Rb = 20e6;
gamma_r=1;
[BERaPPM]= cp0901_MUIBER_2PPM(ebno,erx0,erxMUIa,...
   pulse,Rb,PPMshift,fc,gamma_r);
[BERbPPM]= cp0901_MUIBER_2PPM(ebno,erx0,erxMUIb,...
   pulse,Rb,PPMshift,fc,gamma_r);
[BERcPPM]= cp0901_MUIBER_2PPM(ebno,erx0,erxMUIc,...
   pulse,Rb,PPMshift,fc,gamma_r);
[BERaPAM] = cp0901_MUIBER_2PAM(ebno,erx0,erxMUIa,...
   pulse,Rb,fc,gamma_r);
[BERbPAM] = cp0901_MUIBER_2PAM(ebno,erx0,erxMUIb,...
   pulse,Rb,fc,gamma_r);
[BERcPAM] = cp0901_MUIBER_2PAM(ebno,erx0,erxMUIc,...
   pulse,Rb,fc,gamma_r);
figure(1)
semilogy(ebno,BERaPPM,ebno,BERaPAM)
figure(2)
semilogy(ebno,BERbPPM,ebno,BERbPAM)
figure(3)
semilogy(ebno,BERcPPM,ebno,BERcPAM)
```

The above command lines store vectors `BERaPPM`, `BERaPAM`, `BERbPPM`, `BERbPAM`, `BERcPPM`, and `BERcPAM` in memory. These vectors contain the values of the probability of error for the three scenarios and the two modulation schemes under consideration. The above commands also provide the graphical outputs shown in Figures 9–4, 9–5, and 9–6 after the introduction of axes labels and a change of style for the different curves.

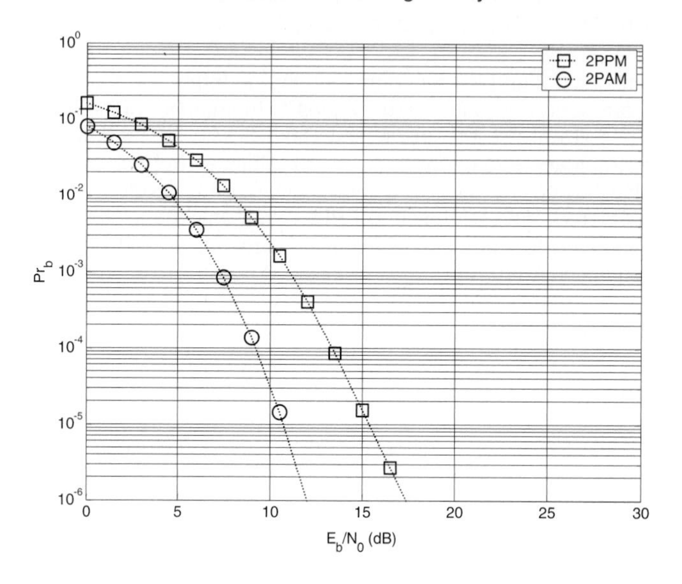

Figure 9–4 Probability of error Pr_b vs. E_b/N_0 for binary orthogonal PPM (squares) and binary antipodal PAM (circles) in *Case A* (five users).

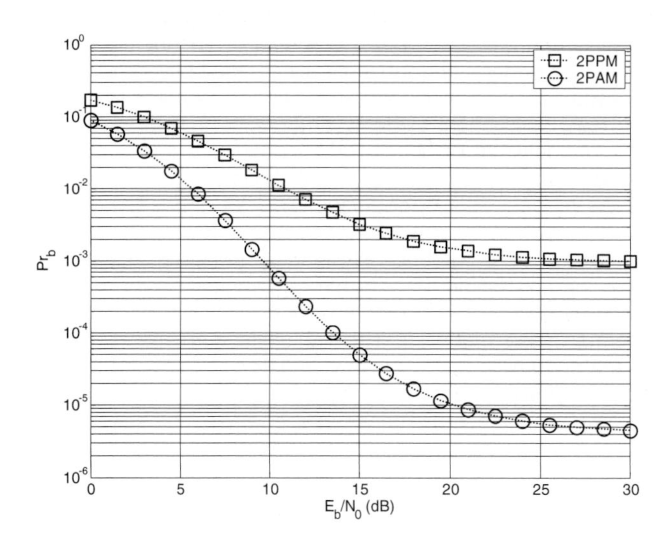

Figure 9–5 Probability of error Pr_b vs. E_b/N_0 for binary orthogonal PPM (squares) and binary antipodal PAM (circles) in *Case B* (20 users).

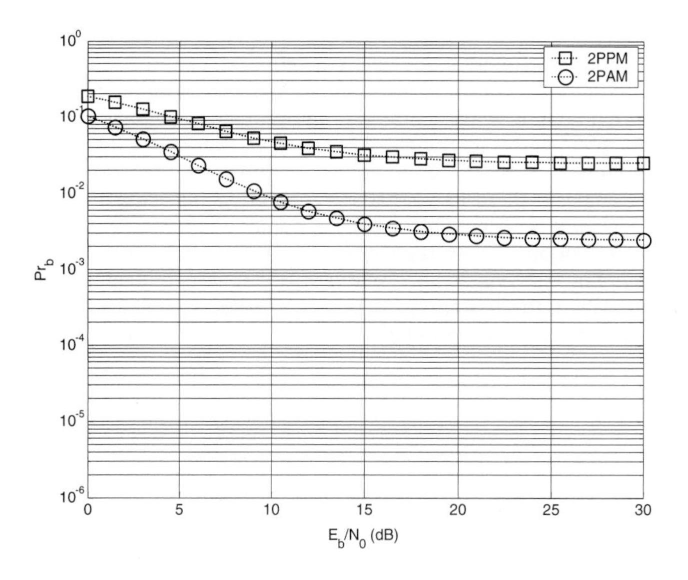

Figure 9–6 Probability of error Pr_b vs. E_b/N_0 for binary orthogonal PPM (squares) and binary antipodal PAM (circles) in *Case C* (50 users).

Figure 9-4 shows Pr_b versus E_b/N_0 for binary orthogonal PPM and binary antipodal PAM when five interfering users are present (*Case A*). We observe that in this scenario, MUI seems to have no effect on system performance, since for both PPM and PAM, the probability of error decreases with E_b/N_0 with the same trend of Figure 8–22. In other words, the only contribution that seems to affect system performance is the presence of thermal noise at the receiver. The gap between PPM and PAM is nearly constant and on the order of 3 dB for values of Pr_b higher than 10^{-4}. It increases slightly for lower Pr_b values.

Figure 9–5 shows the probability of error versus E_b/N_0 for binary orthogonal PPM and binary antipodal PAM when 20 interfering users are present (*Case B*). In this case, we observe that the probability of error for both PPM and PAM tends asymptotically to a constant value, leading to the conclusion that system performance in the presence of high E_b/N_0 values is determined by only MUI. We can thus identify two states of operation for all systems affected by thermal noise and MUI. The first state corresponds to the situation where E_b/N_0 is low, and the probability of error is mainly determined by the thermal noise. In such a situation, we can improve performance by allowing all devices to increase the transmitted energy per pulse, or increase the transmitted power. When the transmitter power increases, the ratio E_b/N_0 at the receiver also increases, with a corresponding decrease in the Pr_b value. According to Figure 9–5, however, the relationship between the increase of power (in dB) and the decrease in the probability of error is not linear, since greater amounts of additional power are required for an equal decrease in the probability of error as we move towards smaller Pr_b values. At a certain point, the probability of error reaches a value that cannot be further decreased by increasing the transmitted power. This value is called the Pr_b floor, and characterizes the second state of the system where performance is limited by MUI. Note that the Pr_b floor is different for PPM and PAM. In particular, PAM seems to be more robust, since it presents a Pr_b floor that is more than two orders of magnitude smaller than for PPM. This is a general result that derives from the loss of 3 dB in the SIR value suffered by PPM. Due to the non-linear relationship that exists between SNR_{ref} and Pr_b (see Eq. (9–15)), the effect on performance of such a loss is not the same in all scenarios. For example, we can observe the plot in Figure 9–6, which compares the probability of error versus E_b/N_0 for binary orthogonal PPM and binary antipodal PAM when 50 interfering users are present (*Case C*). Here, the gap between the Pr_b floors reduces to one order of magnitude.

The comparative analysis of PPM versus PAM in the three examined scenarios is summarized in Figure 9–7, which collects all curves presented in Figures 9–4, 9–5, and 9–6. Figure 9–7 shows that orthogonal PPM results in a loss of performance with respect to antipodal PAM, in agreement with (Vojcic and Pickholtz, 2003). In particular, we observe that in the case under examination, the performance of binary antipodal PAM with 50 interfering users is nearly the same as binary orthogonal 2PPM with 20 interfering users.

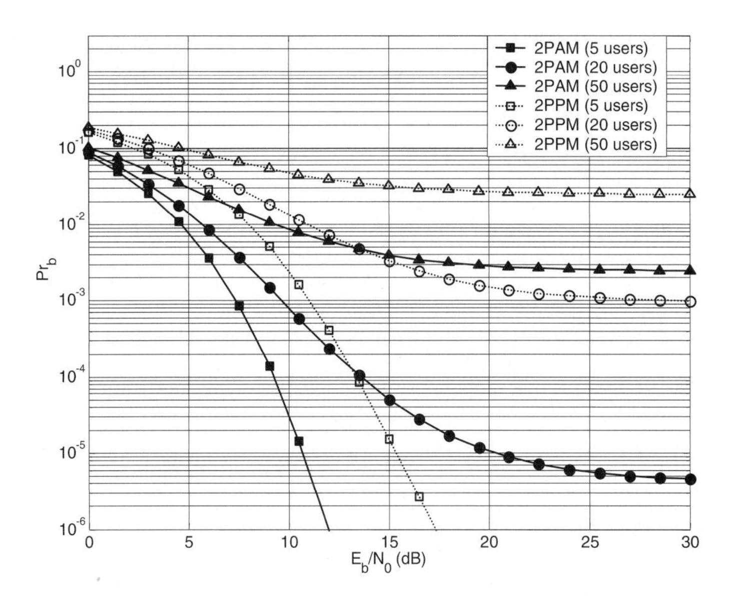

Figure 9–7 Comparison of performance for binary orthogonal PPM (white markers) and binary antipodal PAM (black markers). Three scenarios are considered: *Case A* with five users (squares), *Case B* with 20 users (circles), and *Case C* with 50 users (triangles).

In the second part of this checkpoint, we will simulate the activity of an UWB receiver in the presence of multiple users to compare simulation versus theoretical curves based on the analytical derivations of Section 9.2. In particular, we are interested in evaluating the range of validity of the SGA for the estimation of the probability of error when both thermal noise and MUI are present at the receiver. In this analysis, we will consider the case of binary orthogonal PPM. An identical approach, however, can be followed for the PAM case. The MATLAB functions required for the simulation were already introduced in previous checkpoints.

Two scenarios are considered for this simulation: The first (*Case 1*) is characterized by the presence of six interfering users, the second (*Case 2*) by the presence of nine interfering users. In both cases, perfect power control is assumed. The reference user generates a stream of bits with bit period T_b = 18 ns, leading to an approximate bit rate R_b = 55.55 Mbits/s. Each bit period is organized into three frames of duration T_s = 6 ns, that is, three pulses are transmitted for each bit. Each frame is further subdivided into six slots with length T_c = 1 ns, that is, the TH code can assign one out of six possible positions to the single pulse within each pulse repetition period T_s. We assume that all users transmit with the same signal format.

We start with *Case 1*, where a total number of seven asynchronous users transmit using same power.

The first step of the simulation is the generation of the transmitted signals for the reference user and the interfering user. This task is performed by using Function 2.6 with the following parameters: Pow=-30, fc=0.5e11, numbits=50000, Ts=6e-9, Ns=3, Tc=1e-9, Nh=6, Np=150000, Tm=0.5e-9, tau=0.2e-9, dPPM=0, 5e-9, G=0. The following commands are executed:

```
[bits,THcode,Stx,ref]=cp0201_transmitter_2PPM_TH;
Nmui = 6;
for n_int = 1 : Nmui
[bitsint,THcodeint,Stxint,refint] = ...
   cp0201_transmitter_2PPM_TH;
Smui(n_int,:) = Stxint;
clear Stxint bitsint refint THcodeint
end
fc = 0.5e11;
Ns = 3;
Ts = 6e-9;
numbits = length(bits);
dPPM = 0.5e-9;
numpulses = numbits*Ns;
```

The above commands generate vector Stx, which represents the useful signal, and matrix Smui, which contains the interfering signals, one signal per row. Note that all generated signals are synchronized; all have slots starting at the same time. The next step is thus the introduction of asynchronism among users. This task can be performed by using Function 8.14 (see Appendix 8.A). Through Function 8.14, we can, in fact, introduce a random time shift on each interfering signal:

```
for n_int = 1 : Nmui
MUI = Smui(n_int,:);
delay = (Ts)*rand;
MUId = cp0804_signalshift(MUI,fc,delay);
Smui(n_int,:)=MUId;
end
clear MUI MUId
```

The above commands apply a cyclic shift on each row of matrix Smui. For each row, the amount of the time shift to be applied is randomly extracted within the range $(0,T_s)$. With the next step, the different interfering signals are summed up to generate a global interfering signal RXMUI:

```
RXMUI=zeros(1,length(Stx));
for n_int = 1 : Nmui
RXMUI = RXMUI + Smui(n_int,:);
end
clear Smui
```

At this point of the simulation, we are ready to introduce thermal noise at the receiver. This task is performed by executing Function 8.2, which introduces white Gaussian noise according to a target vector of E_b/N_0 values:

```
ebno = [0 2 5 10 15 20 25 30];
RXnoise = cp0801_Gnoise1(Stx,ebno,numbits);
clear Stx
```

The above commands generate matrix `RXnoise`. Each row of this matrix represents the reference signal at the receiver after the introduction of Gaussian noise at different energy levels. The next step consists of summing up thermal noise and MUI:

```
for j = 1 : length(ebno)
RX(j,:)=RXnoise(j,:)+RXMUI;
end
clear RXnoise RXMUI
```

The above commands generate matrix `RX`. Each row of this matrix represents the signal at the receiver after the introduction of MUI and Gaussian noise at a different energy level. The following step is the simulation of the receiver activity. Since we are considering the case of propagation over an AWGN channel, we will use the same functions that were introduced in Chapter 8, "Propagation over a Multi-Path-Affected UWB Radio Channel," Checkpoint 8–1:

```
[mask] = cp0801_PPMcorrmask(ref,fc,numpulses,dPPM);
[RXbits,Prb1] = ...
   cp0801_PPMreceiver(RX,mask,fc,bits,numbits,Ns,Ts);
```

After the above commands, we obtain vector `Prb1`, which represents the probabilities of error for all the E_b/N_0 values considered in the simulation. The result of the simulation can be visualized as follows:

```
semilogy(ebno,Prb1)
```

Figure 9–8 shows the probability of error versus E_b/N_0 for the case under examination.

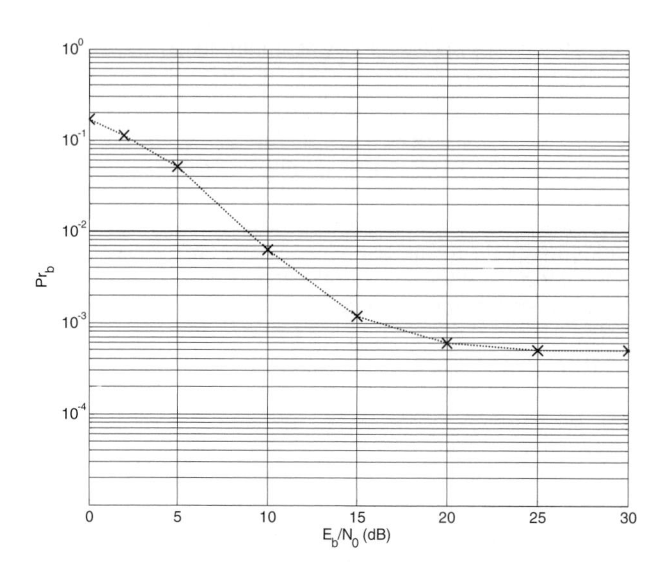

Figure 9–8 Probability of error Pr_b vs. E_b/N_0 for binary orthogonal PPM in *Case 1* (six interfering users).

Note that the curve shown in Figure 9–8 strongly depends on the initial positions of the interfering users, and does not take into account the fact that different delays among users can lead to different performance at the receiver. With different initial conditions, in fact, the Pr_b floor that is measured at the end of the simulation could differ from the one of Figure 9–8. In the presence of MUI, it is important to measure the average probability of error by averaging the results of different simulations. Figure 9–9 compares results of five different simulations. Each simulation was performed by executing the same commands indicated above. As anticipated, the Pr_b floors, which are measured at the end of the simulations, differ by more than one order of magnitude.

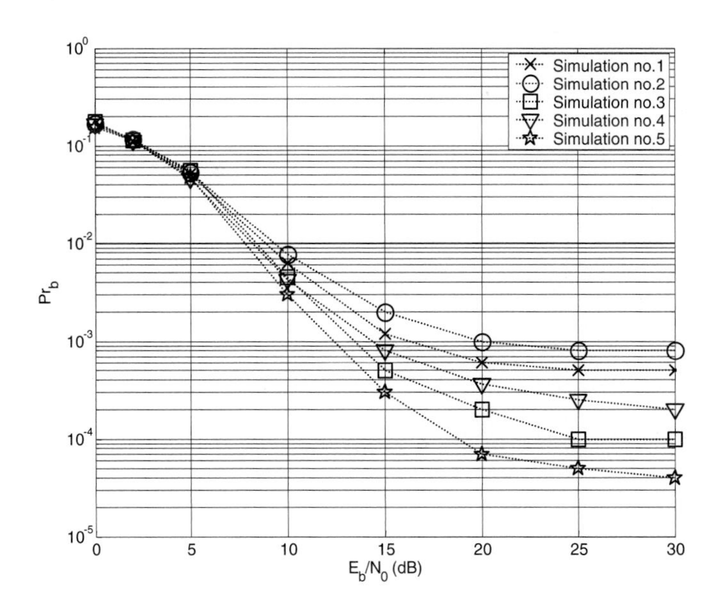

Figure 9–9 Probability of error Pr_b vs. E_b/N_0 for binary orthogonal PPM in *Case 1* (six interfering users). The plot shows the results deriving from five different simulations.

In Figure 9–10, we compare the average probability of error versus E_b/N_0, which is obtained by averaging the results of the five simulations represented in Figure 9–9, with the theoretical curve based on the result in Eq. (9–31). This curve was obtained by applying Function 9.3 with $R_b = (1/18) \cdot 10^9$ bits/s and six interfering users having the same energy per pulse as the useful user.

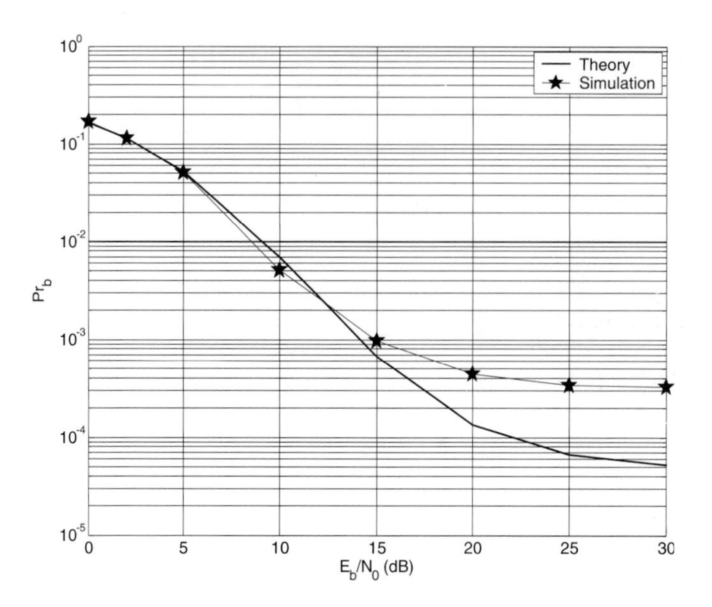

Figure 9–10 Comparison between theoretical results (solid line) and simulation results (dashed line) in *Case 1* (six interfering users).

Figure 9–10 shows that the theoretical model used for evaluating Pr_b underestimates the effect of MUI. In other words, the probability of error measured by simulation is higher than the one predicted under the SGA hypothesis. This result indicates that in the case under examination (*Case 1*), the number of pulses traveling over the air interface is not sufficient for guaranteeing the validity of the Gaussian approximation for the interference noise. The gap between theory and simulation, however, should decrease when considering the second scenario of the simulation (*Case 2*), where a higher number of interferers is considered, that is, nine interferers instead of six. The code for the simulation of *Case 2* is the same as *Case 1*, except for the command `Nmui = 9` in the first block of commands. Once again, we perform five simulations to partially average the effect of the initial condition. Results of the simulation are shown in Figures 9–11 and 9–12. Figure 9–11 compares the results of the five simulations for *Case 2*. Note that the difference among the different Pr_b floors is reduced with respect to *Case 1* (see Figure 9–9). Figure 9–12 compares the average probability of error versus E_b/N_0, which is obtained by averaging the results of the simulations of Figure 9–11, with the theoretical curve based on Eq. (9–31). As expected, the average Pr_b floor is higher than the previous case, and the gap between theory and simulation is reduced.

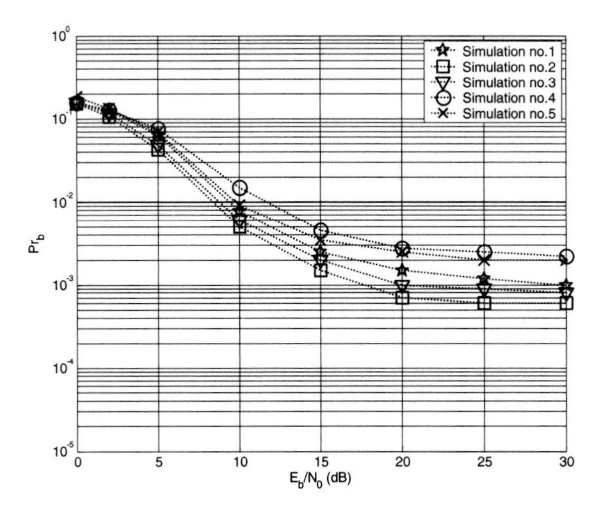

Figure 9–11 Probability of error Pr_b vs. E_b/N_0 for binary orthogonal PPM in *Case 2* (nine interfering users). The plot shows the results deriving from five different simulations.

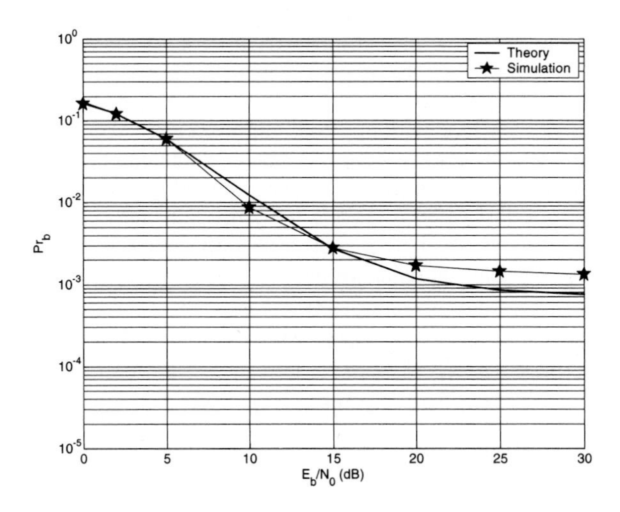

Figure 9–12 Comparison between theoretical results (solid line) and simulation results (dashed line) in *Case 2* (nine interfering users).

CHECKPOINT 9–1

9.3 Multi-User Interference Modeling Based on Packet Collision

In this section, we will analyze MUI from a different perspective, which attempts to take into account the way information is structured and conveyed. Information bits are usually grouped into larger information units called packets. Typically, this reorganization of information flows is made by the MAC module, which generates information units commonly indicated as MAC-Protocol Data Units (MAC-PDUs). It is these packets, after appropriate coding, that are sent over the air interface.

The packet information structure further weakens the hypothesis that interference occurring in one pulse frame is independent of past and future events. MUI can be re-analyzed under this perspective by observing that interference corresponds to packet collision. In this section, we will focus the analysis on IR systems and observe that interference is provoked by collisions occurring between pulses belonging to different transmissions. The time occupied by a single pulse T_M is defined as in Chapter 7, "The Pulse Shaper," as the time interval typically centered on the main lobe in which most of the energy of the pulse at the receiver is concentrated. Typical values for T_M lie between 70 ps and 20 ns, depending on transmitted pulse shape and channel behavior.

We consider here an asynchronous network such as an ad-hoc network in which asynchronous users transmit information in an uncoordinated manner. For asynchronous networks, it is reasonable to suppose that the packet inter-arrival process follows a Poisson distribution (Bertsekas and Gallager, 1992). Each packet contains a set of pulses, each subset of N_s pulses carrying the information of one bit. The pulse inter-arrival process is a complex phenomenon that depends on modulation, dithering, codes, and so on. To try to express the probability of pulse collision in a closed form, let us assume that the pulse inter-arrival process is a Poisson process itself. In this case, the probability that one or more pulses will collide with the useful pulse when N_U active users are transmitting pulses over the air interface can be compared to the Aloha collision probability when the information unit is the pulse instead of the packet. This collision probability is thus expressed as follows:

$$\Pr_{PulseCollision} = 1 - e^{\left(-2(N_U - 1)\frac{T_M}{T_s}\right)} \tag{9–48}$$

Assuming that a pulse collision causes a random decision at the receiver, the pulse error probability can be expressed as:

$$\Pr_{PulseError} = 0.5\Pr_{PulseCollision} \tag{9–49}$$

Consider that each bit is encoded into N_S pulses. We assume an error on the bit when more than $N_S/2$ pulse errors occur. This corresponds to hard receiver detection (see Chapter 8). Bit error probability is thus expressed by:

$$\mathrm{Pr}_b = \sum_{i=\lceil \frac{N_S}{2} \rceil}^{N_S} \binom{N_S}{i} Pr_{PulseError}{}^{i} \left(1 - Pr_{PulseError}\right)^{N_S - i} \tag{9–50}$$

The probability of correct detection is therefore:

$$\mathrm{Pr}_{correctbit} = 1 - \mathrm{Pr}_b = 1 - \sum_{i=\lceil \frac{N_S}{2} \rceil}^{N_S} \binom{N_S}{i} \mathrm{Pr}_{PulseError}{}^{i} \left(1 - Pr_{PulseError}\right)^{N_S - i} \tag{9–51}$$

Equation (9–50) provides the bit error probability in the general hypothesis of asynchronous users with no additional hypothesis on either dithering or MA strategy, which might be either THMA or DS.

Consider, for example, a network in which 100 users generate packets at a rate of G = 10^3 packets/s using a packet length L = 1000 bits. The data rate R_b is set to 10 Mbits/s. Each station therefore transmits at an average rate of 1 Mbits/s. In average, this is equivalent to have $N_U = 10$ active users at a given time. For example, if T_M = 80 ps and $N_S = 11$, then Pr_b is $5.16 \cdot 10^{-5}$.

Here, we assume that a packet is corrupted if at least one bit error occurs.

The average probability to transmit a packet successfully Pr_{Succ} is thus given by:

$$\mathrm{Pr}_{Succ} = \left(1 - \mathrm{Pr}_b\right)^L \tag{9–52}$$

Pr_{Succ} depends on the number of packets P_U and data rate R_b as displayed in Figure 9–13 (full lines). Figure 9–13 drawn from (Di Benedetto et al., 2004), also shows simulation data (dashed line) for the specific case of a rate R = 10 Mbits/s. The simulator used to derive the above data implements the whole transmitter/receiver chain at the physical layer. The effect of thermal noise was neglected in these simulations. Both analytically derived and simulated values were obtained with the following settings: L = 50 bits, T_M = 80 ps, and N_S = 11. Note that the simulated values closely follow the analytical prediction and provide encouraging support for the analytical model. Also note that for low data rates in particular, the probability of packet error is extremely low (< 10^{-4}), up a number of packets in the air interface as large as about 50.

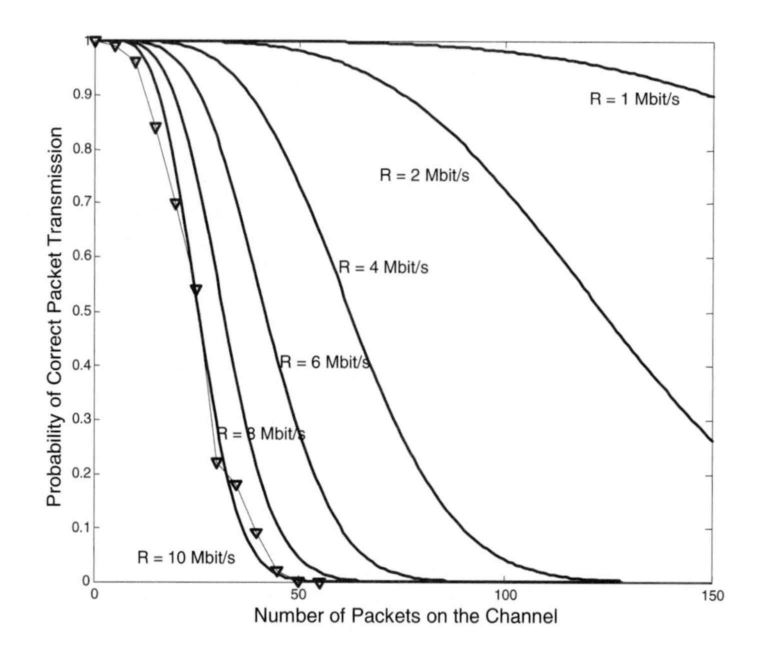

Figure 9–13 Probability of correct packet transmission vs. number of packets on the channel for different values of bit rate R. Solid lines represent theoretical performance; dashed line shows the simulation performance in the case R = 10 Mbits/s (by permission from *Mobile Networks and Applications Journal*, Kluwer Academic Publishers).

CHECKPOINT 9–2

In this checkpoint, we will use computer simulation to compare the probability of error that results from Eq. (9–50) with the Pr_b values deriving from the Gaussian model and simulation. We will focus on IR-UWB systems implementing binary orthogonal PPM with TH multiple access. The same analysis, however, can be easily repeated with other transmission schemes.

A new MATLAB function is required for the evaluation of the probability of error expressed by Eq. (9–50).

Function 9.5 (see Appendix 9.A) evaluates the probability of error in Eq. (9–50) when considering a scenario with `Nu` users, `Ns` pulses transmitted per bit, a pulse duration of `Tm` seconds, and an average pulse repetition period of `Ts` seconds. The above parameters are given as input to the function, which returns the bit error probability `Prb`. The command line for executing Function 9.5 is:

```
[Prb] = cp0902_prbcoll(Nu,Ns,Tm,Ts);
```

Function 9.5 evaluates the probability of error by taking into account the expected number of pulse collisions that occur in the presence of MUI. Two pulses are assumed to collide when a portion of the corresponding waveforms overlap at the receiver input. Note that collision is therefore defined independently of pulse shape. Moreover, since we are working with the hypothesis that collisions occur when two pulses overlap even by a small amount, it is reasonable to limit the preserved percentage of energy to the range 70–90%. The time occupied by a pulse T_M is derived accordingly. However, to take into account the effect of using different pulse shapes, it is useful to introduce the concept of "effective pulse duration." This parameter is defined as the time interval that contains a given percentage of the energy of the pulse. The introduction of the effective pulse duration allows to avoid an over-estimation of the effect of pulse overlapping. Function 9.6 evaluates T_M for any given waveform according to a target percentage of energy.

Function 9.6 (see Appendix 9.A), evaluates the effective pulse duration for an input waveform `pulse` sampled at `fc` in Hz, when a target percentage of energy `pE` is considered. The function returns vector `eff_pulse` to represent the truncated pulse waveform that contains the specified percentage of energy and the value `Tm` of the effective time duration. The command line that executes Function 9.6 is:

```
[eff_pulse,Tm] = cp0902_effpulse(pulse,fc,pE);
```

Given Functions 9.5 and 9.6, we can compare the probability of error that results from the application of the different methods presented in this chapter. We consider a common scenario where N_u asynchronous users transmit binary data over the same air interface through binary orthogonal PPM signals. All these signals are characterized by the transmission of five pulses per bit ($N_s = 5$), with an average pulse repetition period of $T_s =$ 3 ns. Each T_s period is organized in 3 slots of 1 ns ($N_h = 3$; $T_c = 1$). According to the above parameters, each signal conveys a binary stream with $R_b \approx 66.66$ Mbits/s. We further assume that the basic pulse for all the transmitted signals is the second derivative Gaussian waveform with shape factor $\alpha = 0.2$ ns. The PPM shift is assumed to be $\varepsilon = 0.5$ ns. To evaluate the probability of error that derives from the interference model based on Pulse Collision (PC), we must first determine T_M through Function 9.6. In particular, we consider three different values for the percentage of energy that must be contained within T_M, that is, `pE = 0.9`, `pE = 0.8`, and `pE = 0.7`.

```
tau=0.2e-9;
T = 1e-9;
fc = 50e9;
[pulse]= cp0201_waveform(fc,T,tau);
pE = 0.9;
[eff_pulse,Tm1] = cp0902_effpulse(pulse,fc,pE);
pE = 0.8;
[eff_pulse,Tm2] = cp0902_effpulse(pulse,fc,pE);
pE = 0.7;
[eff_pulse,Tm3] = cp0902_effpulse(pulse,fc,pE);
Tm1
>> Tm1 = 2.0000e-010
Tm2
>> Tm2 = 1.6000e-010
```

```
Tm3
>> Tm3 = 1.2000e-010
```

As expected, the effective time duration decreases as `pE` decreases. In particular, we obtain T_M values of 0.2 ns, 0.16 ns, and 0.12 ns corresponding to 90%, 80%, and 70% of the original energy of the pulse, respectively. We can now use Function 9.5 for evaluating Pr_b as expressed by Eq.(9-50). In particular, we can determine the bit error probability as a function of the number of users for different values of T_M:

```
Nu = (1:1:200);
Ns = 5;
Ts = 3e-9;
for n = 1 : length(Nu)
PC1(n) = cp0902_prbcoll(Nu(n),Ns,Tm1,Ts);
PC2(n) = cp0902_prbcoll(Nu(n),Ns,Tm2,Ts);
PC3(n) = cp0902_prbcoll(Nu(n),Ns,Tm3,Ts);
end
```

The above commands store vectors `PC1`, `PC2`, and `PC3` in memory, which contain the Pr_b values for different numbers of users and T_M. The result of the above simulation is shown in Figure 9–14. As expected, the probability of error quickly increases with the number of users due to the higher number of PCs. For a given number of users, the probability of error decreases when considering smaller T_M values.

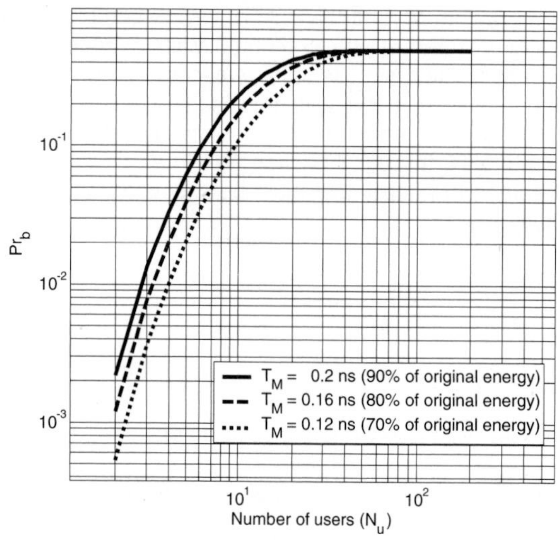

Figure 9–14 Probability of error Pr_b vs. number of users N_u according to the interference model based on PC. Different values of the pulse duration are considered: $T_M = 0.2$ ns (solid line), $T_M = 0.16$ ns (dashed line), and $T_M = 0.12$ ns (dotted line).

We can compare the result in Figure 9–14 with that resulting from the application of the SGA for modeling MUI. This task can be performed using Function 9.3:

```
PPMshift = 0.5e-9;
ebno = Inf;
erx0 = 1;
gamma_r=1;
Rb = 1/(15e-9);
Nu = (2:1:200);
for n = 1 : length(Nu)
erxMUI = ones(1,Nu(n)-1);
PSGA(n) = cp0901_MUIBER_2PPM(ebno,erx0,erxMUI,...
   pulse,Rb,PPMshift,fc,gamma_r);
end
```

The above commands store vector `PSGA` in memory. This vector contains the Pr_b values derived through the SGA for different numbers of users. Note that the probability of error that results from the above commands takes into account only the presence of MUI at the receiver (the E_b/N_0 value is set equal to infinity). Figure 9–15 shows the values of vectors `PC1`, `PC2`, `PC3`, and `PSGA`, that is, the Pr_b values resulting from the interference models based on PC (with different T_M values) and the SGA. The interference model based on the concept of PC overestimates the effect of MUI with respect to the interference model based on the SGA.

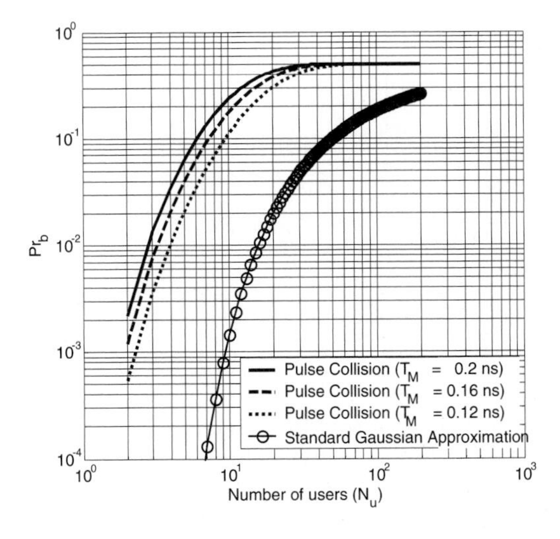

Figure 9–15 Comparison of the probability of error Pr_b vs. number of users N_u derived from PC and SGA models. Solid line refers to PC with T_M = 0.2 ns; dashed line refers to PC with T_M = 0.16 ns; dotted line refers to PC with T_M = 0.12 ns; circles refer to the SGA.

At this point of the analysis, we can evaluate the probability of error that derives from the simulation of the whole transmission chain. This task requires the execution of the same set of codes that was introduced in Checkpoint 9–1 for evaluating receiver performance through simulation. In the present case, we neglect the presence of thermal noise by setting E_b/N_0 to infinity. In addition, we set the flag `HDSD` within Function 8.5 to 1, that is, we consider hard decision detection. We will not reproduce here the code for simulating the whole transmission chain in the presence of multiple users since we can refer to the simulation presented in Checkpoint 9–1. Results of simulations are presented in Figure 9–16, where measured Pr_b values are compared with those obtained with the theoretical models. Figure 9–16 shows that simulation results are between the Pr_b values provided by the SGA and those given by the PC model. In particular, the SGA seems to provide better accuracy in the presence of several users, while the model based on PC provides better results when a small number of interferers is considered. Note that the SGA curve was determined under the hypothesis of soft detection, while the PC model and the simulation were determined under the hypothesis of hard detection. The SGA curve of Figure 9–16 can therefore be considered as a lower bound for Pr_b since with the Gaussian hypothesis, soft detection outperforms hard detection (see Chapter 8). The PC curve, on the other hand, can be seen as an upper bound to Pr_b given its intrinsic severe approximation that a pulse is lost with probability 0.5 independently of the catastrophic degree of the collision as indicated by the amount of overlap between colliding pulses.

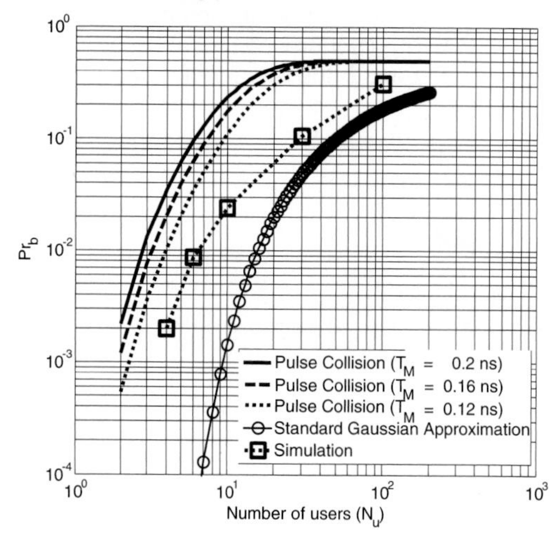

Figure 9–16 Comparison of the probability of error Pr_b vs. number of users N_u derived from theoretical models (PC and SGA) and from simulation. Solid line refers to PC with T_M = 0.2 ns; dashed line refers to PC with T_M = 0.16 ns; dotted line refers to PC with T_M = 0.12 ns; circles refer to the SGA; squares refer to simulation.

CHECKPOINT 9–2

Further Reading

The analyses presented in this chapter are based on the hypothesis of single-user detection. Improved performance can be achieved by using multi-user detection (Verdu, 1998). Multi-user detection is addressed in (Durisi et al., 2003), which also compares system performance with and without RAKE reception for a variety of UWB schemes. The reading of this paper is recommended for the formalism adopted, which provides a unified description of the different UWB radio schemes, as first suggested by (LeMartret and Giannakis, 2002).

Specific families of codes rather than pseudorandom noise coding for TH-UWB have been analyzed in (Iacobucci and Di Benedetto, 2001; Iacobucci and Di Benedetto, 2002). Maggio and colleagues proposed the interesting alternative of pseudo-chaotic codes (Maggio et al., 2001; Laney et al., 2002; Maggio et al., 2002). Pseudo-chaotic encoding is also addressed in (Erseghe and Bramante, 2002). (Jones et al., 2003) analyzed the problem of code assignment in a DS-UWB system based on the DS-CDMA commonly adopted Gold and Kasami codes. (Baccarelli and Biagi, 2003) focused on channel coding and proposed a coding strategy for improving Pr_b in UWB asynchronous networks.

Performance analyses of coded schemes for THMA-UWB are reported in (Fourouzan et al., 2002).

Following the seminal paper by (Pursley, 1977), the characterization of MUI in DS-CDMA systems has been extensively investigated. Among the published literature, we refer the interested reader to (Geraniotis and Pursley, 1982) and (Lehnert and Pursley, 1987).

References

Baccarelli, E., and M. Biagi, "An adaptive codec for multi-user interference mitigation for UWB-based WLANs," *IEEE International Conference on Communications*, Volume: 3 (May 2003), 2020–2024.

Benedetto, S., G. De Vincentiis, and A. Luvison, "Application to Gauss Quadrature rules to Digital Communication problems," *IEEE Transactions on Communications*, Volume: 21, Issue: 10 (October 1973a; legacy pre-1988), 1159–1165.

Benedetto, S., G. De Vincentiis, and A. Luvison, "Error Probability in the Presence of Intersymbol Interference and Additive Noise," *IEEE Transactions on Communications*, Volume: 21, Issue: 10 (March 1973b; legacy pre-1988), 181–190.

Crespo, P.M., M.L. Honig, and J.A. Salehi, "Spread-Time Code-Division Multiple Access," *IEEE Transactions on Communications*, Volume: 43, Issue: 6 (June 1995), 2139–2148.

Di Benedetto, M.-G., and B.R. Vojcic, "Ultra Wide Band (UWB) Wireless Communications: A Tutorial," *Journal of Communication and Networks, Special Issue on Ultra-Wideband Communications*, Volume: 5, Issue: 4 (December 2003), 290–302.

Di Benedetto, M.-G., L. De Nardis, M. Junk, and G. Giancola, "(UWB)^2: Uncoordinated, Wireless, Baseborn medium access for UWB communication networks," in press in *Mobile Networks and Applications special issue on WLAN Optimization at the MAC and Network Levels* (2nd quarter 2004).

Durisi, G., and G. Romano, "On the Validity of Gaussian Approximation to Characterize the Multiuser Capacity of UWB TH-PPM," *IEEE Conference on Ultra Wideband Systems and Technologies* (May 2002), 157–161.

Durisi, G., and S. Benedetto, "Performance Evaluation of TH-PPM UWB Systems in the Presence of Multiuser Interference," *IEEE Communications Letters*, Volume: 7, Issue: 5 (May 2003), 224–226.

Durisi, G., A. Tarable, J. Romme, and S. Benedetto, "A General Method for Error Probability Computation of UWB Systems for Indoor Multiuser Communications," *Journal of Communication and Networks, Special Issue on Ultra-Wideband Communications*, Volume: 5, Issue: 4 (December 2003), 354–364.

Erseghe, T., and N. Bramante, "Pseudo-Chaotic Encoding Applied to Ultra-Wide-Band Impulse Radio," *IEEE Vehicular Technology Conference*, VTC2002-Fall, Volume 3 (September 2002), 1711–1715.

Foerster, J.R., "The performance of a Direct-Sequence Spread Ultra-Wideband System in the Presence of Multipath," *IEEE Conference on Ultra Wideband Systems and Technologies* (May 2002), 87–91.

Fourouzan, A.R., M. Nasiri-Kenari, and J.A. Salehi, "Performance Analysis of Time-Hopping Spread-Spectrum Multiple-Access Systems: Uncoded and Coded Schemes," *IEEE Transactions on Wireless Communications*, Volume: 1, Issue: 4 (October 2002), 671–681.

Geraniotis, E.A., and M.B. Pursley, "Error Probability for Direct-Sequence Spread Spectrum Multiple-Access Communications-Part II: Approximations," *IEEE Transactions on Communications*, Volume: 30, Issue: 5 (May 1982), 985–995.

Giancola, G., L. De Nardis L., and M.-G. Di Benedetto, "Multi User Interference in Power-Unbalanced Ultra Wide Band systems: Analysis and Verification," *IEEE Conference on Ultra Wideband Systems and Technologies* (November 2003), 325–329.

Golub, G.H., and J.H. Welsh, "Calculation of gauss quadrature rules," *Math. Comp.*, Volume: 23 (April 1969), 221–230.

Gui, X., and T.S. Ng, "Performance of asynchronous orthogonal multicarrier CDMA system in frequency selective fading channel," *IEEE Transactions on Communications*, Volume: 47, Issue: 7 (July 1999), 1084–1091.

Hanzo, L., M. Münster, B.J. Choi, and T. Keller, *OFDM and MC-CDMA for Broadband Multi-User Communications, WLANs and Broadcasting*, Chichester, West Sussex, England: John Wiley and Sons, Inc. (2003).

Hu, B., and N.C. Beaulieu, "Precise Bit Error Rate of TH-PPM UWB Systems in the Presence of Multiple Access Interference," *IEEE Conference on Ultra Wideband Systems and Technologies* (November 2003), 106–110.

Hu, B., and N.C. Beaulieu, "Exact Bit Error Rate Analysis of TH-PPM UWB Systems in the Presence of Multiple-Access Interference," *IEEE Communications Letters*, Volume: 7, Issue: 12 (December 2003), 572–574.

Iacobucci, M.S., and M.-G. Di Benedetto, "Multiple Access Design for Impulse Radio Communication Systems," *ICC 2002, IEEE International Conference*, Volume: 2 (2002), 817 – 820.

Iacobucci, M.S., and M.-G. Di Benedetto, "Computer Method for Pseudo-Random Codes Generation," National Italian Patent RM2001A000592 (2001).

IEEE Std 802.15.3-2003, *MAC standard*, Available at *http://www.ieee.org/, 2003*.

Jones, R.A., D.H. Smith, and S. Perkins, "Assignment Spreading Codes in DS-CDMA UWB Systems," *IEEE Conference on Ultra Wideband Systems and Technologies* (November 2003), 359–363.

Laney, D.C., G.M. Maggio, F. Lehmann, and L. Larson, "Multiple access for UWB Impulse Radio with Pseudochaotic Time Hopping," *IEEE Journal on Selected Areas in Communications*, Volume: 20, Issue: 9 (December 2002), 1692–1700.

Lehnert, J.S., and M.B. Pursley, "Error Probabilities for Binary Direct-Sequence Spread Spectrum Communications with Random Signature Sequences," *IEEE Transactions on Communications*, Volume: 35, Issue: 1 (January 1987), 87–98.

LeMartret, C.J., and G.B. Giannakis, "All Digital Impulse Radio with Multiuser Detection for Wireless Cellular Systems," *IEEE Transactions on Communications*, Volume: 50, Issue: 9 (September 2002), 1440–1450.

Maggio, G.M., N. Rulkov, and L. Reggiani, "Pseudo-chaotic Time Hopping for UWB Impulse Radio," *IEEE Transactions on Circuits and Systems I: Fundamental Theory and Applications*, Volume: 48, Issue: 12 (December 2001), 1424–1435.

Maggio, G.M., D. Laney, F. Lehmann, and L. Larson, "A multi-access scheme for UWB radio using pseudo-chaotic time hopping," *IEEE Conference on Ultra Wideband Systems and Technologies* (May 2002), 225–229.

Pursley, M.B., "Performance evaluation for Phase-Coded Spread Spectrum Multiple-Access Communication-Part I: System analysis," *IEEE Transactions on Communications*, Volume: 25, Issue: 8 (August 1977), 795–799.

Ravi, K.V., "Comparison of multiple-accessing schemes for mobile communication systems," *IEEE International Conference on Personal Wireless Communications* (August 1994), 152–156.

Sabattini, M., E. Masry, and L.B. Milstein, "A Non-Gaussian Approach to the Performance Analysis of UWB TH-BPPM Systems," *IEEE Conference on Ultra Wideband Systems and Technologies* (November 2003), 52–55.

Scholtz, R.A., "Multiple access with Time-Hopping Impulse Modulation," *IEEE Military Communications Conference*, Volume: 2 (October 1993), 447–450.

Sunay, M.O., and P.J. McLane, "Sensitivity of a DS-CDMA System with Long PN Sequences to Synchronization Errors," *IEEE International Conference on Communications*, Volume: 2 (June 1995), 1029–1035.

Sunay, M.O., and P.J. McLane, "Calculating error probabilities for DS-CDMA systems: when not to use the Gaussian approximation," *IEEE Global Telecommunications Conference*, Volume: 3 (November 1996), 1744–1749.

Verdu, S., *Multiuser Detection*, 1st Edition, New York: Cambridge University Press (1998).

Vojcic, B., and R.L. Pickholtz, "Direct-Sequence Code Division Multiple Access for Ultra-Wide Bandwidth Impulse Radio," *Proc. of MILCOM2003, Boston, Massachusetts* (Oct. 13–16, 2003), 898–902.

Win, M.Z., and R.A. Scholtz, "Ultra-wide Bandwidth Time-Hopping Spread-Spectrum Impulse Radio for Wireless Multiple Access Communications," *IEEE Transactions on Communications*, Volume: 48, Issue: 4 (April 2000), 679–691.

APPENDIX 9.A

Function 9.1 Evaluation of σ_m^2 for PPM Systems

Function 9.1 is composed of one single step, which basically evaluates the σ_M^2 term by executing the double integration expressed in Eq. (9–25). In the first part of the code, the function normalizes in energy the input pulse and determines the correlation mask $v(t)$. This operation is performed by using Function 8.14 (see Appendix 8.A), which shifts the original pulse by a quantity equal to the PPM shift. Then, both vectors representing the pulse and correlation mask are extended in time by zero padding. This operation is necessary for computing the cross-correlation between the mask and a time-shifted version of the pulse. The double integration in Eq. (9–25) is performed in two steps. First, a loop is introduced for computing and storing in memory the different results generated by the internal integration when all the possible positions in time of the pulse $p_0(t)$ are considered. Then, the external integration is performed by summing up all the values resulting from the above loop.

```
%
% Function 9.1: "cp0901_sm2_PPM"
%
% Evaluates the term SIGMAm^2 ('sm2') for an input 'pulse'
% sampled at 'fc'
% 'PPMshift' is the value in seconds of the PPM shift
%
% Programmed by Guerino Giancola
%

function [sm2] = cp0901_sm2_PPM(pulse,PPMshift,fc)

% ----------------------------------
% Step One - Evaluation of SIGMAm^2
% ----------------------------------

dt = 1 / fc;
PPM_samples = floor(PPMshift/dt);
pulse_samples = length(pulse);


% energy normalization
Ep = sum((pulse.^2).*dt);
```

```
pulse = pulse./sqrt(Ep);


p0 = zeros(1,PPM_samples+pulse_samples);
p0(1:length(pulse)) = pulse;
p1 = cp0804_signalshift(p0,fc,PPMshift);
% single pulse correlator mask for PPM
v = p0-p1;

LM = length(v);
LS = LM + 2*pulse_samples;

pa = zeros(1,LS);
pb = zeros(1,LS);

pa(1:pulse_samples) = pulse;
pb(1:LM) = v;


for tau = 0 : (LS-1)

    pc = cp0804_signalshift(pa,fc,tau*dt);

    % result of internal integration
    I(tau+1) = (sum((pc.*pb).*dt))^2;

end % for tau = 0 : (LS-1)

sm2 = sum(I.*dt);
```

Function 9.2 Evaluation of σ_m^2 for PAM Systems

Function 9.2 is composed of one single step, which basically evaluates the σ_M^2 term by executing the double integration expressed in Eq. (9–43). In the first part of the code, the function normalizes in energy the input pulse and generates two copies of the resulting signal, which are extended in time by zero padding. This operation is necessary for computing the autocorrelation of the pulse. The double integration in Eq. (9–43) is performed in two steps. First, a loop is introduced for computing and storing in memory the different results generated by the autocorrelation when all possible positions in time of the shifted pulse $p_0(t)$ are taken into account. Then, the external integration is performed by summing up all the values resulting from the above loop.

```
%
% Function 9.2: "cp0901_sm2_PAM"
%
% Evaluates the term SIGMAm^2 ('sm2') for an input 'pulse'
% sampled at 'fc'
%
% Programmed by Guerino Giancola
%

function [sm2] = cp0901_sm2_PAM(pulse,fc)

% ----------------------------------
% Step One - Evaluation of SIGMAm^2
% ----------------------------------

dt = 1 / fc;                         % sampling period
pulse_samples = length(pulse);

% energy normalization
Ep = sum((pulse.^2).*dt);
pulse = pulse./sqrt(Ep);

LS = 3*pulse_samples;

pa = zeros(1,LS);
pb = zeros(1,LS);

pa(1:pulse_samples) = pulse;
pb(1:pulse_samples) = pulse;
```

```
for tau = 0 : (LS-1)

    pc = cp0804_signalshift(pa,fc,tau*dt);

    % result of internal integration
    I(tau+1) = (sum((pc.*pb).*dt))^2;

end % for tau = 0 : (LS-1)

sm2 = sum(I.*dt);
```

Function 9.3 Probability of Error for 2PPM-TH-UWB with MUI and AWGN

Function 9.3 is composed of three steps. Step One evaluates all the constant terms that are required for the computation of $\Pr_b$ as expressed by Eq. (9–31): the value of the autocorrelation $R_0(t)$ evaluated in correspondence to the PPM shift, the σ_M^2 value, and the result of the summation of the ratios between interfering energies and useful energy at the receiver. Step Two uses the above terms for evaluating the different values of SNR_n and SIR. Finally, Step Three applies the expression in Eq. (9–31) for determining $\Pr_b$ for all the input E_b/N_0 values.

```
%
% Function 9.3 : "cp0901_MUIBER_2PPM"
%
% Evaluates the theoretical probability of error
% for a 2PPM system in AWGN channels under the
% standard Gaussian approximation
%
% 'ebno'      is a vector containing the values in dB of the
%             ratio Eb/No
% 'erx0'      is the energy of the useful signal
% 'erxMUI'    is a vector containing the received energies
%             of the interfering users
% 'pulse'     is the waveform of the basic pulse
% 'Rb'        is the user bit rate [b/s]
% 'PPMshift'  is the value of the PPM shift [s]
% 'fc'        is the sampling time [Hz]
% 'gamma_r'   represents the ratio (Ts/(Tb/Ns))
%

function [BER] = cp0901_MUIBER_2PPM(ebno,erx0,erxMUI,...
   pulse,Rb,PPMshift,fc,gamma_r)

% ----------------------------------------------------
% Step One - Evaluation of the required constant terms
% ----------------------------------------------------

% R0(epsilon) [R]

  dt = 1 / fc;                          % sampling period
  PPM_samples = floor(PPMshift/dt);
  pulse_samples = length(pulse);
  Ep = sum((pulse.^2).*dt);
```

```
    pulse = pulse./sqrt(Ep);            % energy normalization
    p0 = zeros(1,PPM_samples+pulse_samples);
    p0(1:length(pulse)) = pulse;
    p1 = cp0804_signalshift(p0,fc,PPMshift);
    R = sum((p0.*p1).*dt);

  % (Sigma_m)^2 [sm2]

    sm2 = cp0901_sm2_PPM(pulse,PPMshift,fc);

  % MUI energy summation [EMUI]

    EMUI = sum(erxMUI./erx0);

  % ----------------------------------------
  % Step Two - Evaluation of SIR and SNRn
  % ----------------------------------------

  SIR = (((1-R)^2)*gamma_r)/(sm2*Rb*EMUI);

  EBN0 = 10.^(ebno/10);
  SNRn = EBN0.*(1-R);

  % --------------------------------------
  % Step Three - Performance evaluation
  % --------------------------------------

  SNRref = 1./((1./SNRn)+(1/SIR));

  BER = 0.5.*erfc(sqrt(SNRref./2));
```

Function 9.4 Probability of Error for 2PAM-TH-UWB with MUI and AWGN

Function 9.4 is composed of three steps. Step One evaluates all the constant terms that are required for the computation of Pr_b as expressed by Eq. (9–46): the σ_M^2 value, and the result of the summation of the ratios between interfering energies and useful energy at the receiver. Step Two uses the above terms for evaluating the different values of SNR_n and the SIR. Finally, Step Three applies the expression in Eq. (9–46) for determining Pr_b for all the input E_b/N_0 values.

```
%
% Function 9.4 : "cp0901_MUIBER_2PAM"
%
% Evaluates the theoretical probability of error
% for a 2PAM system in AWGN channels under the
% standard Gaussian approximation
%
% 'ebn0'     is a vector containing the values in dB of the
% ratio Eb/No
% 'erx0'     is the energy of the useful signal
% 'erxMUI'   is a vector containing the received energies
% of the interfering users
% 'pulse'    is the waveform of the basic pulse
% 'Rb'       is the user bit rate [b/s]
% 'fc'       is the sampling time [Hz]
% 'gamma_r'  represents the ratio (Ts/(Tb/Ns))
%


function [BER] = ...
   cp0901_MUIBER_2PAM(ebno,erx0,erxMUI,pulse,Rb,fc,gamma_r)

% ----------------------------------------------------
% Step One - Evaluation of the required constant terms
% ----------------------------------------------------

% (Sigma_m)^2 [sm2]

  sm2 = cp0901_sm2_PAM(pulse,fc);

% MUI energy summation [EMUI]

  EMUI = sum(erxMUI./erx0);
```

```
% ------------------------------------
% Step Two - Evaluation of SIR and SNRn
% ------------------------------------

SIR = (gamma_r)/(sm2*Rb*EMUI);

EBN0 = 10.^(ebno/10);
SNRn = EBN0.*2;

% ------------------------------------
% Step Three - Performance evaluation
% ------------------------------------

SNRref = 1./((1./SNRn)+(1/SIR));

BER = 0.5.*erfc(sqrt(SNRref./2));
```

Function 9.5 Probability of Error Based on PC Probability

Function 9.5 is composed of one single step, which first computes PC probability and then bit error probability. Note that the binomial coefficients in Eq. (9–50) are evaluated by using the MATLAB function `nchoosek(n,k)`, which computes the number of combinations of `n` elements taken in groups of `k` elements.

```
%
% Function 9.5: "cp0902_prbcoll"
%
% Evaluates the probability of error
% on the basis of the probability
% of PC
%
% 'Nu' is the number of users
% 'Ns' is the number of pulses per bit
% 'Tm' is the time duration of the pulse
% 'Ts' is the average pulse repetition period
%
% 'Prb' is the estimated bit error probability
%
% Programmed by Guerino Giancola
%

function [Prb] = cp0902_prbcoll(Nu,Ns,Tm,Ts)

% ----------------------------------------------------
% Step One - Evaluation of the bit error probability
% ----------------------------------------------------


% probability of collision
pr_pulsecollision = 1 - exp((2-2*Nu)*Tm/Ts);

% probability of error for the single pulse
pr_pulseerror = 0.5 * pr_pulsecollision;

% probability of error for the bit
n_min = ceil(Ns/2);
n_max = Ns;
Prb = 0;
for n = n_min : n_max
```

```
    tmp = nchoosek(Ns,n)*(pr_pulseerror^n)*...
     ((1-pr_pulseerror)^(n_max-n));
    Prb = Prb + tmp;

end % for n = n_min = n_max
```

Function 9.6 Effective Pulse Duration

Function 9.6 evaluates the effective pulse duration `Tm` for an input waveform `pulse` sampled at `fc` in Hz. The `Tm` value is determined by first evaluating the new waveform `eff_pulse`, which contains only a fraction `pE` of the original energy, and then by measuring its time duration. Function 9.6 is composed of one single step, which evaluates `Tm` by iteratively decreasing the length of the original waveform until the target energy value is reached.

```
%
% Function 9.6: "cp0902_effpulse"
%
% Evaluates the effective pulse duration of the
% input waveform 'pulse' sampled with frequency 'fc'
% 'pE' is the fraction of the original energy, which must
% be considered for determining the effective duration
%
% The function returns the waveform 'eff_pulse',
% which contains the fraction 'pE' of the original energy
% and the time duration 'Tm' in seconds of 'eff_pulse'
%
% Programmed by Guerino Giancola
%

function [eff_pulse,Tm] = cp0902_effpulse(pulse,fc,pE)

% ----------------------------------------------------
% Step One - Evaluation of the effective pulse duration
% ----------------------------------------------------

dt = 1/fc;
T = length(pulse);
E = sum((pulse.^2).*dt);
Eeff = E*pE;

eff_pulse = pulse;
E0 = E;
T0 = T;
while (E0>Eeff)&(T0>2)

    T0 = T0 - 2;
    tmp = eff_pulse(2:length(eff_pulse)-1);
    eff_pulse = tmp;
    E0 = sum((eff_pulse.^2).*dt);
```

```
end % while E0 > Eeff

Tm = length(eff_pulse)*dt;
```

CHAPTER 10

UWB Ranging and Positioning

Ranging and positioning are very likely to play a major role in the advanced design of wireless communication networks in the coming years. This chapter is devoted to the analysis of ranging and positioning algorithms and related protocols. We will review basic principles and present a few examples of positioning systems and protocols, such as GPS. We will end the chapter with an application example consisting of networks of sensors provided with positioning information called MAgiC world.

10.1 RANGING VS. POSITIONING

The terms "ranging" and "positioning", as well as "localization", are used in current literature in a very flexible manner; there is no common understanding or convention on the exact meaning of these terms. It is thus important to establish a set of definitions that will be used from this moment on regarding, in particular, ranging, node-centered positioning, relative positioning, and absolute or geographical positioning.

Ranging is defined as the action of computing the distance of a target node from a reference node. A reference node wanting to obtain ranging information regarding a target node in the network can acquire this information by establishing a peer-to-peer communication link with the target node. This communication link is used for evaluating parameter values that are used at the reference node for estimating its distance from the target node. The parameters are typically based either on the evaluation of channel attenuation or on delay of propagation, as will be further described in Section 10.2. Figure 10–1 shows an example of ranging, and introduces the following notation: assuming N2 as the reference node, we indicate the ranging information for target nodes N1 and N3 as $RANG_{N2}(N1)$, $RANG_{N2}(N3)$.

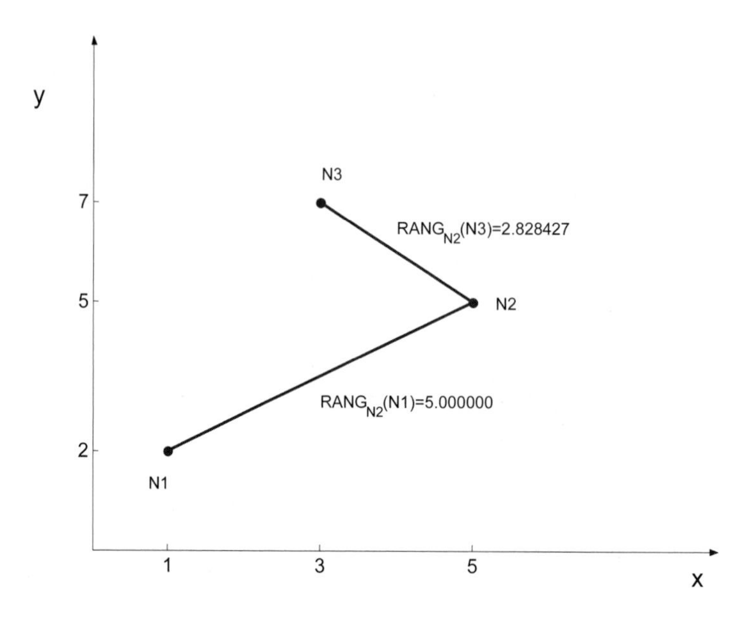

Figure 10–1 Example of ranging with node N2 as reference node.

Node-centered positioning is defined as the action of computing the positions of a set of target nodes with respect to a reference node. Node-centered positioning can also be obtained based on peer-to-peer connections, provided that both distance and angle information about each target node is obtainable at the reference node. Note that any node can play the role of reference node and compute the position of other nodes in its own reference system, but each node associates a different set of coordinates to the same target node. Figure 10–2 shows an example of node-centered positioning with two different reference nodes.

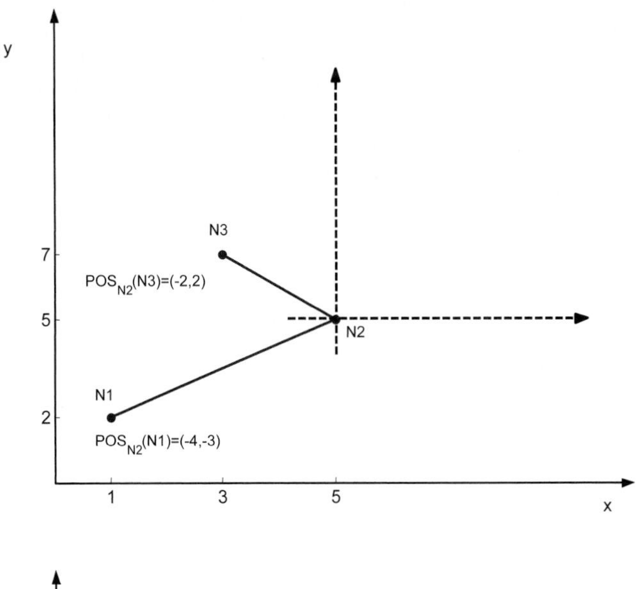

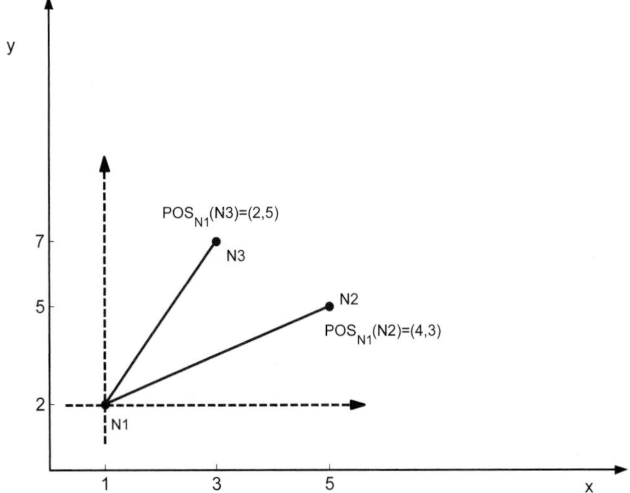

Figure 10–2 Example of node-centered positioning with N1 and N2 as reference nodes on the upper and lower plot, respectively.

The following notation is introduced: assuming N2 as reference node, we indicate the node-centered positioning information related to target nodes N1 and N3 as $POS_{N2}(N1)$, $POS_{N2}(N3)$.

Relative positioning indicates the action of computing the position of a set of nodes with respect to a common system of coordinates. The key difference with node-centered positioning is that all nodes share the same reference system and thus each node is univocally associated with a unique set of coordinates. The adoption of the same coordinate system in all nodes requires organization of the set of nodes into a network, allowing information exchange among more than two nodes, following the rules set by dedicated algorithms for the selection of the common coordinate system and translation of the coordinates of each node. Figure 10–3 gives an example of relative positioning.

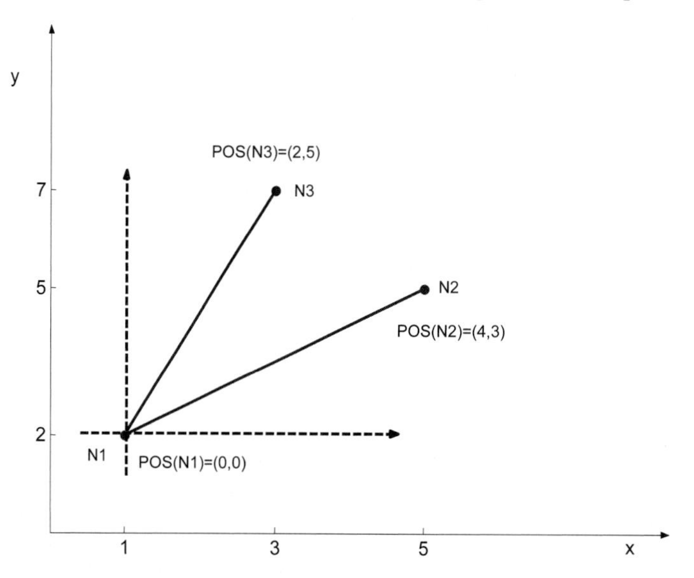

Figure 10–3 Example of relative positioning with a reference coordinate system centered on N1.

The origin of the reference coordinate system can be chosen arbitrarily, and may not coincide with the position of a node in the network. In most cases, however, the reference coordinate system is obtained by evolving from the node-centered coordinate system. As an example, in Figure 10–3 the reference coordinate system coincides with the node-centered coordinate system of node N1, which assumes coordinates POS(N1) = (0,0). Protocols for the selection of the reference coordinate system will be presented in Section 10.4.

A special case of relative positioning corresponds to the adoption of a reference coordinate system coincident with the global coordinate system, with coordinates given in terms of latitude and longitude: This specific solution is referred to as *absolute* (or *geographical*) *positioning* since the coordinates associated to each node are unique worldwide.

Both node-centered and relative positioning require a prior ranging procedure for retrieving distances. The degree of accuracy in distance estimation thus has an impact on the accuracy with which positioning can be achieved. The distance estimation technique must be selected according to requirements imposed by the application layer.

10.2 RANGING

Ranging consists of estimating the distance D between a transmitter and a receiver. Given a transmitted signal $s(t)$, the corresponding received signal writes:

$$r(t) = h(t) * s(t) + n(t) \tag{10–1}$$

where $h(t)$ is the channel impulse response and $n(t)$ is thermal noise. Let us assume for now that the signal propagates over a perfect channel, characterized, as illustrated in Chapter 8, "Propagation over a Multi-Path-Affected UWB Radio Channel," by an impulse response given by:

$$h(t) = A(D)\delta(t - \tau(D)) \tag{10–2}$$

The received signal can be written as follows:

$$r(t) = A(D)s(t - \tau(D)) + n(t) \tag{10–3}$$

Equation (10–3) indicates that distance D can be estimated from either attenuation $A(D)$ or delay $\tau(D)$. The use of $A(D)$ versus $\tau(D)$ determines the method for estimating distance: Received Signal Strength Indicator (RSSI) versus Time of Arrival (TOA).

The RSSI technique is based on the emission at the transmitter side of a signal using a fixed reference power known at the receiver. The receiver measures the power of the received signal and derives the distance from the measured attenuation. Since the relation between distance and attenuation depends on channel behavior, an accurate propagation model is required to reliably estimate distance. Terminal mobility and unpredictable variations in channel behavior can be real problems (Hallberg et al., 2003). An additional source of error with RSSI is introduced by the limitations of the hardware. The transmitter, in particular, must be finely tuned on signal emissions at the reference power; a discrepancy between transmitted and reference power reflects in a systematic bias in distance estimation (Hightower et al., 2001). In conclusion, RSSI is not a very accurate method, and its adoption is confined to applications requiring coarse ranging.

The TOA technique computes distance based on the estimation of the propagation delay between transmitter and receiver. TOA is the most commonly used distance estimation method in the radar field, and for this reason, the terms "TOA" and "ranging" are often interchanged.

Delay estimation is a key topic in wireless communications since it is required for achieving symbol synchronization between transmitter and receiver as discussed in Chapter 8. Several solutions to this problem are available in the literature. Most of these derive from the ML estimator, which is defined as follows.

Under the hypothesis of AWGN $n(t)$, the ML estimate of delay τ, called $\hat{\tau}_{ML}$, corresponds to the τ value, which minimizes the ML function. The $\hat{\tau}_{ML}$ can be expressed as follows (Proakis, 1995):

$$\hat{\tau}_{ML}(r) = \underset{\tau \in \mathbb{R}}{\operatorname{argmin}} \left(e^{-\frac{1}{N_0} \int_{T_{obs}} (r(t) - s(t-\tau))^2 dt} \right) \tag{10–4}$$

where N_0 is the bilateral PSD of the noise and T_{obs} is the observation interval over which the estimation is performed.

A common synchronization scheme that approximates the ML estimator is the so-called early-late gate synchronizer. To understand how this estimator works, let us consider a rectangular pulse of generic duration T, called $s(t)$, as represented in Figure 10–4.

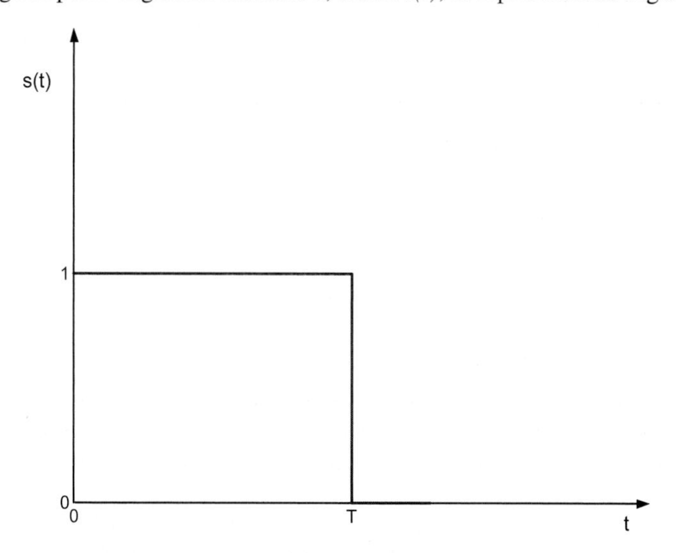

Figure 10–4 Rectangular pulse of duration *T*.

As is well-known, the output $R_s(\xi)$ of a filter matched to an input $s(t)$ is a time-shifted version of the autocorrelation of $s(t)$, and has its maximum value in $\xi = T$, as shown in Figure 10–5.

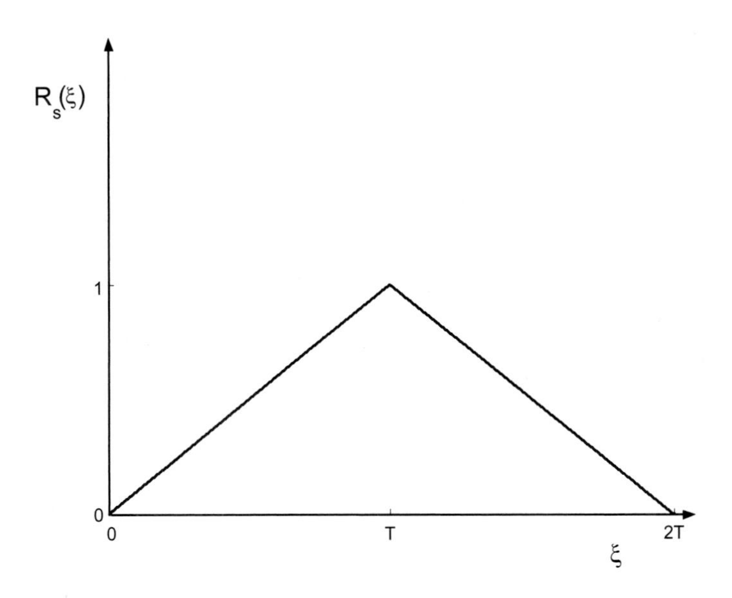

Figure 10–5 Output of the matched filter $R_s(\xi)$ for the signal of Figure 10–4.

Note that $R_s(\xi)$ can be represented as the output of the correlation receiver introduced in Chapter 8. The peak is then obtained by sampling $R_s(\xi)$ at $\xi = T$.

The early-late gate synchronizer exploits the symmetry of $R_s(\xi)$, that is, $R_s(T\text{-}\delta) = R_s(T+\delta)$. The synchronizer extracts two values from $R_s(\xi)$ at symmetrical positions around the expected peak value by sampling the output of the integrator of the correlation receiver at $\xi = T\text{-}\delta$ and $\xi = T+\delta$, respectively. The synchronizer then evaluates the following quantity:

$$\Delta R = R_s(T-\delta) - R_s(T+\delta) \tag{10–5}$$

When perfect synchronization is achieved, the two values $R_s(T\text{-}\delta)$ and $R_s(T+\delta)$ are identical, and $\Delta R = 0$. Conversely, when an unknown delay τ is present in the received signal, the synchronizer extracts two values that are not equal, and $\Delta R \neq 0$. This effect is represented in Figure 10–6.

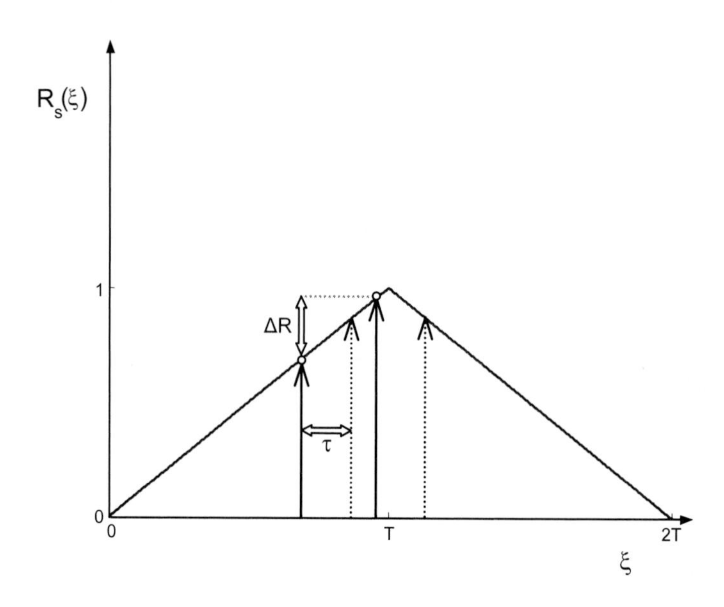

Figure 10–6 Effect of an unknown delay τ on the output of the early-late gate synchronizer ΔR.

In this last case of imperfect synchronization, the synchronizer introduces an additional delay and computes ΔR again. The additional delay is then adjusted until $\Delta R = 0$, using a closed control loop that changes the delay value based on the output of Eq. (10–5). The final additional delay represents the desired estimation of the random delay τ.

Note that when the above scheme is used for synchronization purposes, it compensates for any delay between the received and reference signal. Possible sources of delay include both propagation and misalignments between transmitter and receiver clocks. When applying such a scheme to ranging, it is fundamental to achieve a common time reference between transmitter and receiver, to isolate the delay due to propagation. In the case of radar applications, the common time reference is inherently provided by the fact that transmitter and receiver are physically co-located, since the range to the target is evaluated based on signal echoes. Conversely, for ranging acquisition between different communication devices, a ranging handshake protocol is required between transmitter and receiver to achieve synchronization under a specified accuracy constraint. This topic is widely addressed for the specific case of UWB systems in (Fleming and Kushner, 1997), where several schemes for achieving a common time reference between two UWB devices are proposed, as shown in Figure 10–7 for a selected case.

In Figure 10–7 the reference node N1 sends a packet at time t_0. The target node N2 receives the packet at time $t_0 + \tau$ and collaborates in the ranging procedure by replying with an identical packet sent at time $t_0 + \tau + \Delta$, where Δ is a fixed time delay. Δ is large enough to cover delays due to signal processing at the target node, and is known to both transmitter and receiver. The reference node N1 receives the reply packet at time $t_1 = t_0 + \tau + \Delta + \tau$, and can derive the propagation delay as follows:

$$\tau = \frac{t_1 - t_0 - \Delta}{2} \tag{10–6}$$

Note that this procedure provides ranging only to N1 and not to N2. Either an additional ranging packet from N1 to N2 or an explicit communication of the measured distance by N1 is required to provide N2 with the range.

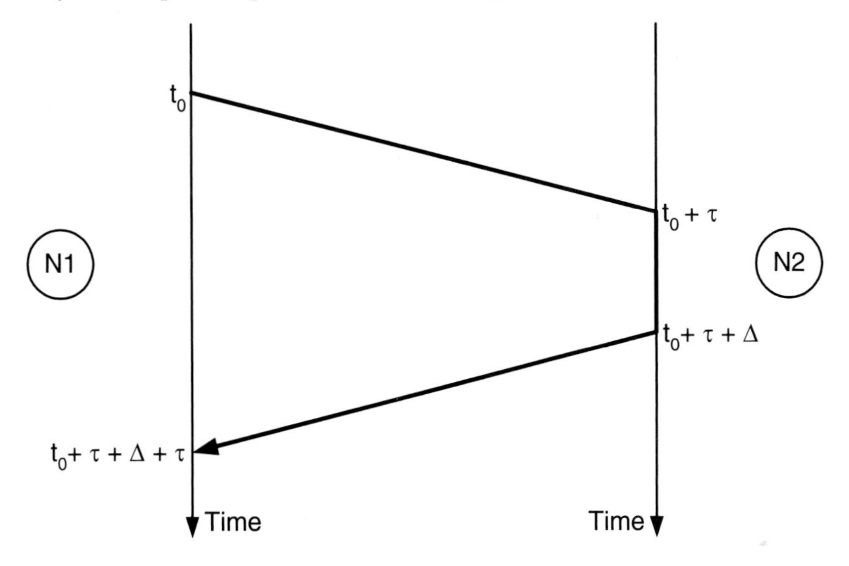

Figure 10–7 Ranging handshake exchange for UWB positioning devices, as proposed in (Fleming and Kushner, 1997).

(Fleming and Kushner, 1997) show that the TOA technique is particularly suited for UWB radio, thanks to the ultra-wide bandwidth. The accuracy of the TOA estimation expressed by the variance of the TOA estimation error $\sigma_{\hat{\tau}}^2$ is in fact related to the bandwidth of the signal and SNR at the receiver. According to the general theory of ML estimators, the lower limit for $\sigma_{\hat{\tau}}^2$ in presence of AWGN is given by the Cramer-Rao lower bound (Urkowitz, 1983):

$$\sigma_{\hat{\tau}}^2 = \frac{N_0}{2\int_{-\infty}^{+\infty} (2\pi f)^2 \left|P(f)\right|^2 df} \tag{10–7}$$

The meaning of this limit and the value it assumes in the case of UWB signals will be investigated in Checkpoint 10–1.

CHECKPOINT 10–1

In this checkpoint, we will further investigate the Cramer-Rao lower bound on the delay estimation error given by Eq. (10–7), and we will evaluate its value for an UWB signal.

Let us suppose that the pulse *p(t)* has a constant bilateral ESD $|P(f)|^2$, that is:

$$|P(f)|^2 = \begin{cases} G_0 & for\ f \in [f_L, f_H] \cup [-f_H, -f_L] \\ 0 & outside \end{cases} \quad (10\text{–}8)$$

Eq.(10–7) can thus be written as follows:

$$\begin{aligned} \sigma_{\hat{\tau}}^2 &= \frac{N_0}{8\pi^2 \int_{-\infty}^{+\infty} f^2 |P(f)|^2 df} \\ &= \frac{N_0}{8\pi^2 2 \int_{f_L}^{f_H} f^2 G_0 df} \\ &= \frac{N_0}{8\pi^2 2G_0 \left[\frac{f^3}{3} \right]_{f_L}^{f_H}} \\ &= \frac{N_0}{\frac{8}{3}\pi^2 2G_0 \left(f_H^3 - f_L^3 \right)} \end{aligned} \quad (10\text{–}9)$$

And one can write:

$$\sigma_{\hat{\tau}}^2 = \frac{N_0}{\frac{8}{3}\pi^2 2G_0\left(f_H - f_L\right)\left(f_H^2 + f_H f_L + f_L^2\right)} = \frac{N_0}{\frac{8}{3}\pi^2 2G_0 B\left(f_H^2 + f_H f_L + f_L^2\right)} \quad (10\text{–}10)$$

Eq. (10–10) shows that the variance in delay estimation is inversely proportional to the signal monolateral bandwidth occupation B and to a term that depends on the lower and upper frequencies f_H and f_L. It can be easily shown that, for a fixed bandwidth B, this term increases as f_H increases.

It is interesting to compute the lower bound given by Eq. (10–10) for an UWB pulse. Consider a pulse that fully exploits the frequency band [3.1–10.6] GHz with the maximum PSD allowed by the FCC and has duration $T_M = 1.33 \cdot 10^{-10}$ s. We obtain the following values: B = 7.5 GHz, f_H = 10.6 GHz, f_L = 3.1 GHz, $2G_0 = 9.86 \cdot 10^{-24}$ Joule/Hz, and $N_0 \cong 2 \cdot 10^{-20}$ W/Hz for a noise temperature T_s equal to FT_0 with F = 7 dB. The limit given by Eq. (10–10) then writes:

$$\sigma_{\hat{\tau}}^2 = 6.63 \cdot 10^{-29} \quad (10\text{–}11)$$

This corresponds to a lower bound average distance estimation error equal to $c\sigma_{\hat{\tau}} = 2.44 \cdot 10^{-6} m$.

The above result provides only a theoretical bound for delay estimation error. Receiver hardware limitations, reduced efficiency in the generation of the transmitted signal, and the presence of multi-path and MUI lead to a far lower accuracy in delay estimation and thus in ranging procedure (Lee and Scholtz, 2003), as will be further discussed in Section 10.5.

CHECKPOINT 10–1

10.3 Positioning

The ranging procedure provides an estimation of distances between pairs of nodes of a given network. Each node Ni knows its distance from all the other nodes. Among these, Ni can choose k reference nodes (N1, …, Nk) to form a reference system, in which it estimates its position as described here.

In the ideal case of free-of-error distance estimations, the *spherical positioning* technique, also known as TOA positioning, is a viable solution. Here, based on the observation that in a tridimensional space (x, y, z), each distance $RANG_{Nj}(Ni)$ between Ni and the reference Nj determines a sphere of radius $D_{ji} = RANG_{Nj}(Ni)$ centered in Nj, position (X_i, Y_i, Z_i) = POS(Ni) of Ni is determined by the intersection of the k spheres of radii $(D_{1i}, \ldots, D_{ki})$ centered in the reference nodes (N1, …, Nk). Since the intersection of four spheres is required for determining a single point in the tridimensional space, at least four reference nodes are required in tridimensional positioning. Note that the introduction of additional reference nodes is not necessary for position computation under the hypothesis of perfect distance estimation, but proves to improve performance in the non-perfect case, which we analyze below. The intersection between the k spheres can be computed by solving the following system of equations:

$$\begin{Bmatrix} \sqrt{(X_1 - X_i)^2 + (Y_1 - Y_i)^2 + (Z_1 - Z_i)^2} \\ \sqrt{(X_2 - X_i)^2 + (Y_2 - Y_i)^2 + (Z_2 - Z_i)^2} \\ \cdots \\ \sqrt{(X_k - X_i)^2 + (Y_k - Y_i)^2 + (Z_k - Z_i)^2} \end{Bmatrix} = \begin{Bmatrix} D_{1i} \\ D_{2i} \\ \cdots \\ D_{ki} \end{Bmatrix} \tag{10–12}$$

with $k \geq 4$.

The same approach can be applied in a bidimensional space (x, y), or a plane. In this case, the position of node Ni, that is, (X_i, Y_i) = POS(Ni), is determined by the intersection of three circles and can be computed by solving the following system of equations:

$$\begin{Bmatrix} \sqrt{(X_1 - X_i)^2 + (Y_1 - Y_i)^2} \\ \sqrt{(X_2 - X_i)^2 + (Y_2 - Y_i)^2} \\ \cdots \\ \sqrt{(X_k - X_i)^2 + (Y_k - Y_i)^2} \end{Bmatrix} = \begin{Bmatrix} D_{1i} \\ D_{2i} \\ \cdots \\ D_{ki} \end{Bmatrix} \tag{10–13}$$

with $k \geq 3$. Figure 10–8 shows an example of spherical positioning in a bidimensional space.

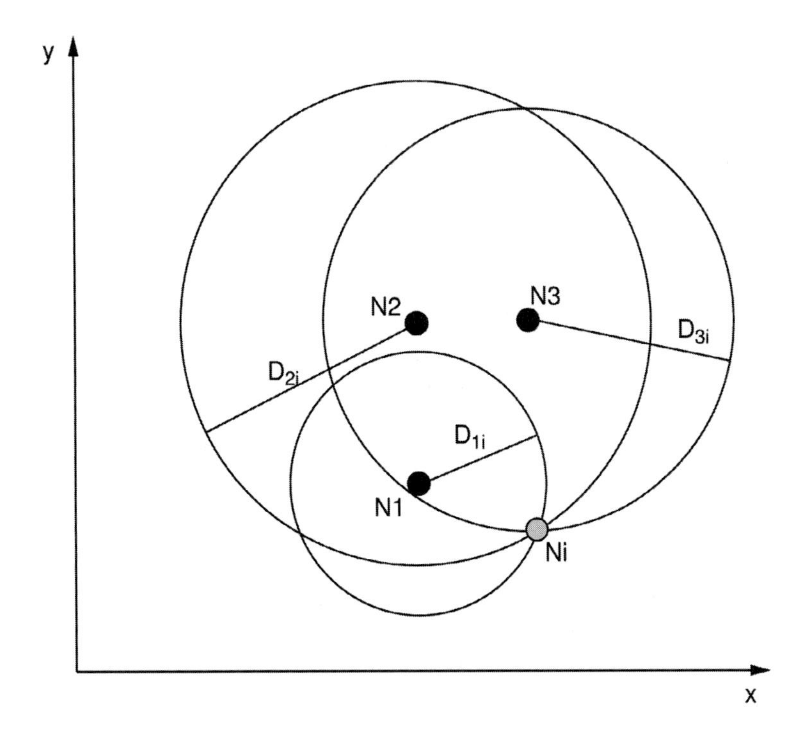

Figure 10–8 Example of spherical positioning of Ni in a bidimensional space with reference nodes N1, N2, and N3.

Spherical positioning can be used only when a common time reference is available to Ni and all reference nodes, that is, a perfect distance estimation can be obtained. This is unfortunately not the case in many practical situations, where misalignments and clock drifts introduce random delays between clocks. Computing exact positioning is still possible provided that a common time reference is available among at least the k reference nodes. The *hyperbolic positioning* technique, also known as Time Difference of Arrival (TDOA), determines the position of Ni based on the difference between times of arrival from the k reference nodes and Ni. Let us assume that the k reference points share a common time reference, and that the clock at Ni is delayed by a time δ with respect to the common time reference. The key point here is that subtraction between times of arrival from different reference nodes removes the delay δ. This can be shown by observing that for a pair (Nn, N(n-1)) of reference terminals, one has:

$$D_{ni} - D_{(n-1)i} = c\left(\tau_{ni} + \delta\right) - c\left(\tau_{(n-1)i} + \delta\right) = c\left(\tau_{ni} - \tau_{(n-1)i}\right) \quad (10\text{–}14)$$

The position of Ni in a tridimensional space is then determined as the intersection of hyperboloids in space, as described by the following equations:

$$\begin{Bmatrix} \sqrt{(X_2-X_i)^2+(Y_2-Y_i)^2+(Z_2-Z_i)^2}-\sqrt{(X_1-X_i)^2+(Y_1-Y_i)^2+(Z_1-Z_i)^2} \\ \sqrt{(X_3-X_i)^2+(Y_3-Y_i)^2+(Z_3-Z_i)^2}-\sqrt{(X_2-X_i)^2+(Y_2-Y_i)^2+(Z_2-Z_i)^2} \\ \cdots \\ \sqrt{(X_k-X_i)^2+(Y_k-Y_i)^2+(Z_k-Z_i)^2}-\sqrt{(X_{k-1}-X_i)^2+(Y_{k-1}-Y_i)^2+(Z_{k-1}-Z_i)^2} \end{Bmatrix} = \begin{Bmatrix} D_{2i}-D_{1i} \\ D_{3i}-D_{2i} \\ \cdots \\ D_{ki}-D_{(k-1)i} \end{Bmatrix} \quad (10\text{–}15)$$

with $k \geq 4$.

In a bidimensional space, one has:

$$\begin{Bmatrix} \sqrt{(X_2-X_i)^2+(Y_2-Y_i)^2}-\sqrt{(X_1-X_i)^2+(Y_1-Y_i)^2} \\ \sqrt{(X_3-X_i)^2+(Y_3-Y_i)^2}-\sqrt{(X_2-X_i)^2+(Y_2-Y_i)^2} \\ \cdots \\ \sqrt{(X_k-X_i)^2+(Y_k-Y_i)^2}-\sqrt{(X_{k-1}-X_i)^2+(Y_{k-1}-Y_i)^2} \end{Bmatrix} = \begin{Bmatrix} D_{2i}-D_{1i} \\ D_{3i}-D_{2i} \\ \cdots \\ D_{ki}-D_{(k-1)i} \end{Bmatrix} \quad (10\text{–}16)$$

with $k \geq 3$. Figure 10–9 shows an example of hyperbolic positioning in a bidimensional space.

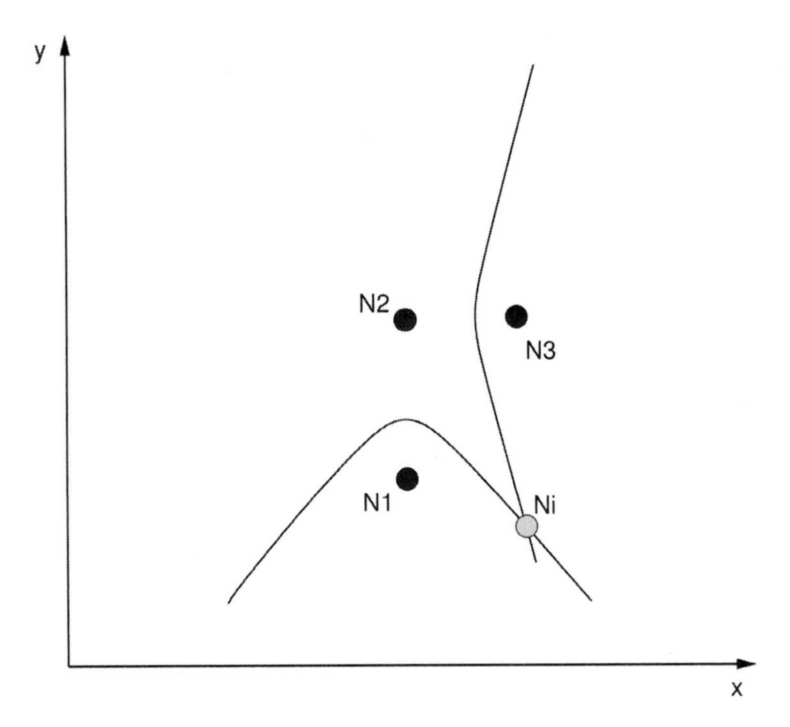

Figure 10–9 Example of hyperbolic positioning of node Ni in a bidimensional space with reference nodes N1, N2, and N3.

Hyperbolic positioning requires an accurate common time reference between the reference nodes, but does not rely on precise synchronization between reference nodes and the target node. It is particularly indicated for infrastructure-based networks, in which coordination between infrastructure nodes is relatively easy to achieve and maintain. For this reason, this technique was proposed for positioning purposes in cellular networks, where base-stations can play the role of reference nodes and allow a mobile node to derive its own position (Drane et al., 1998).

The spherical and hyperbolic positioning techniques presented require error-free ranging information to provide a solution. As shown in Checkpoint 10–1, however, thermal noise can introduce errors in ranging, leading to imperfect estimations of distance between nodes. In this case, the analytical solution of Eqs. (10–12) and (10–15) may not exist, as shown in Figure 10–10 for spherical positioning.

The effect of errors in ranging estimations on the accuracy of positioning can be reduced by adopting minimization procedures such as the Least Square Error (LSE). For this purpose, following the approach presented in (Savarese, 2001), it is convenient to rewrite Eq. (10–12) into a set of equations that are linear in POS(Ni):

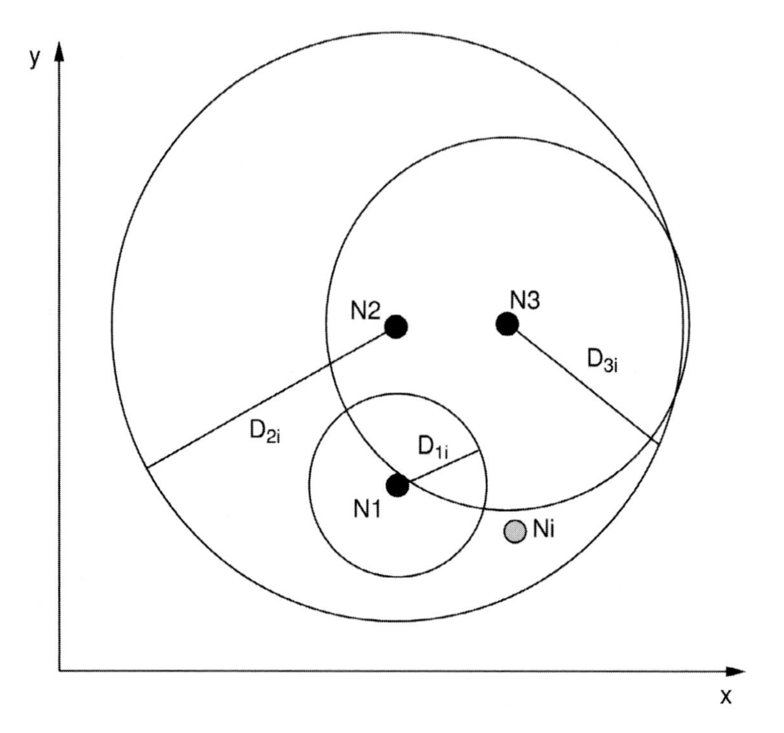

Figure 10–10 Effect of ranging error on spherical positioning of node Ni in a bidimensional space with reference nodes N1, N2, and N3.

$$\overline{A}\overline{I} = \overline{b} \tag{10–17}$$

with

$$\overline{A} = -2\begin{bmatrix} (X_1 - X_k) & (Y_1 - Y_k) & (Z_1 - Z_k) \\ (X_2 - X_k) & (Y_2 - Y_k) & (Z_2 - Z_k) \\ \cdots & \cdots & \cdots \\ (X_{k-1} - X_k) & (Y_{k-1} - Y_k) & (Z_{k-1} - Z_k) \end{bmatrix} \tag{10–18}$$

$$\overline{I} = \begin{bmatrix} X_i \\ Y_i \\ Z_i \end{bmatrix} \tag{10–19}$$

and

$$\bar{b} = \begin{bmatrix} D_{1i}^2 - D_{ki}^2 - X_1^2 + X_k^2 - Y_1^2 + Y_k^2 - Z_1^2 + Z_k^2 \\ D_{2i}^2 - D_{ki}^2 - X_2^2 + X_k^2 - Y_2^2 + Y_k^2 - Z_2^2 + Z_k^2 \\ \dots \\ D_{(k-1)i}^2 - D_{ki}^2 - X_{(k-1)}^2 + X_k^2 - Y_{(k-1)}^2 + Y_k^2 - Z_{(k-1)}^2 + Z_k^2 \end{bmatrix} \quad (10\text{–}20)$$

with $k \geq 4$.

The system defined by Eq. (10–17) can be solved in the sense of LSE minimization.

As already anticipated, in the case of positioning with errors in estimating of ranges, the adoption of a redundant set of ranging measurements is helpful in reducing the variance of the positioning error, as will be shown in the following checkpoint.

CHECKPOINT 10–2

In this checkpoint, we will evaluate the accuracy of LSE for the problem defined by Eq. (10–17) as a function of the degree of error in ranging measurements and of the number of reference nodes. We will analyze the problem in bidimensional space to allow graphical representation of results. Errors are modeled as values extracted from a white Gaussian process with zero mean.

Three MATLAB functions are required: Function 10.1, which generates the positions of the nodes; Function 10.2, which selects the target node and reference nodes; and Function 10.3, which determines the LSE solution.

Function 10.1 (see Appendix 10.A) takes as inputs the number of nodes `N`, the side of the square `area_side`, and a flag `G`, which activates the graphical output. The function returns two outputs: a matrix `positions` containing the positions of the nodes, and a second matrix `ranges` with the exact distances between each pair of nodes in the network. The command line for executing Function 10.1 is:

```
[positions, ranges] = ...
    cp1002_create_network(N,area_side,G);
```

Function 10.2 (see Appendix 10.A) receives as input the number of nodes `N` and the desired number of reference nodes `k`. The function returns an integer `Nx` representing the target node, and a vector `Ref` of dimension `k` containing the IDs of the reference nodes. Function 10.2 is invoked with the command:

```
[Nx,Ref] = cp1002_select_nodes(N,k);
```

Function 10.3 (see Appendix 10.A) operates three basic tasks. First, it introduces errors in the range estimations with error values extracted from a zero mean Gaussian process. Then, it estimates the position of the target node by solving the positioning problem in the linear form of Eq. (10–17). Finally, it determines the positioning error defined as the Euclidean distance between estimated and exact positions. Function 10.3 takes as inputs a matrix `positions` containing the positions of the nodes, a matrix `ranges` containing the distances between each pair of nodes, the ID of the target node `Nx`, a vector `Ref` containing the IDs of the reference nodes, the value `sigma_2` of the noise variance, and a flag `G`, which enables the graphical output. Function 10.3 returns matrix `PosNx` representing the estimated position of the target node and the value of the error `ErrNx`. The command line that executes Function 10.3 is:

```
[PosNx,ErrNx] = cp1002_find_LSE_position(positions, ...
   ranges, Nx, Ref, sigma_2, G);
```

Let us start by defining a set of 10 nodes located in random positions in a square of area 50x50 m^2. This task is performed by executing the following commands:

```
N = 10;
area_side = 50;
G = 1;
[positions, ranges] = ...
   cp1002_create_network(N,area_side,G);
```

Figure 10–11 shows the position of the nodes in our example.

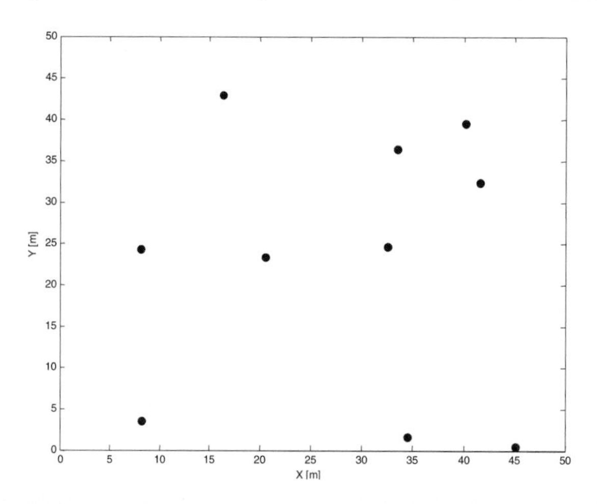

Figure 10–11 Set of 10 nodes in an area 50x50 m^2 generated by Function 10.1

The next step is to choose a target node Nx and k reference nodes in the set of N nodes to test the accuracy of the estimation of POS(Nx) based on the distances from the reference nodes. We start our analysis by considering k = 3, that is, the minimum number of reference nodes required to compute POS(Nx). The case of redundant reference nodes will be analyzed in the last part of the checkpoint.

The selection of Nx and the reference nodes is performed by means of Function 10.2:

```
[Nx,Ref] = cp1002_select_nodes(N,k);
```

We now have to define the amount of error in range measurements. Let us assume for now that error is absent, corresponding to a variance $\sigma^2 = 0$:

```
sigma_2 = 0;
```

The position of the target node is evaluated by invoking Function 10.3:

```
[PosNx,ErrNx] = cp1002_find_LSE_position(positions, ...
   ranges, Nx, Ref, sigma_2, G);
```

The graphical output provided by the function is depicted in Figure 10–12. Here, the nodes that are not involved in the minimization process are represented by empty circles, while the reference nodes are represented by squares. The figure also shows the estimated position of the target node, identified by a star. Since no error is assumed, the estimated and correct positions coincide.

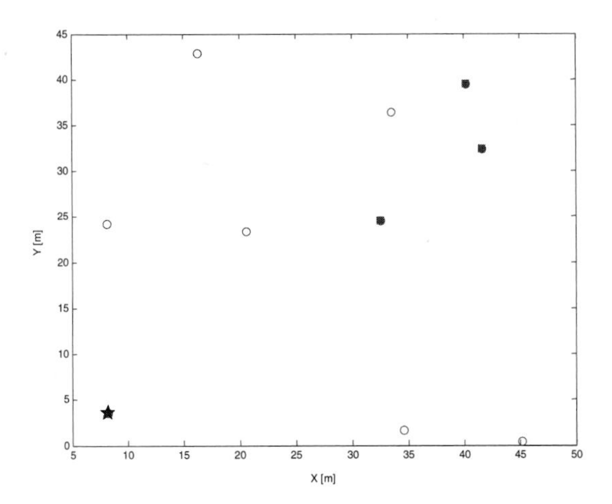

Figure 10–12 Position estimation by means of LSE without ranging errors (squares: reference nodes; circles: other nodes; star: estimated and effective position of target node).

Let us analyze the effect of errors in the range measurements. Figure 10–13 shows the results obtained with the same set of nodes when an error with variance $\sigma^2 = 5$ is introduced. The effective position of the target node is represented by a triangle:

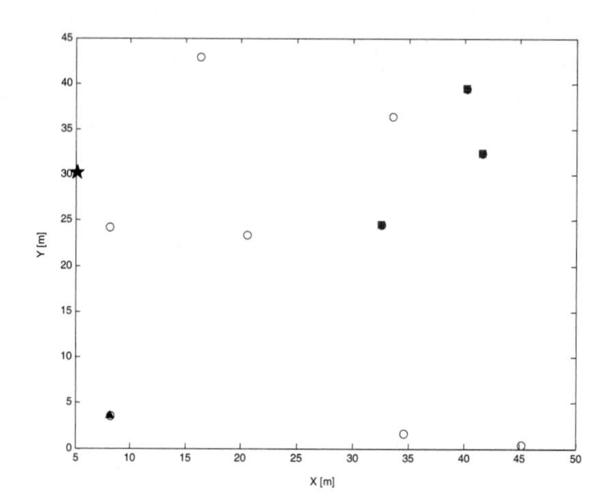

Figure 10–13 Position estimation by means of LSE with ranging errors, $\sigma^2 = 5$ (squares: reference nodes; circles: other nodes; star: estimated position of target node; triangle: effective position of target node).

Figure 10–13 shows that the introduction of a ranging error caused a positioning error of about 25 m. A relation between the ranging error and the positioning error can be obtained by iterating the steps above with different values of σ^2 and by averaging over a large number of trials. Figure 10–14 shows the positioning error for N = 10, k = 3 and σ^2 varying between 0 and 10 with steps of 0.5 for 10,000 averages for each σ^2 value.

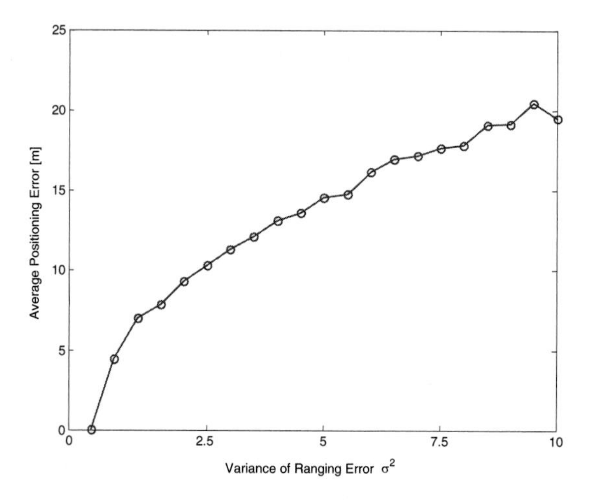

Figure 10–14 Average positioning error as a function of the variance of ranging errors.

Figure 10–14 shows that when the ranging error increases, the position error increases, thus reducing the positioning accuracy. As already anticipated, however, the error in position estimation can be reduced by introducing redundancy in the number of estimated distances. This redundancy can be introduced by increasing the number of reference nodes k, thus leading to an over-determined minimization problem. The effect of an increased k on the positioning error is shown in Figure 10–15 for $\sigma^2 = 5$.

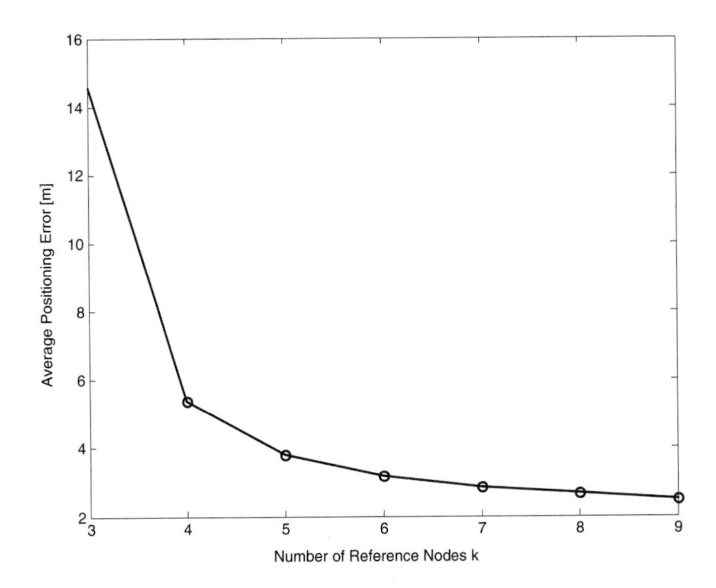

Figure 10–15 Average positioning error as a function of the number of reference nodes k, for variance of ranging errors $\sigma^2 = 5$.

Figure 10–15 clearly shows that a greater number of reference nodes leads to better accuracy in position estimation, even in the presence of large ranging errors.

CHECKPOINT 10–2

10.4 Positioning Protocols

The ranging and positioning techniques illustrated in Sections 10.2 and 10.3 form the basic block in the design of algorithms and related protocols for retrieving positioning in wireless networks of mobile nodes. The aim of a positioning protocol is to provide each node with position information about all other nodes in the network. Such information is then used to perform location-based optimizations at both the local (resource management) and global (routing) levels. Positioning protocols can be roughly subdivided into two categories: anchor-based protocols and anchor-free protocols.

Anchor-based protocols assume that a subset of nodes in the network know their position, either absolute or relative, at the beginning of network operation. Nodes in this subset typically assume the role of reference or anchor nodes, and allow all other nodes to compute their position.

Anchor-free protocols are used when all nodes are in unknown initial positions. In this case, the protocol must first determine a common reference system, and then compute the relative position of each node in this system.

In the case of anchor-based protocols, an external positioning system is required to provide each anchor node with its position. We will now briefly review the basic features of the most important positioning system, GPS (Getting, 1993).

GPS is based on a network of 24 satellites, which act as reference nodes for terrestrial GPS receivers. To achieve a common time reference, each satellite is equipped with an atomic clock with accuracy on the order of 10^{-13} s. Two carrier signals are generated and emitted by each satellite based on this clock, at frequencies L1 = 1575.42 MHz and L2 = 1227.60 MHz. The L2 carrier is modulated by a pseudo-random code with chip time $T_c \approx 10^{-6}$ s, known as Coarse Acquisition (C/A) code. The carrier L1 is modulated by two codes, the C/A code described above and the Precision (P) code with $T_c \approx 10^{-7}$ s.

Ranging in GPS exploits either the C/A or the P code to estimate the delay of the signal between satellite and receiver. The receiver generates a local code and determines the delay required to match the received code. As shown in Checkpoint 10–1, the accuracy in distance estimation depends on signal bandwidth. The C/A code is characterized by a bandwidth B of approximately 1 MHz. The ranging accuracy is roughly evaluated as equal to about 10% of a corresponding wavelength. In the present case, this equals a fraction of 300 m. The P code, thanks to its larger bandwidth of about 10 MHz, can provide an estimation with errors on the order of a meter.

A GPS receiver computes its position based on the estimation of distance by at least four satellites by means of spherical positioning. As indicated in Section 10.3, four equations are required to determine a unique point in a tridimensional space. In the case of GPS, three observations could be enough to determine the approximate tridimensional position of the terminal, since one of the two solutions of the system would normally lead to an impossible position (e.g., below the Earth's surface). Nevertheless, a fourth observation is required to determine the relative delay between the clock of the receiver and the common time reference adopted by the satellites.

The positioning resulting from distance estimations based on the C/A code is the so-called Standard Positioning System (SPS). The SPS specifications foresee 100-m horizontal accuracy, 156-m vertical accuracy, and 340-ns time accuracy, 95% of the time. The above specifications take into account the presence of an additional noise, known as Selective Availability (S/A), introduced by the U.S. Department of Defence to reduce the accuracy of the ranging based on the C/A code. In most cases, a higher than above accuracy is achieved since the U.S. government decided to turn off the S/A signal in May 2000.

The P code defines the Precise Positioning System (PPS). The PPS provides 22-m horizontal accuracy, 27.7-m vertical accuracy, and 100-ns time accuracy, 95% of the time.

The use of the P code is restricted to the U.S. government and its allies, and is not allowed for civilian use.

Accuracy of GPS positioning and ranging can be further increased by adopting techniques that exploit coordination among several GPS receivers, such as carrier phase estimation and differential GPS. The interested reader can find additional information in the "Further Reading" section at the end of the chapter.

The availability of GPS in a network of mobile nodes offers several advantages: a synchronized clock in all terminals, absolute positioning, and low signaling traffic related to the positioning protocols, since only positions of the terminals have to be exchanged in the network, and no additional information is needed to build a network map.

The development of anchor-based protocols was suggested because of the high cost of GPS receivers. In these protocols, in fact, only a subset of expensive nodes is equipped with GPS hardware, while the protocol takes charge of determining the positions of standard, non-GPS equipped nodes.

An example of the anchor-based positioning protocol was proposed in the framework of the PicoRadio Project (Savarese et al., 2001; Savarese, 2002).

Savarese (2002) identifies the main challenges for an anchor-based positioning protocol in a wireless network as: a) the sparse anchor node problem; and b) the range error problem. The sparse anchor node problem appears in large multi-hop networks, where physical connectivity between each node and at least four anchor nodes (for tridimensional positioning) cannot be guaranteed. This requires the information about anchor node positions to be spread all over the network by the positioning protocol to allow nodes that are not in physical connectivity with anchor nodes to compute their position in the reference system. The range error problem is caused by errors in range measurements, which can lead to large errors in position estimation, especially when the position is estimated by means of an iterative approach.

The approach proposed in (Savarese, 2002) is based on two algorithms addressing the two above problems: the Hop Triangulation via Extended Range and Redundant Association of Intermediate Nodes (Hop-TERRAIN) and the refinement algorithm.

The Hop-TERRAIN protocol works as follows:

1. Each anchor node Nk starts a flooding procedure by sending a broadcast packet, containing its own position, with hop count set to 0.
2. A node receiving a packet originated in Nk with hop count set to n checks its actual distance in hops from Nk, h_{Nk}. If no previous distance h_{Nk} was stored, or the hop count n in the packet is such that $n < h_{Nk}-1$, the node sets $h_{Nk} = n+1$ and retransmits the packet with hop count set to n+1. If h_{Nk} is already stored and $n \geq h_{Nk}-1$, the node discards the packet. The procedure expires when all nodes already received a packet from Nk through the path with a minimum number of hops.

When a node collects its distance in hops from the required number of anchor nodes, it estimates its position using the LSE minimization described in Section 10.3. To do so, the node converts the distance expressed in hops h_{Nk} into a physical distance d_{Nk}. The conversion is performed by multiplying the distance in hops by the average distance per hop, approximated with the maximum radio range of the node.

After the application of the Hop-TERRAIN algorithm, each node is provided with an approximate estimation of its position in the reference system. The refinement algorithm is in charge of improving estimation accuracy. The algorithm is based on the exchange of position estimates and ranging information at the local level, that is, between nodes in physical connectivity. At the beginning of each iteration, a node broadcasts its position and range estimations to its neighbors, and receives the same information from its neighbors. Based on the new information, the node recomputes its position by LSE minimization and moves to the next iteration. The algorithm ends when the update in the position falls below a given threshold.

Most of the protocols proposed in the literature assume the availability of anchor nodes equipped with a GPS receiver. This assumption simplifies the positioning task since there is no need for electing reference nodes and building a reference system; GPS-enabled terminals can assume the role of reference nodes and provide a global coordinate system.

A GPS-free approach can provide, however, additional features such as indoor coverage and better miniaturization, which is limited in GPS devices due to the antenna size.

A GPS-free positioning protocol, called the Self-Positioning Algorithm (SPA), has been proposed in the framework of the Terminodes project (Capkun et al., 2001). The protocol relies on the TOA method to obtain ranging measurements that are used to build a relative network map. The SPA is composed of two steps. First, each node in the network attempts to build a node-centered coordinate system, and to determine the position of its neighbors in this system. Second, the node-centered coordinate systems converge to a global network coordinate system.

Each node Ni tries to build a node-centered coordinate system by:

1. Detecting the set of its one-hop neighbors K_{Ni}, by using beacons.
2. Evaluating the set of distances from its neighbors D_{Ni} by means of TOA estimation.
3. Broadcasting D_{Ni} and K_{Ni} to its one-hop neighbors.

Since all nodes perform the above procedure, node Ni knows the distance from all its one-hop neighbors, the IDs of its two-hop neighbors, and a subset of the distances between the one-hop neighbors and the two-hop neighbors. This information is used by Ni to build its own node-centered coordinate system.

The determination of the node-centered coordinate system in a bidimensional space requires coordination between Ni and at least other two nodes, Np and Nq, belonging to K_{Ni} and not aligned with Ni. Ni must also know the distance between Np and Nq. This imposes an additional constraint in the choice of Np and Nq. Np must also belong to K_{Nq}, which also implies $Nq \in K_{Np}$ if same transmission ranges are assumed for all nodes.

The node-centered coordinate system is decided by Ni, and the coordinates of Ni, Np, and Nq are as follows:

$$\begin{cases} X_i = 0, Y_i = 0 \\ X_p = d_{ip}, Y_p = 0 \\ X_q = d_{iq}\cos\gamma, Y_q = d_{iq}\sin\gamma \end{cases} \tag{10–21}$$

where γ is the angle $\angle$(Np,Ni,Nq), that is, the angle formed by vectors $\overrightarrow{NpNi}$ and $\overrightarrow{NiNq}$, defined as:

$$\gamma = \arccos\frac{d_{iq}^2 + d_{ip}^2 - d_{pq}^2}{2d_{iq}d_{ip}} \tag{10–22}$$

Ni may eventually evaluate the position of a subset of its one-hop neighbors formed by the nodes for which Ni knows the distances from Ni, Np, and Nq. The subset is called Local View Set ($LVS_{Ni} \subseteq K_{Ni}$). The position of a node Nj belonging to this subset is given by:

$$\begin{cases} X_j = d_{ij}\cos\alpha_j \\ Y_j = \begin{cases} d_{ij}\sin\alpha_j & if\ \beta_j = |\alpha_j - \gamma| \\ -d_{ij}\sin\alpha_j & if\ \beta_j = \alpha_j + \gamma \end{cases} \end{cases} \tag{10–23}$$

where α_j is the angle $\angle$(Np,Ni,Nj) and β_j is the angle $\angle$(Nq,Ni,Nj), and both angles are given by:

$$\begin{cases} \alpha_j = \arccos\dfrac{d_{ij}^2 + d_{ip}^2 - d_{pj}^2}{2d_{ij}d_{ip}} \\ \beta_j = \arccos\dfrac{d_{ij}^2 + d_{iq}^2 - d_{qj}^2}{2d_{ij}d_{iq}} \end{cases} \tag{10–24}$$

All node-centered systems must evolve into a relative coordinate system by adopting the same orientation of the x and y axes. Iteratively, starting from the selection of one reference coordinate system, pairs of nodes rotate and align their coordinate systems. Using this distributed approach, readaption converges after a few steps to a relative coordinate system.

Note that nodes that are not able to build their own node-centered coordinate system can obtain their position in the relative coordinate system if they are connected with three nodes that already received the network coordinate system.

The approach followed by SPA is a typical example of an anchor-free protocol. This approach requires the exchange of a greater amount of information between nodes compared to anchor-based since information must be exchanged to select the reference nodes and convert the node-centered reference systems into a global, network-wide reference system. The resulting network map is formed by the relative positions of the nodes in the network, without any relationship to a global coordinate system; this knowledge is sufficient for several location-based applications, such as location-based routing and resource management. Anchor-free protocols are thus a viable solution for scenarios in which GPS cannot be exploited, either because LOS to satellites is not available or because the accuracy provided by GPS is not adequate. This is the case, for example, of WPAN applications (range < 10 m), where the accuracy on the order of meters provided by GPS is too coarse. In this case, positioning techniques capable of providing higher accuracy must be adopted; UWB is an appealing candidate thanks to its high potential accuracy provided by the large bandwidth.

In the next section, an example of an application of UWB to a small-scale, position-based scenario will be presented.

10.5 An Application: A MAgiC World

Positioning can play a major role in the advanced design of wireless communication networks. Resource sharing, as well as routing, to cite a few, can be optimized when positioning information is available throughout the network. Consider, for example, a mobile node that is capable of using directive antennas, and thanks to positioning, knows how to emit power in privileged directions, or a mobile node that selects how to route information to destination based on its awareness of the topology of the network.

Knowing how to locate and track static as well as moving objects with high precision is an extremely appealing feature per se, and is likely to introduce important changes in the way humans represent their visual perception into information bits.

Accurate positioning together with the deployment of a large number of small communication devices, such as sensors, opens the door to a variety of applications ranging from static locationing (think of locating a book in the Library of Babel) to the detection of movements and their conversion into maps of trajectories.

In this section, we will illustrate the example of devising a high-tech desk, the MAgiC desk, which incorporates new concepts of peripherals to support a human to machine interface. Suppose MAgiC desk to be designed as a distributed network of objects (sensors and devices called *sensible-areas*), which can be localized and identified using wireless communication; one mobile device (called the *pen*); and one element that manages communication with a computer (called the *control-element*). This design is shown in Figure 10–16.

The system should allow a user to interact with common computer applications (control of the pointer, of the peripherals, digital design, writing, etc.) by using the pen. The interaction consists of the movement of the pen inside a *macro-area* studded with a network of wireless sensors (the *desk*). The sensors are organized to define *micro-areas* of the desk (the sensible-areas). Each sensible-area corresponds to a particular application, which becomes active when the pen locates in that area. When a sensible-area becomes active, a local function to extract pen position information is started. It is the evolution of this information that should allow the control of related computer applications, as moving the pointer when the pen interacts with the mouse-pad, or writing a document when the pen interacts with the notebook. The accuracy in tracking the position of the pen depends on the application (high level of accuracy for digital design and handwriting, medium accuracy to control the pointer, and low accuracy to control the peripherals). Positioning information is obtained by the interaction of the pen/sensible-area and by the interaction of the sensible-area/control-element. The control-element can be wired to the computer and should therefore concentrate computational power.

Figure 10–16 The MAgiC desk.

Let us attempt to formalize the definition of requirements:

1. Rate of the update of positioning information according to the required precision in the description of the movements of the pen.
2. Constraints imposed on the receiver.

Suppose that the pen is moving at speed v in m/s and that pen positioning is determined on the basis of TOA of pulses sent by the pen according to the criterion illustrated in the previous sections. We suppose IR communication between the pen and the sensors located over the sensible-area on which the pen is moving. Pulses have a standard duration ranging from a few hundreds of picoseconds to one nanosecond.

Call M the trajectory of the pen we want to track, as shown in Figure 10–17.

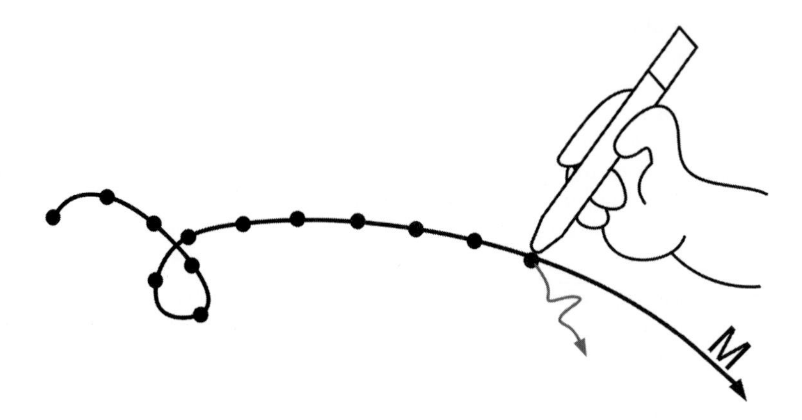

Figure 10–17 The pen is moving at speed v m/sec, tracing trajectory M.

The degree of spatial definition, or the degree of accuracy in the description of trajectory M, depends on the rate of emission of the pulses sent for positioning purposes. Call *R* this transmission rate. If the required degree of definition is *dM*, the following condition on rate *R* is imposed:

$$R \geq \frac{v}{dM} \tag{10–25}$$

When the moving object is a pen, we can reasonably suppose that v is about 1 m/s. Suppose that the application refers to tracking a handwritten text and that this imposes a required spatial definition of about 1 mm. The minimum required rate *R* for tracking the pen trajectory with a degree of accuracy *dM* is thus 1 kHz, corresponding to 1 kpulses/s.

Suppose that at start $t = t_0$, that is, when the pen is at rest, the pen-transmitter and sensor-receiver have synchronized using a specific synchronization phase.

The receiver must be capable of distinguishing between a pulse received at a nominal instant of time multiple of $T = 1/R$ and a pulse delayed or anticipated by δ from nominal time *T*. A positive delay δ indicates that the pen is moving away from the receiver, while $-\delta$ indicates that the pen is moving toward the receiver (see Figure 10–18).

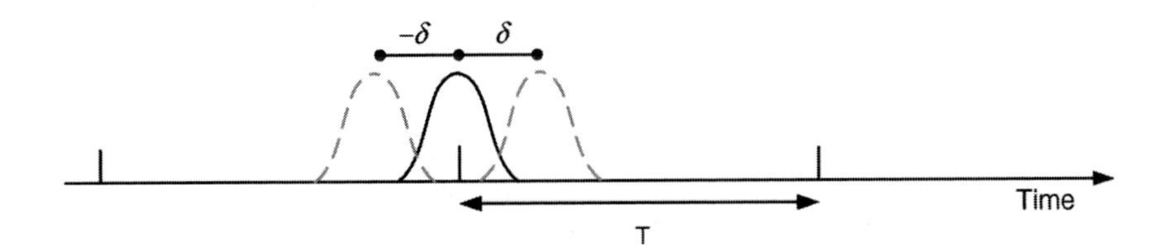

Figure 10–18 Received pulse is delayed by δ or anticipated by δ with respect to nominal time T depending on whether the pen is moving toward or away from the receiver.

Suppose that the receiver is the usual correlation receiver with a correlation mask as shown in Figure 10–19, followed by an integrator. The decision rule on the output of the integrator Z is thus:

$$Z = \begin{cases} A & \textit{if pen moves toward receiver} \\ \textit{zero} & \textit{if pen is still} \\ -A & \textit{if pen moves away from receiver} \end{cases} \qquad (10\text{–}26)$$

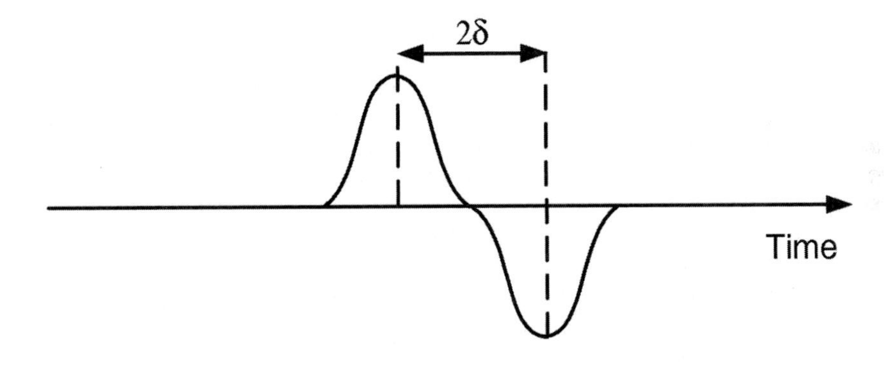

Figure 10–19 Receiver correlation mask.

Delay δ depends on the velocity of propagation of the pulse, which we suppose to be $c \approx 3 \cdot 10^8$ m/s. In correspondence to a degree of definition dM, delay δ is thus:

$$\delta = \frac{dM}{c} \qquad (10\text{–}27)$$

For $dM = 1$ mm, δ is about 3 ps, which indicates that the degree of definition dM imposes a capability of detecting pulses delayed by 3 ps on the receiver. This is far beyond what can be achieved with current technology. Therefore, the theoretical lower bound introduced in Section 10.4 (see Eq. (10–7)) cannot be realistically reached. Note that the receiver must also be capable, after receiving each pulse, of adjusting the correlation mask and centering it on the time of arrival of the current pulse since trajectory tracking is based on the computation of the current position relative to the position determined on the immediately preceding pulse.

Technology constraints may limit the degree of accuracy with which a pulse delayed by δ can be correctly identified as such by the receiver and the degree by which the correlator is capable of time shifting the correlator mask. These limitations impose in turn a constraint on the degree of spatial definition of the trajectory that a receiver is capable of providing. The design of a system based on positioning should therefore take into account the following guidelines:

1. Determine the minimum delay δ, which the receiver is capable of implementing with good accuracy in the correlator mask.
2. Determine the minimum shift of the correlator mask, which can be implemented by the receiver in a time corresponding to the interval between two pulses.
3. Choose the highest between delay δ and the minimum shift of the correlator.
4. Compute dM from Eq. (10–27), where δ is the time defined at Step 3.
5. Determine the rate at which pulses used for positioning must be transmitted based on Eq. (10–25) as a function of the speed of the object.

Equation (10–25) also indicates the effect of a change of speed of the object on the degree of definition dM. If dM must be constant, then rate R must vary linearly with speed. Since a variable rate might introduce complexity in synchronization between transmitter and receiver, a possible way out is to adopt a much higher rate than the minimum required for average speed. This way, depending on the speed of the object, the transmitted pulses may be partly discarded if the difference between their time of arrival is beyond the receiver's capabilities.

As a final comment, note that the degree of definition needed in the description of the trajectory is closely related to the relationship between signs and symbols and what they represent, that is, the semantics of the trajectory.

In the case of a pen writing a text, the degree of accuracy required by the positioning module depends on the sophistication of an interpretation module, which should analyze the tracked trajectory and interpret it as linguistic information. A highly sophisticated text recognition module based on complex algorithmic structures might significantly reduce the required degree of accuracy in positioning. The interpretation module should in fact model the complex action of the human brain, which before decoding an image or shape, sets its status as being in the condition of interpreting a sequence of graphemes conveying linguistic information, and does so by activating the set of neurons dedicated to this task.

FURTHER READING

The topics of ranging and positioning have been widely addressed in the scientific literature. (Drane, 1992) provides a good starting point for the analysis of general ranging and positioning problems. (Rappaport et al., 1996) introduce the fundamentals of positioning, and present a review of the main positioning algorithms. A complete survey of the existing positioning systems, with a comparison in terms of accuracy, is provided in (Hightower and Borriello, 2001).

Regarding ranging based on UWB radio signals, interesting solutions can be found in several U.S. patents dealing with UWB-based ranging and positioning devices, such as (McEwan, 1995) and the already cited (Fleming and Kushner, 1997). Regarding scientific publications on this topic, a TOA-based ranging scheme for UWB signals is proposed in (Lee and Scholtz, 2002), where performance of ranging based on UWB is evaluated in the presence of multi-path and obstacles to propagation. The accuracy of ranging with UWB signals is further investigated in (Shimizu and Sanada, 2003), where the effect of timing jitter between clocks at the transmitter and receiver sides is analyzed, and in (Denis et al., 2003), which deals with the case of NLOS propagation.

Significant efforts have been dedicated to the derivation of theoretical results for positioning algorithms, as for example in (Fang, 1990), where a simple solution to the problem of hyperbolic positioning based on geometrical considerations is given, and in (Deffenbaugh et al., 1996), where the relationship between spherical and hyperbolic positioning is investigated.

Regarding practical implementations of positioning systems, a survey of the GPS system can be found in (Getting, 1993), but the interested reader can find a deeper insight in one of several books dedicated to GPS, such as (Kaplan, 1996) and (Logsdon, 1999). Several methods for improving the accuracy of ranging and positioning provided by GPS have been proposed. The most important is differential GPS, which combines two GPS receivers. This technique is described in (Morgan-Owen and Johnston, 1995).

Positioning in cellular networks is another hot topic, in particular in the United States, where the FCC requests that mobile network operators have the capability of determining the position of a customer during an emergency call. Analyses on positioning capabilities in both TDMA and CDMA cellular networks can be found in (Spirito, 2001; Hepsaydir, 1999).

Positioning in indoor environments has also been investigated. In particular, RSSI-based positioning in 802.11b systems was proposed (Bahl and Padmanabhan, 2000). This approach overcomes the inherent limitations of RSSI by previously tracing a power map of the area of interest and then performing a ML estimation of a device's position based on the received power at each radio access point.

The availability of GPS has also triggered significant research efforts in the field of distributed positioning protocols based on the availability of GPS-equipped anchor nodes. Apart from the aforementioned Hop-TERRAIN approach (Savarese, 2002), it is worth mentioning the work presented in (Niculescu and Nath, 2001), further refined in (Niculescu and Nath, 2003) with the introduction of information on angles between nodes. Other noticeable contributions in the field of positioning protocols can be found in (Doherty et al., 2001) and (Di Stefano et al., 2003).

Regarding UWB-based positioning systems, a system based on hyperbolic positioning with receivers in fixed locations, capable of positioning accuracy in the order of 30 cm, is described in (Fontana et al., 2003), while the possible combination of UWB and GPS for positioning is investigated in (Opshaug and Enge, 2002).

REFERENCES

Bahl, P., and V.N. Padmanabhan, "RADAR: An In-Building RF-based User Location and Tracking System," *Nineteenth Annual Joint Conference of the IEEE Computer and Communications Societies*, Volume: 2 (March 2000), 775–784.

Capkun, S., M. Hamdi, and J.P. Hubaux, "GPS-free positioning in mobile Ad-Hoc networks," *Hawaii International Conference On System Sciences* (January 2001), 3481–3490.

Deffenbaugh, M., J.G. Bellingham, and H. Schmidt, "The relationship between spherical and hyperbolic positioning," *MTS/IEEE OCEANS '96 'Prospects for the 21st Century,"* Volume: 2 (September 1996), 590–595.

Denis, B., J. Keignart, and N. Daniele, "Impact of NLOS Propagation upon Ranging Precision in UWB Systems," *IEEE Conference on UWB Systems and Technologies* (November 2003), 379–383.

Di Stefano, G., F. Graziosi, and F. Santucci, "Distributed positioning algorithm for ad-hoc networks," *International Workshop on Ultra Wideband Systems* (June 2003).

Doherty, L., K.S.J. Pister, and L. El Ghaoui, "Convex position estimation in wireless sensor networks," *Twentieth Annual Joint Conference of the IEEE Computer and Communications Societies*, Volume: 3 (April 2001), 1655–1663.

Drane, C., *Positioning Systems: A Unified Approach*, New York: Springer-Verlag (1992).

Drane, C., M. Macnaughtan, and C. Scott, "Positioning GSM telephones," *IEEE Communications Magazine*, Volume: 36, Issue: 4 (April 1998), 46–54, 59.

Fang, B.T., "Simple solutions for hyperbolic and related position fixes," *IEEE Transactions on Aerospace and Electronic Systems*, Volume: 26, Issue: 5 (September 1990), 748–753.

Fleming, R.A., and C.E. Kushner, "Spread Spectrum Localizers," U.S. Patent No. 6,002,708 (1997).

Fontana, R.J., E. Richley, and J. Barney, "Commercialization of an Ultra Wideband precision asset location system," *IEEE Conference on UWB Systems and Technologies* (November 2003), 369–373.

Getting, I.A., "Perspective/navigation-The Global Positioning System," *IEEE Spectrum*, Volume: 30, Issue: 12 (December 1993), 36–38, 43–47.

Hallberg, J., M. Nilsson, and K. Synnes, "Positioning with Bluetooth," *10th International Conference on Telecommunications*, Volume: 2 (February 2003), 954–958.

Hepsaydir, E., "Mobile Positioning in CDMA Cellular Networks," *IEEE Vehicular Technology Conference*, Volume: 2 (September 1999), 795–799.

Hightower, J., and G. Borriello, "Location systems for ubiquitous computing," *IEEE Computer*, Volume: 34, Issue: 8 (August 2001), 57–66.

Hightower, J., C. Vakili, G. Borriello, and R. Want , "Design and Calibration of the SpotON Ad-Hoc Location Sensing System," Available at *www.cs.washington.edu/homes/jeffro/pubs/hightower2001design/hightower2001design.pdf* (August 2001).

Kaplan, E. (ed.), *Understanding GPS: Principles and application*, Norwood, MA: Artech House (1996).

Lee, J.-Y., and R.A. Scholtz, "Ranging in a dense multipath environment using an UWB radio link," *IEEE Journal on Selected Areas in Communications*, Volume: 20, Issue: 9 (December 2002), 1677–1683.

Logsdon, T., *The Navstar Global Positioning System*, New York: Van Nostrand Reinhold (1992).

McEwan, T.E., "Short range locator system," U.S. Patent 5,589,838 (1995).

Morgan-Owen, G.J., and G.T. Johnston, "Differential GPS positioning," *Electronics & Communication Engineering Journal*, Volume: 7, Issue: 1 (February 1995), 11–21.

Niculescu, D., and B. Nath, "Ad Hoc Positioning System (APS)," *IEEE Global Telecommunications Conference*, Volume: 5 (November 2001), 2926–2931.

Niculescu, D., and B. Nath, "Ad Hoc Positioning System (APS) Using AOA," *Twenty-Second Annual Joint Conference of the IEEE Computer and Communications Societies*, Volume: 3 (March 2003), 1734–1743.

Opshaug, G.R., and P. Enge, "Integrated GPS and UWB Navigation system: (Motivates the Necessity of Non-Interference)," *IEEE Conference on UWB Systems and Technologies* (May 2002), 123–127.

Proakis, J.G., *Digital Communications*, 3rd Edition, New York: McGraw-Hill International Editions (1995).

Rappaport, T.S., J.H. Reed, and B.D. Woerner, "Position location using wireless communications on highways of the future," *IEEE Communications Magazine*, Volume: 34, Issue: 10 (October 1996), 33–41.

Savarese, C., "Robust Positioning Algorithms for Distributed Ad-Hoc Wireless Sensor Networks," Master Thesis. at *http://bwrc.eecs.berkeley.edu/Research/Pico_Radio/docs/Savarese_MS_Thesis_FINAL.pdf* (2002).

Savarese, C., J.M. Rabaey, and J. Beutel, "Location in distributed ad-hoc wireless sensor networks," *IEEE International Conference on Acoustics, Speech, and Signal Processing*, Volume: 4 (2001), 2037–2040.

Shimizu, Y., and Y. Sanada, "Accuracy of Relative Distance Measurement with Ultra Wideband System," *IEEE Conference on UWB Systems and Technologies* (November 2003), 374–378.

Spirito, M.A., "On the accuracy of cellular mobile station location estimation," *IEEE Transactions on Vehicular Technology*, Volume: 50, Issue: 3 (May 2001), 674–685.

Urkowitz, H., *Signal Theory and Random Processes*, Dedham, MA: Artech House (1983).

APPENDIX 10.A

Function 10.1 Network Creation

```
%
% FUNCTION 10.1 : "cp1002_create_network"
%
% This function generates the positions of a set of nodes
% in a square area and computes the distance between each
% pair of nodes
% The function receives as input:
% - The number of nodes N
% - The side length of the square area
% - A flag G to enable/disable the graphical output
% The function returns:
% - A matrix Nx2 containing the (X,Y) positions of each
%   node
% - A matrix NxN containing the distances between each pair
%   of nodes
%
% Programmed by Luca De Nardis
%

function [positions, ranges] = ...
   cp1002_create_network(N,area_side,G)
for i = 1:N
    positions(i,1)=rand*area_side;
    positions(i,2)=rand*area_side;
    j=1;
    for j=1:(i-1)
        ranges(i,j)= sqrt((positions(i,1) -...
           positions(j,1))^2 + (positions(i,2) -...
           positions(j,2))^2);
        while(ranges(i,j)==0)
                X(i)=rand*50;
                Y(i)=rand*50;
                ranges(i,j)=sqrt((positions(i,1)-...
                   positions(j,1))^2+(positions(i,2)-...
                   positions(j,2))^2);
        end
```

```
            ranges(j,i)=ranges(i,j);
        end
    end
    if G
        scatter(positions(:,1),positions(:,2),'filled');
        axis([0 area_side 0 area_side]);
        xlabel('X [m]');
        ylabel('Y [m]');
        box on;
    end
```

Function 10.2 Selection of Target Node and Reference Nodes

```
%
% FUNCTION 10.2 : "cp1002_select_nodes"
%
% This function selects a target node and k reference nodes
% from a set of nodes
% The function receives as input:
% - The total number of nodes N
% - The number of reference nodes k
% The function returns:
% - The ID of the target node Nx
% - A vector Refs of length k, containing the ID of the
%   reference nodes
%
% Programmed by Luca De Nardis
%
function [Nx,Ref] = cp1002_select_nodes(N,k);

% extraction of the target node
Nx = ceil(N*rand);

Check=zeros(1,N);

for i=1:k
    Ref(i) = Nx;
    % check if Ref(i) is different from Nx
    while((Ref(i)==Nx)||(Ref(i)==0))
        Ref(i)= ceil(N*rand);
        % check if Ref(i) is already selected
        if(Check(Ref(i)))
            Ref(i) = 0;
        else
            Check(Ref(i))=1;
        end
    end
end
```

Function 10.3 Evaluation of LSE Position

```
%
% FUNCTION 10.3 : "cp1002_find_LSE_position"
%
% This function determines the LSE solution to a
% positioning problem in a bidimensional space
% The function receives as input:
% - A matrix Nx2 containing the (X,Y) positions of each
%   node
% - A matrix NxN containing the distances between each pair
%   of nodes
% - The ID of the target node Nx
% - A vector Refs of length k, containing the ID of the
%   reference nodes
% - The value of sigma_2
% - A flag G to enable/disable the graphical output
%
% The function returns the estimated position of
% the target node Nx and the error with respect to the
% exact position ErrNx
%
%
%Programmed by Luca De Nardis
%

function [PosNx, ErrNx] = ...
   cp1002_find_LSE_position(positions, ranges, Nx, Ref,...
   sigma_2, G);

% adding errors to the range estimation

N = size(ranges,1);
err_ranges = ranges + sqrt(sigma_2)*randn(N);

% defining the linear problem
% matrix A
k = length(Ref);
for i=1:(k-1)
    A(i,1) = positions(Ref(i),1) - positions(Ref(k),1);
    A(i,2) = positions(Ref(i),2) - positions(Ref(k),2);
end
A=-2*A;
```

```
% matrix b
b=zeros(2,1);
for i=1:(k-1)
    b(i) = err_ranges(Ref(i),Nx)^2 -...
      err_ranges(Ref(k),Nx)^2 - positions(Ref(i),1)^2 +...
      positions(Ref(k),1)^2 - positions(Ref(i),2)^2 +...
      positions(Ref(k),2)^2;
end


% solving the problem
PosNx=A\b;

% computing the error
ErrNx = sqrt((PosNx(1)-positions(Nx,1))^2+(PosNx(2)-...
   positions(Nx,2))^2);

% graphical output
if G
    scatter(positions(:,1),positions(:,2));
    xlabel('X [m]');
    ylabel('Y [m]');
    box on;
    hold on;
    scatter(PosNx(1), PosNx(2), 200, 'filled', 'k','p');
    scatter(positions(Nx,1),positions(Nx,2),200,...
       'filled','^');
    for i=1:k
        scatter(positions(Ref(i),1),positions(Ref(i),2),...
           'filled','r','s');
    end
    hold off;
end
```

CHAPTER 11

UWB Networks: Principled Design of MAC

As a general principle, the role of the MAC module is to allow multiple users to share a common resource. The definition of a resource, and of the procedures by which access to the medium is granted, depends on the adopted transmission and multiple access techniques. As mentioned in Chapter 9 "Multi-User Wireless Communications," the flows of information bits are reorganized in the MAC into larger information packets called MAC-PDUs, which, after appropriate coding, are sent over the air interface.

In a layered architecture for data networks, the MAC is generally considered the bottom part of the Data Link Control (DLC) layer. The service offered by the MAC to the upper DLC is to provide a bit pipe, preventing or resolving contentions in the access to the medium. The functions executed in the MAC should be defined without taking into account the underlying physical layer, which is seen by the MAC as a black box offering the service of transferring bits in the form of signals appropriate for the channel. In this view, the adoption of a novel transmission technique such as UWB does not affect the design of the MAC and existing solutions typically designed for wireless networks can be directly incorporated into the design of an UWB network. In this chapter, we will review examples of MAC implementations for a few popular wireless networks: IEEE 802.11b (Crow et al., 1997), Bluetooth (Haartsen, 2000), and IEEE 802.15.3 (IEEE 802.15.3, 2003). The last was originally developed for traditional narrowband physical layers in the Industrial, Scientific, and Medical (ISM) 2.4 GHz band, but given the current physical layer proposals that focus on UWB, it should also serve in the UWB case. All these are examples of MAC implementations that basically abstract from UWB-specific features.

The design of an efficient MAC often requires an accurate knowledge of the physical layer. In the case of UWB systems, this is a crucial issue where typical UWB features such as the need for operating in low-power conditions versus a rather peculiar precise ranging capability may enable the definition of novel MAC functions, as well as lead to a drastically different implementation of more traditional MAC functions. Cross-layer design strategies for system optimization are at the core of the conception of future wireless communication systems.

Following the principle of cross-layer design, we will introduce at the end of the chapter an UWB-tailored MAC that attempts to take into account UWB-specific MUI and synchronization issues as analyzed in Chapters 8, "Propagation over a Multi-Path Affected UWB Radio Channel," and 9, and incorporates the capability of providing the network of nodes with ranging information. This information can be exploited as explained in Chapter 10, "UWB Ranging and Positioning," to introduce positioning for the design of power-aware and location-based routing strategies (De Nardis et al., 2003).

This chapter does not contain checkpoints with MATLAB simulations as did previous chapters. Rather, specific reference cases are included in the form of examples incorporated into the text. Simulation of MAC algorithms would in fact require the joint implementation of physical and MAC modules for which MATLAB might not be the most appropriate tool. A network simulator such as ns-2 developed at the University of California at Berkeley (Breslau et al., 2000) or OMNeT++ (Varga, 2002) would form the best ground for studying by simulation the behavior of a cooperative network of independent nodes, but this investigation is beyond the scope of this book.

11.1 MAC: General Principles

(Chandra et al., 2000) defined a set of parameters that characterizes a MAC independently of the underlying transmission technique and multiple access method. These parameters are:

- Throughput, defined as the percentage of channel capacity used during data transmission.
- Delay, that is, the average time spent by a packet in a MAC queue.
- Degree of fairness, which states that access to the medium is fair if all nodes have a similar chance of obtaining medium access.

Additional evaluation parameters are usually related to specific MAC protocols, such as the degree of flexibility regarding the possibility of an asymmetrical bandwidth user allocation to downlink versus uplink streams (Choi and Moayeri, 2000). Key objectives are therefore to maximize throughput and to guarantee an acceptable delay and fair access to the channel to all terminals. The MAC should be capable of fulfilling these goals in a dynamic environment. It should be flexible in adapting to different channel behaviors, traffic characteristics, and local network topologies.

Network architecture determines whether and how two nodes in a network can exchange information. The role of the algorithms ruling the network module is therefore to tend to a network topology in which each node can exchange information with all other nodes, that is, the topology of the network is fully connected. Note that connectivity here has a logical meaning and should not be confused with physical connectivity. Two nodes might exchange information, although not in mutual visibility, by using routes that either hop on other nodes (multi-hop connections) or go through special nodes connected to the fixed wired infrastructure (radio access points).

Routing and flow control are functions commonly associated with the network module since it is at this layer that a unique addressing format is created to connect nodes belonging to underlying sub-networks that are possibly heterogeneous. Figure 11–1 shows in a schematic way how different sub-networks can interact thanks to an overlaying structure, the network, which unifies the addressing format. To achieve its task, the network module must work by combining information available for all communication links of the network. Cooperation among all peer processes operating at the transmitter and receiver side of each communication link must be ensured, and is obtained by implementing network functions in the form of distributed algorithms.

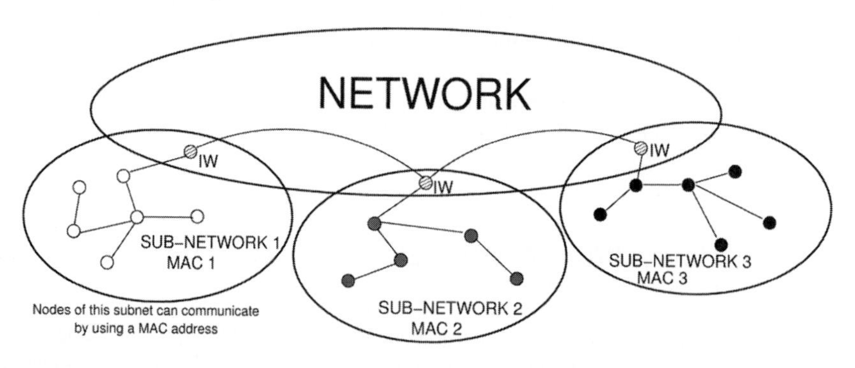

Figure 11–1 Interaction between different MAC subnets.

Routing, flow control, and addressing can be defined as functions of the MAC module in the case of a homogenous network of nodes. This is the typical case of wireless networks, which are not connected to the fixed wired infrastructure, or of independent MACs of independent subnets. Note that also in such a case, subnets can be integrated by adding Inter-Working (IW) units toward the network. These units enable a subnet to connect with other entities in the world (see Figure 11–1). Note that even in the presence of IW units, subnet internal routing and flow control may continue to be managed by the MAC module of the subnet.

Therefore, we need to analyze the MAC architecture. Contrary to network architecture, MAC architecture is related to the topology of potentially active links, that is, those links that the MAC has selected based on the set of all possible physical links to allow information exchange between nodes in a coordinated manner (see Figure 11–2).

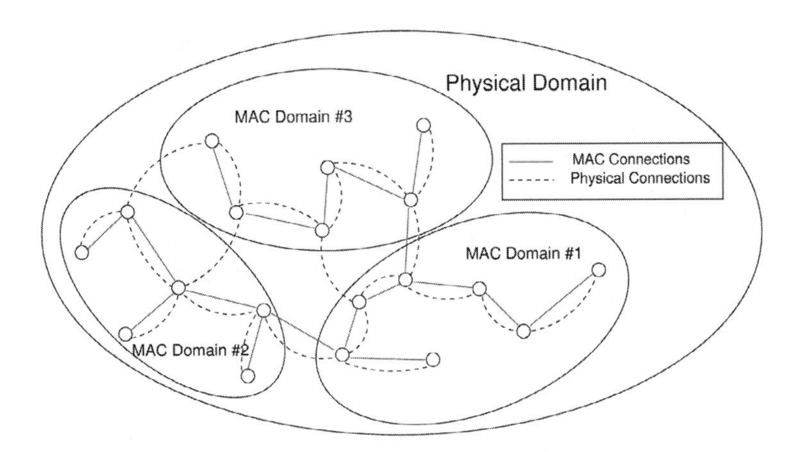

Figure 11–2 The MAC domain concept.

UWB radio is in principle a suitable physical transmission technique for all kinds of applications. Given the strong power constraints that have been set in the United States, and are likely to be adopted in many other parts of the world, UWB is emerging as a particularly appealing transmission technique for applications requiring either high bit rate over short range or low bit rate over medium-to-long ranges. The high bit rate/short range case includes WPANs for multimedia traffic, cable replacement applications (such as wireless USB and DVI), and wearable devices (e.g., wireless hi-fi headphones). The low bit rate/medium-to-long range case applies to long-range sensor networks (e.g., indoor/outdoor distributed surveillance systems), non-real-time data applications (e.g., e-mail and instant messaging), and in general, all data transfers compatible with a transmission rate on the order of 1 Mb/s over several tens of meters. A recent standard release of the IEEE 802.15.4 for low rate WPANs (IEEE 802.15.4, 2003) has increased attention and interest toward low bit rate applications. In this regard, the first commercial UWB precision asset location system named PAL650, operating at an average rate of 70 bits/s over a couple of hundred meters in outdoor links, was recently presented (Fontana et al., 2003).

The above applications refer to networks that commonly adopt the self-organizing/ad-hoc principle and are usually referred to as distributed network architectures, meaning a group of terminals in a limited size area that need to communicate through the radio interface in a fixed-wired, infrastructure-free fashion, that is, without any central coordinating unit or base station. Communication routes or paths are set up in a dynamic way and are reconfigurable. This paradigm can be viewed as the opposite of the cellular networking concept, in which nodes typically communicate by establishing connections using single-hop links with a central coordinating unit, the radio access point, which interfaces the wireless nodes with the fixed wired infrastructure.

In this chapter, we will focus on distributed network architectures such as ad-hoc and sensor networks, which promise to be key scenarios for the UWB technology. We will consider a homogenous network of nodes and therefore assume that routing, addressing, and flow control functions are implemented in the MAC module.

11.2 MAC Functions

Typically, the tasks or functions that must be achieved by the MAC are (De Nardis and Di Benedetto, 2003):

1. Medium sharing — This function determines how terminals access the medium to transmit packets.
2. MAC organization — This function deals with the organization of the MAC, or how terminals coordinate themselves in resource sharing.
3. Admission control — This function is used to regulate the access of traffic sources in the network and avoid congestion.
4. Packet scheduling — When multiple traffic flows are present at the same terminal, packet scheduling is used to select the next packet to be transmitted.
5. Power control — Power control aims at optimizing power utilization in the network.

Overlaying the above functions is Quality of Service (QoS) management. QoS involves most of the functions defined above, and can be seen as a horizontal function. We will briefly address the general problem of introducing QoS in distributed wireless networks in Section 11.3.

11.2.1 Medium Sharing

Several of the existing MAC modules for distributed networks are based on the hypothesis that users share a single channel. Two possible choices are available: either terminals contend to gain channel control (random access), or channel control may be granted by a control unit based on a specific resource assignment protocol (scheduled access). While random access is appropriate for bursty traffic, scheduling allows a more efficient utilization of the channel when continuous streams of data packets must be transferred. Note that even in the case of a scheduled approach, an initial random access phase is necessary since the scheduling sequence is typically unavailable at network startup.

Typical random access solutions for wireless networks are: Aloha (Abramson, 1977), Carrier Sensing Multiple Access (CSMA) (Kleinrock and Tobagi, 1975), and out-of-band signaling (Tobagi and Kleinrock, 1975).

The main advantage of Aloha is simplicity. It only adds a Cyclic Redundancy Code (CRC) field to each data packet before transmission. If collision occurs, a backoff procedure is activated, and the corrupted packet is scheduled for retransmission. Aloha behaves well for low traffic load, while performance decreases abruptly when traffic load increases and packet length grows (Bertsekas and Gallager, 1992). Slotted Aloha adopts a slotted time axis, which forces terminals to attempt transmission only at the beginning of a time slot. Overall performance is slightly improved over the pure Aloha scheme, but is still insufficient to use the Aloha principle beyond the specific case of short, rare packet transmission such as with control packets.

For a heavy traffic load, a higher throughput can be obtained by means of CSMA, which is based on a channel sensing period performed by each terminal before starting transmission. Performance of CSMA is strongly limited by two phenomena: the well-known "hidden terminal" and "exposed terminal" problems. Suppose that, as shown in Figure 11–3, a source terminal TX1 has gained access to the radio interface and has started transmitting to a given receiver RX at $t = t_0$. Suppose in addition that a third terminal TX2, in the neighborhood of RX, has initiated a transmission procedure, but is hidden to TX1 due to distance or obstacles. Terminal TX2 sounds the channel, considers it as free since it does not receive TX1, and starts emitting power at $t = t_0 + T$, eventually causing a collision in the receiving node RX. In CSMA systems, this collision is destructive due to the CSMA hypothesis of using only one channel.

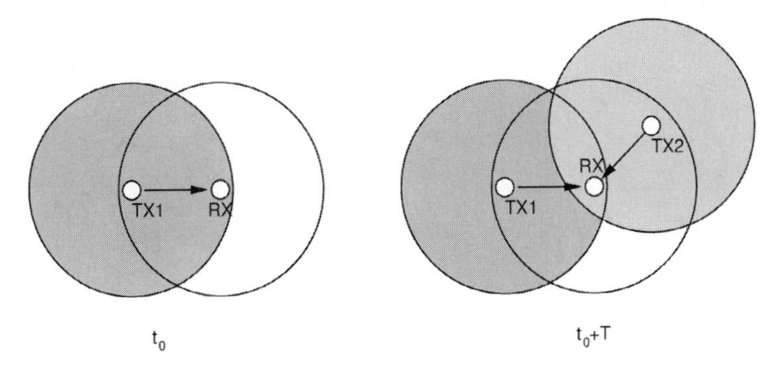

Figure 11–3 Hidden terminal condition.

The exposed terminal problem is represented in Figure 11–4. Suppose that terminal TX1 starts transmitting to RX1, and suppose in addition that terminal TX2 wants to start transmitting to RX2. When TX2 sounds the channel, it detects the transmission from TX1 to RX1 (in this sense, it is exposed to this transmission), assumes the channel is busy, and postpones transmission. Transmission from TX2 to RX2 was, however, harmless, since RX1 and TX2 are out of range.

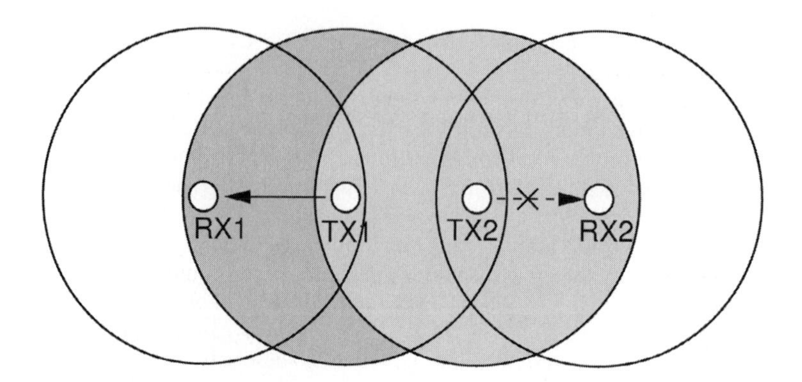

Figure 11–4 Exposed terminal condition.

To solve the hidden and exposed terminal problems, CSMA can be modified by introducing a three-way handshake, as in Multiple Access with Collision Avoidance (MACA), which is described in (Karn, 1990). The handshake actually substitutes the carrier sensing procedure. In the Floor Acquisition Multiple Access (FAMA) algorithm (Fullmer and Garcia-Luna-Aceves, 1995) a combination of handshake and carrier sensing was proposed. FAMA gave rise to a family of algorithms commonly referred to as Carrier Sensing Multiple Access with Collision Avoidance (CSMA-CA), among which was the Distributed Foundation Wireless MAC (DFWMAC), which was adopted for the MAC of IEEE 802.11b standard (Crow et al., 1997), as further illustrated in Example 11–1.

Example 11–1 : The 802.11b IEEE MAC

An example of CSMA-CA is the DFWMAC which was adopted for the MAC layer of IEEE 802.11b standard (Crow et al., 1997).

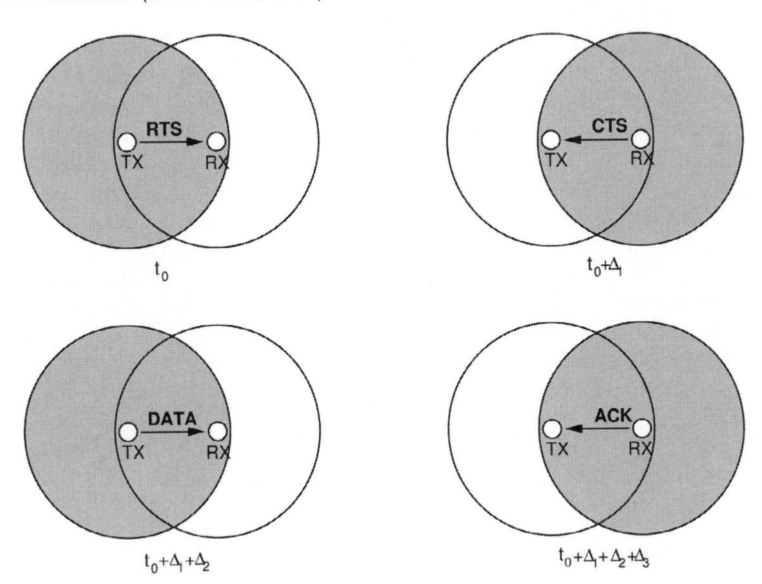

Figure 11–5 DFWMAC setup procedure

Figure 11–5 shows the set-up procedure of DFWMAC. After the emission of the Request To Send (RTS) packet, TX waits for a given time interval before starting transmission. If, within this time, no Clear to Send (CTS) is received, TX assumes that the RTS packet was lost, sends it again, and increments the timeout interval. Similarly, RX waits for a certain time after sending the CTS. If there is no DATA packet reception before timeout, RX sends the CTS packet again, and then waits for a longer time.

The above signaling procedure reduces collisions on data packets. When a terminal detects an RTS or CTS emitted by another terminal, it avoids emitting packets for a given period of time. Collisions are thus restricted to the signaling RTS packets in case two or more terminals initiate a transmission procedure at the same time. On the other hand, this plain CSMA-CA, based on the carrier sensing concept, still suffers from both the hidden and exposed terminal problems.

To address these issues, DFWMAC implements an enhanced version of CSMA-CA, in which the RTS/CTS control packets are used to broadcast information regarding the duration of the scheduled transmission, allowing the introduction of the *virtual carrier sensing* concept. Each terminal builds and updates a Network Allocation Vector (NAV), which contains information regarding the scheduled channel occupation by other terminals. A terminal starts transmission only if it senses an idle channel, and if according to the NAV, there are no other scheduled transmissions. By including the duration information on both RTS and CTS packets, this approach solves the hidden terminal problem and guarantees collision-free transmission of data packets. However, the procedure still suffers of possible collisions among control packets.

EXAMPLE 11–1

Out-of-band signaling MAC algorithms solve the hidden terminal problem by using a dedicated channel to signal the beginning of a transmission. The overall bandwidth available for communication is split into two channels: a data channel, used for data packet exchange, and a narrowband signaling channel on which sinusoidal waveforms, referred to as busy tones, are asserted by terminals transmitting or receiving to avoid interference produced by hidden terminals. The approach in its original form (Tobagi and Kleinrock, 1975) was intended for networks of mobile terminals communicating with a central station; whenever a terminal needs to transmit a packet, it senses the busy tone channel and starts transmission only if no busy tone is detectable (Busy Tone Multiple Access, BTMA). The central station asserts a busy tone whenever a terminal transmits data on the channel, thus preventing all other terminals from transmitting. Note that the extension of this procedure to a distributed network requires that each terminal sensing a transmission emits a busy tone. This operation blocks all nodes in an area of 2R around the transmitting node, where R is the radio coverage radius, and the exposed terminal problem is thus amplified (Chandra et al., 2000; Tobagi and Kleinrock, 1975), (as shown in Figure 11–6). The out-of-band signaling concept is further analyzed in Example 11–2.

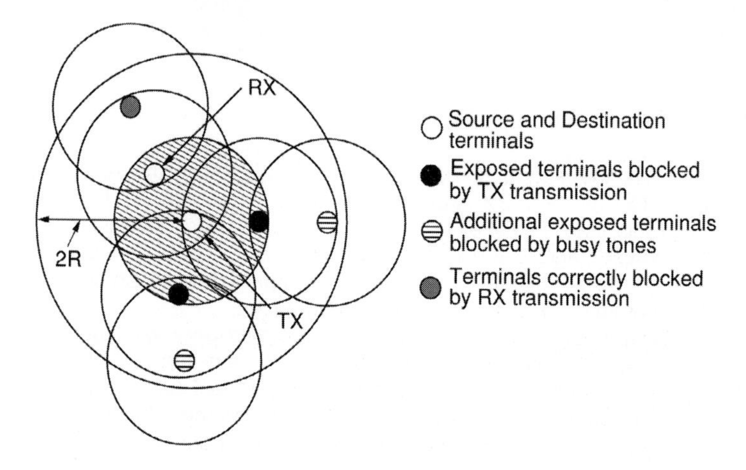

Figure 11–6 Amplification of the exposed terminal problem in BTMA (by permission from *Journal of Communications and Networks*, ISSN 1229-2370, © Korean Institute of Communication Sciences).

Example 11–2 : DBTMA

In Dual Busy Tone Multiple Access, or DBTMA (Deng and Haas, 1998), two different busy tones are used in combination with three-way handshaking, similar to MACA. The two busy tones, BT_r and BT_t, are used during reception vs. transmission. This reduces the exposed terminal problem if compared to one tone.

DBTMA can be described as shown in Figure 11–7. Before transmitting a packet, TX senses the signaling channel. If TX detects a BT_r tone, that is, a receiving node lies within its range, it postpones transmission to avoid interference at the receiving node RX. If no BT_r tone is detected, TX starts a control handshake by sending an RTS packet (Figure 11–7a). The destination RX senses the signaling channel, and if a BT_t is detected, the handshake is aborted since the transmission in the range of RX, which is already active, would interfere with the reception of packets from TX. If no BT_t is detected, RX replies with a CTS packet, and by asserting the BT_r tone, signals to its neighbors its receiving status (Figure 11–7b). Finally, TX asserts the BT_t tone and starts transmitting the data packet (Figure 11–7c).

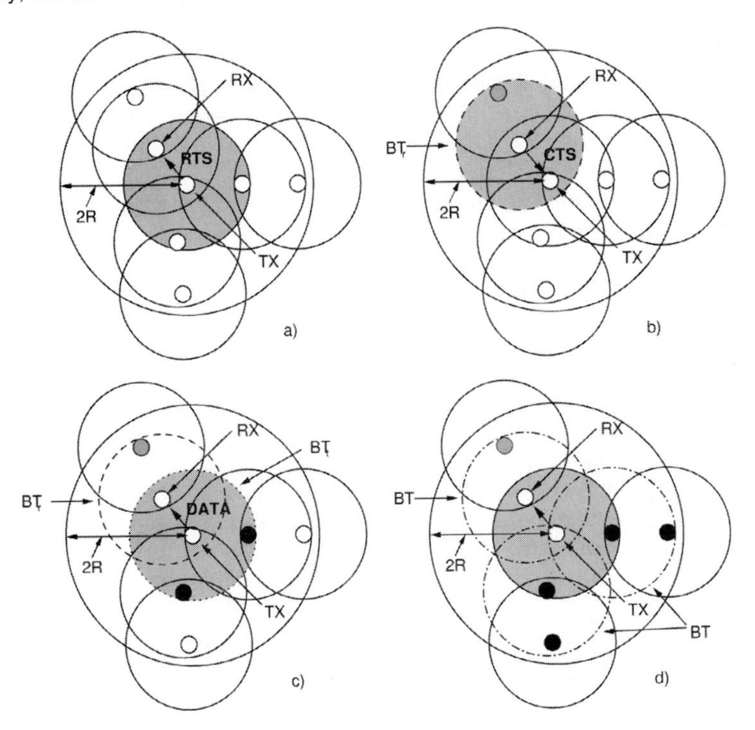

Figure 11–7 DBTMA vs. BTMA.

Figure 11–7d shows BTMA for the same node configuration. Note that the adoption of DBTMA eliminates the exposed terminal problem by blocking only transmissions in the range of the receiver (gray nodes in Figure 11–7c) and receptions in the range of the active transmitter (black nodes in Figure 11–7c). Note that a node that was prevented from receiving can still transmit because it detects the BT_t but not the BT_r, and is therefore not exposed. On the contrary, in the case of BTMA (Figure 11–7d), all nodes that detect either the data sent by the transmitter or a BT signal are blocked, leading to a high number of exposed terminals (black nodes in Figure 11–7d). On the other hand the solution adopted by DBTMA and similar methods, such as the receiver-initiated BTMA (Wu and Li, 1987), require the receiver to decode the entire RTS packet before setting the BT_r, with the consequence of an increase in the probability of collision when compared to the BTMA.

Comparison by simulation of DBTMA versus MACA (i.e., handshaking without busy tones and carrier sensing) indicates that the adoption of busy tones increases network performance in heavy traffic conditions by significantly decreasing the percentage of packet collisions. Note that signaling and data channels are located on different frequency intervals, which might be characterized by different path losses, and therefore different coverage areas for data and control information should be considered.

EXAMPLE 11–2

Multi-channel access methods are the other possible option and have been widely investigated; the achievable throughput is significantly increased (Polydoros and Silvester, 1987). In the case of multi-channel, the overall available resource is partitioned into channels. As discussed in Chapter 9, a channel usually corresponds to either a time slot, as in TDMA, a frequency band, as in FDMA, or a code, as in CDMA.

An UWB system can adopt any of the above resource partitioning criteria. The IEEE 802.15.TG3a proposal, which was examined regarding the physical layer and channel in preceding chapters, proposes a TDMA MAC for UWB (IEEE 802.15.3, 2003), as further described in Example 11–4. Time-Hopping Impulse Radio (TH-IR) UWB provides, however, a straightforward partition of the resource in channels based on TH codes. The design of a multi-channel CDMA MAC algorithm therefore forms a natural basis for TH-IR UWB.

Multi-channel CDMA MAC algorithms, commonly referred to as multi-code, have been intensively investigated for CDMA networks. Among them all, we cite random CDMA access (Raychaudhuri, 1981), and more recently multi-code spread-slotted Aloha (Dastangoo, Vojcic, and Daigle 1998). Note, however, that although in the last years most of the research efforts were focused on DS CDMA, Frequency-Hopping (FH) CDMA and TH CDMA also provide viable solutions.

The performance of multi-code MAC algorithms is limited by two factors:

- MUI, caused by simultaneous transmission of packets from different users on different codes
- Collisions caused by the selection of the same code by two different transmitters in reach of each other

Robustness to MUI is determined by the cross-correlation properties of the codes. The lower the cross-correlation between different codes, the higher the number of simultaneous transmissions that can be allowed.

The effect of code collisions can be mitigated by means of appropriate code selection strategies. The task of assigning codes to different transmitters in the same radio coverage area is a challenging issue in distributed networks. (Sousa and Silvester, 1988) provided a thorough overview of possible code assignment strategies and proposed the following schemes:

- *Common code* — All terminals share the same code, relying on phase shifts between different links for avoiding code collision
- *Receiver code* — Each terminal has a unique code for receiving, and the transmitter tunes on the code of the intended receiver for transmitting a packet
- *Transmitter code* — Each terminal has a unique code for transmitting, and the receiver tunes on the code of the transmitter for receiving a packet
- *Hybrid* — A combination of the above schemes

In the common code scheme, multi-code capability is not exploited. If phase shifts are small, this solution collapses into the single-channel Aloha. The receiver code scheme has the main advantage of reducing receiver complexity, since a terminal must only listen to its receiving code. On the other hand, multiple transmissions involving the same receiver may likely result in collisions, since the same code is adopted by all transmitters. Conversely, the transmitter code scheme avoids collisions at the receiver, since each transmitter uses its own code, but requires that a receiver listen to all possible codes in the network.

Hybrid schemes allow a tradeoff among the above conditions. A hybrid scheme could foresee the use of either the receiver or the common code scheme for transmitting signaling information, in which the receiver can read the code that will be used for sending data. A transmitter code scheme is then used for data.

When the set of codes is limited, however, even the transmitter code scheme may suffer from collisions due to possible reassignments of the same code. Specific code assignment strategies are required, such as the method presented in (Garcia-Luna-Aceves and Raju, 1997), in which distributed code assignment for CDMA multi-hop networks is proposed. In this protocol, if code C is used by terminal T, code C is never selected within a two-hop range from T.

11.2.2 MAC Organization

In a distributed scenario, terminals cooperate to build the network. Two main approaches are possible for network self-organization at the MAC layer: domain-dependent (clustered) MAC and domain-independent (flat) MAC.

Most of the MAC protocols proposed in the literature and adopted in Wireless Local Area Network (WLAN) standards rely on the explicit definition of a MAC domain leading to a clustered network architecture, where each cluster corresponds to a MAC domain. A clustered architecture simplifies resource management within each cluster by allowing a centralized approach. Two examples of domain-dependent MAC protocols are Bluetooth and IEEE 802.15.3, as illustrated in Examples 11–3 and 11–4.

Example 11–3 : The Bluetooth MAC

In Bluetooth (Haartsen, 2000), a Frequency-Hopping Code Division Multiple Access (FH-CDMA) scheme is adopted, and each MAC domain, called a piconet, is associated with a FH sequence. Terminals in a given area self-organize into piconets composed of a maximum of eight terminals. Each terminal has an address composed of three bits. Medium access control within each piconet is centralized and follows a master slave paradigm, in which the terminal that sets up the piconet plays the role of the master. Each master can control up to seven slaves. The FH sequence used in the piconet is determined on the basis of the master ID. The set of independent piconets in the area is called the scatternet.

The piconet setup is started by a terminal during a neighbor discovery phase, achieved by means of a dedicated scan procedure, which allows the terminal to collect information on neighboring devices. The terminal assumes the role of piconet master. The master pages one of the discovered neighbors (slave) to start the piconet. If other devices must be included in the piconet, an individual paging by the master is required for each of them. Alternatively, a device can decide to join an active piconet by paging the master. To add the new device as a slave in the piconet, however, an additional master/slave switch procedure is needed; otherwise, the paging device automatically becomes master of the piconet. The master slave switch procedure is executed on request.

Bluetooth MAC is an example of a scheduled access method. Each slave uses the same FH sequence, which is calculated based on the master ID and adapts its own internal clock by adding a temporal phase to the clock of the master, which provides a time reference within the piconet. Radio access is controlled by the master and is based on a slotted time axis. Each time slot has a duration T_{SLOT} = 625 μs. A different chip of the FH sequence corresponds to each time slot. Any terminal transmitting in that slot makes use of the specific FH chip, that is, of the frequency determined by the FH chip value.

Two types of links can be activated in a piconet:

- Synchronous Connection-Oriented (SCO)—Point-to-point links between the master and a single slave
- Asynchronous Connectionless (ACL)—Point-to-point and point-to-multi-point links between the master and all slaves in the piconet

No direct slave-to-slave communication is allowed in Bluetooth.

Slots are assigned alternatively to the master (TX slots) and to one of the slaves (RX slots).

In a SCO link, the master reserves two consecutive slots to exchange data with a slave on a periodic time basis T_{SCO}. A SCO link is appropriate for latency-sensible applications such as, for example, speech.

In an ACL link, access is based on polling: The master grants an RX slot to a slave by sending the slave ID in the immediately preceding TX slot.

Each packet in a SCO link is transmitted within a single slot, while in ACL links, longer packets, that is three-slots- or five-slots-long, are allowed. A terminal that transmits a long packet does not change frequency of operation during transmission. At the end of the transmission of a long packet, the terminal skips a number of FH chips equal to the number of slots forming the long packet (minus one). Figure 11–8 shows the combination of a SCO link (master – slave #7) with period $T_{SCO} = 8\ T_{SLOT}$, and several ACL links.

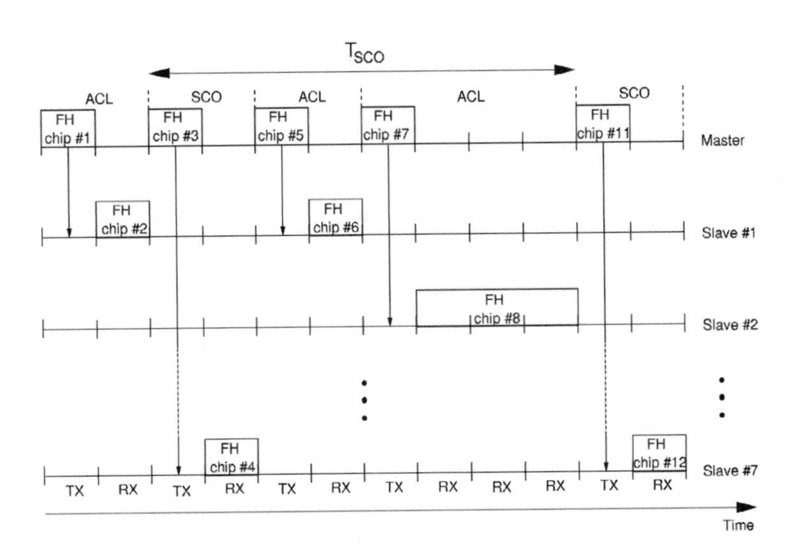

Figure 11–8 Contention-free access in a Bluetooth piconet. Different slots correspond to different FH chip values, that is, different frequencies of transmission.

The centralized approach avoids collisions between terminals in the same piconet. Terminals connected to different piconets can, however, collide and provoke interference. This problem is mitigated by adopting different FH codes in different piconets. Since the FH sequence to be used in a piconet is extracted from the master ID, multiple channels are therefore obtained in an uncoordinated fashion, while inside a piconet, the channel is shared between terminals thanks to a centralized master controlling unit. Regarding inter-piconet communications, note that a terminal willing to participate in *n* different piconets must manage *n* different non-synchronized channels (one for each master), by periodically switching from one to the other. This raises a major issue of scalability, especially regarding traffic scheduling and routing in scatternets composed of numerous piconets.

EXAMPLE 11–3

The IEEE 802.15.3 standard (IEEE 802.15.3, 2003) is another example of domain-dependent MAC. This standard was originally developed for traditional, narrowband physical layers in the ISM band. In 802.15.3, as much as in Bluetooth, medium access is controlled in a centralized fashion within each MAC domain, as further illustrated in the following example.

Example 11–4 : IEEE 802.15.3 MAC

In IEEE 802.15.3 (IEEE 802.15.3, 2003), a MAC domain is, as in Bluetooth, called a piconet. A piconet is controlled by a master called the Piconet Controller (PNC) which grants access to the medium on a TDMA basis. Up to 256 terminals are allowed in each piconet based on an eight-bit addressing format.

Different from Bluetooth, the PNC role in 802.15.3 is assigned to the terminal with the highest PNC potentials. PNC potentials are defined based on a set of parameters, including the transmitter's power level and the maximum transmission rate characterizing the terminal. If a newly associated terminal is better suited than the current PNC for the master role, a specific PNC handover procedure is applied. The PNC maintains piconet global timing thanks to the definition of a superframe composed of:

- A beacon period — Used by the PNC to broadcast information to all terminals in the piconet
- A Contention Access Period (CAP) — Used by terminals to send small amounts of data to other terminals, or to ask the PNC for reserved Channel Time Allocations (CTAs)
- A Contention-Free Period (CFP) — Dedicated to transmissions during the CTAs announced in the beacon period

The PNC sets both global frame duration and durations of different periods, depending on piconet size and traffic conditions.

During the CAP, terminals can send either data or Channel Time Request (CTR) messages. They do so by using a CSMA-CA approach.

In the CFP, terminals that requested reserved CTAs send packets directly to the destination terminals. This sets a key difference between 802.15.3 and the Bluetooth MAC architecture. In Bluetooth, direct communication between two different terminals is not allowed, and the master is in charge of relaying all traffic through the piconet. The piconet topology is, in this case, a typical star, as shown in Figure 11–9a. Conversely, the PNC in 802.15.3 schedules CTAs but is not involved in data packet exchange. Thus, although piconet management is fully centralized, data is transferred in a pure ad-hoc manner (Figure 11–9b).

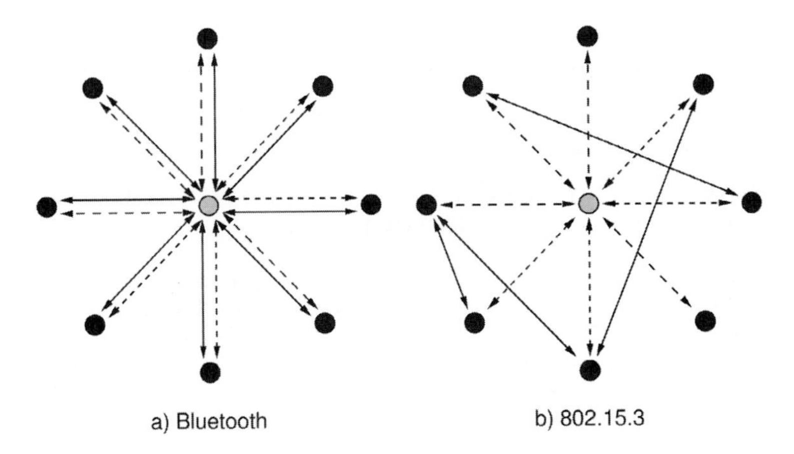

Figure 11–9 Logical piconet topologies: control traffic shown with dashed arrows and data traffic shown with filled arrows (by permission from *Journal of Communications and Networks*, ISSN 1229-2370, © Korean Institute of Communication Sciences).

As in Bluetooth, the IEEE 802.15.3 MAC does not define procedures for the interconnection of independent piconets. As a consequence, the maximum size of a piconet is a first bound for maximum network size and network scalability. Compared to Bluetooth, IEEE 802.15.3 has larger piconets, 256 versus 8 terminals, and it includes the possibility of extending network scale by introducing "child" and "neighbor" piconets as described in the standard (IEEE 802.15.3, 2003).

EXAMPLE 11–4

11.2.3 Admission Control

Admission control is required when congestion must be avoided to meet network performance requirements. Admission control is mandatory in QoS-aware networks, in which unregulated access might easily provoke violation of performance guarantees. Best-effort networks do not require admission control, but can benefit from its introduction.

Admission control is typically implemented with centralized schemes, as in cellular networks (Mouly and Pautet, 1992) and centralized wireless networks (Khun-Jush et al., 2000). A few proposals for distributed schemes are, however, also available in the literature. These schemes rely on the cooperation of terminals in evaluating the impact of additional traffic flows on network throughput, and eventually rejecting requests causing unacceptable performance degradation typically due to MUI generated by potential new entries (Bambos, 1998; Valaee and Li, 2002; Yang and Kravets, 2003).

This approach is suitable for UWB networks, which rely on TH-CDMA for multiple access. A distributed admission control scheme for an UWB network is proposed in (Cuomo et al., 2002), based on (Bambos, 1998), and is illustrated in the following example.

Example 11–5 : A Distributed Admission Control Function

We describe in this example the general concept that forms the basis for an UWB-specific admission control function proposed by (Cuomo et al., 2002) for the particular case of flows of traffic with no specific requirements, similar to the best-effort service of IP networks.

The system foresees at the start the presence of N_u active links between N_u pairs of UWB terminals. The admission control function for a new link is based on the maximization of global network throughput and reflects the following strategy: A new link is admitted if the effect of increased interference that the new link provokes is overlooked by an increment of global throughput. The presence of a new link in the system generates increased MUI and forces active links to reduce their transmission rate to keep verifying the required specifications (SIR). If the overall decrease in throughput due to the decrease of the transmission rates is compensated for by the increase in throughput generated by the new link, the new link is accepted.

Suppose each terminal has a limited power of emission P_{max}. If no constraint is imposed on cumulative emitted power, it can be demonstrated that the emitted power of each single terminal's throughput is maximized when all terminals transmit at P_{max}. The overall system throughput, before introduction of a new link, can be expressed as a function H(**r**) of vector **r** representing bit rates over the N_u active links at the start, or:

$$H(\mathbf{r}) = \sum_{i=1}^{N_u} R_i \quad (11\text{–}1)$$

The SIR_i of each link can be directly derived from the analysis of Chapter 9 (see Eqs. 9–30 and 9–45), after releasing the hypothesis that all terminals transmit at the same rate. For orthogonal PPM, for example, if R_i indicates the transmission rate over link i, one has:

$$SIR_i = \frac{N_s^{(i)} T_s}{\sigma_M^2} \frac{E_{RX}^{(i)}}{\sum_{\substack{n=1 \\ n \neq i}}^{N_u} E_{RX}^{(n)}} = \frac{\gamma_R}{\int_{-T_M}^{T_M} R_0^2(\tau)\, d\tau} \frac{1}{R_i \sum_{\substack{n=1 \\ n \neq i}}^{N_u} \frac{E_{RX}^{(n)}}{E_{RX}^{(i)}}} \quad (11\text{–}2)$$

Given a target SIR_{ispec} to be satisfied on each link, one must have:

$$SIR_i \geq SIR_{ispec} \quad for \quad i = 1, \ldots, N_u \qquad (11\text{–}3)$$

Note that, according to Eq. (11–2), R_i for each i can be expressed as follows:

$$R_i \leq \frac{\gamma_R}{SIR_{ispec} \int_{-T_M}^{T_M} R_0^2(\tau)\, d\tau} \frac{1}{\sum_{\substack{n=1 \\ n \neq i}}^{N_u} \frac{E_{RX}^{(n)}}{E_{RX}^{(i)}}} \qquad (11\text{–}4)$$

Overall system throughput reaches a maximum when all rates R_i have the maximum value defined by Eq. (11–4), that is:

$$H_{N_u} = \sum_{i=1}^{N_u} \frac{\gamma_R}{SIR_{ispec} \int_{-T_M}^{T_M} R_0^2(\tau)\, d\tau} \frac{1}{\sum_{\substack{n=1 \\ n \neq i}}^{N_u} \frac{E_{RX}^{(n)}}{E_{RX}^{(i)}}} \qquad (11\text{–}5)$$

To derive the admission rule, let us consider an initial situation in which N_u already active links satisfy the condition specified in Eq. (11–5).

Suppose a request for activating a new link labeled 0 is presented. In case the new link was admitted, each active receiver would experience a higher interference noise, which would require a reduction of the bit rate on that link to maintain the requested SIR. One would thus have for each N_u+1 link:

$$SIR_i = \frac{\gamma_R}{\int_{-T_M}^{T_M} R_0^2(\tau)\, d\tau} \frac{1}{R_{i,new} \sum_{\substack{n=0 \\ n \neq i}}^{N_u} \frac{E_{RX}^{(n)}}{E_{RX}^{(i)}}} \qquad (11\text{–}6)$$

The hypothetical new throughput value would be:

$$H_{N_u+1} = R_0 + \sum_{i=1}^{N} R_{i,new} \qquad (11\text{–}7)$$

The admission rule is, therefore, to admit the new link if:

$$H_{N_u+1} > H_{N_u} \qquad (11\text{–}8)$$

The R_0 value will eventually be zero, in which case, the new link is not activated.

A straight application of this admission rule introduces unfairness in the system since a new link might never be able to gain access. This effect can be mitigated by introducing the concept of a weighted admission function (Cuomo et al., 2002).

EXAMPLE 11–5

11.2.4 Packet Scheduling

The packet scheduling algorithm determines the order in which buffered packets are selected for transmission. In wired networks, this function has two main objectives: 1) to guarantee fair access to the available capacity to all the flows, and 2) to support QoS if different traffic classes are present. The simplest solution is the First Come First Serve (FCFS) algorithm, in which packets are sent in the same order in which they are buffered. This solution, however, provides no protection against ill-behaving sources, that is, those sources that have very high emission rates and would capture a high percentage of the available bandwidth. To increase fairness, a round robin scheme adopted to serve each traffic flow is proposed in (Nagle, 1987). Fair access, however, is not guaranteed, since packets of different lengths can be present in each queue. The weighted fair queueing algorithm (Demers et al., 1989) addresses this issue by assigning a weight to each queue, with the aim of emulating a bit-per-bit round robin between different flows. In this case, the introduction of QoS in the scheduling strategy is straightforward, since the weights can be easily adjusted to account for QoS classes.

Efficient packet scheduling in wireless networks cannot ignore the status of the wireless channel. Several wireless scheduling algorithms sensitive to channel status have been proposed. These are based on either a simple on/off Markov channel model (Lu et al., 1999) or on more sophisticated channel models leading to accurate evaluation of the SNR (Aida et al., 2000) and external interference (Golmie et al., 2001).

To this end, UWB does not present any relevant difference from other radio transmission techniques and the above methods are directly applicable to the specific UWB case.

11.2.5 Power Control

Due to the broadcast nature of the wireless medium, the achievable performance in wireless networks strictly depends on the capability of minimizing the undesired effects of each radio transmission on neighboring receivers. Power control thus leads to optimization of emitted power levels and achieves three desirable effects (Bambos, 1998): 1) minimization of power consumption, leading to longer autonomy; 2) reduction of interference; and 3) adaptation of emitted power to link variations due to channel modifications and mobility.

Power control has received significant attention in the last few years, in conjunction with the introduction of third-generation cellular networks based on CDMA, since it mitigates the near-far phenomenon in which a transmitter close to the receiver shadows the signal of a further transmitter. The centralized structure of cellular networks, however, simplifies the solution to this problem, since the presence of a base station significantly helps the implementation of efficient power control algorithms. The issue is far more complicated in a distributed network architecture, in which several independent links may be set up at the same time without any central controller. Nevertheless, power control should be a key property of distributed MAC protocols since it allows a significant increase in network capacity (Gupta and Kumar, 1998). A distributed power control protocol for CDMA ad-hoc networks jointly with a power-related admission control function is proposed in (Bambos, 1998).

Power control is important in the case of UWB networks as well, at least for two reasons: 1) UWB networks are affected by the near-far effect, although it can be expected that the high processing gain provided by TH-IR can partially mitigate this phenomenon, and 2) the low power levels allowed for UWB communication networks impose efficiency in the use of power.

11.3 MAC with QoS Management

Limitations in terms of available bandwidth and user terminal capabilities have confined in the past the transfer of data in data networks to small real-time data amounts (e.g. telnet sessions) or larger amounts without any real-time requirement (e.g., FTP sessions). A best-effort approach as adopted in IP-based networks was found to be suitable.

Technological progress and social modifications, however, have made data networks increasingly appealing as a universal way to transfer all kinds of information such as voice, multimedia, streaming video, and in general, real-time traffic.

Voice and multimedia traffic is characterized by requirements that are not present in non-real-time data traffic, and in particular, the need for transferring bit streams at a minimum bit rate (determined by the application generating the traffic) with an upper bound on the end-to-end delay. The fulfillment of the above requirements offers a guarantee to the end-user who perceives the offered service with the requested quality: QoS is thus defined as the performance that must be guaranteed by the network to meet user expectations. QoS is a typical and fundamental issue of network layer design and impacts the MAC, as will be briefly discussed in this section.

11.3.1 Traffic Classes

To guarantee a requested QoS, two different solutions can be foreseen:

- Maintain a pure best-effort approach and increase the available resource (in terms of bandwidth and processing power) by over-dimensioning the whole system with respect to expected traffic loads, and guaranteeing that each service has sufficient resources to satisfy QoS constraints.

- Adopt strategies that modify network behavior depending on traffic characteristics and QoS constraints, preserving the best-effort strategy only for low-priority traffic.

While the first solution is simpler from a conceptual point of view, it is not applicable in most cases for economic reasons; furthermore, it cannot be accepted for traffic such as telemedicine applications or air traffic control due to the imprecise control of network behavior and the lack of guarantee in the fulfillment of QoS requirements. Traffic-dependent network strategies are therefore necessary for the evolution of data networks.

The first step in the design of such strategies is the definition of a set of parameters defining QoS at the network level. Among typical network-level QoS parameters are bandwidth, end-to-end delay, jitter, bit error rate, and packet loss. Note that application-level QoS parameters such as resolution and frame rate for video services versus sample rate and sample size for audio services are all mapped into a unique set of network-level parameters.

According to the considered service, the admissible values for each of the network-level parameters vary. A few examples are reported in Table 11–1 (Nahrstedt and Steinmetz, 1995).

Table 11–1 Mapping Services onto Network-Level QoS Parameters

Service	QoS parameter	Range
Audio (telephone speech)	Bandwidth	16 kb/s
	End-to-end delay	$\leq$ 400 ms
	Packet loss	$\leq 10^{-2}$
Video (HDTV, lossy compression)	Bandwidth	20 Mb/s
	End-to-end delay	$\leq$ 250 ms
	Packet loss	$\leq 10^{-6}$
Data	Bandwidth	0.2–10 Mb/s
	End-to-end delay	$\leq$ 1 s
	Packet loss	$\leq 10^{-11}$

11.3.2 QoS at the MAC Layer

In a network supporting QoS traffic, the MAC must translate the QoS requirements coming from the network layer into MAC layer parameters, to reserve sufficient resources to fulfill requirements for a QoS traffic flow. The strategies to map a set of QoS requirements on a corresponding amount of resources is strictly dependent on the characteristics of the underlying physical layer: In the case of wireless networks, the radio link poses a severe issue on the capability of meeting QoS requirements due to the intrinsic low reliability of this transmission medium.

As a general concept, the MAC cannot guarantee at all times the fulfillment of requirements. All transmission channels are characterized by an *outage probability*. The key difference between wired and wireless networks is in the value assumed by such outage probability. In wired networks, the probability of having a link down is low enough and the upper layers can simply overlook this event. If a link or router is out of order, the problem is solved by selecting a new route. Radio networks are characterized by frequent link failures that impact the MAC and introduce the need for mechanisms that quickly recover errors on the link, such as Forward Error Correction (FEC) codes or Automatic Repeat on request (ARQ). In spite of these mechanisms, however, the fulfillment of QoS requirements can only be promised by the MAC with a given probability of failure in meeting QoS. The weakness of the physical medium translates at the MAC layer into an out-of-service probability.

11.4 An Example of an UWB-Tailored MAC Algorithm: $(UWB)^2$

Impulse Radio features with respect to MUI and synchronization, as analyzed in Chapters 8 and 9, form the basis for the definition of an UWB-tailored MAC algorithm — the uncoordinated, wireless, baseborn medium access for UWB communication networks $(UWB)^2$ described in (Di Benedetto et al., 2004) — which will be described in this section.

$(UWB)^2$ takes advantage for data transmission of the multiple access capabilities offered by TH codes and relies for the access to a common channel on the high MUI robustness provided by the processing gain of IR. MUI modeling based on packet collision, as analyzed in Section 9.3, shows that for TH-IR, the probability of successful packet transmission is fairly high, for uncoordinated transmission of several users and in the presence of MUI. Based on this result, $(UWB)^2$ adopts a pure Aloha approach and makes use of a synchronization scheme that foresees the presence of a synchronization trailer in each transmitted packet.

$(UWB)^2$ applies a multi-code concept to the specific case of a TH-IR UWB system and adopts a hybrid scheme based on the presence of a common control channel, provided by a common TH code, and of dedicated data channels associated to transmitter TH codes. The adoption of a hybrid scheme can be justified as follows:

1. It simplifies the receiver structure since data transmissions (and corresponding TH codes) are first communicated on the control channel.
2. It provides a common channel for broadcasting. Broadcast messages are, for example, required for routing and distributed positioning protocols.

Regarding code assignment, a unique association between MAC ID and transmitter code can be obtained by adopting the algorithm described in (Iacobucci and Di Benedetto, 2002) by which codes are generated on the basis of the MAC ID, thus avoiding the implementation of a distributed code assignment protocol.

$(UWB)^2$ does not assume that synchronization between transmitter and receiver is available at the beginning of packet transmission because of clock drifts in each terminal. As a consequence, a synchronization trailer long enough to guarantee the requested synchronization probability is added to the packet. The length of the trailer depends on current network conditions and is provided to the MAC by the synchronization logic.

$(UWB)^2$ also exploits the ranging capability offered by UWB. Distance information between transmitter and receiver is collected during control packet exchange. Such information can enable optimizations of several MAC features, and allows the introduction of new functions such as distributed positioning.

Performance of $(UWB)^2$ was evaluated by simulation for a fully connected graph of nodes. The number of terminals varied between 25 and 100. Each terminal sent MAC-PDUs containing L = 2000 bits to other terminals in the network following a Poisson process characterized by an average inter-arrival time T_{PDU}. Three different T_{PDU} values were considered: 2.5 s, 1.25 s, and 0.3125 s, corresponding to data rates of 800 bits/s, 1600 bits/s, and 6400 bits/s, respectively.

Regarding UWB physical layer parameters: The pulse rate was set to 1 Mpulses/s, $N_s = 1$, and $T_M = 1$ ns. We assumed all terminals to adopt the same synchronization sequence.

Packet error probability Pr_p was also computed. No correction capability on the packet was considered, and we assumed all bits in a packet to be correct for a packet to be correct. During simulations, a real-time evaluation of the number of active users N_u, as defined in Chapter 9 (rather than the average N_u value) was adopted for computing the probability of pulse collision. Pr_p was evaluated on a modified form of Eq. (9–52), which highlights the dependency of Pr_b on N_u as reflected by a possible variation from one bit to the other, and is expressed as follows:

$$\Pr_p = 1 - \prod_{i=0}^{L-1} \left(1 - \Pr_b(i)\right) \tag{11–9}$$

where $Pr_b(i)$ is the error probability for the i-th bit in the packet.

The measured values for Pr_p are presented in Figure 11–10. Figure 11–10 shows that for the three considered data rates, Pr_p remains below $1.6 \cdot 10^{-3}$ for as many as 100 users.

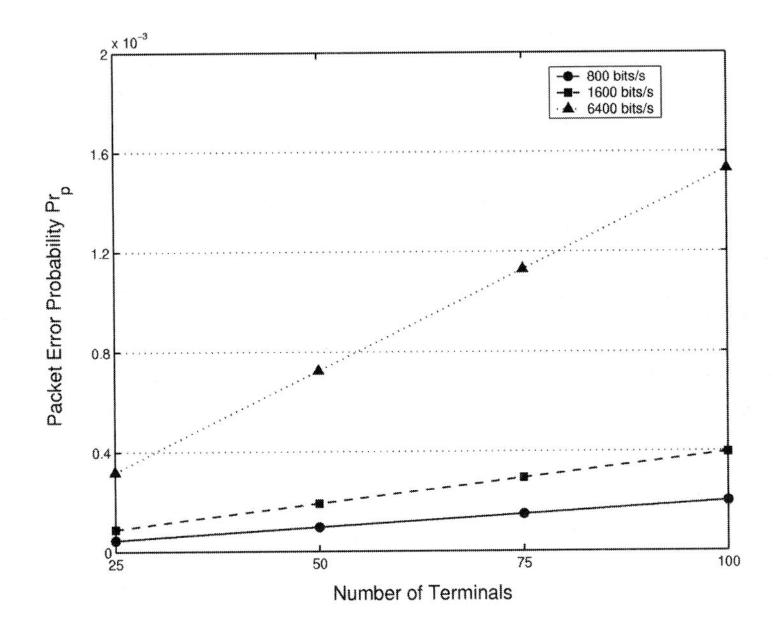

Figure 11–10 Packet error probability as a function of number of terminals for different data bit rates: circle: 800 bits/s; square: 1600 bits/s; triangle: 6400 bits/s (by permission from *Mobile Networks and Applications Journal*, Kluwer Academic Publishers).

Results from simulations indicate that $(UWB)^2$ can be successfully applied when the number of users spans from 20–40 to about 100 for data rates ranging from a few thousands to a few hundreds of bits per second. Results show that data transfers experience a Pr_p lower than about 0.2%, and confirm that $(UWB)^2$ is a suitable and straightforward solution for large networks of terminals transmitting at relatively low bit rates.

Further Reading

Regarding the MACA protocol, several variants have been proposed, among which are MACAW by (Bharghavan et al., 1994) and MACA-By Invitation (MACA-BI) by (Talucci and Gerla, 1997).

Bluetooth's inter-piconet interference problem has received increased interest. An estimate of network performance in a multi-piconet Bluetooth environment can be found in (Lin et al., 2003).

In Example 11–5, we presented a simplified version of the Cuomo et al. admission control function for the case of best-effort traffic. The case of QoS-guaranteed traffic is thoroughly described in (Cuomo et al., 2002) and (Cuomo et al., 2003; Cuomo and Martello, 2003).

The general problem of using positioning to reduce routing overhead in ad-hoc networks is addressed in (Ko and Vaidya, 1997). Optimal routing in UWB networks is presented in (Baldi et al., 2002). The model described in (Di Benedetto and Baldi, 2001) considers a network of cooperative nodes, which, based on ranging information, tend to minimize a power-aware UWB network cost.

REFERENCES

Abramson, N., "The Throughput of Packet Broadcasting Channels," *IEEE Transactions on Communications*, Volume: COM-25, Issue: 1 (January 1977), 117–128.

Aida, H., Y. Tamura, Y. Tobe, and H. Tokuda, "Wireless packet scheduling with signal-to-noise ratio monitoring," *Proceedings of IEEE Conference on Local Computer Networks* (November 2000), 32–41.

Baldi, P., L. De Nardis, and M.-G. Di Benedetto, "Modeling and optimization of UWB communication networks through a flexible cost function," *IEEE Journal on Selected Areas in Communications*, Volume: 20, Issue: 9 (December 2002), 1733–1744.

Bambos, N., "Toward power-sensitive network architectures in wireless communications: Concepts, issues, and design aspects," *IEEE Personal Communications*, Volume: 5, Issue: 3 (June 1998), 50–59.

Bertsekas, D., and R. Gallager, *Data Networks*, 2nd Edition, Upper Saddle River, NJ: Prentice Hall (1992).

Bharghavan, V., A. Demers, S. Shenker, and L. Zhang, "MACAW: A medium access protocol for wireless LANs," *Proceedings of the conference on Applications, Technologies, Architectures and Protocols for Computer Communication* (August 1994), 212–225.

Breslau, L., D. Estrin, K. Fall, S. Floyd, J. Heidemann, A. Helmy, P. Huang, S. McCanne, K. Varadhan, Y. Xu, and H. Yu, " Advances in Network Simulation," *IEEE Computer*, Volume: 33, Issue: 5 (May 2000), 59–67.

Chandra, A., V. Gummalla, and J.O. Limb, "Wireless Medium Access Control Protocols," *IEEE Communication Surveys* (2nd quarter 2000), 2–15.

Choi, H., and N. Moayeri, "Evaluation procedure for 802.16 MAC Protocols," Available at *www.ieee802.org/16/tg1/mac/pres/802161mp-00_16.pdf* (April 2000).

Crow, B. P., I. Widjaja, J. G. Kim, and P.T. Sakai, "IEEE 802.11: Wireless Local Area Networks," *IEEE Communications Magazine*, Volume: 35, Issue: 9 (September 1997), 116–26.

Cuomo, F., C. Martello, A. Baiocchi, and F. Capriotti, "Radio Resource Sharing for Ad Hoc Networking with UWB," *IEEE Journal on Selected Areas in Communications*, Volume: 20, Issue: 9 (December 2002), 1722–1732.

Cuomo, F., C. Martello, and S. Caputo, "An interference-controlled admission control scheme for QoS support in distributed UWB networks," *Proceedings of the IST Mobile & Wireless Communications Summit* (June 2003), 508–512.

Cuomo, F., and C. Martello, "Improving Wireless Access Control Schemes via Adaptive Power Regulation," *Proceedings of Personal Wireless Communications 8th International Conference* (September 2003), 114–127.

Dastangoo, S., B.R. Vojcic, and J.N. Daigle, "Performance Analysis of Multi-Code Spread Slotted ALOHA (MCSSA) System," *IEEE Global Telecommunications Conference*, Volume: 3 (November 1998), 1839–1847.

Demers, A., S. Keshav, and S. Shenker, "Analysis and Simulation of a Fair Queueing Algorithm," *Proceedings of the Conference on Applications, Technologies, Architectures and Protocols for Computer Communication* (September 1989), 1–12.

Deng, J., and Z.J. Haas, "Dual Busy Tones Multiple Access (DBTMA): A New Medium Access Control for Packet Radio Networks," *IEEE International Conference on Universal Personal Communications*, Volume: 2 (October 1998), 973–977.

De Nardis, L., G. Giancola, and M.-G. Di Benedetto, "A Position Based Routing Strategy for UWB Networks," *IEEE Conference on Ultra Wideband Systems and Technologies* (November 2003), 200–204.

De Nardis, L., and M.-G. Di Benedetto, "Medium Access Control design in UWB networks: review and trends," *Journal of Communication and Networks, Special Issue on Ultra-Wideband Communications*, Volume: 5, Issue: 4 (December 2003), 386–393.

Di Benedetto, M.-G., and P. Baldi, "A model for self-organizing large-scale wireless networks," *Invited paper, Proceedings of the International Workshop on 3G Infrastructure and Services* (July 2001), 210–213.

Di Benedetto, M.-G., L. De Nardis, M. Junk, and G. Giancola, "$(UWB)^2$: Uncoordinated, Wireless, Baseborn medium access for UWB communication networks," in press in *Mobile Networks and Applications special issue on WLAN Optimization at the MAC and Network Levels* (2nd quarter 2004).

Fontana, R.J., E. Richley, and J. Barney, "Commercialization of an Ultra Wideband Precision Asset Location System," *IEEE Conference on Ultra Wideband Systems and Technologies* (November 2003), 369–373.

Fullmer, C. L., and J. J. Garcia-Luna-Aceves, "Floor acquisition Multiple Access (FAMA) for Packet Radio Networks," *Proceedings of the conference on Applications, Technologies, Architectures and Protocols for Computer Communication* (September 1995), 262–273.

Garcia-Luna-Aceves, J. J., and J. Raju, "Distributed Assignment of codes for multihop packet-radio networks," *Proceedings of IEEE Military Communications Conference*, Volume: 1 (November 1997), 450–454.

Golmie, N., N. Chevrollier, and I. ElBakkouri, "Interference aware Bluetooth packet scheduling," *Proceedings of IEEE Global Telecommunications Conference*, Volume: 5 (November 25–29 2001), 2857–2863.

Gupta, P., and P.R. Kumar, "The capacity of wireless networks," *IEEE Transactions on Information Theory*, Volume: 46, Issue: 2 (March 2000), 388–404.

Haartsen, J.C., "The Bluetooth Radio System," *IEEE Personal Communications*, Volume: 7, Issue: 1 (February 2000), 28–36.

Iacobucci, M.S., and M.-G. Di Benedetto, *Computer method for pseudo-random codes generation*, National Italian patent number RM2001A000592 (2001).

IEEE 802.15.3-2003, "IEEE standard for information technology - telecommunications and information exchange between systems - local and metropolitan area networks - specific requirements part 15.3: wireless medium access control (MAC) and physical layer (PHY) specifications for high rate wireless personal area networks (WPANs)" (September 2003).

IEEE 802.15.4-2003, "IEEE standard for information technology - telecommunications and information exchange between systems - local and metropolitan area networks specific requirements part 15.4: wireless medium access control (MAC) and physical layer (PHY) specifications for low-rate wireless personal area networks (LR-WPANs)" (October 2003).

Karn, P., "MACA-A new Channel Access Protocol for Packet Radio," *Proceedings of the ARRL/CRRL Amateur Radio Ninth Computer Networking Conference* (September 1990), 134–140.

Khun-Jush, J., G. Malmgren, P. Schramm, and J. Torsner, "Overview and performance of HIPERLAN type 2-a standard for broadband wireless communications," *Proceedings of IEEE 51st Vehicular Technology Conference*, Volume: 1 (May 2000), 112–117.

Kleinrock, L., and F.A. Tobagi, "Packet Switching in Radio Channels: Part I–Carrier Sense Multiple-Access Modes and Their Throughput-Delay Characteristics," *IEEE Transactions on Communications*, Volume: 23, Issue: 12 (December 1975), 1400–1416.

Ko, Y.B., and N.H Vaidya., "Using location information to improve routing in ad hoc networks," *Technical Report 97-013, CS Dept, Texas AM University* (December 1997).

Lin, T.-Y., Y.-K. Liu, and Y.-C. Tseng, "An Improved Packet collision analysis for Multi-Bluetooth Piconets considering Frequency-Hopping Guard Time Effect," *IEEE Vehicular Technology Conference Proceedings* (October 2003), 577–581.

Lu, S., V. Bharghavan, and R. Srikant, "Fair scheduling in wireless packet networks," *IEEE/ACM Transactions on Networking*, Volume: 7, Issue: 4 (August 1999), 473–489.

Mouly, M., and M.B. Pautet, *The GSM System for Mobile Communication*, Palaiseau, France: CELL&SYS (1992).

Nagle, J., "On Packet Switches with Infinite Storage," *IEEE Transactions on Communications*, Volume: 35, Issue: 4 (April 1987), 435–438.

Nahrstedt, K., and R. Steinmetz, "Resource Management in Networked Multimedia Systems," *IEEE Computer*, Volume: 28, Issue: 5 (May 1995), 52–63.

Polydoros, A., and J. Silvester, "Slotted Random Access Spread-Spectrum Networks: An Analytical Framework," *IEEE Journal on Selected Areas in Communications*, Volume: 5, Issue: 6 (July 1987), 989–1002.

Raychaudhuri, D., "Performance Analysis of Random Access Packet-Switched Code Division Multiple Access Systems," *IEEE Transactions on Communications*, Volume: 29, Issue: 6 (June 1981), 895–901.

Talucci, F., and M. Gerla, "MACA-BI (MACA By Invitation): A wireless MAC protocol for high speed ad hoc networking," *IEEE 6th International Conference on Proceedings of Universal Personal Communications*, Volume: 2 (October 1997), 913–917.

Tobagi, F.A., and L. Kleinrock, "Packet Switching in Radio Channels: Part II – The Hidden Terminal Problem in Carrier Sense Multiple Access and the Busy Tone Solution," *IEEE Transactions on Communications*, Volume: 23, Issue: 12 (December 1975), 1417–1433.

Valaee, S., and B. Li, "Distributed Call Admission control for Ad-hoc networks," *Proceedings of IEEE 56th Vehicular Technology Conference*, Volume: 2 (September 2002), 1244–1248.

Varga, A., "OMNeT++" in the column "Software Tools for Networking," *IEEE Network Interactive*, Volume: 16, Issue: 4, Available at *www.comsoc.org/ni/Public/2002/Jul/index.html* (July 2002).

Yang, Y., and R. Kravets, "Contention-Aware Admission Control for Ad Hoc Networks," *University of Illinois in Urbana-Champaign Technical Report UIUCDCS-R-2003-2337* (April 2003).

Wu, C.S., and V.O.K. Lee, "Receiver Initiated Busy Tone Multiple Access in Packet Radio Networks," *Proceedings of ACM SIGCOMM'87* (1987), 336–342.

APPENDIX

Current Trends in UWB Standardization Activities

The last two years have witnessed an increased interest in both chip manufacturing companies and standardization bodies in UWB. UWB's appealing features, such as flexibility and robustness, as well as its high-precision ranging capability, have polarized attention and made UWB an excellent candidate for applications regarding short-range/high-speed wireless communications and sensor networks requiring accurate localization and tracking. Low bit rate, location-aware applications for connecting simple devices that consume minimal power at short distances are an additional area in which UWB has recently started to play a leading role for future standards.

The purpose of this appendix is to briefly review current trends of UWB standardization, and to provide the reader with a roadmap for locating information related to standards in the book.

PHYSICAL LAYER

The development of the first UWB radio device for communication purposes dates back to the early 1990s. During those years, a few small- and medium-sized enterprises started promoting the innovative idea of wireless services operating on an unlicensed basis. The UWB technology developed at that time was based on the IR concept, or on the transmission of virtually carrier less and extremely short pulses.

On February 14, 2002, the FCC approved the first guidelines allowing, at least in the United States, the intentional emission of UWB signals contained within specified emission masks (FCC, 2002). According to the FCC rules, however, the UWB concept is not limited to pulsed transmission, but can be extended to continuous-like transmission techniques, provided that the occupied bandwidth of the transmitted signal is greater than 500 MHz. The effect of the FCC release was twofold. On one side, the FCC regulation of UWB emissions raised the interest of major chip manufacturers such as Texas Instruments, Motorola, IBM, and Intel. On the other hand, discussions were triggered around the advantages of the original IR scheme versus the traditional, carrier-based, continuous transmission alternative.

The above lack of agreement is reflected in the current diatribe on UWB standardization, in particular in the United States in the framework of the IEEE 802.15.3a Task Group. This group was formed in late 2001 with the task of investigating innovative solutions for the development of high-speed and low-power WPANs. Currently (March 2004), two different proposals for a physical layer based on UWB are under consideration: a MB approach combining FH with OFDM (Batra et al., 2003), and a second approach using DS-UWB, which preserves the original UWB pulsed nature (Roberts, 2003). The reader can find a detailed analysis of the generation of **MB-OFDM-**modulated signals in Section 2.3. Spectral properties of OFDM signals are discussed in Chapter 5; specifically, the MB approach is illustrated in Section 5.2 and developed further in Checkpoint 5–1. With reference to the IR approach, the reader can refer to Section 2.1 for the analysis of **PPM-TH-UWB** signals, to Section 2.2 for the analysis of **PAM-DS-UWB** signals, and to Section 4.2 for the analysis of **PAM-TH-UWB** signals. The spectral analysis of UWB signals based on the transmission of trains of pulses is presented in Chapters 3 and 4.

To evaluate and compare the different physical layer proposals that were submitted to the IEEE, the 802.15.3a Study Group formed a subcommittee devoted to the definition of a standard UWB channel model. In February 2003, a final report summarizing the work of the channel modeling subcommittee was released (IEEE, 2003). In this report, the IEEE proposed a channel model for indoor UWB propagation and provided recommendations on how the model should be used for evaluating performance of the physical layer. A description of the **IEEE UWB channel model** for indoor UWB propagation is provided in Section 8.2.3. In addition, Checkpoint 8–2 introduces the MATLAB code that can be used for implementing such a model. The roadmap is summarized in Figure A–1.

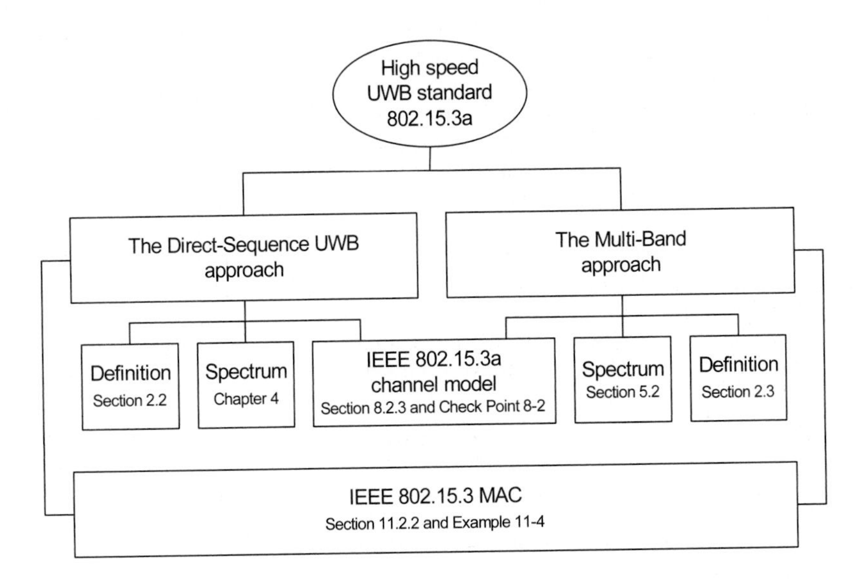

Figure A–1 Location in the book of the information related to the IEEE 802.15.3 standardization activities.

Regarding the adoption of UWB for low-rate, location-enabled applications, standardization is taking its first steps within the IEEE 802.15.4a Task Group with a first meeting in May 2004. The main interest is in providing communications with high-precision ranging and localization, low power emission and consumption, and a low cost.

Outside the United States, we cite the activity carried out in Europe by the TG31A group of the European Telecommunications Standards Institute (ETSI), which has the task of investigating and developing ETSI radio standards for short-range devices using UWB. Currently, the task group is about to deliver a first draft of an ETSI standard for UWB. Still in Europe, research and promoting activities around UWB will be developed in the next few years within the 6th IST EU Framework Integrated Project PULSERS. Project PULSERS (*www.pulsers.net*), which started January 1, 2004 and has gathered over 30 European and international partners, is taking the lead on research and development activities concerning UWB in Europe.

MAC LAYER

A multiplicity of MAC and network protocols for wireless LANs and PANs have been proposed and standardized in the last years. Among the several we cite are the IEEE 802.11 and HIPERLAN/2 standards for wireless LANs up to 54 Mbit/s, the Bluetooth standard for short range and low bit rate wireless communications, and the most recent IEEE 802.15.3

for short range and high bit rate wireless PANs. Within the book, the reader can find a description of the **IEEE 802.11b** MAC protocol in Example 11–1 located within Section 11.2.1. The **Bluetooth** MAC is introduced in Example 11–3 within Section 11.2.2. In the same section, Example 11–4 illustrates the features of the **IEEE 802.15.3 MAC**. The **IEEE 802.15.3 MAC** although originally developed for traditional narrowband physical layers, is currently proposed to be used in combination with an UWB physical layer.

REFERENCES

Batra, A. et al., *Multi-band OFDM Physical Layer Proposal for IEEE 802.15 Task Group 3a*, Available at *www.Multi-Bandofdm.org/papers/15-03-0268-01-003a-Multi-band-CFP-Document.pdf* (September 2003).

Federal Communications Commission, "Revision of Part 15 of the Commission's rules Regarding Ultra-Wideband Transmission Systems: First report and order," *Technical Report FCC 02-48* (adopted February 14, 2002; released April 22, 2002).

IEEE 802.15.SG3a, "Channel modeling Sub-committee Report Final," *IEEE P802.15-02/490r1-SG3a* (February 2003).

Roberts, R., *XtremeSpectrum CFP Document*, Available at *grouper.ieee.org/groups/802/15/pub/2003/Jul03/03154r3P802-15_TG3a-XtremeSpectrum-CFP-Documentation.pdf* (July 2003).

List of MATLAB Functions

List of Acronyms

ACL	Asynchronous Connection Less
ARQ	Automatic Repeat on reQuest
BF	Base Function
BPSK	Binary Phase Shift Keying
BTMA	Busy Tone Multiple Access
CAP	Contention Access Period
CDMA	Code Division Multiple Access
CFP	Contention Free Period
CRC	Cyclic Redundancy Check
CSMA	Carrier Sensing Multiple Access
CSMA-CA	Carrier Sensing Multiple Access with Collision Avoidance
CTA	Channel Time Allocation
CTR	Channel Time Request
CTS	Clear To Send
DARPA	Defense Advanced Research Projects Agency
DBTMA	Dual Busy Tone Multiple Access
DFT	Discrete Fourier Transform
DLC	Data Link Control
DLL	Delay Locked Loop
DS-SS	Direct Sequence Spread Spectrum
DS-SS-UWB	Direct Sequence Spread Spectrum Ultra Wide Band
DS-UWB	Direct Sequence Ultra Wide Band
DVI	Digital Visual Interface
EGC	Equal Gain Combining
EIRP	Effective Isotropic Radiated Power
ESD	Energy Spectral Density
FAMA	Floor Acquisition Multiple Access
FCC	Federal Communications Commission
FCFS	First Come First Serve
FDMA	Frequency Division Multiple Access
FEC	Foward Error Correction
FFT	Fast Fourier Transform
FH	Frequency Hopping
FH-CDMA	Frequency Hopping Code Division Multiple Access
FH-SS	Frequency Hopping Spread Spectrum
GPS	Global Positioning System

HI-FI	High-Fidelity
IDFT	Inverse Discrete Fourier Transform
IEEE	Institute for Electrical and Electronics Engineers
IFFT	Inverse Fast Fourier Transform
IR	Impulse Radio
IR-UWB	Impulse Radio Ultra Wide Band
ISI	Inter Symbol Interference
ISM	Industrial Scientific and Medical
IW	Inter Working
LCR	Large Current Radiator
LOS	Line of Sight
LSE	Least Square Error
MAC	Medium Access Control
MACA	Multiple Access with Collision Avoidance
MACA-BI	Multiple Access with Collision Avoidance By Invitation
MAC-PDU	Medium Access Control Protocol Data Unit
mb	measured bandwidth
MB	Multi Band
MB-OFDM	Multi Band Orthogonal Frequency Division Multiplexing
MC-CDMA	Multi Carrier Code Division Multiple Access
MCSSA	Multi Code Spread Slotted Aloha
MRC	Maximal Ratio Combining
MUI	Multi User Interference
NAV	Network Allocation Vector
NLOS	Non Line of Sight
OFDM	Orthogonal Frequency Division Multiplexing
PAM	Pulse Amplitude Modulation
PAM-TH-UWB	Pulse Amplitude Modulation Time Hopping Ultra Wide Band
PDP	Power Delay Profile
PHY	Physical layer
PLL	Phase Locked Loop
PN	Pseudo Noise
PNC	PicoNet Controller
PPM	Pulse Position Modulation
PPM-TH-UWB	Pulse Position Modulation Time Hopping Ultra Wide Band
PPS	Precise Positioning System
PSD	Power Spectral Density
QoS	Quality of Service
QPSK	Quadrature Phase Shift Keying
RF	Radio Frequencies
RTS	Request To Send
RX	Receiver
SCO	Synchronous Connection Oriented
SD	Selection Diversity
SGA	Standard Gaussian Approximation
SIR	Signal to Interference Ratio

SNR	Signal to Noise Ratio
SPS	Standard Positioning System
TDMA	Time Division Multiple Access
TH-IR	Time Hopping Impulse Radio
TH-UWB	Time Hopping Ultra Wide Band
TX	Transmitter
USB	Universal Serial Bus
UWB	Ultra Wide Band
WLAN	Wireless Local Area Network
WPAN	Wireless Personal Area Network

Index

A

B

C

D

E

F

G

N

O

P

T

U

For The Curious Reader

```
%                  - a l p h a -                    %
%              [ type : alpha_omega  ]               %
%                                                    %
%                                                    %
Pow = 0;  fc = 50e9;  numbits = 1; Ts = 1e-9;
Ns = 3 ;  Np = 3 ;  Tm = 1e-9 ; tau = 0.4e-9;
bits =                     cp0201_bits(numbits);
r1 =                    cp0201_repcode(bits,Ns);
DScode =                          cp0202_DS(Np);
[P1,D1] =     cp0202_2PAM_DS(r1,fc,Ts,DScode);
power =                     (10^(Pow/10))/1000;
Ex =                                power * Ts;
w0 =              cp0201_waveform(fc,Tm,tau);
wtx =                           w0 .* sqrt(Ex);
Sa =                              conv(P1,wtx);
L =                 (floor(Ts*fc))*Ns*numbits;
Stx =                                  Sa(1:L);
ebno =                        linspace(40,8,50);
Eb =                (1/numbits)*sum(Stx.^2);
EbNo =                          10.^(ebno./10);
No =                                Eb ./ EbNo;
nstdv =                            sqrt(No./2);
for j =                        1 : length(EbNo)
NS(j,:) =    nstdv(j) .* randn(1,length(Stx));
OUT(j,:) =                   NS(j,:) + Stx; end
F  =                                 figure(1);
set(F,'Color','white');          S = surf(OUT);
set(S,'LineWidth',[2]);                SX = gca;
set(S,                      'MeshStyle','row');
set(SX,                     'Xcolor','white');
set(SX,                     'Ycolor','white');
set(SX,                     'Zcolor','white');
set(SX,     'CameraPosition',[909 -278 2.14]);
%                                                    %
%       M.G. Di Benedetto - G. Giancola              %
%              Rome, Italy, 2004                     %
%                - o m e g a -                       %
```

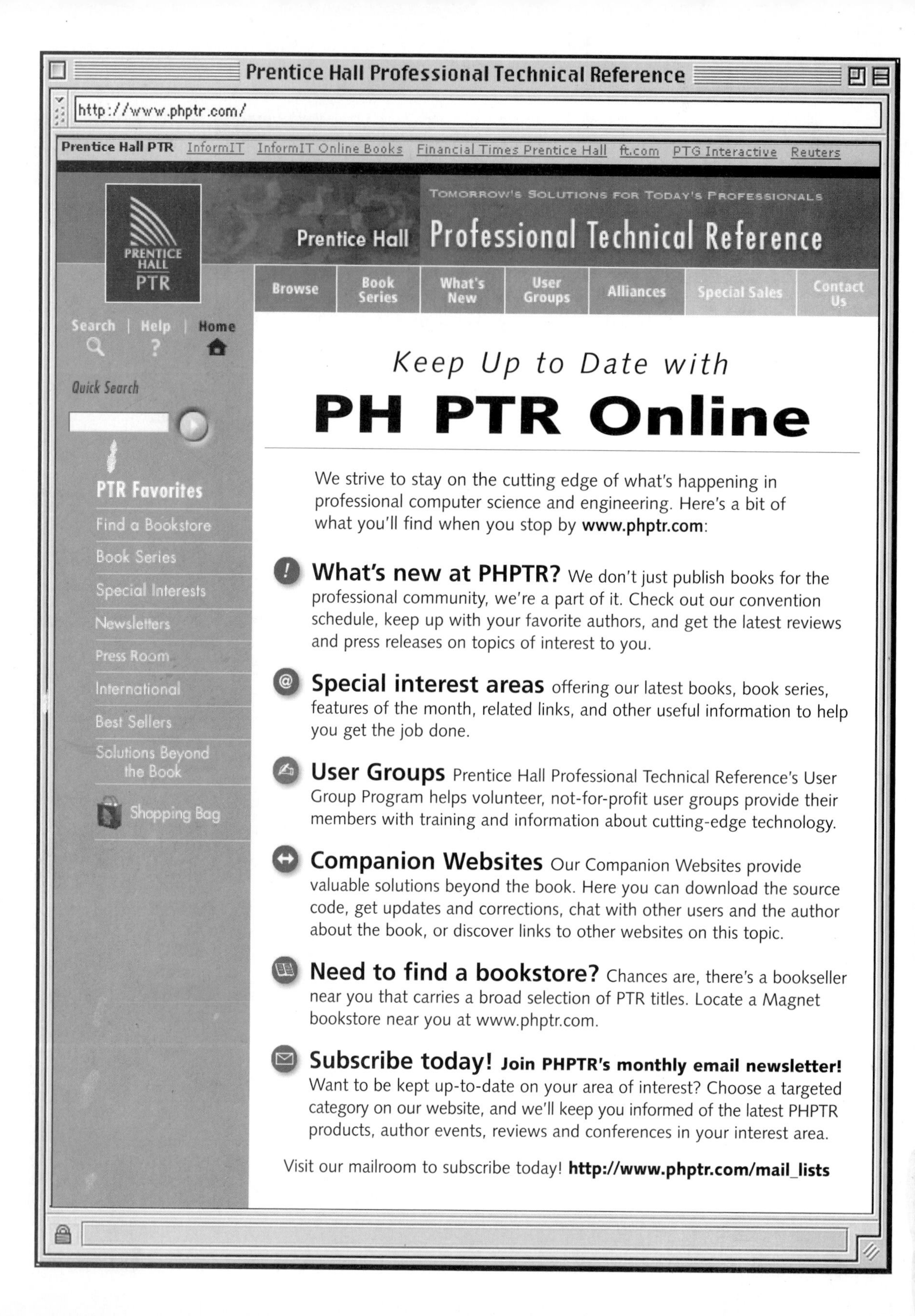
Prentice Hall Professional Technical Reference
http://www.phptr.com/
Prentice Hall PTR
InformIT
InformIT Online Books
Financial Times Prentice Hall
ft.com
PTG Interactive
Reuters
Tomorrow's Solutions for Today's Professionals
PRENTICE HALL PTR
Prentice Hall Professional Technical Reference
Browse
Book Series
What's New
User Groups
Alliances
Special Sales
Contact Us
Search | Help | Home
Quick Search
PTR Favorites
Find a Bookstore
Book Series
Special Interests
Newsletters
Press Room
International
Best Sellers
Solutions Beyond the Book
Shopping Bag
Keep Up to Date with
PH PTR Online
We strive to stay on the cutting edge of what's happening in professional computer science and engineering. Here's a bit of what you'll find when you stop by www.phptr.com:
What's new at PHPTR? We don't just publish books for the professional community, we're a part of it. Check out our convention schedule, keep up with your favorite authors, and get the latest reviews and press releases on topics of interest to you.
Special interest areas offering our latest books, book series, features of the month, related links, and other useful information to help you get the job done.
User Groups Prentice Hall Professional Technical Reference's User Group Program helps volunteer, not-for-profit user groups provide their members with training and information about cutting-edge technology.
Companion Websites Our Companion Websites provide valuable solutions beyond the book. Here you can download the source code, get updates and corrections, chat with other users and the author about the book, or discover links to other websites on this topic.
Need to find a bookstore? Chances are, there's a bookseller near you that carries a broad selection of PTR titles. Locate a Magnet bookstore near you at www.phptr.com.
Subscribe today! Join PHPTR's monthly email newsletter!
Want to be kept up-to-date on your area of interest? Choose a targeted category on our website, and we'll keep you informed of the latest PHPTR products, author events, reviews and conferences in your interest area.
Visit our mailroom to subscribe today! http://www.phptr.com/mail_lists